TRANSFER PHENOMENA IN MAGNETOHYDRODYNAMIC AND ELECTROCONDUCTING FLOWS

FLUID MECHANICS AND ITS APPLICATIONS
Volume 51

Series Editor: R. MOREAU

MADYLAM
Ecole Nationale Supérieure d'Hydraulique de Grenoble
Boîte Postale 95
38402 Saint Martin d'Hères Cedex, France

Aims and Scope of the Series

The purpose of this series is to focus on subjects in which fluid mechanics plays a fundamental role.

As well as the more traditional applications of aeronautics, hydraulics, heat and mass transfer etc., books will be published dealing with topics which are currently in a state of rapid development, such as turbulence, suspensions and multiphase fluids, super and hypersonic flows and numerical modelling techniques.

It is a widely held view that it is the interdisciplinary subjects that will receive intense scientific attention, bringing them to the forefront of technological advancement. Fluids have the ability to transport matter and its properties as well as transmit force, therefore fluid mechanics is a subject that is particulary open to cross fertilisation with other sciences and disciplines of engineering. The subject of fluid mechanics will be highly relevant in domains such as chemical, metallurgical, biological and ecological engineering. This series is particularly open to such new multidisciplinary domains.

The median level of presentation is the first year graduate student. Some texts are monographs defining the current state of a field; others are accessible to final year undergraduates; but essentially the emphasis is on readability and clarity.

For a list of related mechanics titles, see final pages.

Transfer Phenomena in Magnetohydrodynamic and Electroconducting Flows

Selected papers of the PAMIR Conference held in Aussois, France 22–26 September 1997

Edited by

A. ALEMANY
Ph. MARTY

and

J. P. THIBAULT

Industrial and Geophysical Flows Laboratory,
Institute of Mechanics Grenoble,
France

SPRINGER SCIENCE+BUSINESS MEDIA, B.V.

A C.I.P. Catalogue record for this book is available from the Library of Congress.

ISBN 978-94-010-6002-8   ISBN 978-94-011-4764-4 (eBook)
DOI 10.1007/978-94-011-4764-4

Printed on acid-free paper

Conference Chairman

A.Alemany

Scientific Committee

A.Alemany (France)
F.Bark (Sweden)
H.Branover (Israel)
G.Cognet (France)
P.Davidson (UK)
G.Gerbeth (Germany)
I.Kirillov (Russia)
O.Lielausis (Latvia)
J.Leorat (France)
S.Martemianov (France)
Ph. Marty (France)
J.Meng (USA)
R.Moreau (France)
U.Muller (Germany)
J.P.Thibault (France)
J.Walker (USA)
A.Wragg (UK)

Local Organizing Committee

B.Collovati (Secretary)
A.Alemany, Ph.Marty (Scientific Secretary), J.P.Thibault

Sponsoring Organizations and Companies

Association Universitaire de Mécanique, France
European Commission, DG XII, Brussels, Belgium
CNRS, France
EDF-CLI Electricité de France, Lyon
IRSID-Usinor Sacilor, Metz, France
Ministère de l'Enseignement Supérieur et de la Recherche, Paris
SIMULOG, Grenoble, France
LEGI Laboratory, Grenoble, France
Université Joseph Fourier, Grenoble, France
Institut National Polytechnique de Grenoble, France

CONTENTS

III - MHD in Metallurgy and Crystal Growth

IV - Energetic Applications

PREFACE

The "Pamir" international conference on "Transfer Phenomena in MagnetoHydroDynamic & Electroconducting Flows" is one of the manifestations associated to "Hydromag". This international organisation aims to coordinate and promote the MHD research in the world. The organiser of this conference is the Pamir group from the French laboratory LEGI (Laboratoire des Ecoulements Geophysiques et Industriels) which has close connections with INPG (Institut National Polythechnique de Grenoble), with the Joseph Fourier University of Grenoble and with the CNRS (Centre National de la Recherche Scientifique). In September 26-30, 1997 this conference was organised for the third time at the Paul Langevin centre (Aussois, France) which belongs to the CNRS. Approximately 15 countries were represented by about 120 participants. Papers included in this volume are those presented at the third Pamir conference after selection by the scientific committee and referee's procedure.

The formal presentations and invited lectures were focused on four main topics related to: Interfacial heat and mass transfer phenomena, Energetic applications, Dynamo effect, and MEHD Phenomena. One of the perspectives of the conference was to promote a productive interaction between the MHD and the chemical engineering research communities. The possibility to use an external magnetic field to improve and control the mass transfer processes in electrochemical systems, sometimes called *magnetoelectrolysis*, was introduced as a new topic of the Pamir conference and is considered as a relatively new and promising branch of MHD research. In particular a wide field of applications in various domains is expected. This new activity could be compared with the metallurgical applications of MHD and a parallel development could be anticipated. Because it concerned an introduction of the subject, the scientific committee decided to limit the number of accepted papers in this field to one third of the total submitted.

The invited lectures were about the Earth Dynamo presented by Doctor Dominique Jault from Grenoble University (France), Magnetoelectrolysis presented by Professor Thomas Fahidy from Waterloo University (Canada), instability problems by Doctor Philippe Marty from Grenoble University (France), Metallurgical Applications of MHD by Doctor Gunter Gerbeth from Dresden (Germany) and a review on Magnetic Fluids was presented by Professor Stuart Charles from the University of Wales (UK). The presentations and discussions showed a wide interest in theoretical and experimental fluid dynamo research. The situation of different experimental sodium facilities under construction (Karlsruhe, Riga) and preliminary projects (French Ampere programme) were particularly important.

The discussion about the field of magnetoelectrolysis was appreciated by the MHD community as well as by electrochemists. The different presentations revealed the relevance of the fundamental studies to possible industrial applications.The problems involved required competences in the field of electrochemistry, fluid mechanics and electromagnetism. Most of the experimental papers were presented by the Japanese community. It was concluded that the numerical computing had to be improved in order to describe the full complexities of the phenomena.

Metallurgical applications of MHD revealed that a strong effort has been made in the direction of realistic experimental and numerical approaches and that crystal growth activities is a main subject of interest. It would be important for this subject to quickly reach the level of real application. An important effort was devoted to MHD turbulence problems at low magnetic Reynolds number. New numerical developments were proposed especially in the range of moderate values of the interaction parameter which corresponds to working conditions of many industrial devices. The necessity to return toward experimental analysis to control the validity of numerical results was also one of the important conclusions on the subject.

In the field of fundamental MHD Flows the main contributions were devoted to phenomena characterised by asymptotic values of parameters (Reynolds number, Hartmann number and interaction parameter). Many contributed papers were orientated toward Fusion problems, using a lithium-lead alloy subjected to a very strong magnetic field, a condition which allows for an analytic or semi-analytic approach.

In the class of energetic applications of MHD, few papers were devoted to Cold Plasma MHD power generation which seems still active in some countries, e.g. Japan, China and India, while the European and American effort on this subject seems decreasing. The former activity on MHD ship-propulsion is also strongly decreasing in most of the previously involved countries. Nevertheless a quite novel and promising activity : electromagnetic seawater flow control, is presently under consideration by the community.

To prepare the topics of the next Pamir conference, which will be held in France in 2000, two specialists in magnetic fluids were asked to present the general aspects, the state of the art and the possible future developments of the subject. Only one of these two presentations is included in the present book.

A. Alemany
Ph. Marty
J.P. Thibault

CONTRIBUTORS LIST

O. Aaboubi
DTI - UFR Sciences
BP 1039
51687 - Reims Cedex 2 FRANCE

T. Alboussiere
Eng. dept.
Trumpington Street
Cambridge U.K.

O. Andreev
Inst. of Physics Latvian Academy of Sciences
32, Miera str.
2169 - Salaspils LATVIA

R. Aogaki
National Research Lab. for Magnetic Science
1-156, Shibasimo, Kawagucchi
333 - Saitama JAPAN

M. Aurousseau
LEPMI
BP 75
38402 - St Martin d'Hères FRANCE

F. Bark
KTH
Faxen Lab. Dept. of Mechanics
100 44 - Stockholm SWEDEN

E. Blums
Latvian Academy of Sciences
2169 - Salaspils-1 LATVIA

P. Boissonneau
LEGI
BP 53
38041 - Grenoble Cedex 9 FRANCE

V. Bojarevics
University of Greenwhich
Sch. of Comput. and Math. Wellington St.
SE18 6PF - London U.K.

R. Bolcato
EPM - Madylam
BP 53
38041 - Grenoble Cedex 9 FRANCE

P. Bontoux
IMFM
1 Rue Honorat
13003 - Marseille FRANCE

J.P. Brancher
LEMTA
2, Av. de la Forêt de Haye
54504 - Vandoeuvre les Nancy FRANCE

I. Bucenieks
Institute of Physics
Miera 32, Latvia
2169 - Salaspils LATVIA

P. Byrne
Royal Institut of Technology - Applied Electrochemistry
10044 - Stockholm SWEDEN

Ph. Cardin
LGIT
BP 53
38041 Grenoble Cedex 9 FRANCE

Y. Cho
Korea Institute of Science and Technology
PO Box 131 Cheongryang
130-650 - Seoul KOREA

J.P. Chopart
DTI - UFR Sciences
BP 1039
51687 - Reims Cedex 2 FRANCE

M. Cross
University of Greenwhich
Sch. of Comput. and Math. Wellington St.
SE18 6PF - London U.K.

S. Cuevas
Centro de Investigacion en Energia UNAM
A.P. 34 Temixco, Mor.
62580 - Mexico MEXICO

L. Davoust
LEGI
BP 53
38041 - Grenoble Cedex 9 FRANCE

F. Debray
EPM - Madylam
BP 53
38041 - Grenoble Cedex 9 FRANCE

Y. Delannoy
EPM - Madylam
BP 53
38041 - Grenoble Cedex 9 FRANCE

O. Devos
DTI - UFR Sciences
BP 1039
51687 - Reims Cedex 2 FRANCE

M. El Ganaoui
IMFM
1 Rue Honorat
13003 - Marseille FRANCE

J. Etay
EPM - Madylam
BP 53
38041 - Grenoble Cedex 9 FRANCE

Y. Fautrelle
EPM - Madylam
BP 53
38041 - Grenoble Cedex 9 FRANCE

B. Fischer
Inst. of Material Science, Dept 6,
University Erlangen, Martensstr. 7,
91058 - Erlangen GERMANY

E. Fontes
Royal Institut of Technology - Applied Electrochemistry
10044 - Stockholm SWEDEN

J. Friedrich
Inst. of Material Science, Dept 6,
University Erlangen, Martensstr. 7,
91058 - Erlangen GERMANY

A. Gailitis
Latvian SSR Academy of Sciences
Institute of Physics
229021 - Riga Salaspila LATVIA

J.P. Garandet
DEM/SESC
BP 85 X
38041 - Grenoble Cedex FRANCE

Y. Gelfgat
Inst. of Physics Latvian Academy of Sciences
32, Miera str.
2169 - Salaspils LATVIA

G. Gerbeth
Research Center Rossendorf INC
PO Box 510119
01314 - Dresden GERMANY

I. Grants
Inst. of Physics Latvian Academy of Sciences
32, Miera str.
2169 - Salaspils LATVIA

Y. Hasegawa
Mechanical Engineering Laboratory
1-2 Namiki, Tsukuba
305 - Ibaraki JAPAN

D. Jault
LGIT
BP 53
38041 Grenoble Cedex 9 FRANCE

S. Kabashima
Dept. of Energy Sciences Tokyo Inst. of Tech.
4259 Nagatsuta, Midori-ku
226 - Yokohama JAPAN

Y. Khine
Detp of Mechanical and Ind. Eng.
University of Illinois 1206 W. Green Str., MC 244
61801 - Urbana-Illinois USA

Yu. Kolesnikov
Inst. of Physics Latvian Academy of Sciences
32, Miera str.
2169 - Salaspils LATVIA

M. Kumita
Dept. of Chemistry & Chemical Engineering
Kanazawa University
2-Chome, Kodatsuno
920 - Kanazawa JAPAN

C. Kupfer
Crystal Growth Laboratory
Inst. of Material Science, Dept 6,
University Erlangen, Martensstr. 7,
91058 - Erlangen GERMANY

P.Lehmann
EPM - Madylam
BP 53
38041 - Grenoble Cedex 9 FRANCE

J. Legrand
IUT Saint Nazaire
Laboratoire de Génie des Procédes CRTT - B.P. 420
44606 - Saint Nazaire cedex FRANCE

K. Loueslati
LEMTA
2, Av. de la Forêt de Haye
54504 - Vandoeuvre les Nancy FRANCE

D.Lucor
Royal Institut of Technology - Applied Electrochemistry
10044 - Stockholm SWEDEN

T. Maeda
Dept. of Energy Sciences Tokyo Inst. of Technology
4259 Nagatsuta, Midori-ku
226 - Yokohama JAPAN

S. Martemianov
E S I P
40, AV. du Recteur Pineau
86022 - Poitiers Cedex FRANCE

L. Martin Witkowski
LEGI
BP 53
38041 - Grenoble Cedex 9 FRANCE

Ph. Marty
LEGI
BP 53
38041 - Grenoble Cedex 9 FRANCE

E. Merienne
DTI - UFR Sciences
BP 1039
51687 - Reims Cedex 2 FRANCE

A. Mezulis
Latvian Academy of Sciences
2169 - Salaspils-1 LATVIA

S. Mitu
University Politechnica of Bucarest
Dept. of Chemical Eng.
1 Polizu street
78126 Bucarest ROMANIA

S. Molokov
Coventry University
School of Mathematical and Information Sciences
Priory Street
CV1 5FB - Coventry UK

R. Moreau
EPM - Madylam
BP 53
38041 - Grenoble Cedex 9 FRANCE

S. Mori
Dept. of Chemistry & Chemical Engineering
Kanazawa University
2-Chome, Kodatsuno
920 - Kanazawa JAPAN

D. Morvan
IMFM
1 Rue Honorat
13003 - Marseille FRANCE

G. Muller
Inst. of Material Science, Dept 6,
University Erlangen, Martensstr. 7,
91058 - Erlangen GERMANY

T. Murakami
Dept. of Energy Sciences Tokyo Inst. of Techn.
4259 Nagatsuta, Midori-ku
226 - Yokohama JAPAN

H.C. Nataf
LGIT
BP 53
38041 Grenoble Cedex 9 FRANCE

C. Nouar
LEMTA - 2, Avenue de la Forêt du Haye
BP 160
54054 Vandoeuvre Cedex FRANCE

T. Okamura
Dept. of Energy Sciences Tokyo Inst. of Techn.
4259 Nagatsuta, Midori-ku
226 - Yokohama JAPAN

Y. Okuno
Dept. of Energy Sciences Tokyo Inst. of Technology
4259 Nagatsuta, Midori-ku
226 - Yokohama JAPAN

A. Olivier
DTI - UFR Sciences
BP 1039
51687 - Reims Cedex 2 FRANCE

O. Ould-Dris
Laboratoire de Mécanique des Fluides
Institut de Physique - USTHB
BP N° 32
16111 - El Alia, Alger ALGERIE

P. Ozil
LEPMI
BP 75
38402 - St Martin d'Hères FRANCE

B. Pascal
EPM - Madylam
BP 53
38041 - Grenoble Cedex 9 FRANCE

K. Pericleous
University of Greenwhich
School of Computing and Mathematics Wellington Str.
SE18 6PF - London U.K.

Y. Polovko
Phys. Techn. Institute
26, Polyteknicheskaya
194021 - St Petersburg RUSSIA

K. Rajan
Coventry University
School of Mathematical and Information Sciences

E. Ramos
Centro de Investigacion en Energia UNAM
A.P. 34 Temixco, Mor.
62580 - Mexico MEXICO

E. Romanova
Phys. Techn. Institute
26, Polyteknicheskaya
194021 - St Petersburg RUSSIA

A. Salem
Laboratoire de Mécanique des Fluides
Institut de Physique - USTHB
BP N° 32
16111 - El Alia, Alger ALGERIE

K. Shinohara
National Research Lab. for Magnetic Science
1-156, Shibasimo, Kawagucchi
333 - Saitama JAPAN

S. Shioda
Dept. of Energy Sciences Tokyo Inst. of Technology
4259 Nagatsuta, Midori-ku
226 - Yokohama JAPAN

D. Simonsson
Royal Institut of Technology - Applied Electrochemistry
10044 - Stockholm SWEDEN

F. Stefani
Research Center Rossendorf INC
PO Box 510119
01314 - Dresden GERMANY

T. Suekane
Dept. of Energy Sciences Tokyo Inst. of Technology
4259 Nagatsuta, Midori-ku
226 - Yokohama JAPAN

A. Sviridov
State Aviation Technology University
27, Petrovka str.
103767 - Moscow RUSSIA

A. Tadano
National Research Lab. for Magnetic Science
1-156, Shibasimo, Kawagucchi
333 - Saitama JAPAN

M. Takeuchi
Dept. of Chemistry & Chemical Engineering
Kanazawa University
2-Chome, Kodatsuno
920 - Kanazawa JAPAN

A. Thess
Zentralinstitut fur Kernforschung
Akademie der Wissenschaften der DDR, Rossendorf, Postfach 19
8051 - Dresden GERMANY

J.P. Thibault
LEGI
BP 53
38041 - Grenoble Cedex 9 FRANCE

P. Trombetta
LEGI
BP 53
38041 - Grenoble Cedex 9 FRANCE

E. Tropp
Phys. Techn. Institute
26, Polyteknicheskaya
194021 - St Petersburg RUSSIA

K. Tsuji
Dept. of Energy Sciences Tokyo Inst. of Tech.
4259 Nagatsuta, Midori-ku
226 - Yokohama JAPAN

V. Uspenski
Institute of Mechanics, Lomonossov Univ.
Moscow RUSSSIA

J. Valdmanis
Institute of Physics
Miera 32, Latvia
2169 - Salaspils LATVIA

D. Vizman
Faculty of Physics
West University of Timisoara,
Bd V. Parvan 4
1900 - Timisoara ROMANIA

J. Walker
Detp of Mechanical and Ind. Eng.
University of Illinois 1206 W. Green Str., MC 244
61801 - Urbana-Illinois USA

O. Widlund
KTH
Faxen Lab. Dept. of Mechanics
100 44 - Stockholm SWEDEN

H. Yamasaki
Dept. of Energy Sciences Tokyo Inst. of
Technology
4259 Nagatsuta, Midori-ku
226 - Yokohama JAPAN

K. Yoshikawa
Dept. of Energy Sciences Tokyo Inst. of
Technology
4259 Nagatsuta, Midori-ku
226 - Yokohama JAPAN

S. Zahrai
KTH
Faxen Lab. Dept. of Mechanics
100 44 - Stockholm SWEDEN

THERMAL DIFFUSION AND PARTICLE SEPARATION IN FERROCOLLOIDS

E. BLUMS, A. MEZULIS
Institute of Physics, University of Latvia
Salaspils-1, LV-2169, Latvia (e-mail: eblums@tesla.sal.lv)

Abstract. Results of experiments on thermal diffusion in ferrocolloids are discussed in the paper. The Soret coefficient is evaluated from measurements of particle separation in thermodiffusion column. To interpret the separation curves measured in the presence of a magnetic field, the column theory is modified taking into account for MHD effects of free convection. It is shown that the Hartmann effect in hydrocarbon based colloids as well in ionic magnetic fluids does not influence significantly the particle separation dynamics. From unsteady separation curves positive values of the Soret coefficient of surfacted particles in tetradecane based colloids are calculated. Such direction of particle transfer toward decreasing temperatures agrees with the slip-velocity theory of thermophoresis of lyophilized particles. An uniform magnetic field oriented normally to the temperature gradient causes an increase in thermal diffusion coefficient. The results agree qualitatively well with hydrodynamic theory of particle thermomagnetophoresis.

1. Introduction

Thermal diffusion may play an important rule in some magnetic fluid and MHD technologies. For example, a change of directional solidification velocity in tin-bismuth alloy [1] has been observed. Intensive thermophoretic transfer of colloidal particles [2;3] may also effect the long-term stability of ferrofluids in devices employing high temperature gradients. From analysis of hydrodynamic Stokes problem accounting for a non-potentiality of thermomagnetic body forces it follows [2] that the thermal diffusion of particles in ferrocolloids might be effected by internal magnetic field gradients. If the external field **B** is oriented along the temperature gradient ∇T, theory predicts that particles could move toward increasing temperatures whereas in the presence of $\mathbf{B} \perp \nabla T$ an opposite direction of particle thermomagnetophoretic motion is expected. Both these effects may be interpreted as a change of the Soret coefficient of particles. Due to difficulties of small concentration difference measurements in thin layers, direct thermal diffusion measurements in liquid metals and in ferrocolloids are extremely difficult. Recently, it was shown [3] that the thermophoretic mobility of nanoparticles in magnetic colloids may be investigated by using an indirect method based on particle

A. Alemany et al. (eds.),
Transfer Phenomena in Magnetohydrodynamic and Electroconducting Flows, 1–14.

separation measurements in thermal diffusion columns. In the present paper some problems of a possible influence of magnetic field on the thermal diffusion coefficient measurements in free convective flows are discussed. Simple analytical dependencies are obtained which allow us to interpret the experimental results on unsteady particle separation in vertical channels and to evaluate the Soret coefficient of nanoparticles in hydrocarbon based ferrofluids under the effect of an external magnetic field [4].

2. Combined Thermal and Concentration Driven MHD Convection in Thermal Diffusion Column

To interpret the particle separation measurements in thermal diffusion columns performed in the presence of an external magnetic field, the column theory must be modified accounting for MHD effects of natural convection. The steady convection in a flat channel in the presence of an uniform magnetic field **B** oriented normally to the vertical walls $x'=\pm a$ of different temperatures $T(-a)=T_1$ and $T(a)=T_2$ may be considered in the classical Bousinesq approximation

$$\rho\nu\frac{d^2u}{dx'^2}-\frac{dp}{dz'}-\sigma B^2u-g(\rho-\rho_0)=0. \tag{1}$$

Here $u=u(x')$ is the vertical convection velocity profile across the channel (the coordinate z' is directed opposite to the acceleration of gravity **g**), dp/dz' is the vertical pressure gradient, ν and σ are the viscosity and the electric conductivity of the fluid. Due to a strong dependence of the colloid density ρ on particle concentration c:

$$\rho=\rho_0+\frac{\partial\rho}{\partial T}(T-T_0)+\frac{\partial\rho}{\partial c}(c-c_0)=-\beta_T\rho_0(T-T_0)+\beta_c\rho_0(c-c_0), \tag{2}$$

the buoyancy force in equation (1) depends not only on temperature but also on nonhomogeneity of particle concentration which develops during the thermodiffusive transfer. The temperature distribution across the channel in steady convection regime is linear:

$$T=T_0+(T_2-T_1)\frac{x'}{2a}=T_0+\Delta T\frac{x'}{2a}. \tag{3}$$

The steady distribution of particle concentration, which corresponds to a zero value of particle flux on unpermeable channel walls

$$j'_x=-D\frac{\partial c}{\partial x'}-D_T\frac{dT}{dx'}c=-D\left(\frac{\partial c}{\partial x'}+\frac{\alpha_T\Delta T}{2aT_0}c\right)=0, \tag{4}$$

follows to an exponential law

$$C=\frac{c}{c_0}=\frac{k}{\sinh k}\exp(-kx), \qquad k=\frac{\alpha_T \Delta T}{2T_0}, \qquad x=\frac{x'}{a}. \tag{5}$$

Here $S=D_T/D$ is the Soret coefficient, D and D_T are the Brownian and the thermal diffusion coefficients of nanoparticles, k is the separation parameter and $\alpha_T=ST_0$ is the thermal diffusion ratio. In conventional thermal diffusion column theories which are developed considering the molecular liquids with $k<<1$, instead (5) a linear dependence $C=1-kx$ usually is employed. In magnetic colloids the parameter k may reach significantly higher values $k\approx 1$, therefore in column theory the exponential concentration profile (5) must be taken into account.

Taking into account the profiles (3) and (5), from equation (1) we obtain the following distribution of the vertical convection velocity $U=ua/D$ across the channel:

$$U=\frac{Gr_T Sc}{2a}\left(x-\frac{\sinh \alpha x}{\sinh \alpha}\right)-$$
$$\frac{Gr_c Sc}{(\alpha^2-k^2)}\left[\frac{\left(1-\frac{k}{\tanh k}\right)}{\left(1-\frac{\tanh \alpha}{\alpha}\right)}\left(\frac{\cosh \alpha x}{\cosh \alpha}-\frac{\tanh \alpha}{\alpha}\right)+\frac{k\sinh \alpha x}{\sinh \alpha}+\frac{k\exp(-kx)}{\sinh k}-1\right]. \tag{6}$$

Here $Gr_T=\beta_T \Delta T g a^3/\nu^2$ and $Gr_c=\beta_c g c_0 a^3/\nu^2$ are the thermal and the concentration Grashoff numbers, $Sc=\nu/D$ is the Schmidt number of colloidal particles and $\alpha=Ba\sqrt{o/\rho\nu}$ is the Hartmann number. For small values of thermal diffusion parameter k a simpler dependence which corresponds to an additive action of thermal and concentration buoyancy forces on free convection is valid:

$$U=\frac{\left(Gr_T+2kGr_c\right)Sc}{2\alpha^2}\left(x-\frac{\sinh \alpha x}{\sinh \alpha}\right). \tag{7}$$

In initial stage of separation process, when the mean particle concentration may be considered being independent on vertical coordinate and equal to $c=c_0$, from equation (6) we obtain the vertical particle flux j'_z (in a non-dimensional form):

$$j_z=\frac{j'_z a}{c_0 D}=\frac{1}{2}\int_{-1}^{+1} CUdx=\frac{k}{2\sinh k}\int_{-1}^{+1} e^{-kx}Udx=$$
$$\frac{Gr_T Sc}{2a^2}\left[\frac{k}{(\alpha^2-k^2)}\left(\frac{\alpha}{\tanh \alpha}-\frac{k}{\tanh k}\right)-\frac{1}{\tanh k}+\frac{1}{k}\right]-\frac{Gr_c Sc}{(\alpha^2-k^2)}\left\{\frac{(1-k/\tanh k)}{(1-\tanh \alpha/\alpha)}\times\right. \tag{8}$$
$$\left.\left[\frac{k}{(\alpha^2-k^2)}\left(\frac{\alpha \tanh \alpha}{\tanh k}-k\right)-\frac{\tanh \alpha}{\alpha}\right]-\frac{k^2}{(\alpha^2-k^2)}\left(\frac{\alpha}{\tanh \alpha}-\frac{k}{\tanh k}\right)+\frac{k}{\tanh k}-1\right\}.$$

Figure 1 represents the solution (8) graphically. If the thermodiffusive transfer across the channel does not effect the buoyancy force, $Gr_c=0$, the direction of particle convective separation in column depends on the sign of parameter k. Particles which are moving toward increasing temperatures ($k<0$) will be collected in the upper separation chamber of column whereas positive Soret coefficients will cause a rise of particle concentration in the lower chamber. Magnetohydrodynamic suppression of convection velocity in electroconducting fluids in the presence of a transversal magnetic field

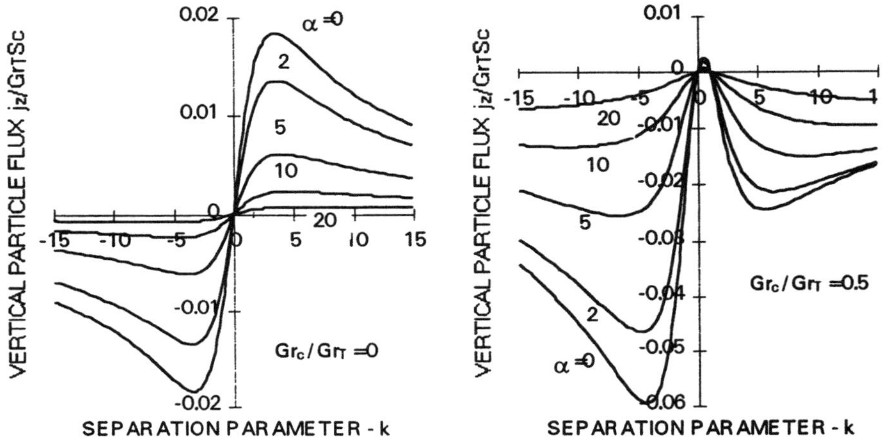

Figure 1. Vertical particle flux $j_z/Gr_T Sc$ in a flat vertical channel in the presence of a transversal magnetic field $\mathbf{B}=B_x=const$.

always causes a significant reduction of the intensity of particle convective separation. In magnetic fluids, usually, the density of particles ρ_p is significantly higher than that of the carrier liquid ρ_0. Therefore, $Gr_c >> 0$, and even at low thermal diffusion coefficients the concentration buoyancy force significantly effects both the convection velocity and the vertical convective transfer of particles. From the expression (8) and from results presented in Fig. 1 it follows that under the conditions of an intensive thermal diffusion the convective particle flux is directed toward vector **g** independently of the sign of parameter k. The particle collection in upper chamber is expected only in a relatively narrow interval of negative values of thermal diffusion parameter $k<Gr_T/2Gr_c$ when the thermogravitational mechanism of free convection in channel still prevails. From results presented in Fig. 1 we can see that this interval of parameter k values practically does not depend on Hartmann number α.

The steady concentration profile (5) is valid only for narrow channels starting some transition time $t=t_0>a^2/D$. In the initial regime of particle separation, $t<t_0$, an unsteadiness of the concentration profile in (1) and (2) has to be taken into account. To simplify the calculations in this regime, we will use an approximate profile

$$\frac{c}{c_0}-1=\pm\frac{km}{3\mp km}(m-1\mp x)^3,\qquad m=\sqrt{12\tau}\,,\qquad \tau=\frac{Dt}{a^2}\,,\tag{9}$$

which corresponds to the equation of an integral mass balance in boundary layer. Here $m=\delta_c/a$ with δ_c being the thickness of the concentration boundary layer near both walls. (The upper signs in (9) correspond to negative $x<0$, the lower ones -- to positive x). This approximation is very close to the concentration profiles of exact solution of the unsteady diffusion equation. For example, from the exact solution it follows that at $\tau<<1$ the gradient of concentration at channel walls is equal to $c'_w=c_0 2\sqrt{\tau/\pi}$, whereas the approximate profile (9) gives a value of the coefficient in this "square root" law equal to $2/\sqrt{3}$.

Let us consider for simplicity only one half of the channel, $x=0\div 1$. For small $km<<3$ the concentration profiles c/c_0-1 near both channel walls are anti-symmetric, therefore $dP/dz=0$. The velocity profile for positive x now is the following:

a) if $|x|<m$:

$$U=\frac{Gr_T Sc}{2\alpha^2}\Big[\Big(x-\frac{\sinh\alpha x}{\sinh\alpha}\Big)-2k\frac{Gr_c Sc}{Ra_T}\Big(\frac{m}{3}+\frac{2}{m\alpha^2}-\frac{2\sinh m\alpha}{m^2\alpha^3}\Big)\frac{\sinh\alpha x}{\sinh\alpha}\Big];\tag{10a}$$

b) if $m\le|x|\le 1$:

$$\begin{gathered}U=\frac{Gr_T Sc}{2\alpha^2}\Big\{\Big(x-\frac{\sinh\alpha x}{\sinh\alpha}\Big)+2k\frac{Gr_c Sc}{Ra_T}\Big[\frac{2\sinh\alpha(1-m)\cdot\cosh\alpha}{m^2\alpha^3}\frac{\cosh\alpha x}{\cosh\alpha}-\\ \Big(\frac{m}{3}+\frac{2}{m\alpha^2}+\frac{2\sinh\alpha(1-m)\cdot\cosh\alpha}{m^2\alpha^3}\Big)\frac{\sinh\alpha x}{\sinh\alpha}-\frac{(1-m-x)^3}{3m^2}-\frac{2(1-m-x)}{m^2\alpha^3}\Big]\Big\}.\end{gathered}\tag{10b}$$

The corresponding mass flux for one-half of the channel is the following:

$$\begin{gathered}j_z=-\frac{kGr_T Sc}{6m^2\alpha^3}\Big\{\Big[\Big(\frac{3m^2}{\alpha^2}+\frac{6}{\alpha^4}\Big)-\frac{1}{\tanh\alpha}\Big(\frac{m^3}{\alpha}+\frac{6m}{\alpha^3}\Big)-\frac{6\sinh\alpha(1-m)}{\alpha^4\sinh\alpha}+\frac{m^4}{4}-\frac{m^5}{20}\Big]-\\ 2k\frac{Gr_c}{Gr_T}\Big[\frac{2\sinh\alpha(1-m)}{m^2\alpha^3\sinh\alpha}\Big(\frac{m^3}{\alpha}+\frac{6m}{\alpha^3}-\frac{6\sinh\alpha m}{\alpha^4}\Big)-\Big(\frac{m}{3}+\frac{2}{m\alpha^2}\Big)\times\\ \Big(\frac{3m^2}{\alpha^2}+\frac{6}{\alpha^4}-\frac{1}{\tanh\alpha}\Big(\frac{m^3}{\alpha}+\frac{6m}{\alpha^3}\Big)-\frac{6\sinh\alpha(1-m)}{\alpha^4\sinh\alpha}\Big)-\frac{m^5}{21}-\frac{2m^3}{5\alpha^2}\Big]\Big\}.\end{gathered}\tag{11}$$

For small Hartmann number values $\alpha\to 0$ instead the (10) and (11) a simpler solution is valid:

$$U = -\frac{Gr_T Sc}{12}(x^3 - x) + kGr_c Sc[\frac{m^3}{60}x - \frac{1}{60m^2}(m-1+x)^5], \tag{12}$$

$$\frac{j_z}{Gr_T Sc} = -\frac{k}{360}(m^3 - \frac{m^4}{2} + \frac{m^5}{14}) - k^2 \frac{Gr_c}{Gr_T}\frac{5}{180 \cdot 36}(m^5 - \frac{9}{25}m^6). \tag{13}$$

Some unsteady convection velocity profiles which correspond to the initial stage of particle separation $Dt/a^2 < 1/12$, when the concentration boundary layer approximation (9) is valid, are shown in Fig. 2. Profiles are plotted in accordance with expressions (10) and (9) for $\alpha = 10$, the curves correspond to equivalent positive and negative values of the thermal diffusion parameter $\pm k$. In colloids having positive thermal diffusion ratio α_T $(k>0)$, the particle transfer across the channel causes a mo-

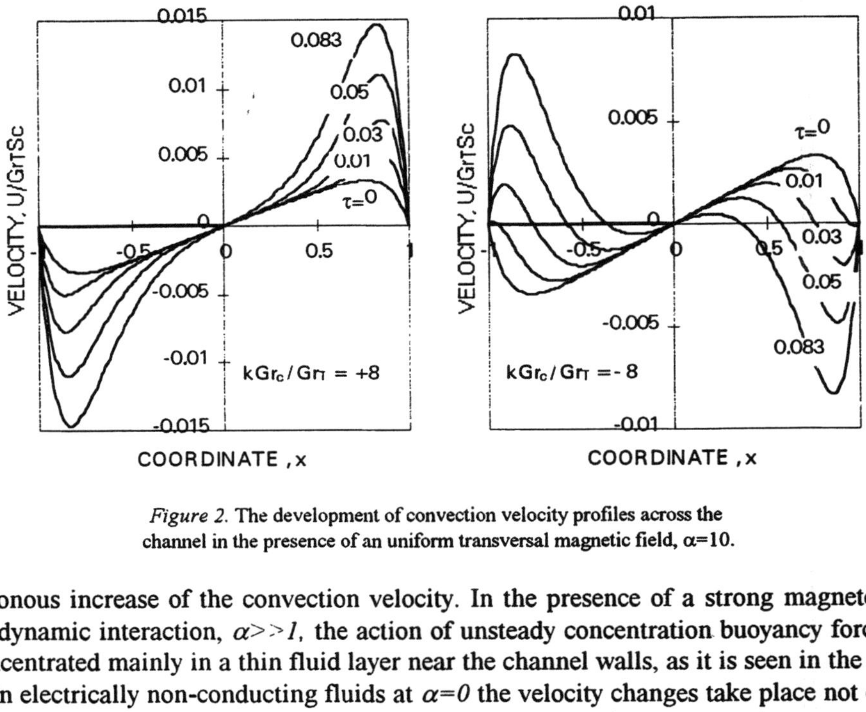

Figure 2. The development of convection velocity profiles across the channel in the presence of an uniform transversal magnetic field, $\alpha = 10$.

notonous increase of the convection velocity. In the presence of a strong magnetohydrodynamic interaction, $\alpha >> 1$, the action of unsteady concentration buoyancy force is concentrated mainly in a thin fluid layer near the channel walls, as it is seen in the Fig. 2. In electrically non-conducting fluids at $\alpha = 0$ the velocity changes take place not only near walls but also in the central part of the channel. If particles are moving toward increasing temperatures, $k<0$, the concentration buoyancy force is directed opposite to the thermogravitation one. Therefore, non-monotonous velocity profiles are developing during the formation of a concentration profile across the channel. Finally, reaching $\tau = \tau_0$ a reverse anti-symmetric velocity distribution develops which corresponds to the steady profile (6) (in the transition period $1/12 < \tau < \tau_0$ the velocity profiles may not be calculated because the boundary layer approximation and the profile (9) are no more valid).

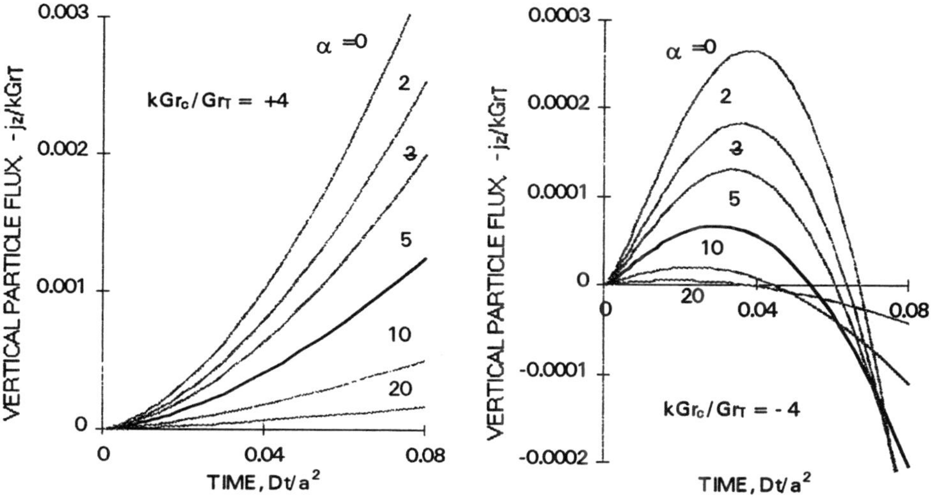

Figure 3. The unsteady vertical particle flux in a flat thermal diffusion channel

The dependence of the vertical particle flux on time for various values of the fluid Hartmann number is shown in Fig. 3. If particles are transferred across the channel toward decreasing temperatures ($k>0$), the concentration buoyancy force causes an increase of vertical particle flux, particles are collected in the lower chamber. If k values are negative, in initial period of time the particles are transferred upward but due to the development of a concentration buoyancy force which now acts opposite to the thermogravitation force, a monotonous reduction of particle flux takes place. When the regime of a reversed convection velocity near channel walls is reached, the vertical mass flux turns to an opposite direction. Therefore, particles like in the previous case $k>0$, are transferred again to the lower separation chamber. Finally, reaching $\tau=\tau_0$ there develops a steady particle flux (8) which corresponds to the velocity profile (6) of a stationary particle distribution across the channel. The magnetohydrodynamic suppression of convection velocity always causes a reduction of the intensity of particle separation in the nonstationary regime.

The non-monotonous velocity profiles of several stagnation points shown in Fig. 2 , obviously, are unstable. The lose of shear flow stability in channel during the reversion of the convection velocity may significantly effect the dynamics of particle separation.

3. Ferrofluid samples and measurement technique

The separation experiments are performed using a vertical flat column of width δ=0.52 mm and of height L=86.5 mm. The heated and cooled walls (copper plates with polished surfaces) are connected with two precise thermostats, keeping the temperatures

T_2 and T_1 constant. The difference $\Delta T=T_2-T_1$ is varied in the interval 2 -18 °C. In order to lower the temperature inhomogeneities caused by heat exchange with the environment, the upper and lower containers of equal volume ($V_c \approx 1$ cm^3) and the outside walls of the channel are made from plastic material with a low heat conductivity.

The thermal diffusion ratio α_T is evaluated from the measured particle separation curves $\Delta c / c_0 = f(\tau)$ ($\Delta c=c_l-c_u$ is the difference between particle concentration in the lower (c_l) and the upper (c_u) chambers). In initial stage of separation, $\tau \ll 1$, the particle concentration changes in column chambers do not influence the vertical particle flux, therefore the separation curves may be analyzed employing a simple relation

$$\frac{\Delta c}{c_0} = -\frac{2Sa}{V_c}\int_0^{\tau} j_z d\tau \tag{14}$$

where S denotes the cross sectional area of the channel. Experiments with hydrocarbon based ferrofluids correspond to $\alpha \ll 1$ even in the presence of strong magnetic fields. From (14) with respect the relation (13) it follows:

$$\frac{\Delta c}{c_0} = \frac{kGr_T Sc}{5400}\frac{a}{L_c} m^5\left[\left(1-\frac{5m}{12}+\frac{5m^2}{98}\right)+k\frac{Gr_c}{Gr_T}\left(\frac{25m^2}{126}-\frac{m^3}{16}\right)\right], \quad m=\sqrt{12\tau} \tag{15}$$

Here $L_c=V_c/S$ is the effective length of the separation chambers (in our experiments $L_c/L=0.682$). In the initial stage of separation process when $m \ll 1$, from (15) it follows that the concentration difference in chambers develops in accordance with a simple relation $\Delta c / c_0 = const \cdot t^{5/2}$. For m close to 1 instead the formula (15) an approximate empirical relation may be used:

$$\frac{\Delta c}{c_0} \approx 0.008kGr_T Sc\frac{2a}{L_c}\tau^2 \ . \tag{16}$$

Starting $\tau=\tau_0$ the vertical particle flux j_z reaches the steady regime (8) (according to Ref. [5] $\tau_0 \approx 0.4$), therefore, the concentration difference Δc starts to develop linearly in time. For $\alpha \ll 1$ and for small k values (k<1) from (14) with the respect (8) we obtain:

$$\frac{\Delta c}{c_0} = \frac{1}{45}\frac{a}{L_c}k(Gr_T + kGr_c)Sc\tau \ . \tag{17}$$

At durable separation, reaching a certain transition time τ_t,, the concentration in column chambers starts to saturate. The smaller the ratio L_c/L or the higher the parameters k and Gr, the sooner the linearity of the saturation curves vanishes. According to Ref. [6], in this regime $\tau > \tau_t$ a certain time interval exists in which the dependence of concentration difference Δc on time τ may be approximated by a square-root law

$$\frac{\Delta c}{c_0} \approx \gamma \sqrt{\frac{7}{40}\frac{L}{L_c}} k\sqrt{\tau}. \qquad (18)$$

The coefficient γ slightly depends on the ratio L/L_c (L is the column height). For devices having nearly equal volumes of the active column V and of the upper and lower containers V_c, from results reported in Ref. [6] it is found the value $\gamma \approx 0.37$ [3].

Particle concentration in both separation chambers is determined by measuring the resonance frequency of an LC oscillator [7]. The inductance coils of the oscillator are mounted inside the both containers. The optimum oscillation frequency in the range of 70-85 kHz is chosen taking into account a compromise between the sensitivity of particle concentration measurements and the errors caused by the relaxation of the colloid's magnetization. The concentration calibration curves of both coils in the presence of an external magnetic field are detected experimentally. For ferrofluids used in the present thermal diffusion experiments, the dependence of LC resonance frequency f on particle concentration is linear, $f=f_0-\beta\cdot c$, the coefficient β is proportional to the colloid magnetic properties. Such a linearity, obviously, is expected only for colloids of low initial magnetic susceptibility $\chi_0 \ll 1$. In the presence of an increasing magnetic field when according to the superparamagnetic nature of magnetic colloids the differential magnetic susceptibility monotonously decreases, a significant reduction of resonance frequency sensitivity to the particle concentration changes takes place.

Experiments are performed employing a tetradecane based ferrofluid sample containing chemically coprecipitated magnetite nanoparticles. The colloid is stabilized by using oleic acid as a surfactant. The saturation magnetization of colloid detected from asymptotics of the sample magnetization curve in strong field is $M_s=10.6$ kA/m [3]. Assuming that the magnetization of particle material is equal to that of a bulk magnetite ($M_s=480$ kA/m) it is found that the value of particle volume fraction is $\varphi_m= 2.2\times10^{-2}$. The mean particle magnetic moment measured by a magnetogranulometry technique [8] is $m=1.4 \times 10^{-19}$ A$\cdot$m^2. Such value of m with the respect of the saturation magnetization M_s corresponds to the mean "magnetic" diameter of particles $d = 7.1 \times 10^{-9}$ m^3. The volume fraction φ_m and the particle diameter d allow us to calculate the particle mean concentration: $c_0=6\varphi_m/(\pi d^3)=1.18\times10^{23}$ m^{-3}. The "magnetic" diameter d is used also to calculate the Brownian diffusion coefficient of particles: $D = k_BT/(3\pi\eta d)$. The dynamic viscosity $\eta = 4.6\times10^{-3}$ N$\cdot$sec/m^2 as well as the thermal expansion coefficient of the colloid $\beta_T=8.3\times10^{-4}$ 1/K are detected experimentally, whereas the coefficient $\beta_\varphi= 5.9$ is calculated by using the known values of density of a magnetite (ρ_p) and of a carrier liquid (ρ_0): $\beta_\varphi \approx \rho_p/\rho_0-1$.

4. Results and discussions

Two series of separation experiments are performed to evaluate the thermal diffusion ratio in a zero magnetic field. In the first series the particle concentration is varied by a dilution of initial sample whereas in the second one the wall temperature difference is changed. Figure 4 represents one of the particle nonstationary separation curve. These measurements are performed by using a colloid of relatively low particle concentration (the initial sample is diluted to 1:4), the temperature difference is $\Delta T=10\ ^{\circ}C$.

From the results presented in Fig. 4 we see that the separation curve confirms both the dependence $\Delta c / c_0 = const \cdot t^{5/2}$ (initial regime, $t<800$ s) and the empirical relation (18) (starting approximately $t=2000$ s). The linear dependence (17) is not seen. Obviously, the volume of column separation chambers V_c is to small to reach the steady vertical particle flux before the concentration difference in column chambers starts to saturate.

Some preliminary measurements are performed using also a water based ionic ferrofluid containing colloidal Mn-Zn ferrite particles. In these experiments an oppo-

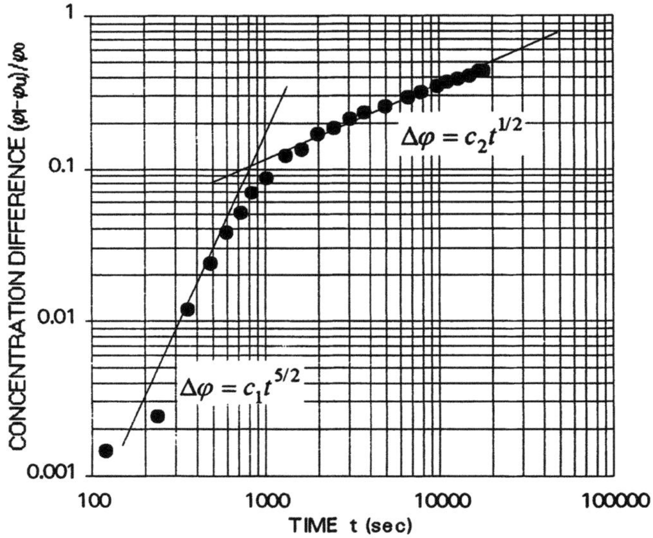

Figure 4. The particle unsteady separation in a vertical column. Diluted sample, $\varphi=5.5\cdot 10^{-3}$, $\Delta T=10\ ^{\circ}C$

site direction of the vertical particle flux is observed. Thus, the electrically stabilized particles in hydrosols are transferred toward increasing temperatures. Unfortunately, in the thermal diffusion column of different copper wall temperature the ionic ferrofluid during the separation experiment loses stability, therefore more detailed measurements are not completed at the present. The difference in the direction of thermodiffusivity of surfacted particles in hydrocarbons and of electrically stabilized particles in ionic ferrofluids has been observed also in forced Rayleigh scattering experiments [10]: from optical signals of dynamic grating in thin ferrofluid layers negative Soret coefficients for electrically stabilized ferrite nanoparticles are calculated.

Analyzing the particle separation curves of various series of experiments we made a conclusion that the thermal diffusion ratio α_T neither depends on particle concentration nor on temperature difference. The average value α_T calculated from the initial part of separation curves is $\alpha_T = +24 \pm 5$. From analysis of measurements in the intermediate interval of times by using formula (18) a practically identical result is obtained, $\alpha_T = +21 \pm 4$.

The thermophoretic transfer of particles in colloidal dispersions is caused mostly by a slip velocity. Exact theories of small particle thermophoretic motion are developed only for gaseous suspensions of hard particles or droplets when the slip characteristics may be calculated by using existing gas theories. Some general ideas of theory of the slip velocity at a solid-liquid interface resulting from tangential temperature gradient is given in Ref. [11]. Analyzing the enthalpy flux carried by forced convection of fluid across a porous barrier and applying Onsager's reciprocal theory for the slip velocity it was predicted that free particles in surfacted colloids would move in the direction of decreasing temperatures. Our experimental results agree with these predictions.

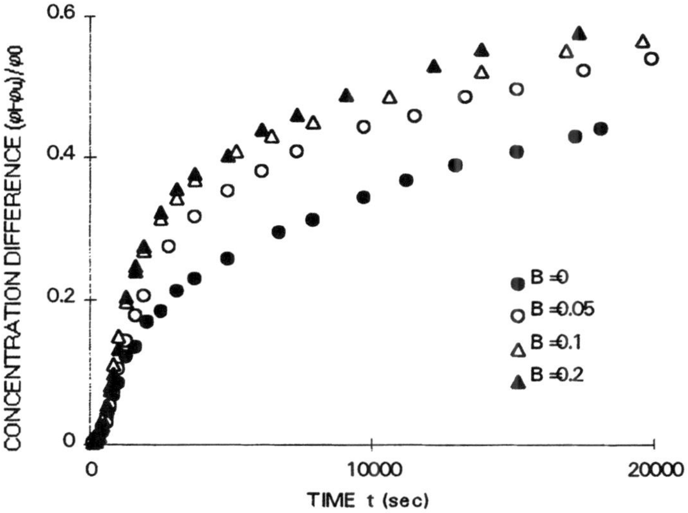

Figure 5. Particle separation curves in the presence of an uniform magnetic field **B** (measured in T) oriented along the heated and the cooled walls, **B**// ∇T.

Figure 5 represents some results of nanoparticle separation measurements performed in the presence of an uniform magnetic field **B** which is oriented horizontally along the heated and the cooled walls of the channel. From the presented separation curves it is seen that the increase of field intensity causes a remarkable intensification of vertical particle transfer. The width of the channel in magnetic field direction b significantly exceed the channel thickness a ($a/b \approx 0.016$). Besides, the Hartmann number in surfacted hydrocarbon based colloids is very small, $\alpha < 1$. Therefore, we may conclude that the MHD hydrodynamic effects in our experiments are not responsible for changes of the mass transfer intensity. Obviously, we have observed an effect of the uniform magnetic field on the thermal diffusion of nanoparticles.

Figure 6 represents the dependence of particle thermophoretic mobility on magnetic field intensity. The results are calculated from unsteady particle separation curves using the expressions (17) and (18) and assuming that only the particle thermal diffusion is effected by external magnetic field. As it is seen, the both methods of analysis give a practically identical result. This result allow us to conclude that we really have observed the effect of magnetic field on the particle thermophoresis but not on the fluid viscosity or the diffusion coefficient. The expressions (17) and (18) depend on ν and on D in a different way, therefore any changes of these coefficients under the effect of a magnetic field would give a divergence of points in Fig. 6 calculated from the separation curves in a different way. Indeed, the effect of a fluid magnetoviscosity is expected to be insignificantly small because the measurements are performed using colloids of a relatively low particle concentration. Moreover, the "square root" dependence (18) does not contain the fluid viscosity at all. Obviously, additional more detailed experiments are needed to evaluate the necessary correction of results by taking into account for a possible effect of the magnetic field on the translation diffusion of particles.

The increase in thermal diffusivity of particles in the presence of the uniform

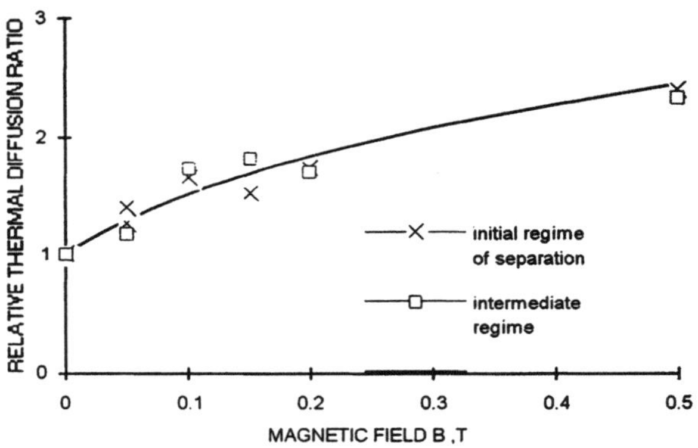

Figure 6 The effect of uniform magnetic field **B**$\perp \nabla T$ on the particle thermal diffusion ratio

magnetic field **B**$\perp \nabla T$ qualitatively agrees well with our previous theoretical predictions [2] which are based on analysis of a modified Stokes problem for magnetic nanoparticles by taking into account for a non-potential thermomagnetic force caused by local magnetic field perturbations and temperature gradients around the particle. According to this theory the magnetic field oriented normally to the temperature gradient causes a particle transfer toward decreasing temperatures. Therefore, in surfacted ferrocolloids in which the ordinary Soret coefficient in a zero field is positive, an increase in α_T in the presence of a transversal magnetic field is expected. It is interesting to note that the experimentally measured thermomagnetophoretic effect presented in Fig. 6 is significantly stronger than the theoretically evaluated one. Obviously, if thermodiffusive transfer of nanometer scaled particles is considered, the hydrodynamic theory [2] must be specified taking into account for slip characteristics and for a temperature jump on the particle solid-liquid interface.

From the hydrodynamic theory [2] and from the measurements of particle dynamic grating in the forced Rayleigh scattering experiment [9] it follows that the Brownian diffusion coefficient of colloidal particles in magnetic fluids also depends on the external magnetic field and on its orientation. The transversal field $\mathbf{B}\perp\nabla T$ causes an monotonous decrease of the translation diffusion coefficient D. Numerical estimates performed on basis of the theory [2] lead to the conclusion that an approximately 20% reduction of D in our experiments in the asymptotic regime of a magnetic saturation of the colloid may be observed. If such effect in our experiments really takes place, the analysis of separation curves by using corrected coefficients in expressions (18) and (19) would give the thermomagnetophoretic effect even stronger than that presented in the Fig. 6.

In the presence of a parallel field ($\mathbf{B}/\!/\nabla T$) under the conditions of a magnetic saturation of the colloid we have observed an oscillatory regime of particle separation in the column. To interpret these results, various reasons of such peculiarities, including the possible change in the direction of particle thermophoretic motion (at negative α_T a loss of the shear flow stability may occur, see Fig. 2) as well as the effect of a thermomagnetoconvective flow instability, have to be taken into account. Our latest separation experiments [12] which are performed by using a temperature sensitive magnetic fluid containing Mn-Zn ferrite nanoparticles of high pyromagnetic coefficients, show a strong reduction of the Soret coefficient under the effect of a parallel magnetic field $\mathbf{B}/\!/\nabla T$. In the regime of magnetic saturation an approximately zero value of α_T or even a change of the direction of particle thermomagnetophoretic transfer is observed.

5. Conclusions

The particle separation measurements in vertical thermal diffusion column indicate a high thermophoretic mobility of magnetic nanoparticles in ferrofluids. Colloidal particles in hydrocarbons stabilized by using surfactants are moving toward lower temperatures. Such direction of thermophoresis agrees with the predictions of thermodynamic theory accounting for a slip velocity of lyophilic solid-liquid boundaries. Some preliminary experiments performed by using a water-based ionic ferrofluids indicate an opposite direction of the thermophoretic transfer of electrically stabilized nanoparticles. Obviously, the temperature non-homogeneity of a double-electric layer is responsible for thermophoresis of the charged particles. Experiments confirm the theoretically predicted effect of an uniform magnetic field on the thermophoretic transfer of nanoparticles in ferrofluids under nonisothermic conditions. The transversal magnetic field $\mathbf{B}/\!/\nabla T$ causes a significant increase of the Soret coefficient of surfacted magnetite particles in hydrocarbons. Hartmann's type MHD effects in thermal diffusion column experiments using both the lyophilized and the electrically stabilized magnetic fluids may be neglected. Strong magnetic field oriented along the temperature gradient may cause a thermomagnetic instability of convective shear flow in vertical channels.

Acknowledgments

The authors are thankful to our colleges G. Kronkalns for providing us with the ferrofluid samples and M. Maiorov for performing the magnetization measurements as well as the magnetogranulometry analysis of polydisperse samples. We are thankful also to J.-C. Bacri and to A. Bourdon for introducing in the problems of optically induced thermodiffusive grating in ionic ferrofluids. The work has been financially supported by the Latvian Science Council (Grant 96.0271).

References:

[1] VanVaerenbergh, S., Coriel, S. R., McFadden, G. B., Muttay, B. T., and Legros, J. C.: Modification of morphological stability of Soret diffusion, *J. Crystal Growth* **147** (1995), 207-214.

[2] Blums, E.: Some new problems of complex thermomagnetic and diffusion driven convection in magnetic colloids, *J. Magn. and Magn. Materials* **149** (1995), 111-115.

[3] Blums, E., Mezulis, A., Maiorov, M., and Kronkalns, G.: Thermal diffusion of magnetic nanoparticles in ferrocolloids: experiments on particle separation in vertical columns, *J. Magn. and Magn. Materials* **169** (1997), 220-228.

[4] Blums, E., and Mezulis A.: The Effect of Magnetic Field on Particle Thermophoresis in Ferrocolloids: Separation Measurements in Thermodiffusion Columns, in *19th International Congress of Theoretical and Applied Mechanics, Kyoto, August 25-31, 1996, Abstracts*, (1996), p. 748.

[5]. Blums, E., and Savickis, A.: Convection in thermomagnetic diffusion column: unsteady effects caused by particle transfer in ferrofluids, in *The 14th International Riga Conference on Magnetohydrodynamics MAHYD'95, August 24-26, 1995, Jurmala, Latvia, Abstracts*, (1995), p. 167.

[6] Blums, E., Kronkalns, G., and Ozols, R.: The characteristics of mass transfer processes in magnetic fluids, *J. Magn. and Magn. Materials* **39** (1983), 142-146.

[7] Mezulis, A., E. Blums, E., Kronkalns, G., and Maiorov, M.: Measurements of thermodiffusion of nanoparticles in magnetic colloids, *Latvian Journal of Physics and Technical Sciences* 1995, 5, 37-50.

[8] Maiorov, M. M.: Magnetization curve of magnetic fluid and distribution of magnetic moment of ferroparticles, in *Proc. 10th Riga MHD Conference, Salaspils*, (1981), pp. 11-18.

[9] Bacri, J.-C., Cebers, A., Bourdon, A., Demouchy, G., Heegard, B. M., Kashevsky, B. M., and Perzynski, R.: Transient grating in a ferrofluid under magnetic field. Effect of magnetic interactions on the diffusion coefficient of translation, *Phys. Rev. E* **52** (1995), 3936-3942.

[10] Langlet, J.: Generation de second harmonique et diffusion Rayleigh forcee dans les colloides magnetiques, *Ph.D. Thesis*, de l'Universite Paris 7 Denis Diderot, Paris, 1996.

[11] Derjaguin, B. V., Churaev, N. V., and Muller, V. M.: *Surface Forces*, Plenum Press, N. Y, 1987.

[12] Blums, E., Odenbach, S., and Mezulis, A.: Soret effect of nanoparticles in ferrocolloids in the presence of a magnetic field, *Phys. Fluids* (in Press).

I - MHD FLOWS AND TURBULENCE

GEODYNAMO AND M.H.D

D. JAULT, P. CARDIN AND H.C. NATAF
Laboratoire de Géophysique Interne et Tectonophysique
BP 53, 38041 Grenoble Cedex 9, France

Abstract. The main part of the geomagnetic field is generated by self-induction in the Earth's molten core. The *geodynamo* mechanism is not yet fully understood as an M.H.D. problem even though we have many constraints coming from observations, as well as from geophysical and geochemical theories. This Earth's science problem combines many of the difficulties we face in physics (turbulence), in applied mathematics (nonlinear equations, boundary layers), in technology/engineering (liquid metal experiments) and in computing sciences (numerical modelling).

1. Introduction

There are already many very good reviews of geodynamo theory (Fearn *et al.*, 1988) (Roberts *et al.*, 1992) (Fearn, 1997). Moreover, a few textbooks include thorough discussions of the application of MHD theory to the Earth's core (Moffatt, 1978) (Roberts, 1987). Thus there is no need for yet another review of geodynamo theory. On the other hand, it may be interesting to summarize what our team investigating the geodynamo mechanisms expects to learn from experiments with liquid metals. A recent and very helpful review discusses the laboratory experiments that have illuminated core dynamics (Aldridge, 1997) but it says very little about experiments with liquid metals. We hope to arouse interest in the geodynamo problem among physicists investigating the magnetohydrodynamics of liquid metals and this paper is intended for them. We confront the experimental approach with the analytical and numerical approaches and we discuss previous relevant experiments. On the other hand, in recent years, most progress has come from numerical calculations of the solution to the MHD equations, of which the limitations are underlined.

A. Alemany et al. (eds.),
Transfer Phenomena in Magnetohydrodynamic and Electroconducting Flows, 17–30.

In the first part of this paper, the geophysical data on the geodynamo is presented. Secondly, we discuss turbulence and rotating fluid theory for the Earth's core. We report also on experiments with precessing flows, which have shed some light on this problem. The third part is devoted to boundary layers and internal shear layers. Such layers are indeed as important for rotating flows as for conducting fluid cavities permeated by a magnetic field. Geodynamo theory has been largely influenced by analytical and numerical analyses of the magnetoconvection problem. Some of their conclusions are discussed in the fourth section. Then, we give a short overview of the recent progress in geodynamo numerical modelling and finally ask the question : how can liquid metal experiments help to understand the geodynamo mechanism?

2. Geophysical constraints on the geodynamo

The Earth's core is a quasi-spherical ball, with radius $R = 3480km$, of iron-rich (90%) materials encased in a rocky mantle. It is liquid, except for the solid inner body at the center of the Earth ($R = 1220km$). Its density ρ (10-13 g/cm^3) and its shape are very well determined by inversion of seismological data together with the mass and moment of inertia of the Earth. From its density, we can deduce the hydrostatic pressure which varies from 130 GPa at the CMB (Core Mantle Boundary) to 360 GPa at the center. The core's composition is inferred from theories of Earth's formation. Finally, the equation of state for iron and iron alloys are being determined in geophysical laboratories using high pressure devices (Poirier, 1991). These studies will tell us how the liquid at the bottom of the outer core freezes at the inner core surface as the Earth cools during its history. Now, even the temperature (from $3500 \pm 500K$ at the CMB to $5500 \pm 500K$ at the center) is not very well known. Compressibility of the fluid core strongly influences heat transfer. However, most compressibility effects can be neglected when deriving the equations for convection in the Earth's core (Braginsky *et al.*, 1995).

Of course, other evidence of the presence of the metallic core within the Earth is the observation of the magnetic field. Since the mantle is insulating, the geomagnetic field can be downward continued from Earth's surface to the CMB where it can be described, at first order, as a dipole field aligned along the axis of rotation. This dipole term apart, the energy spectrum for the lowest degrees (at least up to $l = 13$), in the expansion on the spherical harmonic basis $(Y_l^m)_{lm}$ of the geomagnetic potential (Backus *et al.*, 1996), is almost flat. Changes of the field on timescales from years to centuries (its secular variation) have been monitored in dedicated observatories (Merrill *et al.*, 1996). Unfortunately, these times are very short compared to the magnetic diffusion time for the dipole (40000 years) calculated from the electrical conductivity σ ($7-810^5 Sm$) of the core that is extrapolated from lower pressure measurements (Secco, *et al.1989*). The physical properties of liquid iron in the Earth's core are not thought to be too disparate from the properties of liquid metals used in laboratories, as it can be seen in the table below.

Prandtl number	Pr	ν/κ	0.1
Magnetic Prandtl number	Pm	ν/λ	10^{-6}

Here κ and λ are, respectively, thermal and magnetic diffusivities. $\lambda = 1/\mu\sigma$ where μ is magnetic permeability. Viscosity ν is so small that the spin-up time is similar to the magnetic diffusion time allowing fluctuations in the differential rotation rate between core and mantle. Thus, the torques acting between these two bodies do not cancel each other out. They impart changes in the rotation rate of the solid Earth, which can be observed.

Two important dimensionless numbers allow to understand the difficulties that would be encountered were one to try and mimic in the laboratory the processes taking place inside the core. These are the Ekman number E and the Elsasser number Λ.

Ekman number	E	$\nu/\Omega R^2$	10^{-15}
Elsasser number	Λ	$\sigma B^2/\rho\Omega$	1

Ω is the Earth's rotation rate and B is a typical intensity of the geomagnetic field. In the laboratory, high rotation rate would be necessary to get Ekman numbers as large as $10^{-5}-10^{-7}$ and the magnetic field intensity would have to approach $10^{-2} - 10^{-1}T$. to make the Elsasser number unity.

To conclude this geophysical overview of the core, we would like to underline a recent result which is interesting for this community. Seismologists have shown that the inner core is anisotropic. Travel times of seismic waves are shorter for polar paths than for equatorial paths. This is not yet understood. Amongst the possible explanations, there is crystal growth in the presence of a magnetic field.

3. Turbulence in the Earth's core

The liquid motions are strongly influenced by rotational and magnetic forces. The magnetic field must be of predominantly large scale since otherwise ohmic dissipation of the electrical currents sustaining it would produce excessive quantities of heat. The upper bound on the heat flux entering the bottom of the mantle ($10^{13}W$) is indeed a severe constraint on dynamo models (Brito *et al.*, 1996). The kinetic Reynolds number (10^8) indicates that the motions are strongly turbulent. Braginsky and Meytlis (1990) have developed an heuristic approach of the turbulence in presence of rotation and magnetic field that may give some insight into core turbulence. It is well argued in the thorough discussion of the equations governing convection in Earth's core by Braginsky *et al.* (1995). Motions are decomposed into small scale and large scale motions, denoted respectively $\mathbf{u}$ and $\mathbf{U}$. Magnetic forces and rotation forces are able to constrain heavily the geometry of the small scale motions. These local motions are expected to be invariant in the directions of the rotation vector $\mathbf{\Omega}$ and of the large scale magnetic field $\mathbf{B}$. In an incompressible fluid ($\nabla.\mathbf{u} = 0$), it implies that the small scale motions are confined to the planes defined by these two vectors. Thus, the effect of the small scale motions on large scale quantities $\mathbf{U}, \mathbf{B}, \Theta$ (Θ temperature) can be qualitatively estimated. The term $\mathbf{u}.\nabla\mathbf{u}$ would play a negligible role in the momentum equation whereas the $\mathbf{u}.\nabla\Theta$ term would ensure an efficient and anisotropic turbulent mixing.

Under these conditions, the momentum equation becomes

$$\rho(\frac{\partial \mathbf{U}}{\partial t} + \mathbf{U}.\nabla\mathbf{U} + 2\mathbf{\Omega} \wedge \mathbf{U}) = -\nabla p + \mathbf{J} \wedge \mathbf{B} + \alpha\Theta\rho\mathbf{g} \tag{1}$$

where we have already neglected the turbulent term $\mathbf{u}.\nabla\mathbf{u}$. α is the volume expansion coefficient and $\mathbf{g}$ is the gravity vector. We have no information on the rapid motions (with period on the order of a day) that can arise in such a rotating flow. If their role can be neglected, as it is usually assumed, the momentum equation for the slow motions reduces to a magnetostrophic equilibrium, with archimedean forces as the energy source

$$2\rho\mathbf{\Omega} \wedge \mathbf{U} = -\nabla\mathbf{p} + \mathbf{J} \wedge \mathbf{B} + \alpha\Theta\rho\mathbf{g} \tag{2}$$

where the Boussinesq approximation has been adopted. The curl of (2) eliminates the pressure and yields

$$-2\rho\frac{\partial \mathbf{U}}{\partial z} = \nabla \wedge (\mathbf{J} \wedge \mathbf{B} + \alpha\Theta\rho\mathbf{g}) \tag{3}$$

where (s, ϕ, z) are cylindrical coordinates. This equation can be read as a diagnostic equation for the velocity $\mathbf{U}$ which satisfies the no-penetration condition $\mathbf{U}.\mathbf{n} = \mathbf{0}$ at the boundary ($\mathbf{n}$ outward normal). Taylor (1963) has shown that it has a solution if and only if

$$\int\int_{s=s_0} (\mathbf{J} \wedge \mathbf{B})_\phi d\phi dz = 0 \tag{4}$$

on each cylindrical surface ($s = s_0$) whose axis is the rotation axis. When this condition is not satisfied, geostrophic motions $\mathbf{U}_G$ (rotation of these cylinders about the axis) are accelerated. Recently, we have (Jault, 1995) illustrated with numerical examples Taylor's idea that these motions would change the magnetic field so that eventually the Taylor's condition on the magnetic force is satisfied. One important consequence of the magnetostrophic equilibrium would be that the total magnetic energy stored in the Earth's core predominates by far over the kinetic energy. However, the main weakness of the Taylor's description remains its purely theoretical character. There are no experiments to validate this approach. Since the theory of Taylor's constraint is behind all the numerical models that are being developed, we think that it is urgent to devise such an experiment.

In point of fact, Taylor's solution might be vitiated. Malkus (1968) and Vanyo *et al.* (1995) have demonstrated that the fluid circulation in a precessing ellipsoid can be very different from what is predicted by the model of Poincaré. Poincaré had shown that the response of a fluid cavity to a slow precession of its boundary is a quasi-rigid rotation about an axis in the equatorial plane. As the precession rate is increased, vigourous geostrophic motions are generated that are not yet understood but it is suspected that nonlinear interactions (spawned from the $\mathbf{U}.\nabla\mathbf{U}$ term) of the inertial waves (the very same motions with periods on the order of one day that were neglected above) play an important role (Hollerbach *et al.*, 1995). The theory for these waves has been nicely vindicated by an experiment in which the rotation speed of a rapidly rotating spherical container filled with water was forcibly varied (Aldridge *et al.*, 1969). In his review, Aldridge (1997) discusses the role played by inertial waves in core dynamics. These waves arise in precessing spheroids as a secondary motion driven by the Ekman layer (see next section) where the Poincaré mode adjusts to the velocity of the container (Greenspan, 1968). We are currently investigating these waves and their nonlinear interaction with a numerical code. However, Malkus' experiment demonstrates transition to full turbulence as the precession rate is further increased. This effect may be beyond the scope of numerical modelling. We plan eventually to study experimentally the effect of an imposed magnetic field on the fluid motions in a sphere and in a precessing ellipsoid. Replacing water by a conducting fluid and imposing an external magnetic field would not change the frequency of the inertial waves very much. However, one should observe caution since the properties of the viscous boundary layers may be altered. The imposed magnetic field may also hinder the occurence of geostrophic shear. There has already been a study of hydromagnetic precession in a cylinder, filled with sodium (Gans, 1970). The instrumentation was minimal but the author argued that the magnetic Reynolds number in his system attained 20. Thus, dynamo action may well occur in a precessing system set up in a laboratory. Moreover, building such an experiment appears to be a sensible way to study turbulence in the presence of rotation and magnetic forces.

4. Boundary layers and internal shear layers

In a rotating flow, the viscous boundary layer - the Ekman layer - exerts a control on the fluid interior as does the Hartmann layer in a flow permeated by an imposed magnetic field.

As an example, we have studied numerically the motions generated in a spherical shell by a slight differential rotation $\Delta\omega$ of a conducting inner core with respect to an insulating container. The entire set-up rotates rapidly and is permeated by a dipolar magnetic field of internal origin. Without magnetic field, the solution of Proudman (1956), valid in the asymptotic limit of vanishing viscosity, has been completed by the boundary layer study of Stewartson (1966). The latter proved that the boundary layer attached to the cylinder tangent to the inner sphere does not exert a control on the interior flow in contrast with the Ekman layers attached to the spherical boundaries. Our numerical solution illustrates these asymptotics well. Without rotation of the set-up, viscous boundary layers are of Hartmann type. Magnetic forces tightly couple the inner body to the fluid shell and most of the shear is confined at the outer viscous boundary layer. Since the strength of the magnetic field varies along the boundary, there is a flux of electrical currents out of the boundary layer and hence the interior solution is controlled by the boundary layer. There is thus close similarity between Hartmann and Ekman layers. However, the most interesting feature of this system is the internal shear layer parallel to the magnetic field line tangent to the outer surface. The width of this layer scales as $M^{-1/2}$, where M is the Hartmann number (fig 1). Finally, we have studied the solution with both rotation and magnetic field but only for small $\Delta\omega$. As $\Delta\omega$ is increased, we anticipate growth of three dimensional perturbations. The motions of the boundaries are very efficiently transmitted to the interior of the fluid shell because of the Proudman-Taylor constraint.

In short, rotation together with magnetic field give an important role to the viscous boundary layers, which are able to drive the interior flow.

5. Magnetoconvection

All published numerical dynamo models have used convection as energy source. These convective dynamo problems have followed up linear solutions to the magnetoconvection problem. This problem consists in calculating the critical Rayleigh number Ra for onset of convection as a function of

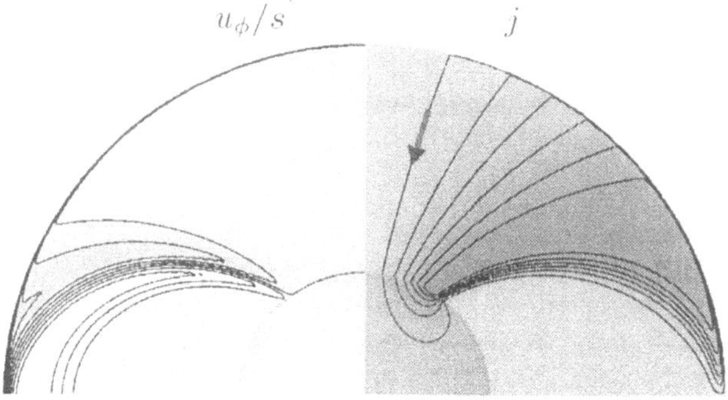

Figure 1. Numerical modelling of the flow driven by the rotation of the inner core. The zonal angular velocity u_ϕ/s and the meridional electric currents j. Meridional section through the sphere, with the North magnetic pole at the top. $M = 1000$.

the strength of an imposed magnetic field, measured by the Elsasser number Λ . In an horizontal plane layer, where the gravity and the rotation vector are vertical and the imposed magnetic field is uniform and horizontal, Ra is minimum when Λ is $O(1)$ (Chandrasekhar, 1961): the inhibitive influence of the magnetic force and of the rotation force cancel each other out. In a sphere, the magnetic field cannot be assumed to be uniform. As a consequence, there are purely magnetic instabilities, which may grow even without a temperature gradient (Zhang, 1992). With small magnetic fields ($\Lambda \leq O(Pr^{-1/3}E^{1/3})$), the solution resembles the non-magnetic solution. Viscous forces are important in the interior and determine the horizontal length scale, which is $O(E^{1/3})$. The solution is strongly dependent on the Prandtl number. The magnetic field is stabilizing for Prandtl numbers smaller than unity. Thus the convective cells are displaced towards regions of weak magnetic field. When the magnetic field is further increased, the solution is large scale and viscous forces become negligible in the interior. The magnetic field becomes destabilizing.

If we use these results as a guide for convective dynamo models, we expect to find two critical Rayleigh numbers, one for the onset of convection, and the other for the onset of dynamo action. We expect the second bifurcation to be supercritical (Busse, 1975). However, when the strength of the field reachs a critical value, which is $O(Pr^{-1/3}E^{1/3})$, the motions are accelerated in presence of the magnetic field, and there is "run-away growth" of the field (Soward, 1979). Saturation is expected for ($\Lambda = O(1)$). These ideas remain theoretical because the small value of the magnetic Prandtl number in the Earth's core implies that this scenario should take place at high values of the Rayleigh number, for which the motion is turbulent and cannot be easily modelled either analytically or numerically.

Because magnetoconvection studies have been so important in building up our intuition, we have found it useful to set up a laboratory experiment to investigate magnetoconvection in a rapidly rotating fluid. It follows up several studies of thermal convection in rapidly rotating fluids (Busse *et al.*, 1976) (Carrigan *et al.*, 1983) (Chamberlain *et al.*, 1986) (Cordero *et al.*, 1992) (Cardin *et al.*, 1994). In all these experiments, centrifugal acceleration takes over the role played by radial gravity in the Earth's core. The fluid (water, $Pr = 7$) is heated at the outer surface and it is also cooled at the inner surface when there is a central core. Busse and his collaborators have concentrated their efforts on the flow structure just above the critical Rayleigh number where analytical (Busse, 1970) and numerical results are available. The two-dimensional rolls with axis parallel to the axis of rotation predicted by the theory have been observed and power-law dependences of the critical Rayleigh number and of the critical azimuthal wave number have been borne out. However, with centrifugal gravity, no purely conductive state is possible because gravity is not perpendicular to equidensity surfaces. There is always a motion, "thermal wind", which hinders accurate comparison between experiments and theory. Cordero *et al.* (1992) have combined centrifugal gravity and Earth's gravity to minimize the baroclinicity of the basic state. It has enabled them to show some agreement for the drift of the columns and to anticipate much better agreement when accurate numerical calculations are available. There has been only one comparison between numerical solutions calculated respectively with centrifugal and radial gravity (Glatzmaier *et al.*, 1993) and only Rayleigh numbers well above critical were studied. This numerical study has legitimated the use of centrifugal gravity in experiments especially for high Rayleigh numbers, where the role played by thermal wind becomes negligible. In our group, E. Dormy has developed a numerical code for convection in a spherical shell. His preliminary results indicate good agreement with experimental measurements (Dormy, 1997). Finally, Cardin *et al.* (1994) have shown in an experiment at high Rayleigh number that motions remain two-dimensional even when they are highly turbulent.

In our magnetoconvection experiment, water is replaced by Gallium ($Pr = 0.025$). The Ekman number $E \simeq 3.10^{-7}$ will be small enough (according to predictions by numerical calculations (Ardes *et al.*, 1997)) to ensure that the motion at the onset is still organized in two-dimensional columnar cells. Onset of convection occurs for a temperature difference between top and bottom less than 10% in the same experiment for water. Thus, it will be difficult to fine tune the temperature difference to study the onset of convection. In presence of a magnetic field, the critical temperature difference can be increased up to tenfold (because Pr is small) (Fearn, 1979). For high values of the imposed magnetic field, we expect small azimuthal wavenumbers to be favoured. At high Rayleigh number, small scale two-dimensional columns may add to this convection pattern, without changing much the geometry of the induced magnetic field, which should remain large scale. This magnetoconvection experiment should allow us to investigate turbulence in presence of both rotation and magnetic field. Finally, we will study whether the constraint (4) is obeyed.

6. Numerical modelling of the geodynamo mechanism

The dynamo models that have helped to plan dynamo experiments were originally analytical models. The Lowes and Wilkinson experiment (Herzenberg *et al.*, 1957) (Lowes *et al.*, 1968) was inspired by the Herzenberg model (Herzenberg, 1958). The experiments that are now set up in Riga and Karlsruhe derive respectively from Ponomarenko and G.O. Roberts models. The solution of these kinematic models in unbounded media can indeed be calculated analytically. However, the solutions for contained fluid models require numerical calculation (Apel *et al.*, 1996) (Rädler *et al.*, 1997). Numerical modelling is now a necessary stage before an experiment is devised. Fortunately, there exists a host of kinematic models for the full sphere (Dudley *et al.*, 1989).

Numerical modelling will also be necessary to complete the scarce measurements that are possible in an opaque fluid. Brito *et al.* (1995) have put a vortex of gallium in a uniform and transverse magnetic field. A spinning crenellated disc at the bottom of a cylinder filled with gallium forces the fluid. This set-up is placed on a turn-table, rotating much more slowly. The main result is that the rotation rate of the vortex is determined by the Elsasser number, in which the rotation rate of the turn-table enters. Thus, it demonstrates that the fluid dynamics are governed by the Elsasser number instead of the Stuart number in non rotating flows (see below). The experiment is completed by a numerical model to map the electrical currents inside the cylinder. There is good agreement with the measurements of the induced magnetic field. However, the numerical model was purely kinematic and unable to describe how the motions are influenced by the magnetic forces. Thus, the actual dependence on the Elsasser number remains unexplained.

Finally, a few recent models are better described as substitutes for laboratory experiments than as prologues to new experiments. These numerical models (Glatzmaier *et al.*, 1995) (Kuang *et al.*, 1997) include the back reaction of the induced magnetic field on the fluid dynamics. Both Glatzmaier *et al.* and Kuang *et al.* have presented Earth-like models, with a dominant dipole. The success of these simulations relies on the significance of the cylinder tangent to the inner core with axis the rotation axis and ultimately on the part played by Coriolis acceleration. However, it turns out that to get converged solutions, "hyperviscosity", which has no physical justifications, has to be used: the smaller the scale of the motions, the larger the viscosity. In the Glatzmaier *et al.* model, the viscosity increases more than the magnetic diffusivity with the length scale. As a consequence, these models will not be useful guides to understand fluid dynamics with small magnetic Prandtl number.

Another way consists in following the conclusions of Braginsky *et al.* (1990): viscous and inertial forces are entirely neglected. Then, a conflict between spherical and cylindrical geometry is met. Spherical geometry is dictated by the vanishing of the magnetic field at infinity. However, the dominance of the Coriolis force in the momentum equation (3) demands the use of cylindrical coordinates. In our group, we are developing such models using both sets of coordinates. Numerical accuracy is difficult to ensure because we have to interpolate at each time step between the two sets. So far, we have only been able to demonstrate that the solution of several kinematic models obey the Taylor's condition (Jault *et al.*, 1998). The robustness of the numerical algorithm, which is used to calculate the geostrophic velocity, has also been tested.

All these three dimensional dynamo models have neglected possible mean field effects. When there is scale separation, small scale motions may indeed induce electrical currents antiparallel to the large scale magnetic field. This α effect has been demonstrated in an experiment where flow of sodium was forced into two sets of perpendicular pipes immersed in a uniform magnetic field aligned along the third direction (Steenbeck *et al.*, 1967). In contrast to the original expectations, this experiment did not lead to an actual dynamo model. Interestingly enough, the strength of the electromotive force changed according to the Stuart number. In a rotating flow, we expect that it would depend on the Elsasser number. Hence, there is here scope for new experiments. On the other hand, we wonder whether such an effect is not hindered by the small value of the magnetic Prandtl number. In any case, such mean field effects cannot be present in the three dimensional models that are being developed because only large scale motions are retained.

In short, numerical simulations are not yet able to uncover dynamo mechanisms.

7. Dynamo experiments

In this paper, we have insisted on the role of experiments. There are still many open questions. In particular, it is very important to resolve whether the core motions are in a magnetostrophic regime, as it is universally assumed today. The ultimate goal, a dynamo experiment in a rotating sphere, seems still difficult to reach. The discussion above has made clear that it is important that the rotation period is very small compared to the magnetic diffusion time. This is a rather demanding constraint.

On the other hand, a dynamo experiment with non rotating flows may *not* be easier. Let us put forward a naïve argument. In a rotating device, growth of the magnetic field may be favoured because it can counteract the constraining effect of the rotation forces. In a non rotating flow, the magnetic field will only try to oppose the motions that generate it. Moreover, the approach of Braginsky *et al.* (1990) does not apply to non-rotating flows and the turbulent term $\mathbf{u}.\nabla\mathbf{u}$ may be very efficient to dissipate the kinetic energy. Finally, studies of kinematic dynamos (with imposed velocity field) show that small changes in $\mathbf{U}$ can lead from strong field generation to catastrophic collapse. In short, the geophysicist is bound to insist on the role of the rotation forces for the Earth's dynamo and also for dynamo experiments in the laboratory.

Let us estimate the Joule dissipation per unit time in a dynamo experiment where magnetostrophic equilibrium is attained

$$P = \lambda \frac{B^2 L^3}{\mu L^2} \tag{5}$$

where L measures the size of the experiment. The condition of magnetostrophic equilibrium yields

$$2\rho\Omega U = \frac{B^2}{\mu L} \tag{6}$$

Thus

$$P = 2\rho\Omega U \lambda L^2 = 2\rho\Omega\lambda^2 R_m L \tag{7}$$

Let us now assume that a power P_{max} is available to drive the experiment, P_{max} being limited also by the heat that we can extract from the system.

$$2\rho\Omega\lambda^2 R_m L \leq P_{max} \tag{8}$$

On the other hand, the rotation period has to be small compared to the magnetic diffusion time

$$\frac{1}{\Omega} \ll \frac{L^2}{\lambda} \tag{9}$$

Then, we derive the condition

$$\frac{2\rho\lambda^3 R_m}{L} \ll P_{max} \tag{10}$$

which seems possible to satisfy in an experiment with liquid sodium ($\lambda \simeq 0.1 m^2 s^{-1}$, $\rho \simeq 10^3 kg.m^{-3}$). This estimate is very crude; we have, for example used L^2/λ as magnetic diffusion time τ , whereas we know than in a sphere

$$\tau \leq \frac{L^2}{\lambda \pi^2} \tag{11}$$

where L is radius. Thus, we should at least multiply the left hand side of (10) by π^4. Moreover, we have not taken into account the fact that more heat can be taken away in a larger system ($P_{max} \sim L^2$). However, this ad hoc calculation suffices to demonstrate that size is particularly important in a dynamo experiment with rotating flows. We believe that a dynamo experiment illustrating magnetostrophic equilibrium is feasible: Values of $\Omega = 20 \quad s^{-1}$ and $L = 1m$. are very reasonable values indeed for a sphere filled with sodium. A simple forcing, like the differential rotation discussed in section 4, may suffice to generate the necessary motions.

References

Aldridge, K. (1997) Perspectives on core dynamics from laboratory experiments *Earth's Deep Interior* pp 65-78 Crossley, D.J. editor Gordon and Breach Science Publishers, Amsterdam.

Aldridge, K.D., Toomre, J. (1969) Axisymmetric inertial oscillations of a fluid in a rotating spherical container *J. Fluid Mech.* **Vol. 37** pp 307-323

Apel, A., Apstein, E., Rädler, K.-H., Rheinhardt, M. (1996) Contributions to the theory of the planned Karlsruhe dynamo experiment *Report Astrophysikalisches Institut Postdam*

Ardes, M., Busse, F.H., Wicht, J. (1997) Thermal convection in rotating spherical shells *Phys. Earth Planet. Inter.* **Vol. 99** pp 55-67.

Backus, G., Parker, R., Constable, C. (1996) *Foundations of Geomagnetism* Cambridge University Press.

Braginsky, S.I., Meytlis, V.P. (1990) Local turbulence in the Earth's core *Geophys. Astrophys. Fluid Dynamics* **Vol. 55** pp 71-87.

Braginsky, S.I., Roberts, P.H. (1995) Equations governing convection in Earth's core and the geodynamo *Geophys. Astrophys. Fluid Dynamics* **Vol. 79** pp 1-97.

Brito, D., Cardin, P., Nataf, H.C., Marolleau, G. (1995) Experimental study of a geostrophic vortex of gallium in a transverse magnetic field (1995) *Phys. Earth Planet. Inter.* **Vol. 91** pp 77-98.

Brito, D., Cardin, P., Nataf, H.C., Olson, P. (1996) Experiments on Joule Heating and the dissipation of energy in the Earth's core (1996) *Phys. Earth Planet. Inter.* **Vol. 91** pp 77-98.

Busse, F.H. (1970) Thermal instabilities in rapidly rotating systems *J. Fluid Mech.* **Vol. 44** pp 441-460

Busse, F.H. (1975) A model of the Geodynamo *Geophys. J. R. Astron. Soc.* **Vol. 42** pp 437-459.

Busse, F.H., Carrigan, C.R. (1976) Convection induced by centrifugal buoyancy *J. Fluid Mech.* **Vol. 62** pp 579-592

Cardin, P., Olson, P. (1994) Chaotic thermal convection in a rapidly rotating spherical shell: consequences for flow in the outer core *Phys. Earth Planet. Inter.* **Vol. 82** pp 235-259.

Carrigan, C.R., Busse, F.H. (1983) An experimental and theoretical investigation of the onset of convection in rotating spherical shells *J. Fluid Mech.* **Vol. 126** pp 287-305

Chamberlain, J.A., Carrigan, C.R. (1986) An experimental investigation of convection in a rotating sphere subject to time varying thermal boundary conditions *Geophys. Astrophys. Fluid dynam.* **Vol. 41** pp 17-41

Chandrasekhar, S. (1961) *Hydrodynamic and hydromagnetic stability* Oxford University Press.

Cordero, S., Busse, F.H. (1992) Experiments on convection in rotating spherical shells *Geophys. Res. Lett.* **Vol. 19** pp 733-736

Dormy, E. (1997) *Modélisation numérique de la dynamo terrestre* Thesis, Institut de Physique du Globe de Paris.

Dormy, E., Cardin, P., Jault, D. (1998) MHD flow in a slightly differentially rotating spherical shell, with conducting inner core, in a dipolar magnetic field *Earth Planet. Sci. Lett.* **Vol. 160** pp 15-30

Dudley, M.L., James, R.W. (1989) Time-dependent kinematic dynamos with stationary flows *Proc. R. Soc. Lond.* **Vol. A 425** pp 407-429

Fearn, D.R. (1979) Thermally driven hydromagnetic convection in a rapidly rotating sphere *Proc. R. Soc. Lond.* **Vol. A 369** pp 227-242

Fearn, D.R. (1997) The Geodynamo *Earth's Deep Interior* pp 79-114 Crossley, D.J. editor Gordon and Breach Science Publishers, Amsterdam.

Fearn, D.R., Roberts, P.H., Soward, A.M. (1988) Convection, stability and the dynamo *Energy Stability and Convection* Galdi, G.P., Straughan, B. editors Pitman Research notes in mathematics series 168 pp 60-324 Longman, Harlow.

Gans, R.F. (1970) On hydromagnetic precession in a cylinder *J. Fluid Mech.* **Vol. 45** pp 111-130

Glatzmaier, G.A., Olson, P. (1993) Highly supercritical thermal convection in a rotating spherical shell: centrifugal vs radial gravity *Geophys. Astrophys. Fluid dynam.* **Vol. 70** pp 113-136

Glatzmaier, G.A., Roberts, P.H. (1995) A three-dimensional convective dynamo solution with rotating and finitely conducting inner core and mantle *Phys. Earth Planet. Inter.* **Vol. 91** pp 63-73

Greenspan, H.P. (1968) *The theory of rotating fluids* Cambridge University Press

Herzenberg, A. (1958) Geomagnetic dynamos *Phil. Trans. R. Soc. Lond.* **Vol. A 250** pp 543-585 Herzenberg, A., Lowes, F.J. (1957) Electromagnetic induction in rotating conductors *Phil. Trans. R. Soc. Lond.* **Vol. A 249** pp 507-584

Herzenberg, A., Lowes, F.J. (1957) Electromagnetic induction in rotating conductors *Phil. Trans. R. Soc. Lond.* **Vol. A 249** pp 507-584

Hollerbach, R., Kerswell, R.R. (1995) Oscillatory internal shear layers in rotating and precessing flows *J. Fluid Mech.* **Vol. 298** pp 327-339

Jault, D. (1995) Model Z by computation and Taylor's condition *Geophys. Astrophys. Fluid Dynamics* **Vol. 79** pp 99-124

Jault, D., Cardin, P. (1998) On dynamic geodynamo models with imposed velocity as energy source *Phys. Earth Planet. Inter.* submitted

Kuang W., Bloxham, J. (1997) An Earth-like numerical dynamo model *Nature* **Vol. 389** pp 371-374

Lowes, F.J., Wilkinson, I. (1968) Geomagnetic dynamo: an improved laboratory model *Nature* **Vol. 219** pp 717-718

Malkus, W.V.R. (1968) Precession of the Earth as the cause of geomagnetism *Science* **Vol. 160** pp 259-264

Merrill, R., McElhinny, M., McFadden, P. (1996) *The magnetic field of the Earth* Academic Press, New-York.

Moffatt, H.K. (1978) *Magnetic Field Generation in Electrically Conducting Fluids* Cambridge University Press.

Poirier, J.P. (1991) *Introduction to Physics of the Earth's Interiors* Cambridge University Press.

Proudman, I. (1956) The almost-rigid rotation of viscous fluid between concentric spheres *J. Fluid Mech.* **Vol. 1** pp 505-516

Rädler, K.-H., Apstein, E., Rheinhardt, M., Schüler, M. (1997) Contributions to the theory of the planned Karlsruhe dynamo experiment - Supplements and Corrigenda - *Report Astrophysikalisches Institut Postdam*

Roberts, P.H. (1987) Origin of the main field: Dynamics *Geomagnetism 2.* pp 251-306 Jacobs J. editor Academic Press, New-York

Roberts, P.H., Soward, A.M. (1992) Dynamo theory *Ann. Rev. Fluid Mech.* **Vol. 24** pp 459-512

Secco, R.A., Schloessin, H.H. (1989) The electrical resistivity of solid & liquid Fe at pressures up to 7. GPa *J. Geophys. Res.* **Vol. 94** pp 5887-5894

Soward, A.M. (1979) Convection driven dynamos *Phys. Earth Planet. Inter.* **Vol. 20** pp 134-151

Stewartson, K. (1966) On almost rigid rotations. Part 2 *J. Fluid Mech.* **Vol. 26** pp 131-144

Taylor, J.B. (1963) The magneto-hydrodynamics of a rotating fluid and the earth's dynamo problem *Proc. R. Soc. Lond.* **Vol. A 274** pp 274-283

Steenbeck, M., Kirko, I.M., Gailitis, A., Klawina, A.P., Krause, F., Laumanis, I.J., Lielausis, O.A. (1967) Der experimentelle Nachweis einer elektromotorischen Kraft längs eines äußeren Magnetfeldes, induziert durch eine Strömung flüssigen Metalls (α-effect) *Monats. Dt. Akad. Wiss.* **Vol. 9** pp 714-719

Vanyo, J., Wilde, P., Cardin, P., Olson, P. (1995) Experiments on precessing flows in the Earth's liquid core *Geophys. J. Int.* **Vol. 121** pp 136-142

Zhang, K. (1992) Spherical shell convection in the presence of a toroidal magnetic field *Proc. R. Soc. Lond.* **Vol. A 448** pp 245-268

VELOCITY PROFILE OPTIMIZATION FOR THE RIGA DYNAMO EXPERIMENT

F. STEFANI AND G. GERBETH
Forschungszentrum Rossendorf
P.O. Box 510119, D-01314 Dresden, Germany

AND

A. GAILITIS
Institute of Physics, Latvian Academy of Sciences
LV-2169 Salaspils 1, Riga, Latvia

Abstract. The Riga dynamo experiment is an attempt to show the self-excitation of a magnetic field in an electrically conducting fluid. Due to the absence of any scale separation effects, the precise shape of the velocity profile has a strong influence on the values of the critical magnetic Reynolds numbers (Rm). Several numerical codes have been developed and used for the determination of these critical Rm for arbitrary velocity profiles. Helicity maximizing profiles are shown to provide smaller critical Rm than the solid body rotation flow. We outline how the velocity profiles have been optimized in an iterative process including flow measurements, constructive improvements and numerical calculations.

1. Introduction

The magnetic fields of planets, stars and even galaxies are believed to be generated by the motion of electrically conducting fluids via the dynamo effect. Whereas the underlying theory has been elaborated in the last decades, experimental efforts are still rare. Notable is the successful experiment of Lowes and Wilkinson [1] with two rotating cylinders in a solid metal block. A first liquid metal experiment was undertaken in Riga in 1987 [2], [3]. Although the self-excitation threshold was not reached for reasons of mechanical instabilities the measured reaction of the system on an imposed magnetic field gave a clear indication of a possible dynamo effect at higher

A. Alemany et al. (eds.),
Transfer Phenomena in Magnetohydrodynamic and Electroconducting Flows, 31–44.

flow rates.
From the viewpoint of flow topology, the present Riga experiment is very similar the the first one. Stimulated by the early work of Ponomarenko [4], the basic component of the facility consists of an axially symmetric helical flow of liquid sodium. In contrast to the Ponomarenko dynamo, this central helical flow is surrounded by a straight coaxial counterflow and an outermost cylinder with sodium at rest. This experimental arrangement enables not only a convective instability for the magnetic field self-excitation (as in the case of the Ponomarenko dynamo) but an absolute instability, too [2]. Calculating the critical Rm for a given geometry under the assumption of solid body rotation in the innermost cylinder, Gailitis [5] had estimated the sodium flow and the power necessary for magnetic field self-excitation.
However, a main problem of the Riga dynamo experiment is connected with the strong dependence of the critical Rm upon the shape of the velocity profile and with the difficulty to generate a desirable profile by means of a propeller and suitable guiding blades[1]. For example, the velocity profiles measured in a water pre-experiment [7] in 1996 differed remarkably from the solid body rotation (SBR)[2]. With these first profiles self-excitation of a magnetic field turned out to be clearly impossible with the installed motor power.
In this situation two problems arose. At first, it was indicated to look theoretically for velocity profiles yielding self-excitation with minimal power. At second, propellers and leading vanes had to be constructed in order to come close to those desirable profiles. The present paper outlines this process of velocity optimization.
In the second section the criterion of optimization is discussed. In order to compute the critical Rm for arbitrary profiles and to look for profiles with critical Rm as low as possible, we carried out a number of numerical simulations. In section 3 we describe a simple finite difference solver for the time-dependent induction equation in one dimension (1D), taking into account only the first azimuthal component and fixed Fourier components in z-direction. Some of the results have been verified by comparison with the outcomes of an eigenvalue equation solver. Additionally, a two-dimensional (2D) code was developed and used in order to check the effects of the finite length of the system and of possible variations of the velocity profiles along the z-axis. This code is presented in section 4. In section 5 the calculated critical Rm are given for several theoretical as well as measured velocity profiles. Helicity maximizing profiles are shown to provide lower critical Rm

[1]This is in contrast to the Karlsruhe experiment, where the precise shape of the velocity profile plays only a negligible role due to the effect of scale separation [6].

[2]The notation "solid body rotation" is defined here by a radial dependence of the axial and the azimuthal velocity components according to $v_z(r) \sim const$ and $v_\phi(r) \sim r$.

than the solid body rotation flow. The process of technical improvements of the facility leading to satisfying velocity profiles is reported.

2. Optimization Criterion

The full optimization of a whole dynamo experiment is a very complex project. Supposing a certain relation of construction costs and operating costs, Gailitis [5] had optimized in a first step the basic geometric relations of the facility. Figure 1 shows the resulting dimensions and the basic components of the experimental setup.

First experiments at a water dummy channel had shown a remarkable deviation of the measured velocity profiles from the solid body rotation profiles supposed in former calculations. At the same time, preliminary computations had indicated a strong dependence of the critical Rm on the shape and the relative strengths of the azimuthal velocity v_ϕ and the axial velocity v_z. It turned out that in order to achieve self-excitation for given power resources a change of the velocity profiles was unavoidable.

The first difficulty for a velocity profile optimization applies to a reasonable definition of the optimization criterion. In a technical sense, it would be desirable to find velocity profiles which require minimal power for the self-excitation of a magnetic field. However, the determination of the necessary power would represent an enormous hydrodynamic task for its own. An easier task is to look for velocity profiles with a minimal critical Rm. But even the definition of Rm is to some extent arbitrary. In the following we use the definition

$$Rm = \mu\sigma \, R_1 \sqrt{\overline{v_z{}^2} + 2\,\overline{v_\phi{}^2}}\,. \qquad (1)$$

By virtue of the averaging

$$\overline{v_*{}^2} = \frac{2}{R_1^2} \int_0^{R_1} v_*{}^2(r)\, r dr \qquad (2)$$

over the radius R_1 of the innermost cylinder, this definition has regard to the integral properties of the flow and not only to the maximal velocity. The stronger weighting of $\overline{v_\phi{}^2}$ is motivated by the fact that for a solid body rotation the major part of the losses comes from the boundary region. For this special case, our definition is equivalent to the often used definition $Rm = \mu\sigma R_1 v_{max}$ based on the maximum value of the velocity. According to the definition of Rm, we define the pitch parameter κ as

$$\kappa \;=\; \sqrt{\frac{\overline{v_z{}^2}}{2\,\overline{v_\phi{}^2}}} \qquad (3)$$

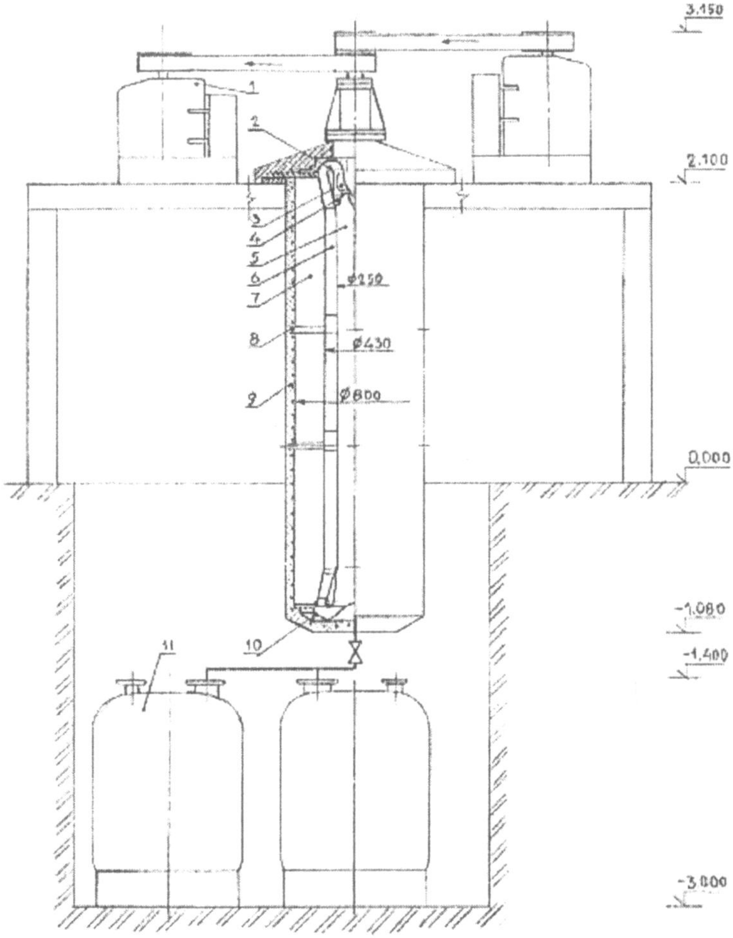

Figure 1. The experiment. 1 – DC motors 2 – pre-propeller vanes, 3 – propeller, 4 – post-propeller vanes, 5 – principal channel with swirling flow, 6 – channel with straight counterflow, 7 – immobile sodium, 8 – mechanical supports with field sensors inside, 9 – heaters with thermo-insulation, 10 – flow reversing guides, 11 – sodium storage

characterizing the relation between the averaged axial and azimuthal velocity. Restricting the following considerations on this Rm-oriented optimization, the power necessary for self-excitation must be kept in mind as the actual quantity to be minimized. Forcing a per se "optimal" velocity profile by means of too much leading blades can have a counterproductive effect.

3. One-dimensional calculations

Being interested only in the onset of the magnetic field self-excitation, we restrict ourselves in the following to the kinematic dynamo problem. Combining Maxwell's equations and Ohm's law we find the induction equation for the magnetic field:

$$\frac{\partial \mathbf{B}}{\partial t} = -\frac{1}{\mu\sigma}\nabla \times (\nabla \times \mathbf{B}) + \nabla \times (\mathbf{v} \times \mathbf{B}) . \tag{4}$$

For the one-dimensional approximation we assume the velocity profile to be axisymmetric and independent of z:

$$\mathbf{v}(\mathbf{r}) = (0, v_\phi(r), v_z(r)) . \tag{5}$$

For the magnetic field we can make the ansatz

$$\mathbf{B}(r,\phi,z,t) = \left(\tilde{B}_r(r,t), \tilde{B}_\phi(r,t), \tilde{B}_z(r,t)\right) exp(im\phi + ikz). \tag{6}$$

Then the induction equation can be written in the following form:

$$\begin{aligned}
\frac{\partial \tilde{B}_r(r,t)}{\partial t} &= \frac{1}{\mu\sigma}\left[\left(\frac{1}{r}\frac{\partial}{\partial r}r\frac{\partial}{\partial r} - \frac{1}{r^2} - \frac{m^2}{r^2} - k^2\right)\tilde{B}_r(r,t) - \frac{2im}{r^2}\tilde{B}_\phi(r,t)\right] \\
&\quad - \left(\frac{im}{r}v_\phi + ikv_z\right)\tilde{B}_r(r,t) \qquad (7) \\
\frac{\partial \tilde{B}_\phi(r,t)}{\partial t} &= \frac{1}{\mu\sigma}\left[\left(\frac{1}{r}\frac{\partial}{\partial r}r\frac{\partial}{\partial r} - \frac{1}{r^2} - \frac{m^2}{r^2} - k^2\right)\tilde{B}_\phi(r,t) + \frac{2im}{r^2}\tilde{B}_r(r,t)\right] \\
&\quad - \left(\frac{im}{r}v_\phi + ikv_z\right)\tilde{B}_\phi(r,t) - \left(\frac{v_\phi}{r} - \frac{\partial v_\phi}{\partial r}\right)\tilde{B}_r(r,t) \qquad (8) \\
\frac{\partial \tilde{B}_z(r,t)}{\partial t} &= \frac{1}{\mu\sigma}\left(\frac{1}{r}\frac{\partial}{\partial r}r\frac{\partial}{\partial r} - \frac{m^2}{r^2} - k^2\right)\tilde{B}_z(r,t) \\
&\quad - \left(\frac{im}{r}v_\phi + ikv_z\right)\tilde{B}_z(r,t) + \frac{\partial v_z}{\partial r}\tilde{B}_r(r,t) . \qquad (9)
\end{aligned}$$

At all interfaces and boundaries the two tangential components of the electric and all three components of the magnetic field are demanded to be continuous.

A central difference scheme is applied for the space dependent part of the differential equation inside the system. For the outer part, modified Bessel functions are used and matched via the boundary conditions to the solution in the inner part. Concerning the time dependence, the 1D code is realized as a Crank-Nicolson scheme.

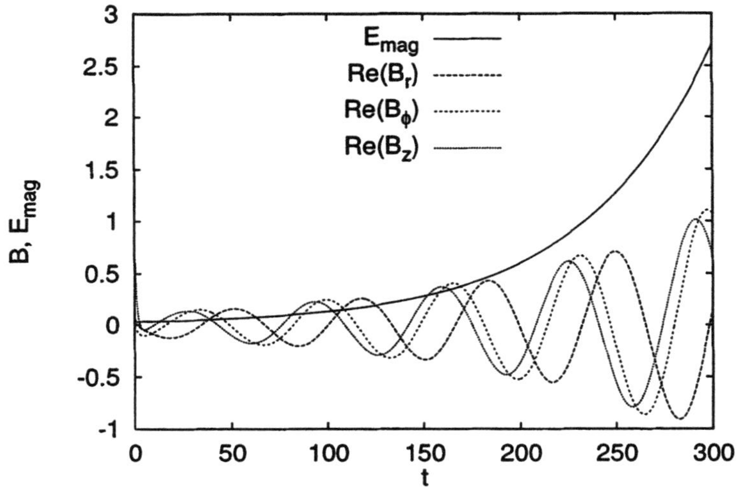

Figure 2. Time dependence of the the magnetic field components and the total magnetic energy (Ponomarenko flow).

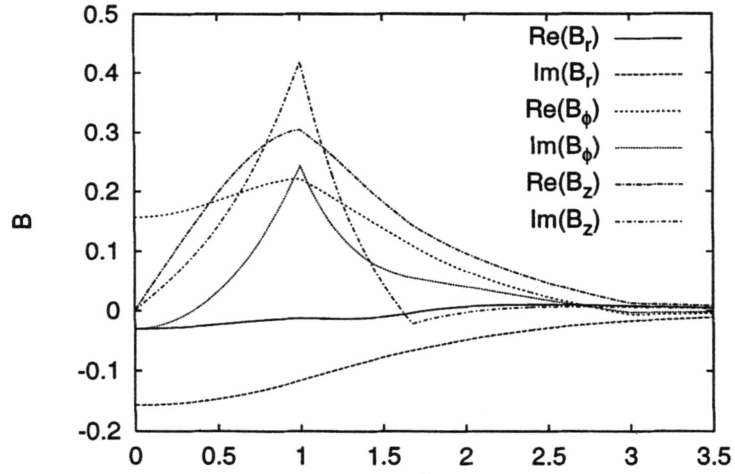

Figure 3. Radial dependence of the magnetic field components (Solid body rotation flow).

For the special case of a Ponomarenko flow figure 2 shows a typical time evolution of the real parts of the magnetic field (at a given r), and of the

total magnetic energy. Rm is chosen slightly above the critical value. Therefore, after the decline of an imposed seed field at the very beginning, an exponentially growing oscillation can be detected.
As an example of the resulting eigenstates, figure 3 shows the radial dependence of the magnetic field components for the solid body rotation profile. The relation of the three cylinder radii is, according to the experiment, $1.0 : 1.68 : 3.0$.
Once the eigenstate of the magnetic field is reached its time evolution can be written in the form

$$\mathbf{B}(r,t) = \mathbf{B}_o(r)\ exp(p(Rm,k)t) \tag{10}$$

with a (complex) growth factor p depending on Rm and the wave number k. Then, the critical Rm for the convective instability, Rm^*, can be determined as the leftmost point of the so-called neutral curve which is the solution of

$$Real(p(Rm,k)) = 0\ . \tag{11}$$

Rm^{**}, the critical Rm for the absolute instability of the infinite length system, can be determined as the Rm where the (complex) group velocity

$$v_{group} = i\ \frac{\partial p(Rm,k)}{\partial k} \tag{12}$$

vanishes (for details see [8]). However, the most important critical Rm is Rm^{***} denoting the absolute instability for the real system of finite length (global instability). In principle, it can be estimated in the framework of a one-dimensional calculation, too [8]. We will mainly use the 2d code for its determination.
The results of this described 1D time evolution code were verified in part by an older 1D code developed by one of the authors (A.G.), which uses modified Bessel functions or polynomial expressions in the various cylinders and solves the nonlinear matching equations for these functions at the inner and outer boundaries.

4. Two-dimensional calculations

In order to check the effect of the finite length of the system and of possible variations of the velocities along the z-axis on the self-excitation threshold it was indicated to complement the one-dimensional results with two-dimensional calculation. In contrast to systems which can be embedded in spheres without major loss of accuracy, the Riga dynamo must be treated as a cylindrical problem due to its high aspect ratio.
At least in the redirection zones, the z-dependence of the velocity as well

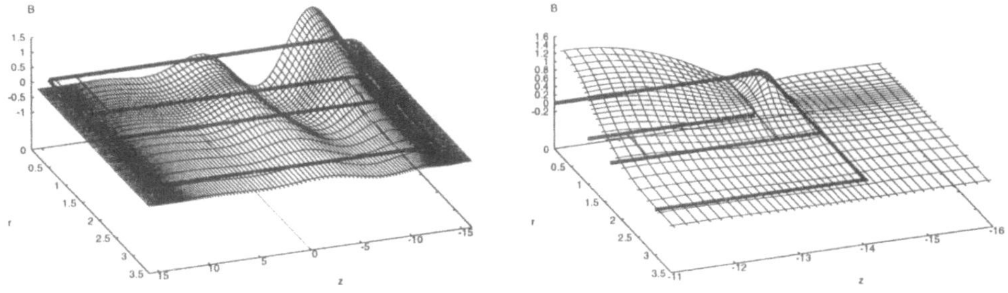

Figure 4. $Real(B_r(r,z))$ for the solid body rotation profile resulting from a 2d calculation. The thick lines delineate different flow zones. The helical flow in the innermost cylinder is located between $r = 0$ and $r = 1$. The propeller is on the left, driving the helical flow to the right. Left: overall picture. Right: details at the lower redirection zone.

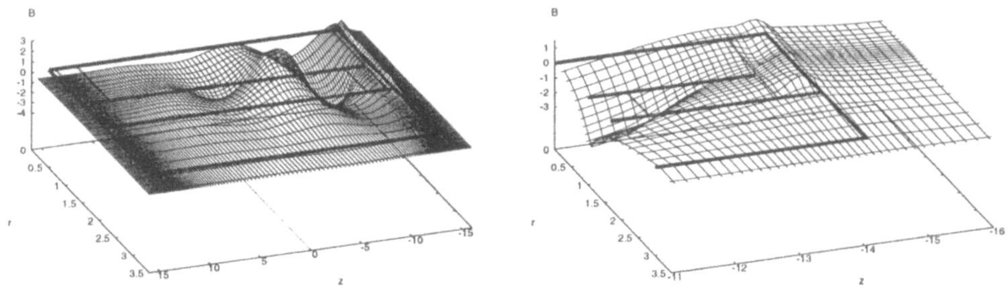

Figure 5. Same as figure 4 for $Real(B_\phi(r,z))$

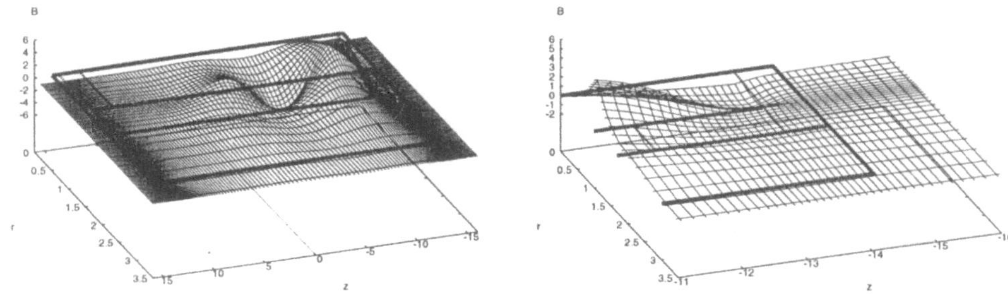

Figure 6. Same as figure 4 for $Real(B_z(r,z))$

as its v_r-component must be taken into account according to

$$\mathbf{v}(\mathbf{r}) = (v_r(r,z), v_\phi(r,z), v_z(r,z)) \ , \tag{13}$$

and the magnetic fields must be treated as z-dependent ones:

$$\mathbf{B}(r,\phi,z,t) = \left(\tilde{B}_r(r,z,t), \tilde{B}_\phi(r,z,t), \tilde{B}_z(r,z,t)\right) exp(im\phi) .$$

We omit the precise form of the induction equation. It can be derived from (7)-(9) replacing the terms $-k^2\tilde{B}_*$ by $\partial^2\tilde{B}_*/\partial z^2$ and adding several terms including v_r and its derivatives.
Again, a central difference scheme is applied for the space dependent parts. Non-uniform grid spacing is used in order to improve the resolution in the critical redirection zones of the flow. A slight rectangular simplification of the real geometry is necessary and the velocity in the redirection zones is approximated in an appropriate manner.
The time evolution is realized as an explicit Adams-Bashforth method of second order. The Laplace equation in the outer part is solved during every time step using a pseudo-transient method [9]. Indeed, this procedure is very time consuming during the first time steps. When the time evolution has reached a state close to the eigenstate of the induction equation, the pseudo-transient method can be speeded up enormously by multiplying the fields in the outer part first at all with a phase factor extracted from the inner part. However, it has to be noted that the total time necessary to reach the eigenstate is much longer than for the one-dimensional code since the large aspect ratio of the system leads to a long characteristic diffusion time.
In order to get an impression of the resulting fields, figures 4-6 show the three field components for the special case of a solid body rotation profile with $\kappa = 1$. The boundaries of the system as well as the interfaces between different flow regions are delineated with thick lines. The propeller is situated at the left. The magnetic fields are oscillating with an amplitude growing in flow direction.

5. Comparison of various velocity profiles

Although a wide variety of theoretical and measured velocity profiles has been considered, we restrict ourselves in the following to five representative profiles. For these profiles the first two rows of table 1 show the shape of the axial and azimuthal velocity components in the innermost cylinder. The relation between the radius R_1 of the innermost cylinder, the radius R_2 of the counterflow cylinder, the radius R_3 of the outermost cylinder and the length L of the apparatus is $R1 : R2 : R3 : L = 1.0 : 1.68 : 3.0 : 25.0$. All the profiles are assumed to be completed by a purely axial and constant counterflow. We tested also a number of counterflow profiles with azimuthal velocity components and a non-constant axial component. However, in view of the difficulties to maintain such profiles along the upward flow we considered a purely axial and constant counterflow as most appropriate for the experiment.

The critical Rm-values are determined either by means of the above mentioned 1D eigenvalue equation solver (a) and/or the 1D time-evolution code (b) or the 2D time-evolution code (c). Additionally, figure 7 shows the de-

TABLE 1. Critical Rm for the considered velocity profiles

	SBR	BEF	NGB	WGB	TGB
v_z	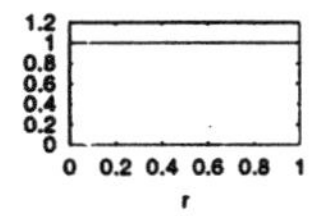	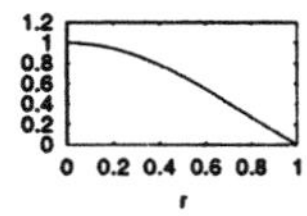	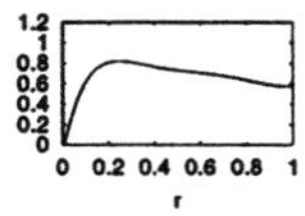	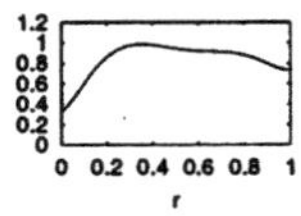	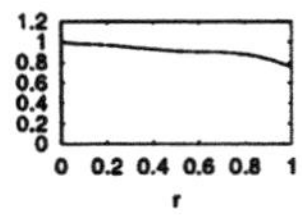
v_ϕ	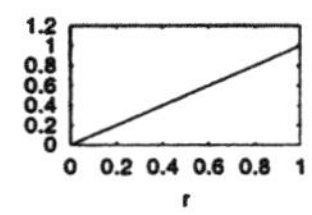	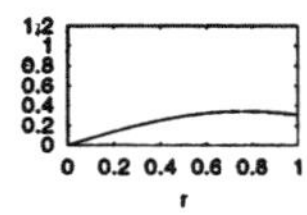	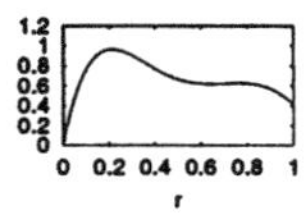	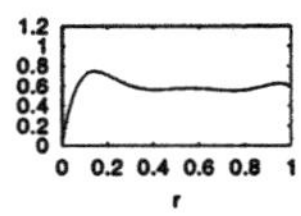	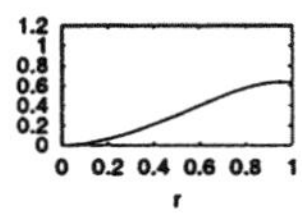
κ	1.00	1.20	0.73	1.10	1.30
Rm^*	$14.9^{a,b}$	$12.0^{a,b}$	25.6^b	17.0^b	14.3^b
Rm^{**}	$16.1^{a,b}$	$13.2^{a,b}$	27.6^b	18.3^b	15.7^b
Rm^{***}	$18.0^{a,c}$	$14.7^{a,c}$	31.0^c	21.5^c	17.6^c

pendence of Rm^* on the parameter κ (computed with method b).

The first profile represents the solid body rotation profile (SBR), which was assumed in former calculations, too. The second profile (BEF) requires some explanatory remarks: It is widely believed that helicity is effective in promoting the dynamo effect. Therefore, it is indicated to test velocity profiles with maximal helicity H for a given Rm (defined in (1)). The helicity is given as

$$\begin{aligned} H &= 2\pi\, L \int_0^R \mathbf{v}(r) \cdot \nabla \times \mathbf{v}(r)\, r dr \\ &= 2\pi\, L \int_0^R \left(v_z(r) \frac{\partial v_\phi(r)}{\partial r} + \frac{v_z(r) v_\phi(r)}{r} - v_\phi(r) \frac{\partial v_z(r)}{\partial r} \right) r dr\,. \quad (14) \end{aligned}$$

Bessel functions are resulting from the calculus of variation. Thus, the v_z- and the v_ϕ-component of the second profile (BEF) are given by the following expressions:

$$v_z(r) = \kappa\, J_0(2.4\, r) \quad (15)$$

$$v_\phi(r) = 1/\sqrt{2}\, J_1(2.4\, r)\,. \quad (16)$$

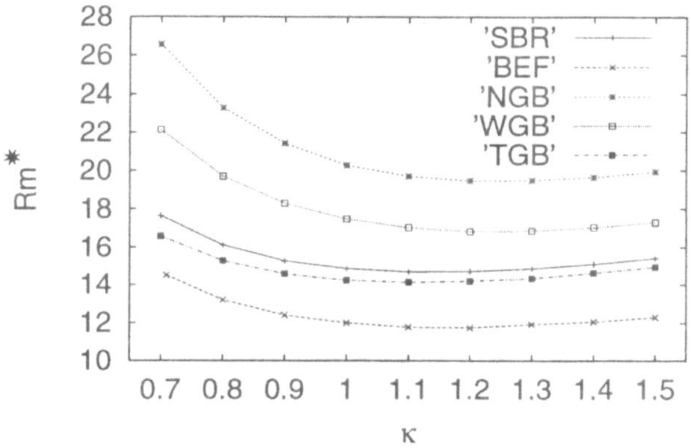

Figure 7. **Dependence of Rm^* on the pitch parameter κ**

The factor $1/\sqrt{2}$ in front of J_1 ensures (for $\kappa = 1$) helicity maximization for our somewhat "skew" expression $\overline{v_z{}^2} + 2\,\overline{v_\phi{}^2}$. The Factor 2.4 in the argument of the Bessel functions is the maximal Lagrangian parameter used in the calculus of variation that ensures positivity of both velocity components in the innermost cylinder. Indeed, the critical Rm-values for these helicity maximizing profiles are the lowest we were able to find. Although the minimum of Rm^* is achieved for $\kappa = 1.2$ rather than for $\kappa = 1.0$, this result underlines the important role of helicity for the dynamo effect.
For the following discussion it is important to note that with the installed motors a Rm of the order of 22 can be achieved (depending slightly on the profile).
The third profile was measured in a first water experiment, where only a propeller and no guiding blades (NGB) were used[3]. All critical Rm are larger than the achievable Rm. The curve NGB in Figure 7 shows that these very high critical Rm are in part due to the small value of the pitch parameter ($\kappa = 0.73$), but the unfavourable shape of the profile is also responsible for that.
In order to avoid the over-rotation, guiding blades were welded into the upper turn-round in front of the propeller. The resulting profile with guiding blades (WGB) yields already much lower critical Rm-values. However, the WGB-profile still showed a serious defect of the axial velocity close to the axis which represented a disadvantage for the self-excitation as can be seen by comparison with the BEF-profile. To overcome this additional blades

[3]The measurements were carried out by means of a pressure probe. It should be noted that the measured values close to the axis may be not very accurate [10].

behind the propeller were necessary.
The fifth profile (two guiding blades: TGB) represents the velocity values

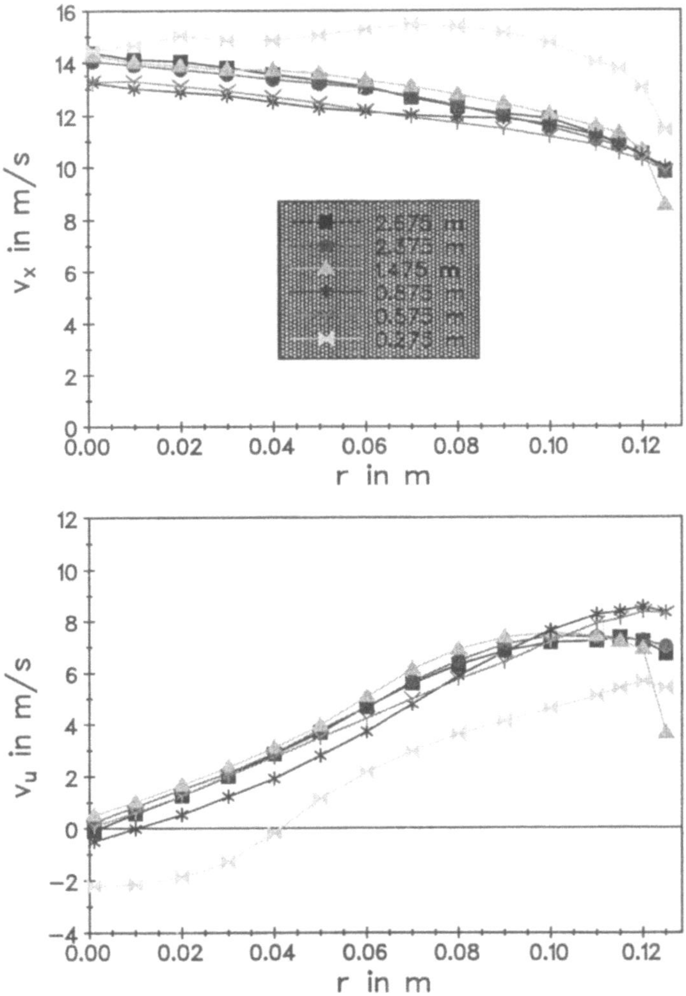

Figure 8. Velocity profiles measured for an propeller rotation rate of 1900 rpm depending on the distance from the propeller. Pre- and post-propeller blades are installed. Above: axial velocity versus radius. Below: azimuthal velocity versus radius.

which were measured after these additional guiding blades had been installed. [10]. As a matter of fact, the profile (TGB) is a mean representative of the really measured velocities at different distances from the propeller (see figure 8) The corresponding Rm-values are even a bit lower than the SBR values.

6. Conclusions

The real shape of the velocity profiles in the Riga dynamo experiment has a serious influence on the critical Rm. For helicity maximizing profiles the critical Rm can be reduced by 20 per cent compared to the corresponding value for the solid body rotation profile. This represents an important feature of the experimental design since the necessary power scales approximately with Rm^3. For all considered profiles, Rm^* has a minimum for κ close to 1.1. Thus, high helicity and a suitable pitch parameter are both important for the optimization of the velocity profiles.
The experimental velocity profiles in a water dummy channel have been improved by constructive changes. Whereas the propeller alone yields profiles which would by no means lead to self-excitation, the improvement by adding pre- and post-propeller blades leads eventually to velocity profiles giving a serious hope that self-excitation can be reached in the Riga sodium experiment.

Acknowledgments

This work was supported by Deutsche Forschungsgemeinschaft under INK 18/A1-1.
We would like to thank Prof. G. Will, M. Christen and H. Hänel from the Technical University of Dresden for designing the propeller and guiding vanes and measuring the velocity profiles.

References

1. Lowes, F. J. and Wilkinson, I (1968) Geomagnetic Dynamo: An improved Laboratory Model, *Nature*, **219**, pp. 717-718
2. Gailitis, A.K., Karasev, B.G., Kirillov, I.A., Lielausis, O.A., Luzhanskii, S.M., Ogorodnikov, A.P. and Preslitskii, G.V. (1987) Liquid metal MHD dynamo model experiment, *Magnitnaya Gidrodinamika* **4**, pp. 3-7
3. Gailitis, A. (1993) Experimental Aspects of a Laboratory Scale Liquid Sodium Dynamo Experiment, in: *Theory of Solar and Planetary Dynamos* (eds. M.R.E. Proctor, P.C. Matthews and A.M. Rucklidge), Cambridge University Press, pp. 91-98
4. Ponomarenko, Yu. B. (1973) On the theory of hydromagnetic dynamos, *Zh. Prikl. Mekh. Tekhn. Fiz.*, **6**, pp. 47-51
5. Gailitis, A. (1996) Project of a liquid sodium MHD dynamo experiment *Magnitnaya Gidrodinamika* **32**, pp. 63-67
6. Stieglitz, R. and Müller, U. (1996) GEODYNAMO - Eine Versuchsanlage zum Nachweis des homogenen Dynamoeffektes, Forschungszentrum Karlsruhe, Wissenschaftliche Berichte, FZKA 5716
7. Christen, M., Hänel, H., Will, G. (1998) Entwicklung der Pumpe für den hydrodynamischen Kreislauf des Rigaer "Zylinderexperimentes", To appear in: *Beiträge zu Fluidenergiemaschinen* **4**, Faragallah-Verlag und Bildarchiv
8. Gailitis, A. K. and Freiberg, J. ZH. (1977) Calculation of helical flux dynamo-instability, Preprint Riga

9. Fletcher, C. A. J. (1991) Computational Techniques for Fluid Dynamics, Vol. I, Springer Berlin-Heidelberg, S. 208
10. Will, G. and Christen, M. (1997) private communication

MAGNETOHYDRODYNAMIC FLOWS AROUND BODIES IN STRONG TRANSVERSE MAGNETIC FIELDS

S.MOLOKOV and K.RAJAN
Coventry University
School of Mathematical and Information Sciences
Priory Street, Coventry CV1 5FB, UK

Abstract
Unconfined magnetohydrodynamic flows around two-dimensional bodies of arbitrary cross-section in a strong magnetic field are considered. The direction of the free-stream velocity, the applied magnetic field, and the axis of the body are perpendicular to each other. The flow is supposed to be inertialess. A detailed asymptotic analysis of the flow at high values of the Hartmann number is presented. The results show that the local shape of the body cross-section in the vicinity of the front and rear stagnation points is very important for both the drag and the overall flow balance. For blunt-nosed bodies, the flow is severely disturbed in the vicinity of the magnetic field lines crossing these stagnation points. The asymptotics of the drag in this case strongly depends on the Hartmann number and the local shape of the cross-section, which means that any small, local deviation of the shape of the cylinder from being circular may significantly affect the drag. Also, previously suggested logarithmic asymptotics of the drag on the circular cylinder with the Hartmann number has been confirmed.

1. Introduction

Recent difficulties with velocity measurements in liquid metal flows in bends [1,2] have renewed interest in magnetohydrodynamic (MHD) flows around obstacles. In the experiments the probes inserted into a liquid metal induce the well-known shading effect (see the discussion in [2]). In essence, owing to MHD interaction a column of fluid with reduced velocity (a wake) is formed along magnetic field lines intersecting the probe surface. This leads to a distorted velocity profile, and thus questions reliability of the experimental results.

Concerning MHD flows around bodies, two different situations must be distinguished, namely confined and unconfined flows. The flow is confined if the wake from the obstacle reaches a solid wall thus giving a possibility for the electric currents to

A. Alemany et al. (eds.),
Transfer Phenomena in Magnetohydrodynamic and Electroconducting Flows, 45–60.

close their loop either within the wall or adjacent Hartmann layers. If the probe is sufficiently small, the wake may not reach the walls of the duct, and the flow may be considered as unconfined.

Unconfined flow around obstacles is one of the oldest subjects in MHD. However, many features of the flow remain unclear. Perhaps, the most controversial situation occurs when the free-stream velocity $\mathbf{v}_0 = -v_0\hat{\mathbf{x}}$ is transverse to the external magnetic field $\mathbf{B} = B_0\hat{\mathbf{y}}$. Here we will be concerned with a two-dimensional situation, when a body is an infinite cylinder with its axis perpendicular to both $\mathbf{v}_0$ and $\mathbf{B}$ (fig. 1). The magnetic field is supposed to be strong in a sense that the Hartmann number $Ha = B_0 a(\sigma/\rho\nu)^{1/2}$ is high. While aiming at developing a more general theory in the future, we assume that the flow is inertialess, and that the induced magnetic field is negligible. In the above σ, ρ, ν are the electrical conductivity, density and kinematic viscosity of the fluid, respectively, while a is half the length of the projection of the cylinder on the x-axis.

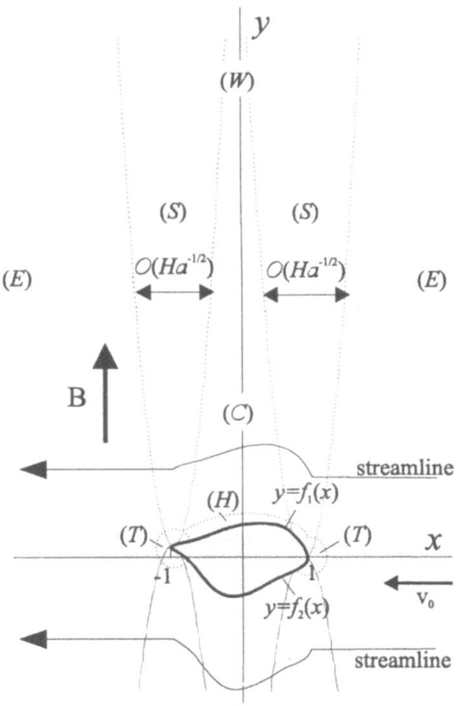

Fig.1 Schematic diagram of the flow around a cylinder of arbitrary shape and flow subregions at high Ha for $y>0$: (C, W) near and far wakes, (H) Hartmann layer, (E) external regions, (S) shear layers, (T) edge layers. For $y<0$ the subregions are similar.

An inviscid flow past a two-dimensional body in a strong transverse field was discussed by Ludford [3]. He employed an asymptotic theory to analyze the flow in the wakes, which spread along the magnetic field lines crossing the body surface. He showed that in major parts of the wakes the flow pattern repeats the shape of the body surface, as qualitatively shown in figure 1. Outside the wakes the fluid flows with the free-stream velocity.

Viscous, inertialess flows past two-dimensional bodies in a strong transverse magnetic field have been analyzed by a number of authors [4]-[7], who use heuristic methods and present contradictory results for the drag on a circular cylinder, see the discussion in [7]. The major point for disagreement is the qualitatively different asymptotics for the drag on the circular cylinder at high Ha, namely $O(1)$ in [4], [5], and $O(Ha^{-1}\ln Ha)$ in [6], [7]. Here the drag is normalized by $\sigma v_0 a^2 B_0^2$.

In this paper we present a complete description of the flow pattern for a cylinder of arbitrary shape and electrical conductivity, σ_c. It turns out that the local shape of the cylinder in the vicinity of the front and rear stagnation points plays a very important role for both the drag and the overall flow balance.

2. Formulation

The steady, two-dimensional, inertialess, inductionless flow of a viscous, conducting, incompressible fluid is governed by the following dimensionless equations:

$$Ha^{-2}\nabla^2 u-(1+u)=\frac{\partial p}{\partial x},\quad Ha^{-2}\nabla^2 v=\frac{\partial p}{\partial y},\quad \frac{\partial u}{\partial x}+\frac{\partial v}{\partial y}=0,\tag{1}$$

where $\nabla=\hat{\mathbf{x}}\,\partial/\partial x+\hat{\mathbf{y}}\,\partial/\partial y$. The fluid velocity $\mathbf{v}=u\hat{\mathbf{x}}+v\hat{\mathbf{y}}$ and the pressure p are normalized by v_0 and $\sigma a v_0 B_0^2$, respectively.

The boundary conditions are the no-slip condition on the cylinder surface and that of vanishing disturbance at infinity. This gives

$$u=0,\ v=0\quad\text{at}\ y=f_{1,2}(x),\tag{2}$$

$$u\to-1,\quad v,p\to 0\quad\text{as}\ x^2+y^2\to\infty.\tag{3}$$

The cross-section of the cylinder is determined by two smooth single-valued functions, $y=f_{1,2}(x)$. The assumption of smoothness is not essential and can be easily relaxed by introducing additional boundary layers at the jumps in slopes of the surface.

The streamfunction, Ψ, may be introduced in a usual way:

$$u=\frac{\partial\Psi}{\partial y},\quad v=-\frac{\partial\Psi}{\partial x}.\tag{4}$$

Since the problem is two-dimensional, both the electric current $\mathbf{j}$ (normalized by $v_0 B_0 \sigma$) and the constant electric field have z-components only. The requirement of vanishing current at infinity gives

$$\mathbf{j} = (-1-u)\hat{\mathbf{z}}, \quad \mathbf{j}_c = -\sigma_c / \sigma \hat{\mathbf{z}}. \tag{5}$$

Here $\mathbf{j}_c$ is the electric current within the cylinder.

The origin of the co-ordinate system (x,y) may be placed in such a way that $f_1(1) = f_2(1) = 0$, and let at the front stagnation point

$$f_1(x) \sim -f_2(x) \sim \beta(1-x)^{\alpha} \quad \text{as } x \to 1-0, \tag{6}$$

where $\alpha>0$ and β are arbitrary parameters. For $\alpha>1$, $\alpha<1$ and $\alpha=1$ one gets sharp-nosed (≻), blunt-nosed (⊃) and wedge-nosed (>) cylinders, respectively. Equation (6) implies that the nose of the cylinder is supposed to be *locally* symmetric with respect to y. Without loss of generality we may assume that asymptotics (6) holds for the rear stagnation point as well.

3. Flow analysis at high Hartmann numbers

At high Hartmann numbers the flow region consists of several subregions shown in fig. 1. Below we analyze the flow in each of these subregions by matched asymptotic expansions.

3.1 NEAR WAKES C

In the near wakes viscous terms are negligible. The reduced equations governing the flow are

$$u_{C,i} = -1 + \frac{\partial p_{C,i}}{\partial x}, \quad \frac{\partial p_{C,i}}{\partial y} = 0, \quad \frac{\partial u_{C,i}}{\partial x} + \frac{\partial v_{C,i}}{\partial y} = 0. \tag{7}$$

Throughout this paper subscripts of u, v , Ψ and p denote the flow region where corresponding limit equations are valid; the index i=1,2 refers to the upper and lower part of the problem, respectively.

From the second equation in (7) the usual result is recovered, namely that the pressure does not vary along the magnetic field lines. The leading terms of the asymptotic solution to the equations (7), which satisfies the non-penetration condition on the surface of the cylinder, are

$$p_{C,i} = Ha^{-1} g_i(x), \quad u_{C,i} = -1 - Ha^{-1} g_i'(x), \quad v_{C,i} = -f'_i(x), \quad \Psi_{C,i} = -y + f_i(x) + c_i, \tag{8}$$

where c_i are arbitrary constants, and $g_i(x)$ are arbitrary functions. The $O(1)$ pressure in the near wakes vanishes because of the inability of the far wakes to accommodate the disturbance of such a magnitude. The $O(Ha^{-1})$-term in the expression for $u_{C,i}$ is retained because it is required for the discussion of the edge layers in §3.6. From the last equation in (8) follows that in the near wakes the streamlines repeat the shape of the cylinder.

From (6) and (8) follows that a singularity develops in $v_{C,i}$ as $x \to 1-0$ for blunt-nosed cylinders ($\alpha<1$). It will be shown that this singularity plays an important role for the overall flow balance.

The dimensions of the near wakes are $O(1)\times O(Ha)$, i.e. they extend very far from the cylinder along magnetic field lines in the region $-1<x<1$.

3.2 EXTERNAL REGIONS *E*

The external regions, in which the flow is also inviscid, are situated outside the strip $|x|<1$. The equations analogous to (7) govern the flow there. It follows that in these regions the fluid flows with the free-stream velocity and the flow is current-free to any order in Ha^{-1}, i.e.

$$p_{E,i}=0, \quad u_{E,i}=-1, \quad v_{E,i}=0, \quad \Psi_{E,i}=-y. \tag{9}$$

The reason for this is that in the shear layers the disturbance decays exponentially into the external regions, see §3.5. An algebraic decay would induce a non-zero disturbance in the external regions. This was the case in the inertial, inviscid counterpart of the present problem discussed by Ludford [3], who discovered an upstream influence of the shear layers governed by the electromagnetic-inertia interaction.

3.3 HARTMANN LAYERS *H*

The Hartmann layers separate the near wakes from the cylinder. Their role is to satisfy the no-slip condition. The standard procedure for the Hartmann layers [8] yields

$$u_{H,i}=-1+E_i, \quad v_{H,i}=f'_i(x)\{-1+E_i\}, \quad p_{H,i}=Ha^{-1}\{g_i(x)-s_i f'_i(x)E_i\}, \tag{10}$$

where $s_1=1$, $s_2=-1$, and

$$y_{H,i}=Ha[y-f_i(x)], \quad E_i(x,y_{H,i})=\exp[-|y_{H,i}|F_i^{-2}(x)], \quad F_i(x)=\left(1+[f'_i(x)]^2\right)^{1/2}.$$

The Hartmann layers have thickness $O(Ha^{-1})$, which varies along the surface of the cylinder. The thickness is minimal at local maxima and minima of $f_{1,2}(x)$ and in general grows as $x\to\pm 1$, where the layers degenerate, and different layers, *T*, are formed.

3.4 FAR WAKES *W*

Far wakes are formed at distances $O(Ha)$ from the surface of the cylinder in the $\pm y$-direction. Their role is to match the conditions at infinity. Introducing a compressed variable $y_{W,i} = Ha^{-1}y$ into the governing equations leads to the following problem for $v_{W,i}$ and the rescaled pressure $\tilde{p}_{W,i} = Hap_{W,i}$:

$$\frac{\partial^2 v_{W,i}}{\partial x^2} = \frac{\partial \tilde{p}_{W,i}}{\partial y_{W,i}}, \qquad \frac{\partial^2 \tilde{p}_{W,i}}{\partial x^2} = \frac{\partial v_{W,i}}{\partial y_{W,i}}, \tag{11}$$

$$\left\{v_{W,i}, \tilde{p}_{W,i}\right\} = \eta(1-|x|)\left\{-f'_i(x), g_i(x)\right\} \quad \text{at } y_{W,i}=0, \tag{12}$$

where $\eta(x)$ is a step function. The conditions (12) are those of matching the near-wake solution. From (11) follows that the far wakes may be viewed as the boundary layers at infinity, where viscosity reappears acting in the x-direction only.

The requirement of limitedness of $v_{W,i}$ and $p_{W,i}$ as $|y_{W,i}| \to \infty$ places a restriction on values (12), whereby the near-wake pressure becomes fully determined (for the details of a similar analysis see [9,10]):

$$p_{C,i}(x) = Ha^{-1}g_i(x) = -Ha^{-1}s_i f_i'(x). \tag{13}$$

The solution to the problem (11)-(13), obtained by applying the forward and inverse Fourier transforms with respect to x, is

$$v_{W,i} = s_i Hap_{W,i} = -\tfrac{1}{2}\pi^{-1/2}\left|y_{W,i}\right|^{-1/2}\int_{-1}^{1} f'_i(t)\exp\left[-\tfrac{1}{4}(x-t)^2\left|y_{W,i}\right|^{-1}\right]dt. \tag{14}$$

It is shown in fig. 2 for a circular ($f_1(x)=-f_2(x)=(1-x^2)^{1/2}$) and a parabolic ($f_1(x)=-f_2(x)=1-x^2$) cylinders.

In regions *W*, *S* and *T* the expressions for u can be easily calculated from the inviscid version of equation (1), namely

$$u = -1 - \frac{\partial p}{\partial x}, \tag{15}$$

and thus are not presented here.

It turns out that the far wake solution is a composite solution for regions *W*, *C*, *E*, and *S*. This remarkable property has been noticed in other problems of flows around obstacles [11], [9], [10].

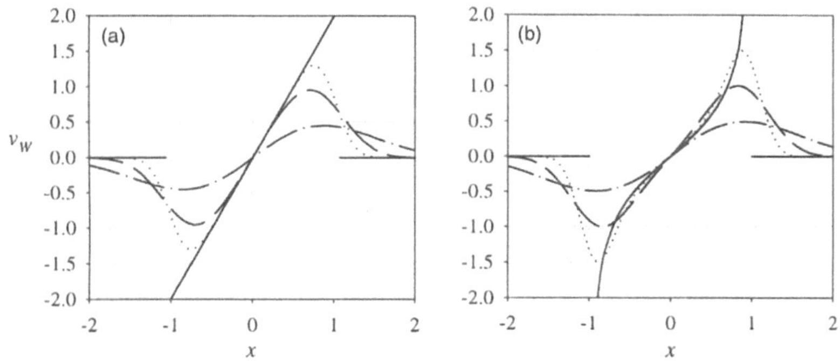

Fig. 2 Transverse component of velocity in the far wake for a parabolic cylinder, $f_1(x)=1-x^2$ (a), and a circular cylinder, $f_1(x)=(1-x^2)^{1/2}$ (b) at different $y_{w,1}$: 0 (solid line), 0.0175 (dotted line), 0.057 (dashed line), 0.25 (dashed-dotted line).

3.5 SHEAR (OR PARALLEL) LAYERS S

Free shear layers separate the near wakes C from the external regions. They are formed along the lines $x=\pm 1$, have the origin at the edge layers, and have thickness $O(Ha^{-1/2})$. Their role is to smooth jumps in the flow variables across the lines $x=\pm 1$. In the inertialess flow the layers have parabolic shape and are governed by a parabolic system of heat equations, which is equivalent to the equations (11).

One way to obtain the solutions for the shear layers is to consider the limiting process of the far-wake solution (14) as $x \to \pm 1$ for $y=O(1)$. Consider, for example, the shear layer at $x=1$, $y>0$. First we rewrite the far-wake solution in terms of the shear-layer variables $x_S=Ha^{1/2}(x-1)$, $y_S=y$. Then the limiting process as $Ha \to \infty$ gives a self-similar solution

$$v_S = Hap_S = Ha^{(1-\alpha)/2}\alpha\beta\Gamma(\alpha)\left(2y_S\right)^{(\alpha-1)/2}(2\pi)^{-1/2}e^{-\eta^2/4}D_{-\alpha}(\eta)\,, \tag{16}$$

where $\eta=x_S/(2y_S)^{1/2}$, $\Gamma(\bullet)$ is the gamma function, and $D_{-\alpha}(\bullet)$ is the parabolic-cylinder function [12]. Parameters α and β in (16) are determined by the local shape of the body in the vicinity of $x=1$, $y=0$ given by (6).

For $x_S<0$ an analytical continuation of the parabolic-cylinder function must be used (see [12]). Then an asymptotic expansion of (16) as $|x_S| \to \infty$ shows that the flow is characterized by a gradual, algebraic decay of the disturbance as $x_S \to -\infty$ (into the near wake), and a fast, exponential decay as $x_S \to \infty$ (into the external region).

Equation (16) shows that there is a jet in the shear layer for $\alpha<1$ (blunt-nosed cylinders). In particular, for a circular cylinder the transverse velocity component is $O(Ha^{1/4})$. The solution (16) for a circular cylinder is shown in fig. 3.

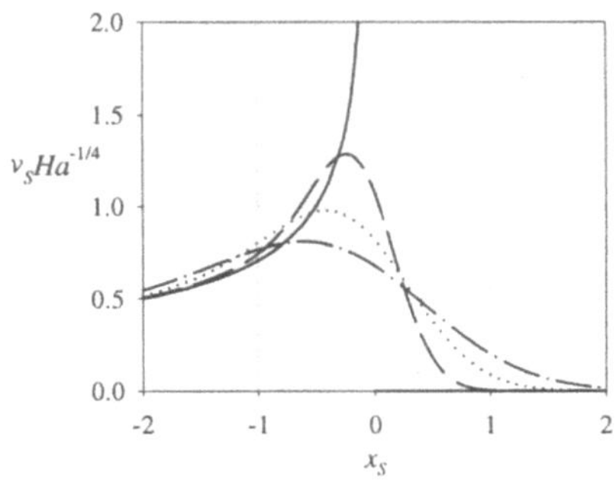

Fig. 3. Normalized transverse component of velocity in the shear layer at $x=1, y>0$ for a circular cylinder at different $y_{s,1}$: 0 (solid line), 0.05 (dashed line), 0.15 (dotted line), 0.32 (dashed-dotted line).

3.6 EDGE (OR TANGENTIAL) LAYERS *T*

These are tiny layers at the front and rear stagnation points. In the vicinity of these points the Hartmann layers degenerate, while the shear layers appear. For sharp-nosed cylinders they need not be analyzed in detail. However, for blunt-nosed cylinders the edge layers determine the drag, and therefore become very important.

The dimensions of the edge layers and the flow pattern there strongly depend on the local shape of the cylinder at the edge points (cf. [11]). The dimensions vary from $O(Ha^{-1})\times O(Ha^{-1})$ for a flat plate aligned with the flow to $O(Ha^{-1/2})\times O(1)$ for the case when the magnetic field is aligned with the surface of the cylinder along a finite interval (e.g. a flat plate transverse to the flow). For a circular cylinder, for example, the dimensions are $O(Ha^{-2/3})\times O(Ha^{-1/3})$, which is similar to a flow in a circular duct [12].

Here we consider a general case. Since it was assumed that at the front stagnation point the body is *locally* symmetric with respect to y (cf. (6)), we consider the upper half plane, $y\geq 0$, only. Let's introduce the stretched variables

$$x_T=Ha^{\gamma}(1-x), \quad y_T=Ha^{\delta}y, \tag{17}$$

where parameters γ and δ are to be determined. Since in the vicinity of $x=1$ the expansion (6) holds, the near-wake solution (8) behaves as follows:

$$u_C\sim -1+\alpha(\alpha-1)\beta x_T^{\alpha-2}Ha^{\gamma(2-\alpha)-1}, \tag{18a}$$

$$v_C\sim \alpha\beta x_T^{\alpha-1}Ha^{\gamma(1-\alpha)}, \tag{18b}$$

$$p_C\sim \alpha\beta x_T^{\alpha-1}Ha^{\gamma(1-\alpha)-1}. \tag{18c}$$

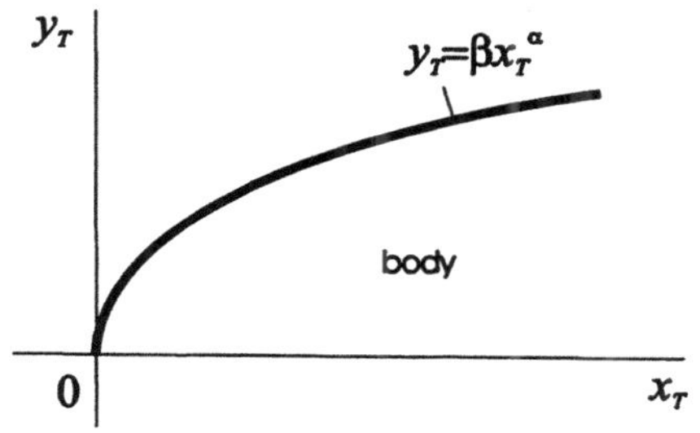

Figure 4. Transformed boundary in the vicinity of the front stagnation point x=1, y=0.

This suggests that in the edge layer, $u_T=Ha^{\gamma(2-\alpha)-1}U$, $v_T=Ha^{\gamma(1-\alpha)}V$, and $p_T=Ha^{\gamma(1-\alpha)-1}P$, where U, V and P are $O(1)$ quantities. The fact that there is an $O(1)$ jump in u across the layer, implies that u_T is also $O(1)$, and hence

$$\gamma=(2-\alpha)^{-1}.$$

The still unknown stretching parameter δ is determined from the limiting process in the governing equations to give

$$\delta=\alpha(2-\alpha)^{-1},$$

whereby these equations become

$$\frac{\partial^2 V}{\partial x_T^2}=\frac{\partial P}{\partial y_T}, \qquad \frac{\partial^2 P}{\partial x_T^2}=\frac{\partial V}{\partial y_T}. \tag{19}$$

Equations (19) are of the same type as those governing the flow in the shear layers and the far wakes, but the shape of the boundary is different now. From (6) and (17) follows that it is a semi-infinite area bounded by the curve $y_T=\beta x_T^{\alpha}$, see figure 4. This makes the problem intractable analytically, and thus a numerical analysis is necessary. For this purpose the flow region in figure 4 is mapped on the quarter-plane shown in figure 5 by means of the following transformation:

$$R=x_T-\left(y_T/\beta\right)^m, \quad \xi=y_T,$$

where m=1/α. Then the equations (19) become

$$\frac{\partial^2 V}{\partial R^2}=-m\beta^{-m}\xi^{m-1}\frac{\partial P}{\partial R}+\frac{\partial P}{\partial \xi}, \qquad \frac{\partial^2 P}{\partial R^2}=-m\beta^{-m}\xi^{m-1}\frac{\partial V}{\partial R}+\frac{\partial V}{\partial \xi}. \tag{20}$$

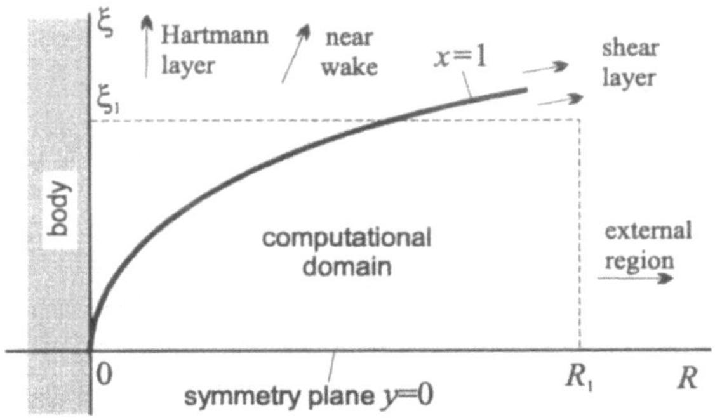

Figure 5. Computational domain for the edge layer on the (R,ξ)-plane.

For α=1/2 this system reduces to the Roberts' equations developed for the tangential layers in a circular duct [13].

From numerical point of view, it is more convenient to work with the equation for the streamfunction

$$\frac{\partial^4\Psi_T}{\partial R^4}=\frac{\partial^2\Psi_T}{\partial\xi^2}-2m\beta^{-m}\xi^{m-1}\frac{\partial^2\Psi_T}{\partial R\partial\xi}+m^2\beta^{-2m}\xi^{2(m-1)}\frac{\partial^2\Psi_T}{\partial R^2}-m(m-1)\beta^{-m}\xi^{m-2}\frac{\partial\Psi_T}{\partial R}. \quad (21)$$

The conditions for Ψ_T are the boundary conditions on the cylinder surface and the matching conditions with the solutions in regions *H, S, C*, and *E*. This gives

$$\Psi_T=0\,,\quad \frac{\partial\Psi_T}{\partial R}=0 \quad \text{at } R=0, \quad (22)$$

$$\Psi_T\to-\xi\,,\quad \frac{\partial\Psi_T}{\partial R}\to 0 \quad \text{as } R\to\infty\,, \quad (23)$$

$$\Psi_T=0\,,\quad \text{at } \xi=0, \quad (24)$$

$$\Psi_T\to\hat{\Psi}_{CHSE} \quad \text{as } \xi\to\infty\,, \quad (25)$$

where $\hat{\Psi}_{CHSE}$ is the composite asymptotic expansion valid in regions *C, H, S* and *E*.

The problem (21)-(25) has been solved numerically with a finite-difference method. Central differences were used throughout. The computational domain is shown in figure 5. The constants R_1 and ξ_1 are chosen sufficiently high for the asymptotics as $R\to\infty$ and $\xi\to\infty$ to establish.

The streamlines in the edge layer for α=1/2 (e.g. circular cylinder) are shown in figure 6. The formation of the shear layer along the line x=1, or $R=\frac{1}{2}\xi^2$, is clearly visible.

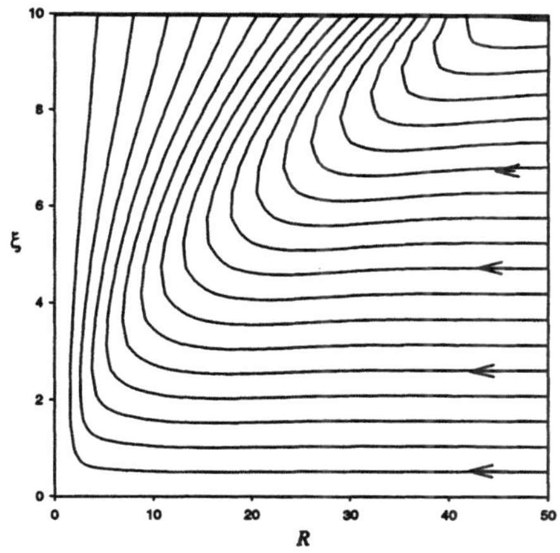

Fig. 6. Streamlines in the edge layer for α=1/2.

4. Drag

The drag on the cylinder, D, normalized by $\sigma v_0 a^2 B_0^2$, consists of two parts. The first part, D_{mech}, results from the contribution of viscous and pressure forces. The second part, D_{EM}, is determined by the interaction of the electric current with the magnetic field within the cylinder. The expressions for the two parts of the drag are

$$D_{mech} = -\sum_{i=1}^{2} s_i \int_{-1}^{1} \left\{ \left(p - 2Ha^{-2} \frac{\partial u}{\partial x} \right) f_i'(x) + Ha^{-2} \left(\frac{\partial u}{\partial y} + \frac{\partial v}{\partial x} \right) \right\}_{y=f_i(x)} dx\,, \tag{26}$$

$$D_{EM} = \iint_{\Omega} \mathbf{j} \times \hat{\mathbf{y}} dxdy = \frac{\sigma_c}{\sigma} \int_{-1}^{1} [f_1(x) - f_2(x)] dx = \frac{\sigma_c}{\sigma} \Omega\,. \tag{27}$$

Here Ω is the cross-sectional area of the cylinder.

The electromagnetic part is independent of the flow outside the cylinder, because the electric current in a two-dimensional flow is in the transverse, z-direction only. This contribution is equal to the product of the relative conductivity of the cylinder and the cross-sectional area. In contrast, the asymptotics of the mechanical drag as $Ha \to \infty$ strongly depends on the shape of the body.

Consider now the mechanical drag. It consists of the contribution of the Hartmann layers and the edge layers. Formal substitution of the respective solutions into (26) and neglecting terms of lower order in Ha^{-1} with respect to those retained gives the following expression for the leading term of the drag:

$$D_{mech} \cong D_{H,frict} + D_{H,pres} + D_{T,pres}, \tag{28}$$

where

$$D_{H,frict} = 4Ha^{-1}, \tag{29}$$

$$D_{H,pres} = 2Ha^{-1} \int_{-1}^{1} \left[f'^2_1(x) + f'^2_2(x) \right] dx, \tag{30}$$

$$D_{T,pres} = 4Ha^{-\frac{\alpha+1}{2-\alpha}} \int_0^\infty P|_{R=0} d\xi \qquad (\alpha<2) \tag{31}$$

are contributions of the Hartmann-layer friction, Hartmann-layer pressure, and edge-layer pressure, respectively. We recall that P is the rescaled, $O(1)$ pressure in the edge layer. While deriving (31) it was assumed that the local shape of the body at the rear stagnation point is the same as at the front, so that (31) represents the contribution of both edge layers.

Several observations are important. First, the drag due to friction is $O(Ha^{-1})$, and is independent of the shape of the cross-section, which is somewhat surprising. Second, the contribution of the Hartmann-layer pressure is also $O(Ha^{-1})$, but it does depend on the shape of the cylinder. Third, in the edge layers the drag is created by the pressure forces only[†].

It should be stressed that the expressions (30) and (31) are formal in a sense that the integrals may either converge or diverge depending on the value of α, which leads to three different situations discussed below.

4. 1 $\alpha > \frac{1}{2}$

For $\alpha > 1/2$ the integrals in (30) converge, while $D_{H,pres}$ remains $O(Ha^{-1})$ for all values of α. Only the Hartmann layers contribute to the drag, since $D_{T,pres}$ is of lower magnitude.

To demonstrate the dependence of $D_{H,pres}$ on α, let's consider a special class of the cylinder cross-sections, defined by

$$f_1(x) = -f_2(x) = (1-x^2)^\alpha, \tag{32}$$

and shown in figure 7a. In this case

$$D_{H,pres} \cong 16Ha^{-1}\alpha^2 B(\tfrac{3}{2}, 2\alpha - 1), \tag{33}$$

[†] For sharp-nosed bodies ($\alpha > 1$) the edge-layer friction gives a comparable contribution to that of pressure, but then both these contributions to the drag are neglegibly small with respect to the Hartmann-layer ones.

where $B(.,.)$ is the beta-function. The results for the total mechanical drag for this specific class are shown in figure 7b. The drag is minimal for $\alpha \approx 1.5$. Higher values of α are of no particular interest in connection with the class (32), because the rise of the drag with further increase in α is caused by a different effect, namely the tendency to the formation of the shear layer at x=0. Indeed, as $\alpha \to \infty$ the cylinder cross section tends to become an infinitely thin cross. At one of the plates, x=0, which comprises the cross, and which is transverse to the flow, there will be an additional shear layer, which is responsible for the rise of the drag. Of course, the analysis can be extended to this case if necessary.

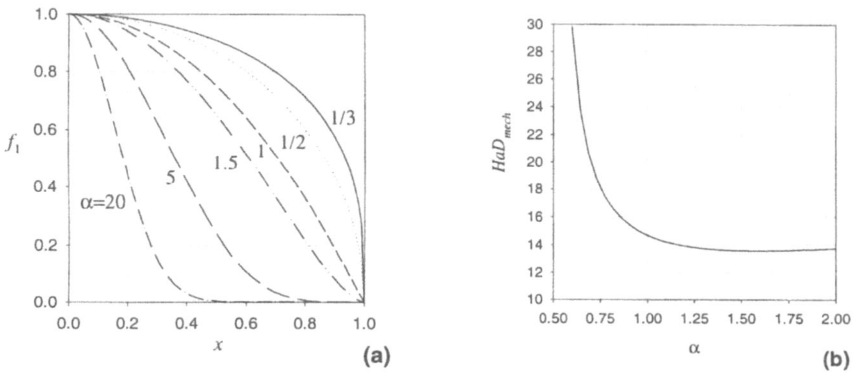

Fig. 7. Shapes of the cross-sections from a special class (32) (a) and variation of the drag with α for this class (b).

An important observation in figure 7b is that D_{mech} tends to infinity as $\alpha \to \frac{1}{2}+0$, i.e. as the shape of the cylinder becomes circular, which indicates a transition to a different asymptotics in Ha^{-1} for α=1/2, see below.

4. 2 $\alpha < \frac{1}{2}$

For α>1/2 the integrals in (30) diverge, while that in (31) converges. However, since the edge-layer pressure matches the Hartmann-layer one, the divergent part of the integrand in (30) is taken into account by the integrand in (31). Since the non-divergent part of (30) remains $O(Ha^{-1})$, it can be neglected, so that (31) represents the total drag. Therefore, the drag in this case is created by the pressure forces at the front and rear stagnation points only. Further, the asymptotics of D_{mech} as $Ha \to \infty$ strongly depends on the value of α. Its maximum value is $O(Ha^{-1/2})$, which is reached as $\alpha \to 0$, e.g. for a rectangular cylinder, or a flat plate transverse to the flow.

4.3. $\alpha = \frac{1}{2}$

The case α=1/2 (γ=2/3, δ=1/3), circular cylinder in particular, is special, because both integrals (30) and (31) diverge, so that special treatment of the pressure term in (26) is required. Let's determine the asymptotics of the following integral

$$I = Ha\int_{-1}^{1} p\Big|_{y=f_1(x)} f_1'(x)dx. \tag{34}$$

First, we split the integral into three, integrating over two edge layers and the Hartmann layer separately. Then we substitute the composite expansion for pressure, $\hat{p}_{HT}$, on the surface of the cylinder

$$\hat{p}_{HT} = p_H + p_T - p_{HT} \quad \text{at } y=f_1(x), \tag{35}$$

where $p_{HT}=Ha^{-1}\beta(1-x)^{-1/2}$ is the common part of p_H and p_T. This gives

$$I = Ha\left(\int_{-1}^{-1+Ha^{-2/3}} + \int_{1-Ha^{-2/3}}^{1}\right)\left[(p_H - p_{HT}) + p_T\right]_{y=f_1(x)} dx$$

$$+ Ha\int_{-1+Ha^{-2/3}}^{1-Ha^{-2/3}} \left[p_H + (p_T - p_{HT})\right]_{y=f_1(x)} dx \equiv I_{12} + I_3. \tag{36}$$

The terms in brackets are combined in a way to cancel the non-integrable singularities.

In I_{12} integration is performed within the edge layers. The term p_H-p_{HT} gives negligible contribution to the integral, so that

$$I_{12} = 2\int_0^1 P\Big|_{R=0} d\xi. \tag{37}$$

In I_3 integration is performed along the Hartmann layers, and the result is

$$I_3 = 2\int_1^{\infty}\left\{P\Big|_{R=0} - \frac{2}{\xi}\right\}d\xi + Ha\int_{-1+Ha^{-2/3}}^{1-Ha^{-2/3}} p_H\Big|_{y=f_1(x)} dx. \tag{38}$$

By differentiating the second integral in (38) with respect to a parameter, Ha, it is easy to show that its asymptotics as $Ha \to \infty$ is $\frac{4}{3}\beta^2 \ln Ha + d$, where d=const. independent of Ha. Combining the expressions (26), (28), (29), (34), (36)-(38) leads to the following asymptotics for the mechanical drag:

$$D_{mech} \cong \tfrac{4}{3}\beta^2 Ha^{-1}\ln Ha + \tilde{d}Ha^{-1}, \tag{39}$$

where a constant $\tilde{d}$, incorporating both the pressure and friction terms on the whole cylinder surface, depends on the shape of the cylinder.

Combining the results (28)-(31) and (39), it may be concluded that the mechanical drag varies from being $O(Ha^{-1})$ to $O(Ha^{-1/2})$.

6. Discussion and conclusions

Consider the flow pattern for $y=O(1)$, when the flow passes through the following sequence of subregions: *E, S, C, S, E.* In the external regions the streamlines are straight lines y=constant. In the near wakes the streamlines repeat the shape of the cylinder surface, analogous to the inviscid flow [3]. There may, or may not be a jet in layers S depending on the value of the parameter α, which determines the local shape of the cylinder surface in the vicinity of the front and rear stagnation points.

For $\alpha>1$ (sharp-nosed cylinder, $\succ$), the transverse component of velocity in the shear layers is $v_S=O(Ha^{(1-\alpha)/2})<<1$, cf. (16). This means that the shear layer does not appear to the main order, and that the flow smoothly passes through $x=1$.

For $\alpha=1$ (wedge-nosed cylinder, $>$, e.g. a parabolic cylinder $f_{1,2}=\pm(1-x^2)$), $v_S=O(1)$.

For $\alpha<1$ (blunt-nosed cylinder, $\supset$, including the circular cylinder), there is a jet in layer S in the y-direction. In the extreme case of $\alpha\to 0$ (e.g. finite flat plate transverse to $\mathbf{v}_0$), $v_S=O(Ha^{1/2})$, and this leads to a finite jump in the value of the streamfunction across the layer.

The mechanical drag due to friction and pressure vary from being $O(Ha^{-1})$ to $O(Ha^{-1/2})$ depending on the shape of the cylinder cross-section. Due to strong dependence of the asymptotics for the drag on α for $\alpha<1/2$, *any small deviation of the shape of the cylinder from being circular, for example, may significantly affect the drag.* This is especially true for small tips of the probes, when a strict circular cross-section is difficult to enforce. Concerning the contradiction concerning the drag on the circular cylinder, discussed in [7], a logarithmic asymptotics of the drag with *Ha*, which was obtained first in [6], can be confirmed.

The results for the flow pattern, as well as for the drag show that sharp-nosed cylinders produce less disturbance to the flow than the blunt-nosed ones. While developing probes for fluid-velocity measurements in MHD experiments these results should be taken into account. A complete picture, however, will emerge only with the analysis of a three-dimensional, inertial flow over a finite body with different orientation of the field and various conditions at infinity.

7. References

1. Reimann, J., Bucenieks, I., Dementiev, S., Flerov, A., Molokov, S., Platnieks, I. *Magnetohydrodynamics* **31** (1995), 338-348.
2. Stieglitz, R., Barleon, L., Bühler, L., Molokov, S. *J. Fluid Mech.* **326** (1996), 91-123.

3. Ludford, G.S.S. *J. Fluid Mech.* **10** (1961), 141-155.
4. Tsinober, A.B., Shtern, A.G. *Magnitnaya Gidrodinamika*, **3** (1967), 146-147 (In Russian).
5. Tsinober, A.B. *MHD flows around bodies*, Zinatne, Riga, 1970 (In Russian).
6. Gailitis, A., Gerbeth, G. *Magnitnaya Gidrodinamika*, **23** (1987), 137-139 (In Russian).
7. Gerbeth, G. *Magnetohydrodynamics*, **26** (1990), 210-215.
8. Moreau, R. *Magnetohydrodynamics*, Kluwer Academic Publishers, Dordrecht, 1990.
9. Molokov, S. *Magnetohydrodynamics*, **28** (1992), 79-86.
10. Molokov, S. *Magnetohydrodynamics*, **29** (1993), 175-180.
11. Chang, I.D. *ZAMP* **14** (1963), 134-147.
12. Bateman, H., Erdélyi, A. *Higher Transcendential Functions, Vol. 2*, McGraw-Hill, New York, 1953.
13. Roberts, P.H. *Proc. Roy. Soc.* **A300** (1967), 94-107.

ON MHD TURBULENCE MODELS FOR SIMULATION OF MAGNETIC BRAKES IN CONTINUOUS STEEL CASTING PROCESSES

OLA WIDLUND

Faxén Laboratory, Royal Institute of Technology
SE-100 44, Stockholm, Sweden

SAID ZAHRAI

ABB Corporate Research
SE-721 78, Västerås, Sweden

AND

FRITZ H. BARK

Faxén Laboratory, Royal Institute of Technology
SE-100 44, Stockholm, Sweden

Abstract. A novel approach is discussed for extending Reynolds stress transport (RST) turbulence modelling to include magnetic Joule dissipation. A scalar dimensionality anisotropy parameter is introduced to carry information about the distribution of energy in spectral space. This information is vital in modelling MHD turbulence, as it determines both magnitude and anisotropy of the Joule dissipation tensor. The Joule dissipation tensor is modelled by a tensor function in the Reynolds stress anisotropy tensor, the unit direction vector of the magnetic field and the scalar dimensionality parameter. A phenomenological transport equation for the dimensionality parameter is proposed for the case of homogeneous turbulence. The model is closed using the pressure–strain model of Sarkar, Speziale and Gatski, and a magnetic destruction term in the standard dissipation equation. Model predictions for the case of decaying turbulence show very good agreement with direct numerical simulations.

A. Alemany et al. (eds.),
Transfer Phenomena in Magnetohydrodynamic and Electroconducting Flows, 61–76.

1. Introduction

In continuous casting of steel, the liquid metal is continuously supplied to a water cooled copper mould via a submerged inlet pipe. The strong momentum of the injected liquid steel will induce recirculation in the mould and this may lead to inclusion of particles and impurities from the slag layer at the surface.

The fluid motion in the mould can be damped and controlled by applying a static magnetic field oriented perpendicular to the direction of flow from the inlet pipe. The magnetic field is provided by a so-called electromagnetic brake (EMBR). This device has become an important tool for improving both quality and productivity in continuous casting operations.

The primary effect of the magnetic field is to brake the mean flow. A secondary effect is magnetic dissipation of turbulence, or Joule dissipation, which will reduce the turbulent transport of heat and momentum. Depending on operating conditions, the magnetic field may well cause laminarization of the melt flow in quiescent regions of the mould. Turbulent conditions are, however, generally sustained by production in the high-velocity entry zone of the jet and its free-shear boundary layers (see e.g. Lehman *et al.*, 1994).

Figure 1 illustrates the effect of a localized magnetic field on the mean flow in the mould. When a magnetic brake is applied, the momentum of the inlet jet is reduced and redirected. The penetration depth of the jet is reduced considerably, as is the entrapment of non-metallic inclusions. The surface of the meniscus will be calmer and the surface temperature higher. The increased stability of the process also leads to a substantial reduction of cracks in the surface of the product.

Manufacturers of magnetic brakes use numerical flow simulations to optimize their designs and verify that they meet performance specifications. The extension of a commercial numerical flow solver to include the effects a magnetic field on the mean flow is relatively straightforward. However, the effect of the magnetic field on the turbulence is much more difficult to incorporate. The magnetic dissipation affects turbulent structures in a very complex way, elongating them in the direction of the magnetic field. Conventional turbulence models cannot capture effects of this kind. A reasonably accurate prediction of turbulence statistics is important, as turbulent diffusion plays a major role in heat transfer and transport of non-metallic inclusions in the mould. Today, the way around this dilemma is to either define anisotropic diffusion coefficients, based on simple scaling arguments and empirical knowledge, or to use transient large eddy simulation (LES) techniques.

The turbulence models normally used for modelling complex flows in

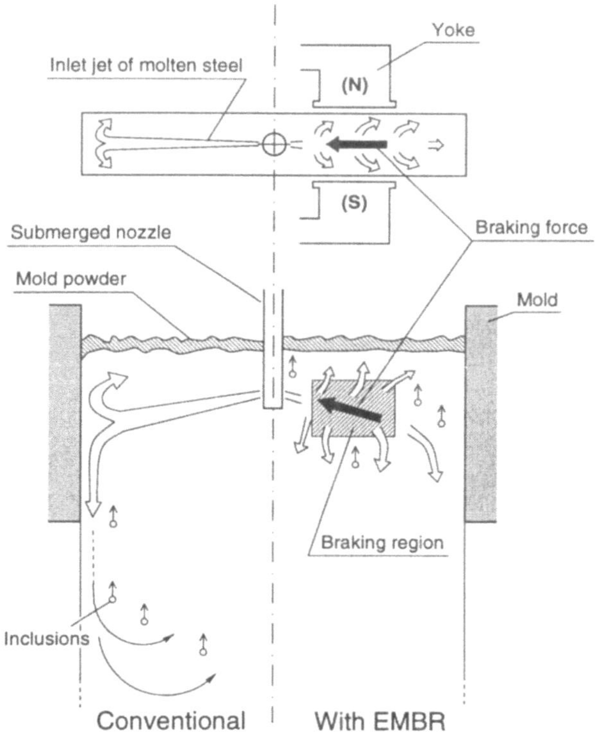

Figure 1. **Outline of flow field and transport of inclusions in continuous slab casting. The left side of the figure is for conventional casting, without magnetic braking. On the right side, an electromagnetic brake (EMBR) has been applied to brake the momentum of the inlet jet.**

industrial applications are so-called one-point closures. The more advanced of these generally introduce transport equations for the individual Reynolds stresses, i.e. the velocity correlations in the one-point limit, and the viscous dissipation, ε. A majority of the different terms in the transport equations must be closed using model assumptions. The classical modelling approach is to use tensor function expansions in the Reynolds stress tensor itself (or rather in its non-dimensional form, the stress anisotropy tensor) and, where applicable, the mean strain rate and mean rotation rate tensors, S_{ij} and Ω_{ij}. The Reynolds stress tensor, however, carry information only about the turbulent velocity components, i.e. the componentality of the turbulence, whereas critical structural information, the dimensionality of the turbulence, is missing (Reynolds & Kassinos, 1995). This is a fundamental obstacle in developing an effective Reynolds stress transport (RST) model for MHD turbulence, in which dimensional anisotropy is a prominent feature.

In the present paper we outline a novel approach to extending conventional one-point turbulence models to include also magnetic effects (see also Widlund *et al.*, 1998). This requires additional structural information about the turbulence, and in Section 2 we identify tensorial and scalar quantities that carry the necessary information. An RST model including this structural information is developed in Section 3 for the case of homogeneous turbulence decaying in a magnetic field. The model is demonstrated in Section 4, and predictions are compared with direct numerical simulations (DNS). Section 5 is a concluding summary.

2. Theory

The Navier–Stokes equations for the instantaneous velocity field $u_i(\vec{x}, t)$ of an incompressible fluid are

$$\frac{\partial u_i}{\partial t} + u_j \frac{\partial u_i}{\partial x_j} = -\frac{1}{\rho}\frac{\partial p}{\partial x_i} + \nu \frac{\partial^2 u_i}{\partial x_j \partial x_j} + f_i, \tag{1}$$

$$\frac{\partial u_i}{\partial x_i} = 0, \tag{2}$$

where $p(\vec{x}, t)$ is the pressure field, ν is the kinematic viscosity, and f_i is a volume force.

In engineering applications, turbulent flows are often analyzed using a statistical approach, in which only averaged effects of the turbulent fluctuations are considered. The so-called Reynolds decomposition is then used to separate the flow field into a mean part and a fluctuating part. Here we define the mean of a quantity v as the ensemble average over an infinite number of realizations,

$$\langle v \rangle = \lim_{N \to \infty} \frac{1}{N} \sum_{n=1}^{N} v_n.$$

The corresponding Reynolds decomposition is defined by

$$v(\vec{x}, t) \equiv \langle v(\vec{x}, t) \rangle + v'(\vec{x}, t) = V(\vec{x}, t) + v'(\vec{x}, t),$$

where we have used a capital letter to denote the mean quantity, and a prime to denote the fluctuating part. By inserting similar expressions for all flow variables into the Navier–Stokes equations (1) and (2), and performing an ensemble averaging, this eventually yields the Reynolds averaged equations

$$\frac{\partial U_i}{\partial t} + U_j \frac{\partial U_i}{\partial x_j} = \frac{1}{\rho}\frac{\partial}{\partial x_j}\left(-P\delta_{ij} + 2\rho\nu S_{ij} - \rho\langle u_i' u_j' \rangle\right) + F_i, \tag{3}$$

$$\frac{\partial U_i}{\partial x_i} = 0, \tag{4}$$

where S_{ij} is the mean strain-rate tensor. We see that the effects of turbulent fluctuations enter the mean momentum equations in (3) through the *Reynolds stresses* $-\rho\langle u_i' u_j'\rangle = -\rho R_{ij}$. Transport equations for the Reynolds stresses may be derived by multiplying (1) by u_i, taking the ensemble average of all terms, and subtracting an expression for the mean kinetic energy of the mean flow (not shown here). See for example the book edited by Hallbäck *et al.* (1996) for more details on Reynolds decomposition and derivation of the exact turbulence equations.

In the context of magnetohydrodynamics, the volume force f_i in (1) corresponds to the magnetic Lorentz force. The instantaneous Lorentz force is (here we use vector notation for readability)

$$\vec{f} = \frac{1}{\rho}\vec{j} \times \vec{B} \tag{5}$$

The electric current $\vec{j}$ is given by Ohm's law,

$$\vec{j} = \sigma(\vec{E} + \vec{u} \times \vec{B}), \tag{6}$$

where $\vec{E}$ is the electric field. Considering a Newtonian, incompressible fluid with constant density ρ, magnetic permeability μ, and electrical conductivity σ, the magnetic Reynolds number is $Rm_L \equiv \mu\sigma u L$, where u is the root-mean-square (r.m.s.) velocity, and L is the integral length scale. The magnetic Reynolds number can be interpreted as the ratio between the time scale of diffusion of the magnetic field, and the turbulence time scale. In our intended application, and indeed in most industrial and laboratory-scale flows, the magnetic Reynolds number is very small, $Rm_L \ll 1$. The physical significance of this condition is that the magnetic diffusion will dominate over convective and local changes of the magnetic field. An induced instantaneous distortion $\vec{b}(\vec{x},t)$ of the magnetic field about the imposed external magnetic field $\vec{B}_0$ will therefore be small, and can be neglected beside $\vec{B}_0$. This quasi-static approximation gives

$$\vec{B}(\vec{x},t) \equiv \vec{B}_0 + \vec{b}(\vec{x},t) \approx \vec{B}_0, \tag{7}$$

which simplifies the Maxwell's equations considerably. The scalar electrostatic potential ϕ is defined by

$$\vec{E} = -\nabla\phi, \tag{8}$$

and it can be shown that ϕ satisfies

$$\nabla^2\phi = \nabla \cdot (\vec{u} \times \vec{B}_0). \tag{9}$$

Given the proper boundary conditions, the Lorentz force f_i is implicitly determined by equations (5) through (9). A Reynolds decomposition may be performed also on these equations, and the averaged equations can be used to solve for the mean Lorentz force F_i appearing in the mean momentum equations (3).

The Lorentz force counteracts also turbulent velocity fluctuations and thus gives rise to a magnetic Joule dissipation term in the Reynolds stress equations. The strength of magnetic damping, relative to non-linear inertial mechanisms, is measured by the interaction number N, which is defined as

$$N \equiv \frac{\sigma B^2}{\rho}\frac{L}{u}.$$

This is the ratio of the turbulence time scale, $u^{-1}L$, and the time scale of the Lorentz force, $\rho\sigma^{-1}B^{-2}$. We will focus our attention on moderate interaction numbers, up to order 10.

For homogeneous turbulence (allowing a homogeneous mean deformation), the Reynolds stress transport equations can be written as

$$\frac{dR_{ij}(t)}{dt} = \left(\frac{dR_{ij}}{dt}\right)_{NS} - \mu_{ij}(t), \tag{10}$$

where μ_{ij} is the Joule dissipation rate tensor. The first term on the right-hand side represents the hydrodynamic contributions from the Navier–Stokes equations,

$$\left(\frac{dR_{ij}}{dt}\right)_{NS} = P_{ij} + \Pi_{ij}^{(r)} + \Pi_{ij}^{(s)} - \varepsilon_{ij}, \tag{11}$$

where P_{ij} is the turbulence production and ε_{ij} is the dissipation rate tensor. The redistributive pressure–strain rate term has here been divided into a rapid part $\Pi_{ij}^{(r)}$, representing the mean deformation–turbulence interaction, and a slow part $\Pi_{ij}^{(s)}$, representing turbulence–turbulence interaction. We will limit our attention to homogeneous turbulence without mean deformation, in which both the production term and the rapid pressure–strain term vanish.

For vanishing magnetic Reynolds numbers, $Rm_L \ll 1$, analysis of MHD effects on the development of the spectral energy tensor in Fourier space, $\hat{E}_{ij}$, yields the following expression for the Joule dissipation tensor (see e.g. Schumann, 1976):

$$\mu_{ij}(t) = \int 2\frac{\sigma}{\rho}\frac{[\vec{k}\cdot\vec{B}]^2}{k^2}\hat{E}_{ij}(\vec{k},t)d^3\vec{k}. \tag{12}$$

The quasi-static approximation is valid, so that $\vec{B}$ henceforth is given by the imposed magnetic field. We rewrite the inner product of wave vector and magnetic field vector using index notation as

$$(\vec{k} \cdot \vec{B})^2 = B^2 N_{pq} k_p k_q,$$

where

$$N_{pq} = \frac{1}{B^2} B_p B_q.$$

The tensor N_{ij} is assumed to be constant under integration so that expression (12) can be written as

$$\mu_{ij}(t) = 2\frac{\sigma B^2}{\rho} N_{pq} \int \frac{k_p k_q}{k^2} \hat{E}_{ij}(\vec{k}, t) d^3\vec{k} \tag{13}$$

$$= 2\frac{\sigma B^2}{\rho} N_{pq} M_{ijpq}, \tag{14}$$

where we have introduced the fourth-rank tensor

$$M_{ijpq} \equiv \int \frac{k_p k_q}{k^2} \hat{E}_{ij}(\vec{k}, t) d^3\vec{k}.$$

In the context of conventional turbulence modelling, we recognize the M-tensor from the formulation of the rapid pressure–strain rate term in the RST equations (11),

$$\Pi_{ij}^{(r)} = 2U_{k,n}(M_{inkj} + M_{jnki}),$$

where $U_{i,j}$ is the mean velocity gradient tensor. In classical turbulence modelling, the M-tensor is modelled in terms of the Reynolds stress tensor alone (or rather in the non-dimensional Reynolds stress anisotropy tensor, a_{ij}). Models of this type seem to do well in many irrotational flows, but fail to predict e.g. the complex behavior of turbulence subjected to rapid rotation. Reynolds (1989) and Reynolds & Kassinos (1995) note that the stress anisotropy tensor alone does not provide an adequate tensorial base for a model of the M-tensor in the case of rapid rotation. They argue that the problem can only be overcome if one lets the model of the M-tensor depend also on other tensors, readily admitting that such a model would be very complex. The new tensors would carry information about the dimensionality and structure of the turbulence. One of the tensors introduced by Reynolds and Kassinos is the structure dimensionality tensor, defined as

$$Y_{ij} \equiv \int \frac{k_i k_j}{k^2} \hat{E}_{nn}(\vec{k}) d^3\vec{k} = M_{nnij}. \tag{15}$$

As we shall see, the dimensionality tensor Y_{ij} contains the additional dimensionality information we need in order to model the Joule dissipation in MHD turbulence.

The dimensionality tensor is symmetric by definition, with trace $2K$. We define the dimensionality anisotropy tensor as

$$y_{ij} = \frac{Y_{ij}}{K} - \frac{2}{3}\delta_{ij}. \tag{16}$$

In analogy with the Reynolds stress anisotropy tensor, this tensor is traceless and the values of its diagonal components are limited to the finite interval $[-2/3, 4/3]$ (see e.g. Reynolds, 1989). A value of y_{11} close to $-2/3$, for example, indicates that the energy-containing eddies are relatively long in the x_1 direction, while a value of $4/3$ would indicate that they are relatively short in the x_1 direction.

Let us now return to the RST equations in (10) and the Joule dissipation tensor as formulated in (13) and (14). Contraction of indices gives us the turbulent kinetic energy equation,

$$\frac{dK}{dt} = \left(\frac{dK}{dt}\right)_{NS} - \frac{1}{2}\mu_{ii} = \left(\frac{dK}{dt}\right)_{NS} - \frac{\sigma B^2}{\rho} N_{pq} M_{iipq},$$

so that the total Joule dissipation is $\mu = \frac{1}{2}\mu_{ii}$. But according to the definition in (15), we can express the total Joule dissipation in terms of the Reynolds dimensionality tensor, or the dimensionality anisotropy tensor (16),

$$\mu(t) = \frac{\sigma B^2}{\rho} N_{ij} Y_{ji} = \frac{\sigma B^2}{\rho} K (N_{ij} y_{ji} + \frac{2}{3}).$$

We now introduce a scalar dimensionality anisotropy parameter, α_μ, and write the total Joule dissipation as

$$\mu(t) \;=\; 2\frac{\sigma B^2}{\rho}\alpha_\mu K, \tag{17}$$

with

$$\alpha_\mu \;\equiv\; \frac{1}{2K} N_{ij} Y_{ji} = \frac{1}{2} N_{ij} y_{ji} + \frac{1}{3}. \tag{18}$$

In the case of initially isotropic homogeneous turbulence decaying in a magnetic field, there is axisymmetry about an axis parallel to the magnetic field vector B_i. $N_{ij} y_{ji}$ is then the axial diagonal component of the dimensionality anisotropy tensor. In the limit of two-dimensional (2D) turbulence, with turbulent structures elongated along the axis of symmetry, $N_{ij} y_{ji} = -2/3$ and consequently $\alpha_\mu = 0$. For isotropic turbulence, $N_{ij} y_{ji} = 0$ and $\alpha_\mu = 1/3$.

In the following section we propose an extension of an RST closure to include a transport equation for the dimensionality anisotropy parameter, α_μ, and a model of the Joule dissipation tensor in terms of α_μ, the stress anisotropy tensor and the unit direction vector of the magnetic field. The space available here permits only a summary of the modelling approach; the interested reader is referred to a paper published elsewhere by the authors (Widlund *et al.*, 1998).

3. Modelling

3.1. THE JOULE DISSIPATION TENSOR

Assuming the Joule dissipation tensor is axisymmetric about an axis parallel with the magnetic field, it seems reasonable to model the tensorial part of the Joule dissipation tensor in (14) in terms of the dimensionality anisotropy tensor y_{ij}, the stress anisotropy tensor a_{ij}, and the unit direction vector n_i of the magnetic field. The latter two quantities are defined as

$$\begin{aligned} a_{ij} &= \frac{R_{ij}}{K} - \frac{2}{3}\delta_{ij}, \\ n_i &= \frac{B_i}{B}. \end{aligned}$$

We propose a coordinate invariant tensorial model of the Joule dissipation tensor on the form

$$\mu_{ij}(t) = 2\frac{\sigma B^2}{\rho} K F_{ij}(a_{kl}, n_m, \alpha_\mu), \tag{19}$$

where F_{ij} is a second-rank symmetric tensor function. In the following we will work with an ansatz which is bilinear in a_{ij} and $n_i n_j$. Further, for μ_{ij} to vanish when $R_{ij} \to 0$, F_{ij} should be of the following form:

$$F_{ij} = G(a_{ij} + \frac{2}{3}\delta_{ij}) + H(\frac{1}{2}[n_i n_k a_{kj} + a_{ik} n_k n_j] + \frac{2}{3} n_i n_j). \tag{20}$$

The coefficients G and H are formally functions of α_μ and $I_{na} \equiv n_i n_k a_{ki}$. The latter is the only relevant tensor invariant of the terms in F_{ij}.

Contraction of (19) and (20) gives the total Joule dissipation as

$$\mu = \frac{1}{2}\mu_{ii} = \frac{\sigma B^2}{\rho} K F_{ii} = \frac{\sigma B^2}{\rho} K \left\{ 2G + H(I_{na} + \frac{2}{3}) \right\}.$$

Comparing this with equation (17) enables us to write

$$G = \alpha_\mu - \frac{H}{2}(I_{na} + \frac{2}{3}). \tag{21}$$

As α_μ is a transported quantity in the present model, it follows from equation (21) that only one of the functions G and H is independent. Assuming H is a function of α_μ only, G will be a function of both α_μ and the invariant I_{na}.

Due to axisymmetry around the direction of the magnetic field, we need to distinguish only between components in directions parallel and perpendicular to the magnetic field. Following the convention of Schumann (1976), we let indices n and p denote components parallel and perpendicular, respectively, to an axis of symmetry $\vec{n}$. The tensor function ansatz in equation (20) reduces to

$$\mu_{nn} = 2\frac{\sigma B^2}{\rho}(G+H)R_{nn}, \qquad \mu_{pp} = 2\frac{\sigma B^2}{\rho}GR_{pp}.$$

In order to obtain appropriate boundary conditions for the function H, we investigate the components of the Joule dissipation tensor μ_{ij} in two extreme states: (instantaneous) isotropic turbulence, and the limit of 2D turbulence. For isotropic turbulence, we may assume an isotropic turbulence spectrum according to Batchelor (1953):

$$\hat{E}_{ij}(\vec{k},t) = \frac{\hat{E}(k,t)}{4\pi k^2}\left(\delta_{ij} - \frac{k_i k_j}{k^2}\right).$$

The spherical integrals in expression (12) can then be computed, which yields the values of G and H in the isotropic limit: $G_{iso} = 2/5$ and $H_{iso} = -1/5$. In the limit of two-dimensionality (where $\alpha_\mu \to 0$), the Joule dissipation becomes zero (Schumann, 1976). For the ansatz (20) this requires $H(\alpha_\mu = 0) = 0$ and $G(\alpha_\mu = 0) = 0$. Furthermore, a linear analysis of the development of an initially isotropic spectral energy tensor over time, neglecting all contributions other than the magnetic dissipation, shows that the ratio of the Reynolds stresses in parallel and perpendicular directions will approach a finite value asymptotically in the limit of 2D turbulence (Schumann, 1976). With some algebra, this condition gives a Neumann condition for $H(\alpha_\mu)$ in the limit of 2D turbulence:

$$\left.\frac{dH}{d\alpha_\mu}\right|_{\alpha_\mu=0} = 0.$$

The simplest function H that fulfills the conditions above is

$$H(\alpha_\mu) \quad = \quad -\frac{9}{5}\alpha_\mu^2, \tag{22}$$

which through equation (21) gives

$$G(\alpha_\mu, I_{na}) = \alpha_\mu + \frac{9}{10}\alpha_\mu^2(I_{na} + \frac{2}{3}). \tag{23}$$

Note that our boundary conditions merely focus on the behavior of the Joule dissipation tensor in the two extreme states of isotropy and 2D turbulence; explicit control of the anisotropy of μ_{ij} for intermediate values of α_μ would require functions G and H of higher order.

3.2. MODEL TRANSPORT EQUATIONS

We propose the following phenomenological model transport equation for the dimensionality anisotropy parameter, α_μ:

$$\frac{d\alpha_\mu(t)}{dt} = \left(\frac{d\alpha_\mu}{dt}\right)_{NS} - \gamma. \tag{24}$$

The first term represents interactions with the mean flow and turbulence, while γ is the magnetic destruction of α_μ. The latter is proposed to be

$$\gamma = 2C_{\alpha 1}\frac{\sigma B^2}{\rho}\alpha_\mu^2. \tag{25}$$

The constant $C_{\alpha 1}$ in (25) can be determined by considering the purely magnetic problem, neglecting viscous dissipation and non-linear effects in the RST and α_μ equations. This gives $C_{\alpha 1} = 3/(5\ln 2)$.

Non-linear inertial mechanisms act to restore an anisotropic turbulence state to isotropy. α_μ has the value 1/3 for isotropic turbulence. For homogeneous shear-free turbulence without production we therefore propose

$$\left(\frac{d\alpha_\mu}{dt}\right)_{NS} = C_{\alpha 2}\frac{\varepsilon}{K}(\frac{1}{3} - \alpha_\mu),$$

in analogy with the linear part of the "return-to-isotropy" term (26) in the RST equations (see below).

For shear-free homogeneous turbulence one can write

$$\left(\frac{dR_{ij}}{dt}\right)_{NS} = \pi_{ij} - \frac{2}{3}\varepsilon\delta_{ij},$$

where π_{ij} is the "return-to-isotropy" tensor, responsible for the relaxation from anisotropy to the isotropic state, and the last term governs the decay of turbulence. In this work we have chosen to model π_{ij} according to the model (for non-conducting fluids) by Sarkar and Speziale (Sarkar *et al.*, 1990, Speziale *et al.*, 1991),

$$\pi_{ij} = -\varepsilon\left(1.7a_{ij} - 1.05\left[a_{ik}a_{kj} - \frac{II_a}{3}\delta_{ij}\right]\right), \tag{26}$$

where $II_a = a_{ik}a_{ki}$ is the second invariant of the Reynolds stress anisotropy tensor, a_{ij}.

We use the conventional model transport equation for the viscous dissipation rate, supplemented by an extra destruction term due to magnetic effects. The Joule destruction term is modelled in analogy with the Joule dissipation term (17) in the turbulent kinetic energy equation, scaled by the turbulent time-scale, $\varepsilon^{-1}K$. The model transport equation for the dissipation rate then reads

$$\frac{d\varepsilon}{dt} = -1.9\frac{\varepsilon^2}{K} - C_{\alpha 3}\frac{\varepsilon}{K}\frac{2\sigma B^2}{\rho}\alpha_\mu K.$$

Like constants for the corresponding terms in the ordinary RST and ε equations, $C_{\alpha 2}$ and $C_{\alpha 3}$ would have to be calibrated against accurate numerical or experimental data. The data available to us at this point is scarce and does not support a serious attempt to calibrate the model. We have found the following values to work well for demonstrating the properties of the new model:

$$C_{\alpha 2} = 0.2, \quad C_{\alpha 3} = 0.5.$$

4. Results

The described model has been applied on the case of decaying homogeneous turbulence. The test case is the same as analyzed by Schumann (1976) using direct numerical simulation (DNS). Schumann integrated the Navier–Stokes equations over a three-dimensional periodic box in Fourier space, with a resolution corresponding to 32^3 points in physical space. The initial turbulent Reynolds number was about 60, as defined by

$$Re_L = \frac{uL}{\nu}, \quad u = \sqrt{\tfrac{2K}{3}},$$

where L is the integral length scale, K is the turbulent kinetic energy, u is the r.m.s. velocity and ν is the kinematic viscosity.

The first time steps ($0 \leq t < t_1$) were integrated with zero magnetic field. A constant magnetic field was switched on in the time interval $[t_1, t_2]$ and the calculations were terminated at $t = t_3$. All results were scaled with $L(t_1)$ and $u(t_1)$, i.e. the integral length scale and r.m.s. velocity (from the turbulent kinetic energy) at $t = t_1$, respectively. In these units, the times at which the magnetic field was switched on (t_1) or off (t_2), and where the computations were terminated (t_3), were (rounded to one decimal)

$$\{t_1, t_2, t_3\} = \{0.4, 1.2, 2.0\}\frac{L_1}{u_1}.$$

Schumann performed simulations with magnetic fields corresponding to initial interaction numbers $N = 0$, 1, 5 and 50.

The RST model simulations used the equations outlined in Section 3. With a hydrodynamic Reynolds number of 60, interaction numbers $N = \{0, 1, 5, 50\}$ (based on turbulence conditions at $t = 0$) and an arbitrary initial turbulent kinetic energy, $K_0 = 10^{-5}$, the initial viscous dissipation is calculated from

$$\varepsilon_0 = \frac{u_0^3}{L_0},$$

where the initial r.m.s. velocity and integral length scale are given by

$$u_0 = \sqrt{\frac{2K_0}{3}}, \quad L_0 = \frac{Re_L \nu}{u_0}.$$

The magnetic time scale was computed from initial conditions and the relation

$$N = \frac{\sigma B^2}{\rho} \frac{L_0}{u_0}.$$

Figure 2 shows the normalized turbulent kinetic energy as a function of time. For $N = 0$, i.e. with no magnetic effects, there is a difference between DNS and RST results that can be attributed to the DNS simulations. This is a case for which conventional RST closures have been calibrated and can be expected to be very accurate. Schumann notes that the DNS simulations will never reach zero kinetic energy, due to the limited resolution of the calculations. This is obvious for $N = 50$ and may affect the result also for $N = 5$.

The development of Reynolds stress anisotropy at on-set of the magnetic field is a characteristic feature of the test case. It is captured very well by the model, both qualitatively and quantitatively. This is illustrated in Figure 3, which shows the ratio between Reynolds stresses in parallel and perpendicular directions. Schumann's DNS data tends towards an anisotropy in the opposite direction already from the start, which is clearly seen for $N = 0$. The trend is likely to affect the results also for higher interaction numbers.

The total Joule dissipation can be calculated directly using equation (17). Figure 4 shows that RST model predictions are in very good agreement with DNS data.

5. Summary and discussion

In this paper we have outlined an approach for extending conventional Reynolds stress closures to include magnetic effects. A scalar dimensionality anisotropy parameter, α_μ, is introduced to carry information about

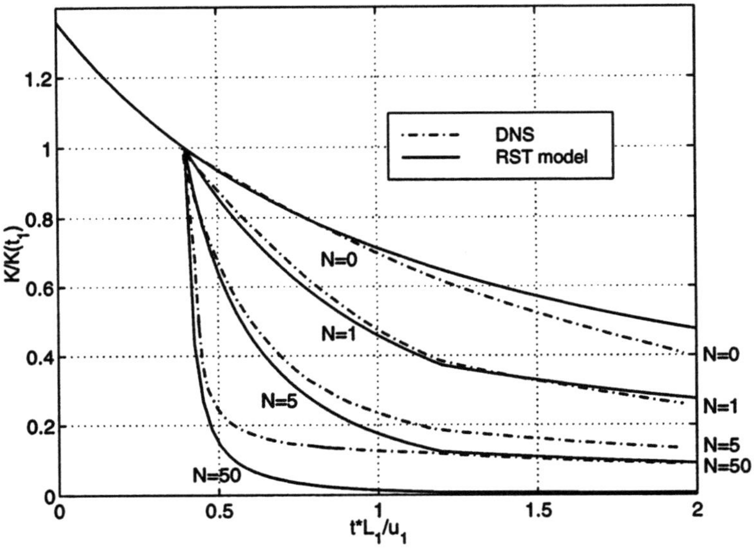

Figure 2. Normalized turbulent kinetic energy as a function of time. Solid lines show results calculated with the new MHD RST closure. Dashed lines represent the corresponding DNS data (Schumann, 1976).

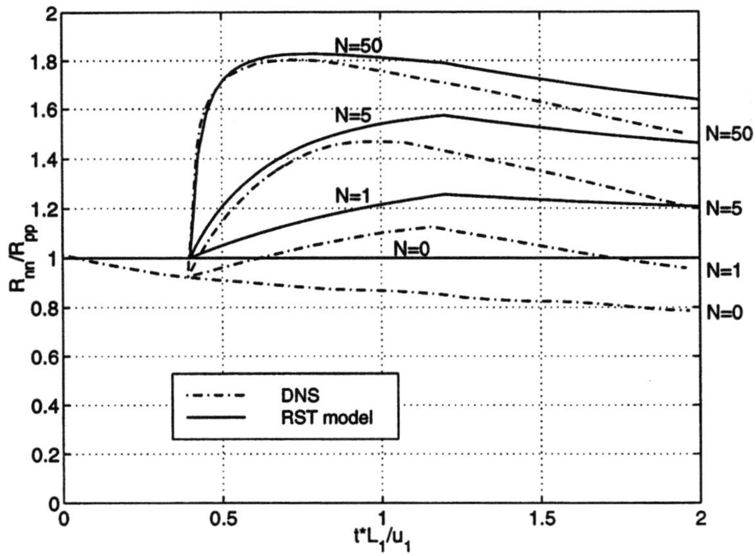

Figure 3. Development of the ratio of energy in normal and parallel Reynolds stress components. Solid lines show results calculated with the new MHD RST closure. Dashed lines represent the corresponding DNS data (Schumann, 1976).

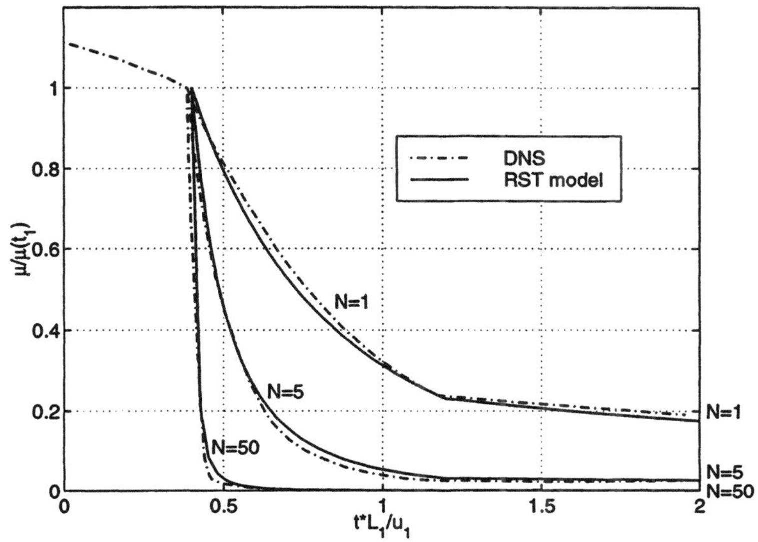

Figure 4. Development of the total Joule dissipation. Solid lines show results calculated with the new MHD RST closure. Dashed lines represent the corresponding DNS data (Schumann, 1976).

the distribution of energy in spectral space. This information is vital for modelling of MHD turbulence, as it determines the strength and anisotropy of the Joule dissipation tensor in the RST equations.

The concept is demonstrated on the case of homogeneous turbulence decaying in a magnetic field. The Joule dissipation tensor is modeled by a tensor function, which is bilinear in the Reynolds stress anisotropy, a_{ij}, and the unit direction vector of the magnetic field, n_i. At present the tensor function coefficients are second-order in α_μ, based on the simplest possible set of boundary conditions applied in the isotropic limit and in the 2D limit. The transport equation governing the development of the dimensionality anisotropy parameter is currently based on phenomenological reasoning.

In order to get a model for practical applications, the MHD RST model must be extended to include effects of mean shear, strain and, long-term, rotation. This will i.a. introduce production terms in the hydrodynamic part of the α_μ equation. To support further development of the presented model, future work will include new DNS experiments, with higher resolution and Reynolds numbers, on both homogeneous decaying turbulence and turbulence with homogeneous mean shear and strain. DNS or LES on channel flow may help to formulate appropriate and effective wall boundary conditions.

The results so far are encouraging and the concept of a scalar dimensionality parameter seems a promising route towards a complete RST closure suitable for engineering MHD applications at low magnetic Reynolds numbers. Engineering applications require simple, robust models that can be implemented into existing commercial CFD codes. The presented model is a strong candidate, as it adds only one extra transport equation and extends existing conventional turbulence models through extra source terms.

References

Batchelor, G.K. (1953) *The Theory of Homogeneous Turbulence.* Cambridge University Press, Cambridge.

Hallbäck, M., Henningson, D.S., Johansson, A.V. and Alfredsson, P.H., editors (1996) *Turbulence and transition modelling.* ERCOFTAC Series. Kluwer Academic Publishers, Dordrecht.

Lehman, A.F., Tallbäck, G.R. and Hackl, H.R. (1994) Fluid flow control in continuous casting using various configurations of static magnetic fields. In *Proceedings of International Symposium on Electromagnetic Processing of Materials, Oct. 25-28, 1994*, pp. 372–377, ISIJ, Nagoya, Japan.

Reynolds, W.C. (1989) Effects of rotation on homogeneous turbulence. In *Proc. 10th Australasian Fluid Mechanics Conference*, University of Melbourne, Melbourne, Australia.

Reynolds, W.C. and Kassinos, S.C. (1990) One-point modelling of rapidly deformed homogeneous turbulence, *Proceedings of the Royal Society of London, Series A*, **451**(1941), 87

Sarkar, S. and Speziale, C.G. (1990) A simple nonlinear model for the return to isotropy in turbulence, *Physics of Fluids A*, **2**(1), 84

Schumann, U. (1976) Numerical simulation of transition from three- to two-dimensional turbulence under a uniform magnetic field, *Journal of Fluid Mechanics*, **74**, 31

Speziale, C.G., Sarkar, S. and Gatski, T.B. (1991) Modelling the pressure-strain correlation of turbulence, *Journal of Fluid Mechanics*, **227**, 245

Widlund, O., Zahrai, S. and Bark, F.H. (1998) Development of a Reynolds stress closure for modeling of homogeneous MHD turbulence. Accepted for publication in *Physics of Fluids*, **10**(8)

ABSOLUTE AND CONVECTIVE MHD STABILITY OF A CAPILLARY LIQUID METAL JET WITH AZIMUTHAL VELOCITY

LOUESLATI K. AND BRANCHER J.P.
Laboratoire LEMTA UMR 7563.
2 avenue de la Forêt de Haye - 54504 Vandoeuvre les Nancy
e-mail : klouesla@ensem.u-nancy.fr

The stability of a circular jet of an inviscid perfectly conductive liquid is investigated. The infinite jet is subject to surface tension , rigid rotation and axial magnetic field for high magnetic Reynolds number . The stability is governed by inertial, capillary and electromagnetic forces. A general analytical solution to the linearised problem is presented. In the first part, the temporal stability is analysed for large values of swirl, Weber and Alfvèn numbers. In the second part, the instability type (convective or absolute) is determined as a function of the different parameters by examining the spatial branch-point singularities of the dispersion relation.

1. Introduction

The temporal stability of an infinite jet was first analysed by Batchelor & Gill [1962]. An inviscid jet is unstable to any disturbance for all azimuthal modes. In the same time, Chandrasekhar [1961] have demonstrated that a uniform magnetic field applied on a capillary jet can stabilise axisymmetric disturbances of all wavelengths. For jet-like flows, the distinction between absolute and convective instability is very significant. The response of the flow to an impulsion shows whether it is convectively or absolutely unstable. If the response grows with time at a fixed location in space, then the flow is absolutely unstable .On the other hand, if the disturbance energy is convected away faster than it is produced one speaks of a convectively unstable flow. So when the instability is convective, the jet is continuous up to a breaking point situated far downstream. In the case of absolute instability the dripping phenomenon occurs at the nozzle opening. This notion can be used effectively, to obtain a qualitative and quantitative description of the spatio-temporal dynamics of the flow. Leib & Goldstein [1986] have carried out a stability analysis for a circular capillary jet to find out whether the transition is absolute or convective. They found a critical Weber of 0.32 when the disturbance is axisymmetric. Monkewitz [1988] has found experimentally that the breakup occurs about We=0.3. Recently, Yakubenko [1995] has analysed the passage between convective and absolute instabilities of current-carrying jet of

A. Alemany et al. (eds.),
Transfer Phenomena in Magnetohydrodynamic and Electroconducting Flows, 77–92.

magnetic liquid. Finally, Loiseleux, Chomaz and Huerre [1998] have studied the instability nature of a Rankine vortex. The originality of this work is the introduction of an azimuthal velocity for the jet.

2. Basic flow

We consider a infinitely long liquid jet, which comes out of a circular opening of radius R_0. The axial velocity U_0 of the flow is assumed to be uniform and the azimuthal velocity $\Omega_0 r$ (solid-body rotation). The conductive fluid used in this study is characterised by a constant density ρ_1, a magnetic permeability μ_0 equal to the void permeability , and a conductivity σ. The outer fluid is non conductive with constant density ρ_2. The two fluids are assumed to be inviscid, the basic flow is steady, incompressible and isothermal. A uniform longitudinal magnetic field $\vec{H}_0$ is applied in all space.

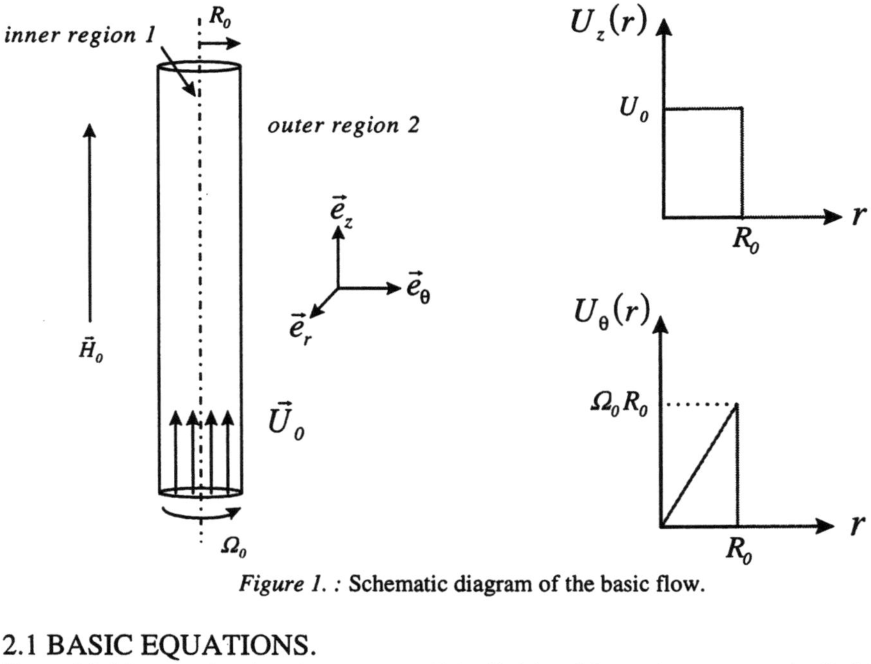

Figure 1. : Schematic diagram of the basic flow.

2.1 BASIC EQUATIONS.

For a highly conducting incompressible fluid subjected to magnetic field, with a large magnetic Reynolds number , the governing equations are :

$$\frac{\partial \vec{U}}{\partial t} + \overline{\overline{\nabla}}\vec{U}.\vec{U} = -\frac{1}{\rho}\vec{\nabla}P + \frac{\mu_0}{\rho}\left(\vec{\nabla} \wedge \vec{H}\right) \wedge \vec{H} \tag{1}$$

$$\frac{\partial \vec{H}}{\partial t} = \vec{\nabla} \wedge \left(\vec{U} \wedge \vec{H}\right) \tag{2}$$

$$\nabla.\vec{U}=0 \tag{3}$$

$$\nabla.\vec{H}=0, \tag{4}$$

where $\vec{U}$ is the velocity field and $\vec{H}$ is the magnetic field. The induction vector $\vec{B}$, the current density $\vec{j}$ and the electric field $\vec{E}$ can be obtained from the Maxwell equations :

$$\begin{cases} \vec{B}=\mu_0\vec{H} \\ \vec{j}=\sigma\left(\vec{E}+\mu_0\left(\vec{U}\wedge\vec{H}\right)\right) \\ \nabla\wedge\vec{E}=-\dfrac{\partial\vec{B}}{\partial t} \end{cases} \tag{5}$$

In the outer region, the fluid resistivity is infinite. The magnetic field satisfies both equation (4) and :

$$\vec{\nabla}\wedge\vec{H}=\vec{0}. \tag{6}$$

2.2 BOUNDARY CONDITIONS AND BASIC SOLUTION

The interface between the two fluids requires the following jump conditions :

- the normal component of the magnetic field must be continuous

$$\left[\vec{H}.\vec{n}\right]=0 \tag{7}$$

- the normal stress, which includes the pressure and the magnetic stress must be balanced by the normal force due to the surface tension.

$$\left[\vec{n}.\left(\overline{\overline{\Sigma}}+\overline{\overline{T}}\right)\vec{n}\right]=\gamma\, div\vec{n} \tag{8}$$

where γ denotes surface tension, $\vec{n}$ is the normal vector at the interface, Σ_{ij} and T_{ij} are the stress and the Maxwell's tensors ($T_{ij}=\mu_0\left(H_iH_j-1/2\,H^2\delta_{ij}\right)$).

- the normal component of the velocity must be continuous

$$\left[\vec{U}.\vec{n}\right]=0 \tag{9}$$

A basic solution of the Euler and magnetohydrodynamic equations for the specified boundary conditions is :

$$\forall\, r\le R_0 \qquad\qquad \forall\, r\ge R_0$$

$$\begin{cases} \vec{U}_1(r,\theta,z)=\left(0,\Omega_0 r,U_0\right) \\ \vec{B}_1(r,\theta,z)=B_0\vec{e}_z \\ \mathrm{P}_1(r,\theta,z)=\mathrm{P}_0+\dfrac{\rho_1\Omega_0^2}{2}\left(r^2-R_0^2\right) \end{cases} \quad \text{and} \quad \begin{cases} \vec{U}_2(r,\theta,z)=\vec{0} \\ \vec{B}_2(r,\theta,z)=B_0\vec{e}_z \\ P_2(r,\theta,z)=P_0 \end{cases} \tag{10}$$

where P_0 is the pressure at infinity.

3. Dispersion relation

In order to study the linear stability, small perturbations are introduced in the basic flow. The velocity, the magnetic and the pressure fields result from small perturbations components in addition to the mean flow :

$$\bar{\vec{U}} = \vec{U}(r) + \vec{u}(r,\theta,z,t)$$

$$\bar{P} = P(r) + p(r,\theta,z,t)$$

$$\bar{\vec{H}} = \vec{H}(r) + \vec{h}(r,\theta,z,t)$$

The perturbations quantities are assumed to have normal-mode form as follow:

$$\vec{u}(r,\theta,z,t) = \begin{pmatrix} \tilde{u}_r(r,\theta,z,t) \\ \tilde{u}_\theta(r,\theta,z,t) \\ \tilde{u}_z(r,\theta,z,t) \end{pmatrix} = \begin{pmatrix} i\, u_r(r) \\ u_\theta(r) \\ u_z(r) \end{pmatrix} e^{i(kz+m\theta+\omega t)},$$

$$p(r,\theta,z,t) = \tilde{p}(r) e^{i(kz+m\theta+\omega t)},$$

$$\vec{h}(r,\theta,z,t) = \begin{pmatrix} \tilde{h}_r(r,\theta,z,t) \\ \tilde{h}_\theta(r,\theta,z,t) \\ \tilde{h}_z(r,\theta,z,t) \end{pmatrix} = \begin{pmatrix} i\, h_r(r) \\ h_\theta(r) \\ h_z(r) \end{pmatrix} e^{i(kz+m\theta+\omega t)} \quad ,$$

where k is the complex axial wavenumber, ω the complex temporal frequency and m an integer azimuthal wavenumber. The perturbed shape of the free boundary is given by :

$$r = R_0 + \eta\, e^{i(kz+m\theta+\omega t)} \quad .$$

3.1 INNER REGION

Introducing these expressions into the governing system (1) and (3) and assuming that the perturbation quantities are sufficiently small to neglect their square, we obtain the following linearised equations for the inner region:

$$\frac{1}{r}\left(\frac{\partial u_{r_1}(r)}{\partial r} + r u_{r_1}(r)\right) + \frac{m}{r} u_{\theta_1}(r) + k u_{z_1}(r) = 0$$

$$\begin{cases} \left(\omega + m\Omega_0 + kU_0 - \dfrac{\mu_0 H_0 \delta k}{\rho}\right) u_{r_1}(r) + 2\Omega_0 u_{\theta_1}(r) = \dfrac{1}{\rho_1}\dfrac{\partial \Pi_1(r)}{\partial r} \\ \left(\omega + m\Omega_0 + kU_0 - \dfrac{\mu_0 H_0 \delta k}{\rho}\right) u_{\theta_1}(r) + 2\Omega_0 u_{r_1}(r) = -\dfrac{m}{\rho_1 r}\Pi_1(r) \\ \left(\omega + m\Omega_0 + kU_0 - \dfrac{\mu_0 H_0 \delta k}{\rho}\right) u_{z_1}(r) = -\dfrac{k}{\rho_1}\Pi_1(r) \end{cases} \tag{11}$$

where $\Pi_1(r)=\tilde{p}_1(r)+\mu_0 h_z(r).H_0$ is the sum of the perturbed pressure and the perturbed magnetic pressure and $\delta=(kH_0)/(\omega+m\Omega_0+kU_0)$. After elimination of the velocity components, a Bessel equation for the pressure is obtained :

$$\Pi_1''(r)+\frac{1}{r}\Pi_1'(r)+\left(\beta^2-\frac{m^2}{r^2}\right)\Pi_1(r)=0,$$

where :

$$\beta^2=\frac{k^2\left(4\Omega_0^2-\alpha^2\right)}{\alpha^2}$$

$$\alpha=\omega+m\Omega_0+kU_0-\frac{\mu_0 H_0 \delta k}{\rho_1}.$$

The solution of this equation is a linear combination of the Bessel functions of first and second kind of order m : $J_m(r)$ and $Y_m(r)$. Taking into account that the pressure remains finished at the center of the jet and that $\lim_{r\to 0} Y_m(r)=+\infty$, the solution is :

$$\Pi_1(r)=A_1 J_m(\beta r), \tag{12}$$

where A_1 is an arbitrary constant. Substituting this expression into the linearised system (11), the velocity field can be obtained :

$$\begin{cases} u_{r1}(r)=-i\dfrac{A_1 k^2}{\rho_1\alpha\beta^2}\left(\dfrac{2\Omega_0 m}{\alpha r}J_m(\beta r)+\beta J_m'(\beta r)\right) \\ u_{\theta 1}(r)=\dfrac{A_1}{\rho_1\left(4\Omega_0^2-\alpha^2\right)}\left(2\Omega_0\beta J_m'(\beta r)+\dfrac{\alpha m}{r}J_m(\beta r)\right) \\ u_{z1}(r)=-\dfrac{kA_1}{\rho_1\alpha}J_m(\beta r) \end{cases} \tag{13}$$

where $J_m'(\beta r)=\left.\dfrac{dJ_m(x)}{dx}\right|_{x=\beta r}$.

Using (2), the magnetic field linearised equation can be expressed as following :

$$\frac{\partial \vec{h}_1}{\partial t}=\overline{\overline{grad\vec{U}_0}}.\vec{h}_1-\overline{\overline{grad\vec{h}_1}}.\vec{U}_0+\overline{\overline{grad\vec{u}_1}}.\vec{H}_0-\overline{\overline{grad\vec{H}_0}}.\vec{u}_1$$

The relation between the disturbed magnetic field and the disturbed velocity field is :

$$\vec{h}_1=\delta\vec{u}_1 \tag{14}$$

Like as the case of the free jet with longitudinal magnetic field, the disturbed magnetic field is directly proportional to the disturbed velocity field.

3.2 OUTER REGION

Outside of the jet, there is no current density because the medium is not conductive. So, the rotational of the magnetic field is null. The magnetic field derives from a magnetic potential ϕ, which is defined by $\vec{h}_2 = \vec{\nabla}(-i\phi)$ and verifies :

$$\Delta\phi = \frac{1}{r}\frac{\partial}{\partial r}\left(r\frac{\partial\phi}{\partial r}\right) + \frac{1}{r^2}\frac{\partial^2\phi}{\partial\theta^2} + \frac{\partial^2\phi}{\partial z^2} = 0$$

By replacing the modal development of ϕ in this expression, a modified Bessel equation is obtained. The solution of this equation is a linear composition of the modified Bessel functions of first and second kind of order m respectively $I_m(r)$ and $K_m(r)$. When $r \to \infty$, $I_m(r)$ diverges, so the solution for the potential is :

$$\phi(r,\theta,z,t) = BK_m(kr)e^{i(kz+\omega t+m\theta)},$$

where B is an arbitrary constant. So, we obtain the expression of the perturbed magnetic field :

$$\vec{h}_2 = B\begin{cases} -i\,K'_m(kr)e^{i(kz+\omega t+m\theta)} \\ \dfrac{mK_m(kr)}{r}e^{i(kz+\omega t+m\theta)} \\ kK_m(kr)e^{i(kz+\omega t+m\theta)} \end{cases} \tag{15}$$

For the non conductive fluid, the Euler equation (1) gives us the following velocity and pressure perturbed fields :

$$\begin{cases} \vec{u}_2 = \vec{\nabla}\left(-iA_2K_m(kr)e^{i(kz+\omega t+m\theta)}\right) = A_2\begin{cases} -i\,K'_m(kr)e^{i(kz+\omega t+m\theta)} \\ \dfrac{mK_m(kr)}{r}e^{i(kz+\omega t+m\theta)} \\ kK_m(kr)e^{i(kz+\omega t+m\theta)} \end{cases} \\ \tilde{p}_2 = -\rho_2 A_2\omega K_m(kr)e^{i(kz+\omega t+m\theta)} \end{cases} \tag{16}$$

Substituting the eigenfunctions into the jump conditions at the free boundary, we obtain the dimensionless dispersion relation :

$$\left(rho\,\omega^2 - Al\,k^2 + We\,k\left(k^2+m^2-1\right)\frac{K'_m(k)}{K_m(k)}\right)\left(2mS + \alpha\beta\frac{J'_m(\beta)}{J_m(\beta)}\right) + (\omega + mS + k)\frac{\alpha^2\beta^2}{k}\frac{K''_m(k)}{K_m(k)} = 0 \tag{17}$$

where :

$$\begin{cases} Al = \mu_0 H_0^2 / \left(\rho_1 U_0^2\right) & \text{the Alfvèn number, ratio of the magnetic to inertial effects} \\ S = \Omega_0 R_0 / U_0 & \text{the swirl number, ratio of the azimuthal to axial velocity} \\ \mathrm{We} = \gamma / \left(\rho_1 R_0 U_0^2\right) & \text{the Weber number, ratio of the surface tension to inertial forces} \\ \mathrm{rho} = \rho_2 / \rho_1 & \text{the ratio of outer to the inner mass density} \end{cases}$$

All variables have been dimensionlessed by using U_0 as characteristic velocity, R_0 as characteristic length, H_0 as magnetic field and ρ_0 as characteristic density. Then :

$$\delta = k/(\omega + mS + k)$$

$$\beta^2 = k^2\left(4S^2/\alpha^2 - 1\right)$$

$$\alpha = (\omega + mS + k - Al\,\delta\, k).$$

The admissible values of k and ω are defined by this non linear relation which depends on the control parameters listed before. By setting *S*, *Al* and *We* to zero the same result as that obtained by Batchelor & Gill is obtained. In the same way, for a non null Alfvèn number, our dispersion relation is identical to the Chandrasekhar's solution.

4. Stability analysis

4.1 TEMPORAL STABILITY

In this analysis, the dispersion equation is solved numerically with a real axial wavenumber *k* and a complex temporal frequency value ω. The sign of the imaginary compound ω_i determines the temporal stability. If ω_i is negative the system is temporally unstable.

Without magnetic field, with $rho = 1$ and for a null surface tension, the jet is always unstable for all axial wavenumbers (Batchelor [1962]). The temporal growth rate increases linearly with the axial wavenumber. Thus it does not exist selection of a wavenumber. In the short wavelength limit $(k >> 1)$, the temporal growth rate is $\omega \approx -k(1 \pm i)/2$, then the small scales give place to amplified waves. The surface tension will reveal a preferential frequency. An explicit solution, for the temporal growth rate, exist :

$$\omega = \frac{-2k \pm \sqrt{-4\, rho\, k^2 L_m(k) + 4(1 + rho L_m(k))(k^2 + m^2 - 1) We \dfrac{k I'_m(k)}{I_m(k)}}}{2(1 + rho L_m(k))} \tag{18}$$

where $L_m(x) = -\dfrac{K_m(x) I'_m(x)}{K'_m(x) I_m(x)}$

One can distinguish several effects of the Weber number according to the number of azimuthal wavenumber and to the axial wavenumber. When the surface tension increase, the system is more and more stabilised, the temporal growth rate decreases. This conclusion is true for all azimuthal wavenumber except for the axisymmetric mode. For this mode the introduction of surface tension has two distinct effects :

if $0 \leq We \leq 1$, the surface tension stabilises the small wavelengths and does not affect large wavelengths similarly to other azimuthal modes (see fig.2).

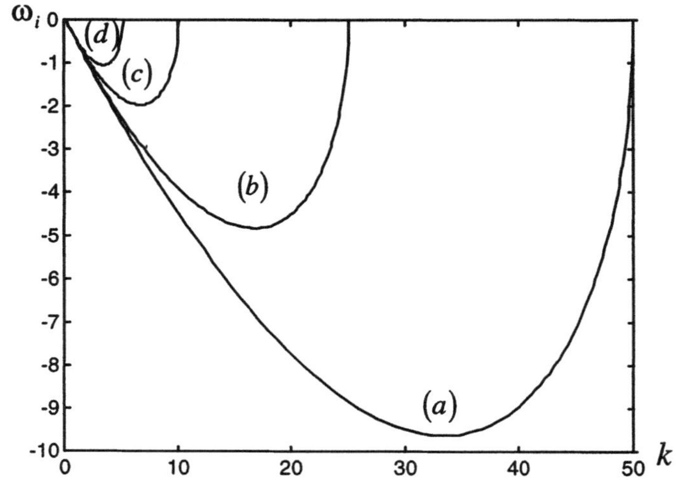

Figure 2. : Values of ω_i plotted as function of k for rho=1, (a) We=10^{-2}(b)We=2.10^{-2} (c) We=5.10^{-2} (d) We=0.1

if $1 \leq We \leq +\infty$, the surface tension makes the jet more unstable for large axial wavelengths, even if the system is stable for axial wavenumber greater than unity (see fig. 3). The axial wavenumber corresponding to the maximum growth rate tends towards a constant value of k_0=0.6769 . This constant value corresponds to a preferential dimensional wavelength of $\lambda_0 = 2\pi/k_0 \approx 4.25R_0$.

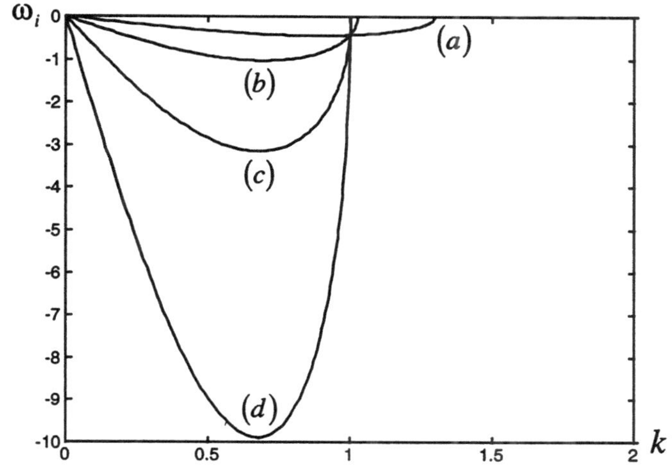

Figure 3. : Values of ω_i plotted as function of k for rho=1, (a) We=1 (b)We=10 (c) We=100 (d) We=1000

In fact, this effect of destabilisation is due to the surface tension energy. This energy is proportional to the coefficient of the surface tension and to the following dimensionless disturbed surface : $S = 2\pi + \pi\eta^2\left(k^2 + m^2 - 1\right)/4$. The second term will be negative when k is smaller than one, and positive when k is larger than one. However, the jet tend towards to minimise the contact surface with the outer fluid. Then, for large axial wavelengths the contact surface decreases, only for the axisymmetric mode, and the jet will move away to the initial configuration. As shown on the figure 4, the critical Weber number is decreasing while the axial wavenumber is increasing.

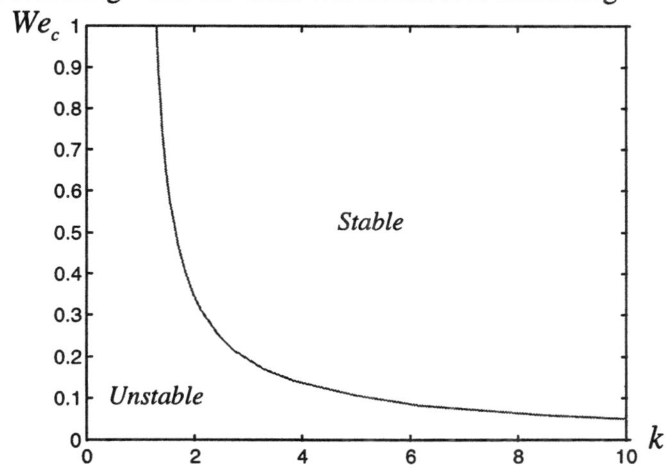

Figure 4. : Critical Weber number as function of k for rho=1, m=0

As shown on figure 5, the density of the jet has a simple effect on the stability; as the density of the jet increases it become more stable.

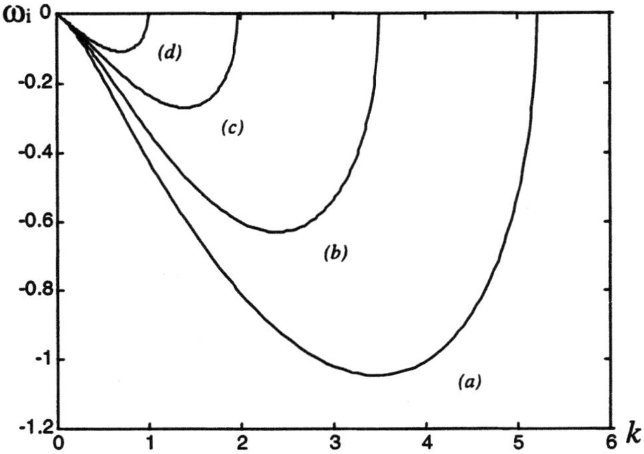

Figure 5. : Values of ω_i plotted as function of k for m=0,We=0.1 without magnetic field and rotation, (a) rho=1 (b) rho=0.5 (c) rho=0.02 (d) rho=10^{-4}

Introducing the magnetic field, the solution takes the following form :

$$\omega = \frac{-k \pm \sqrt{-\, rho\, k^2 L_m(k) + \left(1 + rho L_m(k)\right)\left(Al\, k^2\left(1 + L_m(k)\right) + We\left(k^2 + m^2 - 1\right)\frac{kI'_m(k)}{I_m(k)}\right)}}{\left(1 + rho L_m(k)\right)}. \quad (19)$$

Figure 6 is a characteristic scheme of the magnetic field contribution : the magnetic field has a stabilising effect for all azimuthal wave, swirl, and Weber numbers. One can point out that the short waves are first stabilised by the magnetic field. Increasing the Alfvèn number leads to a stabilisation of the long waves.

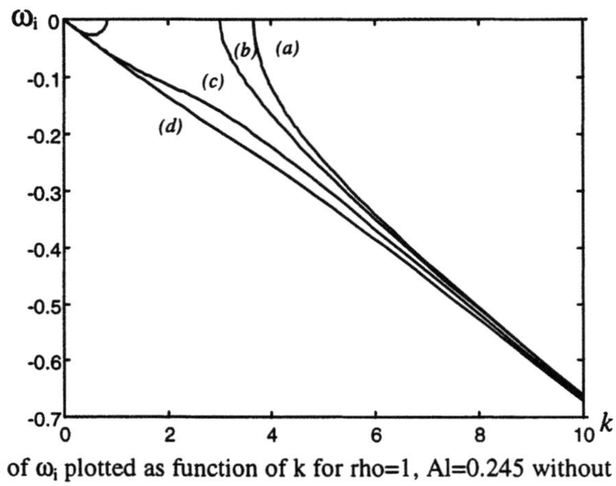

Figure 6. : Values of ω_i plotted as function of k for rho=1, Al=0.245 without surface tension and rotation, (a) m=0 (b) m=±1 (c) m=±2 (d) m=±3

The azimuthal wavenumber has a prevalent importance on stability. Indeed, the magnetic field is more stabilising for the large values of $|m|$. The magnetic field can completely remove the capillary instability of the axisymmetric mode for values of an Alfvèn number higher than the critical value $Al_c = We/2$ (see fig. 7). This critical value correspond to the following magnetic field critical value : $H_{0_c} = \sqrt{\gamma/2\mu_0 R_0}$

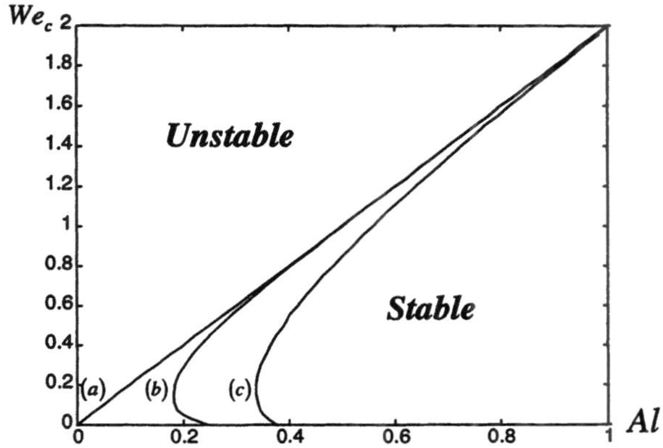

Figure 7. : Critical Weber number as function of Alfvèn number for m=0, (a) rho=0 (b)rho=1 (c) rho=2.5

When a rotation is introduced, the equation becomes more sophisticated and no more analytical solution is available. In order to obtain the roots of the equation, a solver routine of the MATLAB software has been used. For the axisymmetric mode, after the introduction rotation, the temporal branches have always an unstable evolution, however, their linear growth rates simply are attenuated. So, the swirl number can reduce partially the Kelvin-Helmholtz instability.

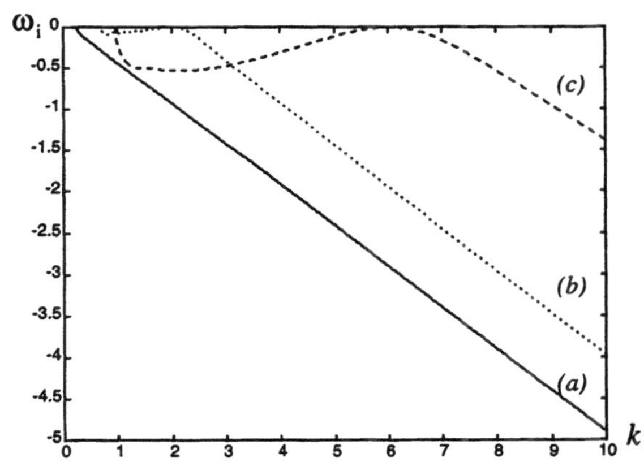

Figure 8. : Values of ω_i plotted as function of k for m=-2,rho=1 without magnetic field and surface tension, (a) S=0.1 (b) S=1.5 (c) S=3

For negative azimuthal wavenumbers (the perturbation turns in opposite direction of the jet), the result is more complex. For $m=-1$ and $m=-2$, the swirl number has a stabilising effect especially for small axial wavenumbers (see fig. 8).

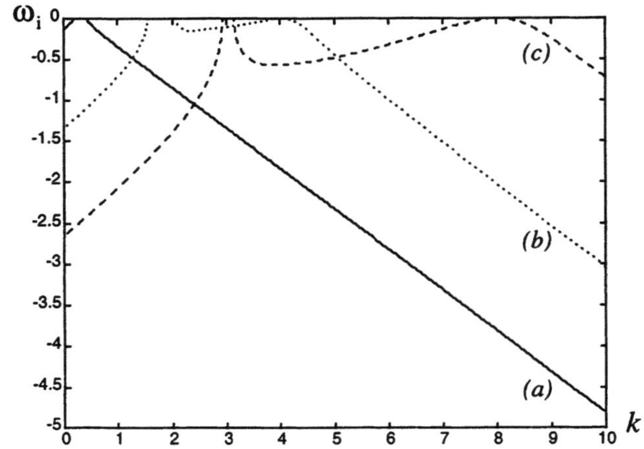

Figure 9. : Values of ω_i plotted as function of k for m=-4, rho=1 without magnetic field and surface tension, (a) S=0.1 (b) S=1 (c) S=2

From $m=-3$, the swirl number have two effects (fig. 9) : stabilisation for short waves and destabilisation for long waves. With the addition of the magnetic field, the system can be made stable for all axial wavenumbers k except for the small wavenumbers. For positive azimuthal wavenumbers, the swirl has a global destabilising effect (see fig. 10). Furthermore, the instability of the rotating jet can be reduced by increasing the Alfvèn number.

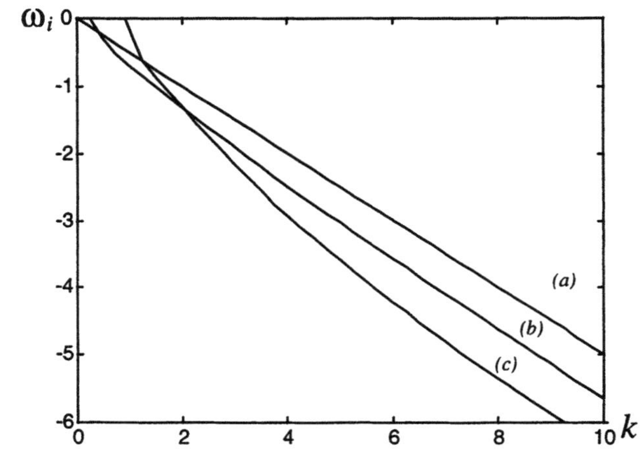

Figure 10. : Values of ω_i plotted as function of k for m=2,rho=1 without magnetic field and surface tension, (a) S=0 (b) S=1 (c) S=2

4.2 SPATIO-TEMPORAL STABILITY.

The time-dependent stability analysis provides the evolution of a perturbation which is spread periodically in space but gives no indication on its nature. Will the perturbation be advected by the flow (convective instability) or will it stay in place and perturb the flow (absolute instability) ? The principle of the spatio-temporal analysis is simple, a concentrated impulsion is introduced at x=0 and t=0, and the response of the system is observed for the long time. The asymptotic response or Green's functions is obtained in Fourier space in terms of k and ω :

$$G(x,t)=\frac{1}{(2\pi)^2}\int_F\int_L\frac{e^{i(kx+\omega t)}}{D(\omega,k,R_i)}\,d\omega dk$$

where $D(\omega,k,R)$ represent the dispersion equation including all control parameters R_i. L and F are the paths of integration, in the complex ω plane and in the complex k plane. The contour L is a straight horizontal line located above all the singularities of the integrand, to satisfy $G(x,t<0)=0$. The path F in the complex axial wavenumber plane is initially taken along the real axis.

When the flow is linearly unstable, two types of impulse response can be distinguished : the basic flow is referred to as *convectively unstable* if

$$\lim_{t\to+\infty} G(x=0,t)=0 \text{ along the ray } x/t=0$$

and it is *absolutely unstable* if

$$\lim_{t\to+\infty} G(x=0,t)=\infty \text{ along the ray } x/t=0.$$

The major problem is the evaluation of the integral. By applying the steepest descent method, one finds that along the ray $x/t=0$ the response is dominated by a specific complex wavenumber k_0 with zero group velocity i.e. $\partial\omega/\partial k(k_0)=0$. So, under this consideration, an absolute frequency ω_0 is defined.

This absolute growth rate characterises the temporal evolution of the wavenumber k_0 observed at a fixed station in the limit $t\to+\infty$.

The sign of the imaginary part of ω_0 determines the absolute/convective nature of the flow :

if $\mathrm{Im}(\omega_0)<0$ the flow is *absolutely unstable*,

if $\mathrm{Im}(\omega_0)>0$ the flow is *convectively unstable*.

The absolute frequency ω_0 is an algebraic branch point of the function $k(\omega)$ in the complex ω–plane or, equivalently, k_0 is a saddle point of $\omega(k)$. This is the last property which are used in our case to determine the nature of the instability. Two different spatial branches $k^+(\omega)$ an $k^-(\omega)$ take part in the pinching process.

The spatial branches with positive and negative indexes correspond, respectively, to waves that propagate downstream and upstream. For a whole of fixed control parameters, the spatial branches $k^+(\omega)$ and $k^-(\omega)$ are located on opposite sides of the k-plane when $\omega_i \rightarrow -\infty$. When ω tends towards the absolute frequency ω_0, the curves move toward each other until the process of pinching for $\omega=\omega_0$ (see fig. 11)

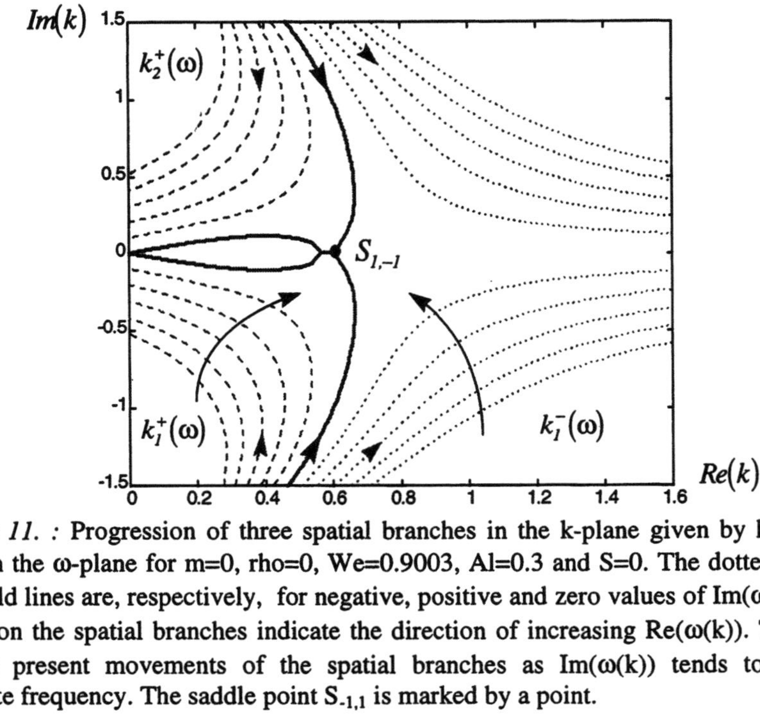

Figure 11. : Progression of three spatial branches in the k-plane given by horizontal lines in the ω-plane for m=0, rho=0, We=0.9003, Al=0.3 and S=0. The dotted, dashed and bold lines are, respectively, for negative, positive and zero values of Im(ω(k)). The arrow on the spatial branches indicate the direction of increasing Re(ω(k)). The other arrows present movements of the spatial branches as Im(ω(k)) tends toward the absolute frequency. The saddle point $S_{-1,1}$ is marked by a point.

On figure 12, the marginal stability curve separating the two type of instability for the axisymmetric mode is plotted Vs Weber and Alfvèn numbers. Large Weber numbers make the flow absolutely unstable. Without magnetic field $(Al = 0)$, the critical Weber number is approximately 0.32, which is the result of Leib and Goldstein for a top-hat jet. For a given Weber number, an increase in Alfvèn number can lead to a convectively unstable capillary jet and can even stabilise the jet. The critical values of the Alfvèn number of the temporal analysis is asymptotically found in the spatio-temporal analysis.

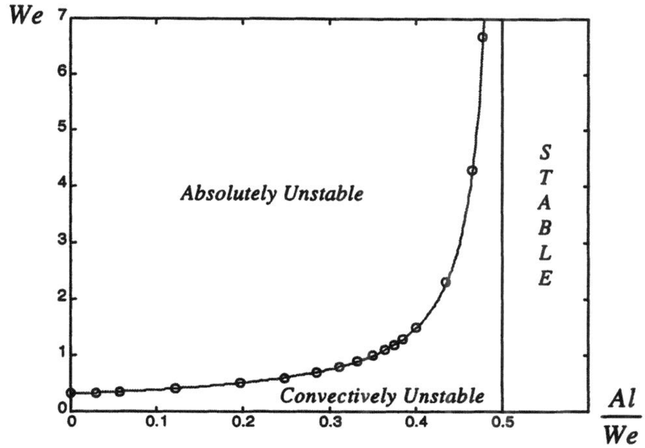

Figure 12. : Domains of absolute and convective instability on the We, Al/We plane for m=0, rho=0

Finally, the presence of an rigid body rotation (see fig. 13) increases the convective instability zone. One must note that if the jet is not rotated, the critical Weber number is higher when the ratio of the density is non null.

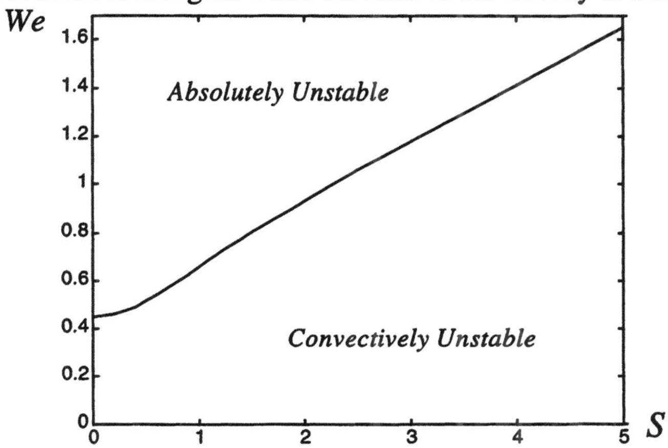

Figure 13. : Domains of absolute and convective instability on the We, S plane for m=0, rho=1

4. Conclusion

A circular, rotating jet of inviscid, incompressible and perfectly conductive liquid is temporally unstable. A uniform axial magnetic field can stabilise the system. The introduction of the surface tension is stabilising except for the axisymmetric mode. The instability type is determined by the combination of the capillary, electromagnetic and inertial instability. The magnetic field and the rotation reduces the domain of absolute instability while higher surface tension enlarges it.

References

BATCHELOR & GILL.(1967), Analysis of the stability of axisymmetric jets. J. Fluid Mech., 14, 529-551

CHANDRASEKHAR (1961) Hydrodynamic and Hydromagnetic stability Clarendon Press, Oxford

LOISELEUX, MONKEWITZ, HUERRE (1998), The effect of swirl on jets and wakes .Phys. Fluid (in press)

LEIB & GOLDSTEIN (1986), The generation of capillary instabilities on a liquid jet. J. Fluid Mech., 168, 479-500

MONKEWITZ, DAVIS., BOJORQUEZ, YU (1988), The breakup of a liquid jet at high Weber number. Bull. Amer. Phys. Soc., 33, 2273

YAKUBENKO (1995), Absolute and convective instability of current-carrying jet of magnetic liquid. Eur. J Mech. B Fluids 14, 823

QUASI-TWO-DIMENSIONAL TURBULENCE IN MHD SHEAR FLOWS : THE "MATUR" EXPERIMENT AND SIMULATIONS

Y.DELANNOY*, B.PASCAL*, T.ALBOUSSIERE**, V.USPENSKI***, R.MOREAU*
** Laboratoire EPM-MADYLAM, CNRS UPR 9033*
ENSHMG, BP 95, 38402 St Martin d'Hères Cedex, France
*** Engineering Department, Trumpington St., Cambridge CB21PZ, UK*
**** Institute of Mechanics, Lomonossov Univ., Moscow, Russia*

Abstract

An experiment and numerical simulations have been carried out to investigate the properties of 2D turbulence in a mercury layer in the presence of a steady magnetic field. Visualisation of the vortices and quantitative measurements of the velocity were recorded, *via* the observation of the free surface and using potential probes located at the bottom of the fluid layer. In parallel, some 2D numerical simulations were issued, using laminar Navier-Stokes equations with a source term modelling the Hartmann friction.

1. Introduction

MHD turbulence exhibits quite specific properties associated with its tendancy to become two-dimensionnal (Sommeria and Moreau, 1982; see also Moreau, 1990). But, so far, most of the experiments performed with electrically conducting fluids in the presence of a uniform magnetic field were devoted to homogeneous turbulence (except within the Hartmann layers), so that the need of knowledge and understanding of the behaviour of quasi-2D turbulence in MHD shear layers is still enormous.

Those shear layers are of special importance because all sorts of transport phenomena take place there. They develop in the direction of the magnetic field along the walls parallel to this field or from any kind of singularity (either geometrical, such as a sharp angle, or electrical, such as a jump in the electrical conductivity). The first observation, purely qualitative, of the turbulent character of these layers is due to Lehnert (1955) and the first quantitative description of their properties is due to Kljukin and Kolesnikov (1989). However, to our knowledge, nothing is known on the transport of a scalar quantity, such as heat, by this kind of turbulent flow, and its dynamics is still far from being well understood.

The first purpose of this paper is therefore to present new experimental results which might be used as a reference to be compared with predictions of adequate modelling. The MATUR equipment (MAgnetohydrodynamic TURbulence) has been

A. Alemany et al. (eds.),
Transfer Phenomena in Magnetohydrodynamic and Electroconducting Flows, 93–106.

realized with a special care to insure the stability and uniformity of all the control parameters and to select a combination of them which guarantees the two-dimensionnal character of this turbulence. A special effort has been made to improve the accuracy of the diagnostic techniques, based on measurements of temperatures and electric potential differences.

In the second part of the paper, we evaluate the pertinence of 2D large eddy simulations for this flow, using source terms derived from Sommeria and Moreau (1982) to model the Hartmann friction. None of the available models for ordinary turbulent shear flows can be applied because of the quasi-2D behaviour of MHD turbulence, but, according to the inverse energy cascade, the small turbulent stuctures (smaller than the size at which the turbulent energy is injected) have little influence. Consequently, we use a commercial code ("Fluent") without subgrid scale model : the dissipation is due to the Hartmann friction, and to the molecular viscosity for the (low-energy) small scales.

2. Apparatus and Measurement Techniques

The working fluid is a layer of mercury. It is located in a horizontal circular cell (internal diameter 22cm) above an electrically insulating bottom plate. The upper boundary may be, either a free surface, to allow to observe the deformation produced by the turbulent eddies, or an insulating horizontal plate, to priviledge the two-dimensionnality when the quantitative measurements are performed. This cell is located in a vertical coil which generates a uniform vertical magnetic field (*Figure 1*).

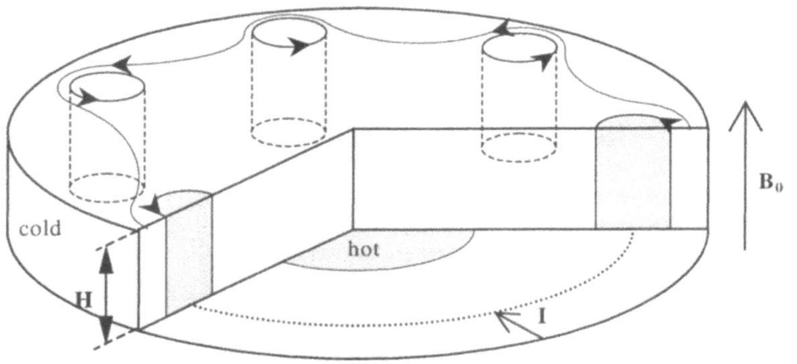

Figure 1 : Structure of the electrically driven 2D turbulent shear flow.
The control parameter I is the intensity of the electric current passing from the ring of electrodes to the external vertical wall, while B_0=0.17T, H=1cm are constants.

Both the depth of the fluid layer and the strength of the magnetic field are adjusted in such a way that the typical time scale to establish a good two-dimensionality, τ_{2D}, is much smaller than the typical turnover time τ_{tu}. These typical times are defined by :

$$\tau_{2D}=\frac{\rho.H^2}{\sigma.B_0^2.l^2} \qquad \text{and} \qquad \tau_{tu}=\frac{l}{u} \tag{1}$$

Here ρ stands for the density and σ for the electrical conductivity of the fluid, H for its depth, B_0 for the magnetic field, l and u for a typical length scale and a typical velocity scale of a turbulent eddy. Typically, with H=1cm and B_0=0.17T, which are the values selected for all the results presented in this paper, $\tau_{2D}\approx10^{-2}$s whereas $\tau_{tu}\approx1$s, so that two-dimensionnality is well achieved. At this stage, it may be noticed that, because the energy transfer is towards the large scales in 2D turbulence, this condition is well satisfied as soon as it is satisfied for the length scale of the forcing mechanism.

On the bottom plate a set of 292 small electrodes are located on a circle of radius 9.3cm and make an anode from which the electric current arrives, whereas the electrically conducting external vertical plate collects this electric current. Because the Hartmann number is significantly larger than one, this radial electric current is passing through the Hartmann layers which behave like an electric motor and drive the fluid in rotation. The Hartmann number (=42 with the choosen values) is defined here as

$$Ha=\sqrt{\frac{\sigma}{\rho.\nu}}.B_0.H \tag{2}$$

From the theory of the Hartmamn layer, if the flow would be laminar, the core velocity between two Hartmann layers should be proportionnal to the total current I and in the circumferencial direction (unit vector $\mathbf{e}_\theta$) :

$$\mathbf{u}_0=\frac{I}{4\pi r\sqrt{\sigma\rho\nu}}.\mathbf{e}_\theta \tag{3}$$

This means that a shear is necessarily established between the rotating external annulus (where I≠0), and the central core region (where I=0), which stays at rest in a laminar regime. The typical maximum velocity which is expected is of the order of 40cm/s. And, if the flow would be laminar, the typical thickness of the circular shear layer above the ring of electrodes, would be of the order of 0.3cm. But, as expected, this circular shear (or mixing) layer is unstable. Two-dimensionnal columnar eddies aligned with the applied magnetic field develop, with an initial wave length which corresponds to the most linearly unstable disturbances (close to 2cm), and interact. These structures remain 2D since any 3D disturbance induces an electric current and a damping Lorentz force (2D eddies do not induce any current). Therefore, an energy transfer towards the large scales feeds larger and larger vortices, until the largest allowed scale which is the cell radius (11cm). This arrangement and the 2D shear flow are sketched on *Figure 1*

A central part of the bottom plate (radius 3.65cm), made of a thick copper piece (but electrically insulated from the mercury), could be heated from below at a given power (0 to 8W), whereas the external vertical wall is maintained at a given and precisely controlled (±0.02°C) temperature. This allows to study how a 2D turbulent shear flow transports a scalar quantity such as heat, as well as angular momentum (*Figure 1*).But, in this paper, which focuses on the velocity field, any data on temperature are disregarded.

To measure the two velocity components in the radial and azimutal directions, we are using electric potential sensors. Between two such sensors close enough to each other, the electric potential difference is equal to the product of the velocity component in the direction perpendicular to the sensors lines times the magnetic field B_0 and the distance between these sensors. We have nevertheless developped a significant originality in these techniques, since our sensors are in the same time thermocouples and velocity probes. Each of them is made of two small circular metallic surfaces, made one of platinium and the other of constantan. The diameter of each is 0.2mm. These two surfaces are located within a cylinder of alumina whose external diameter is 0.9mm. Each probe behaves like a Pt-Cst thermocouple and allows to get the local temperature. And measuring the electric potential between the Pt surfaces of two neighbouring probes gives the local and instantaneous velocity. Platinium was chosen because it has almost the same absolute thermoelectric power as mercury, in order to minimize the error due to any thermoelectric Pt-Hg contribution. All the results presented in this paper have been obtained with a total of 88 such sensors inserted in the insulated bottom plate and located, either on some radial lines (to provide the angular velocity component), or on some circles (to provide the radial velocity component).

Flow visualisations may be realized when a free surface is present, either using tracers moving with the free surface of mercury, or looking at the image of a regular grid reflected by the mercury surface. Each eddy able to locally depress the surface yields a disturbance of the image. On the video shown during the oral presentation of this paper, it is clear that the initial instability of the mixing layer (made of about 30 periodic eddies of typical length scale 2cm), rapidly feeds an inverse cascade. At high values of the electric current (20 or 30A), only a small number of large coherent structures are present (2 or 3 structures with length scales of the order of 10cm), moving around the cell at a velocity smaller than the maximum fluid velocity by about one order of magnitude.

3. Experimental results and interpretation

The mean velocity profiles $V_\theta(r)$ are shown on *Figure 2*. It is remarkable that, whereas the wall shear layer thickness is almost the same as in the laminar regime (≈ 0.3cm), the free shear layer thickness is increased by about one order of magnitude.

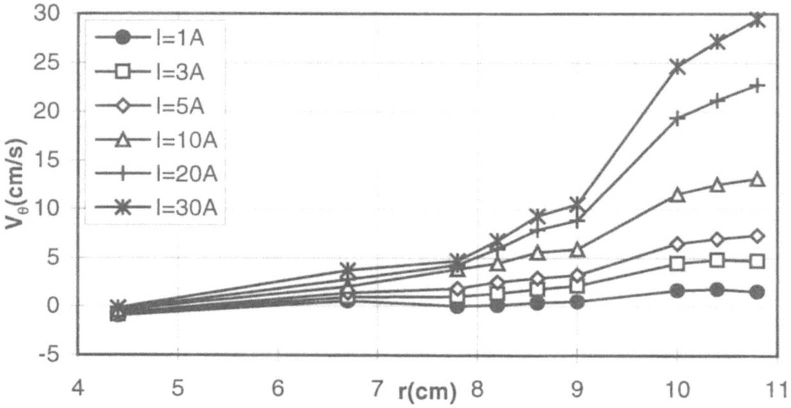

Figure 2 : Profiles of the mean angular velocity component for different values of the control parameter I.

It is also noticeable that the maximum angular velocity is nearly half that predicted for the laminar regime. For instance, the expression $I / (4\pi r\sqrt{\sigma\rho\nu})$ predicts a velocity of 43cm/s for I=20A, but the measured value is close to 23 cm/s. And, consistently, a significant angular momentum is present in the central part of the cell where no electric current at all is passing within the Hartmann layers. The thickness for the turbulent free shear layer derived from the maximum slope of the mean velocity profile, is of the order of 3cm (instead of 0.3cm in a laminar regime) : This illustrates the turbulent transport of angular momentum and suggest that we have a significant turbulent intensity in the free shear layer (between r=8cm and r=11cm). Precisely, the profile of the root-mean-square (rms) of the fluctuations of the two velocity components v'_θ and v'_r are shown in *Figure 3* and *Figure 4*. The first point to notice is the importance of the turbulence level : the ratio v'_θ/V_θ is of the order of 10 to 12% in the free shear layer and goes to values of the order of 35% in the central part. Indeed, the rms of the azimutal velocity fluctuation is almost uniform all over the cell, whatever the value of the mean velocity.

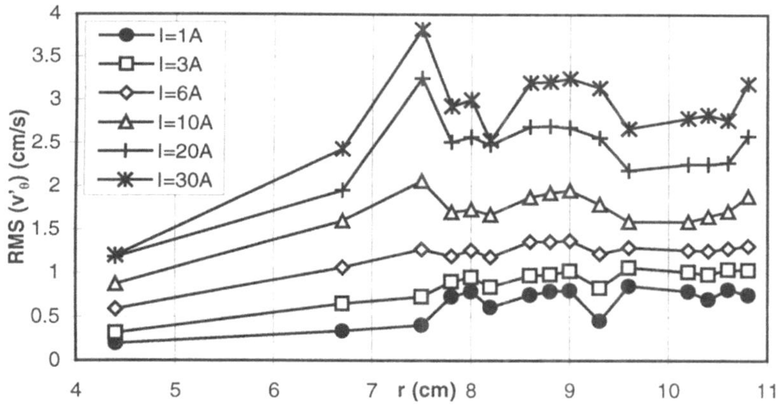

Figure 3 : Profiles of the rms angular velocity fluctuations for different currents I

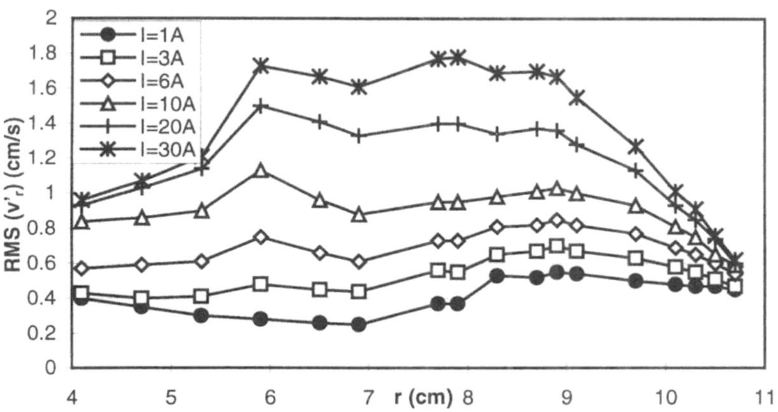

Figure 4 : Profiles of the rms radial velocity fluctuations for different currents I

One may also notice that the rms of the radial velocity fluctuation is about one half that of the angular fluctuation. And, finally, the almost linear increase of v'_r with the distance from the wall towards the centre, seems to be quite consistent with continuity.

Two typical spectra of velocity fluctuations are shown on *Figure 5* (for a moderate electric current I=3.5A) and *Figure 6* (for a high value of the electric current I=30A). Both figures show nice signatures of the large coherent structures illustrated on the video, in the forms of peaks in a range of small wave numbers. The first peak corresponds to a frequency of the order of 0.2Hz for I=3A, or 0.9Hz for I=20A. Using the Taylor hypothesis based on the corresponding mean velocity (3.6 and 19cm/s), one can derive the order of size of the structures : 18cm for 3.5A and 22cm for 30A. Those values exhibit the increase of the size of the large coherent structures (or the reduction of their number) when the kinetic energy of the turbulence increases. However, the velocity of the vortices is only a fraction of the mean velocity, so that the use of the Taylor hypothesis gives only qualitative results.

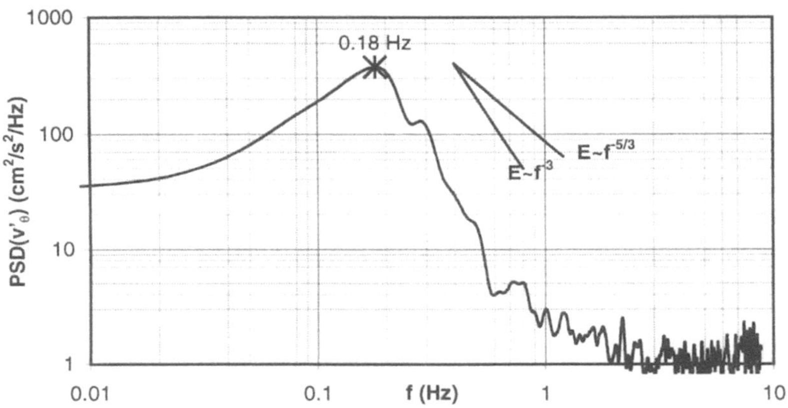

Figure 5 : Power spectral density of v'_θ at r=9.6cm (I=3.5A)

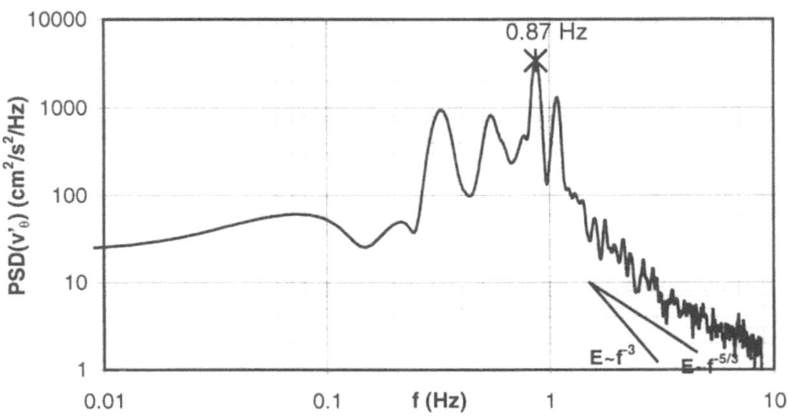

Figure 6 : Power spectral density of v'_θ at r=9.6cm (I=30A)

At larger wave numbers, but before those dominated by noise, an inertial range seems to be well defined, although it starts with a clear reminiscense of the peaks. In both cases (I=3.5A and I=30A), a power law fits well with the measurements, but the exponent should be close to –5/3 at low electric current (or kinetic energy), and close to –3 at high electric current (or high kinetic energy). These results are in agreement with previous measurements. Sommeria *et al.* (1989) had also measured $k^{-5/3}$ spectra in a 2D homogeneous turbulence dominated by an inverse energy cascade. And many authors, in Riga (see Lielausis, 1975), in Purdue (Brouillette and Lykoudis, 1967 ; Garner and Lykoudis, 1971) and in Grenoble (Alemany *et al.*, 1979) had previously reported k^{-3} energy spectra in MHD turbulence, but in quite different conditions (see a discussion of these conditions in Moreau,1990). An important point we wish to stress here is our interpretation of these power laws. We consider the k^{-3} law, which is present in the scales larger than the forcing scale, could be the signature of a quasi-steady equilibrium between the inertial transfer mechanisms and the damping mechanisms. The first one has a typical time scale $\tau_{tr}(k)$, whereas the second one has a typical time scale τ_H, where :

$$\tau_{tr}(k)=\frac{1}{\sqrt{k^3E(k)}} \qquad \text{and} \qquad \tau_H=\frac{H}{2.B_0}\sqrt{\frac{\rho}{\sigma.\nu}} \tag{4}$$

The Hartman time τ_H is k-independent, and represents the damping taking place within the Hartman layers (Sommeria and Moreau, 1982), which results from both viscous and ohmic effects (it results from the Hartmann layer theory that they are equal to each other) : it is the only dissipation suffered by this 2D turbulence. We think that, when the two-dimensionnality is well achieved, the key to determine this spectral law is the ratio between the Hartmann dissipation just mentionned and the dissipation in the shear layer present along the external wall. If the last one is predominant, the energy flux through the cascade must be almost constant and the turbulence exhibits a Kolmogorov inverse cascade with a $k^{-5/3}$ law. On the contrary, if the first one is predominant, it implies that the energy flux through the cascade cannot be constant. Then, the quasi-steady equilibrium between the amount of energy received by each wave number and its dissipation within the Hartmann layers (at both ends of the columnar eddies) applies and explains the k^{-3} law.

4. Numerical model

The present work uses a standard fluid flow code (Fluent) for the large-eddy simulation of a MHD turbulent flow, including simulation of scalar transport properties. The smallest eddies in which energy is injected are supposed to have the same size as the laminar parallel layers present in MHD flows. According to the inverse energy cascade, smaller size have little role in the flow and no subgrid scale model is used : this method is similar to the one of Verron & Sommeria (1987) but we use only the molecular viscosity, without artificial viscosity, and our application deals with a shear layer and not with homogeneous turbulence.

Following Sommeria and Moreau (1982), we consider in this work that the horizontal mercury layer is divided in a core region, where the velocity is constant on a vertical line, and Hartmann layers concentrating the vertical gradients. The Hartmann layers are H/Ha=0.23mm thick, and our model considers that the core flow extends in the whole 1cm layer of mercury. From experimental observations with a free surface, we consider that a fixed oxide skin is always present on the mercury surface (floating particles were not moving during velocity and temperature measurements), so that both horizontal boundary (above and below) impose a no-slip condition, leading to two Hartmann layers on each side of the core flow. Of course, this remains true with a cover plate in place of the free surface.

The laminar Hartmann layer theory gives an analytical expression for the influence of Hartmann layers on the core flow, in the case of an uniform magnetic field and of high Hartmann numbers : the electromagnetic force in the core flow reduces to a force $\mathbf{f}= \rho.(\mathbf{u}-\mathbf{u}_0) / \tau_H$ per unit volume, added to the Navier-Stokes equations :

$$\frac{\partial \mathbf{u}}{\partial t}+(\mathbf{u}.\nabla)\mathbf{u}=-\frac{1}{\rho}\nabla p+\nabla.(\nu\nabla\mathbf{u})+\frac{1}{\tau_H}(\mathbf{u}-\mathbf{u}_0) \tag{5}$$

$\mathbf{u}$ is the (horizontal) velocity in the core flow, $\tau_H=\sqrt{\rho/(\nu.\sigma)}.H/(2B)$ is the Hartmann defined by equation (4) and $\mathbf{u}_0$ is the forcing velocity field, due to the injected current (equation 3) : in our experiment, $\mathbf{u}_0=\mathbf{0}$ for $r<R_{inj}$, and $2\pi.r.\mathbf{u}_0=(I/2)/\sqrt{\sigma.\rho.\nu}$ for $r>R_{inj}$, where R_{inj}=9.3cm is the radius of the anode.

Equation (5) is able to generate shear layers with a thickness representative of parallel MHD layers : its curl, in a steady simple shear layer (where D/Dt=0), is a purely diffusive vorticity equation : $\nu.\Delta\omega= -(1/\tau_H).(\omega-\omega_0)$. When a spatial discontinuity exists for ω_0 (as in our experiment at the injection radius), the characteristic scale of change of ω is δ, defined by $\nu/\delta^2 \sim 1/\tau_H$, i.e. $\delta=(H/2)/\sqrt{Ha/2}$, which is correct order of magnitude for the thickness of parallel shear layers (Moreau, 1990). Considering the velocity field in a stationary simple shear layer (i.e. in laminar conditions), this means that the forcing term $(\mathbf{u}-\mathbf{u}_0)/\tau_H$ and the dissipation term $\nu\Delta\mathbf{u}$ must have the same order of magnitude.

The commercial code Fluent/UNS (1996) uses a classical pressure-velocity formulation on an unstructured mesh to solve equation (5) coupled to the continuity equation $\nabla.\mathbf{u}=0$. We use it in unsteady laminar mode, and with the SIMPLEC algorithm for pressure-velocity coupling : In this implicit pressure correction method, each time step is solved iteratively in a segregated fashion (each scalar equation in turn). There is no Courant-Friedrich-Levy condition on the time step that can be very long, with the penalty of longer iteration times for longer time steps.

A finite volume formulation is used for the spatial discretization, with a 2nd order upwind scheme for convection (convected variable on each face extrapolated from the upwind value and gradient), and a central scheme for diffusion (gradients on the faces calculated by central differences). A limiter is included to prevent spurious oscillations caused by the second order scheme, but it is probably only necessary at the beginning of our calculation, because of the sharp gradients contained in the initial solution. After a few time steps, the discretization is second-order accurate in space but only first-order accurate in time.

13700 quadrilateral cells (*Figure 7*) were used to describe the 11cm diameter disk of the MATUR experiment : the mesh size is 1mm in the r direction except near the wall and is between 1 (axis) and 5.4mm (external wall) in the θ direction, so that the model is able to represent turbulent structures of size similar to, or larger than the laminar parallel layer thickness($\delta_{//}$~3.(H/2)/$\sqrt{Ha/2}$=3.3mm). Near the wall, the radial size decreases up to 0.1mm (at r=R), in order to describe correctly the thin wall layer (~10 cells inside a layer of thickness $\delta_{//}$). The external boundary is a wall with a no-slip condition, and all MHD char urrent, are contained ir routine".

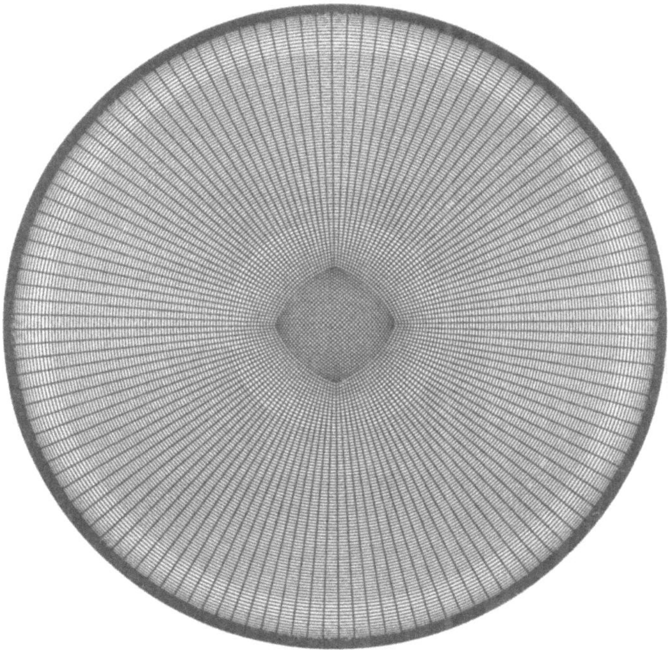

Figure 7 : 2D quadrilateral mesh used for the simulations

Because of the 1st order discretization in time, the time step is chosen very small compared to the characteristic times of the problem, i.e. the Hartmann braking time τ_H, and the travel time τ_0 of a fluid particle around the box, at the "forcing velocity" $u_0(R_{inj})$ (τ_0 is the characteristic time τ_{tu} of the largest possible eddy, if its characteristic velocity is u_0). The time step Δt and the total duration simulated T are shown in *Table 1* for the three simulations presented here. Each simulation represents more than 25 turnover times τ_0 around the box, after dissipation of initial conditions (i.e., neglecting the first τ_H seconds after the beginning of instability). Mean and RMS values are calculated using results at $t > \tau_H$.

I(A)	B(T)	H(mm)	Ha	τ_H(s)	$U_0(R_{inj})$ (cm/s)	τ_0(s)	Δt (s)	T(s)
3	0.17	10	42	9.37	6.35	9.21	0.03	270
10	0.17	10	42	9.37	21.2	2.76	0.01	180
30	0.17	10	42	9.37	63.5	0.92	0.005	75

Table 1 : Experimental and numerical parameters

The calculation starts from a laminar solution, calculated with equation (5) at Ha=42000 using a steady state resolution. Such a laminar solution is plotted together with mean turbulent values in Figure 9, and is very close from the "forcing velocity" distribution $u_0(r)$. At the beginning of the unsteady calculation, the MHD forcing term returns to its normal value (Ha=42), and the unstability appears due to roundoff errors. It has been verified that the addition of a white noise to start the unstabilities give the same statistical results.

5. Numerical results

The instability of the shear layer always starts with small eddies (*Figure 8*, left), then larger and larger eddies are formed, according to the inverse cascade in 2D turbulence. An equilibrium state is reached (*Figure 8*, right), after a time of the order of τ_H, where only a few structures remain in the domain, as in experimental visualisations using the reflection of a grid on the free surface. However, even in the presence of large scale vortices, some concentration of cyclonic vorticity appears above the electrode in the "wake" of the large vortices (*Figure 8*), and can be interpreted as the feeding mechanism providing turbulent energy, at a scale presumed to be the parallel layer thickness (which is the resolution of our mesh). Another concentration of vorticity is produced near the external wall, before each large vortex : at these points (dark zones on the figure) anticyclonic vorticity is injected into the flow from the outer parallel wall layer.

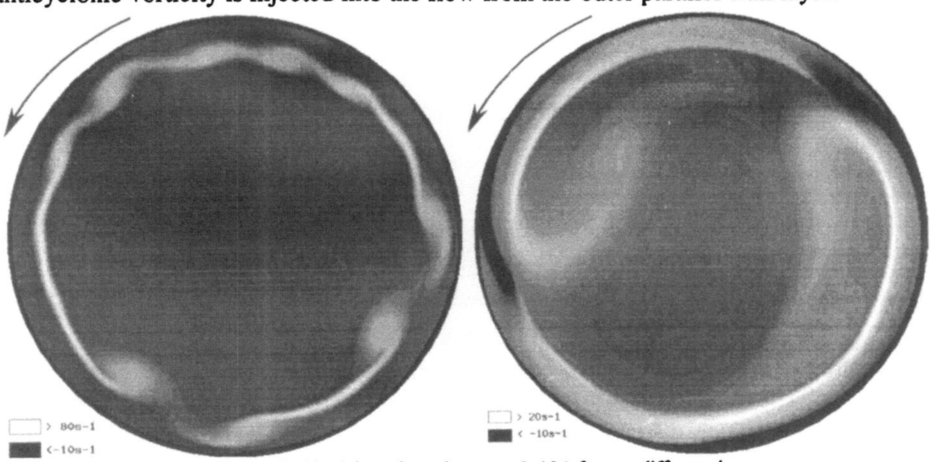

Figure 8 : Vorticity coloured maps at I=10A for two different times :
left : t=1.7s, just after instability ; right : t=180s, in developed turbulence. The arrow indicates the flow direction

The radial distributions of calculated mean velocities and their RMS fluctuations are compared to experimental values in *Figure 9*: the mean distribution of V_θ is reproduced fairly well at 3A and 10A. The well reproduced thickness of the shear layer indicates a correct diffusion of momentum without subgrid scale model. The boundary layer near the external wall is stable in calculations as in experiments, but seems to be thinner in experiments. However, experimental velocity values are calculated from electric potential measurements, using a 2D theory that is not exact in such a parallel layer. At 30A, the mean values are overpredicted by the model : a beginning of instability in Hartmann layers could be responsible of this overprediction (our source term in equation 5 comes from a laminar theory for Hartmann layers).

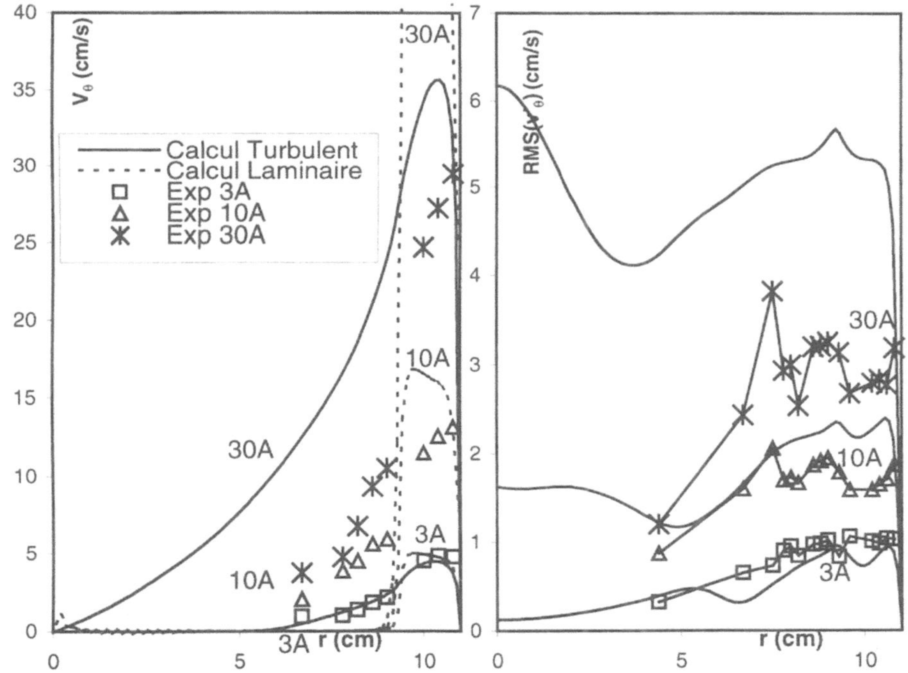

Figure 9 : Radial distribution of the mean circumferencial velocity (left) and its RMS fluctuation (right), compared to experiments and to the laminar solution. I=3A, 10A or 30A.

An important anisotropy of turbulence exists in the shear layer, as shown by the variance of the velocity fluctuations (square of the RMS fluctuation), plotted in *Figure 10* for I=10A. This is due to the large scale of eddies, transporting on a far distance the influence of the wall, whereas in 3D turbulence this anisotropy is confined in a near-wall layer, the thickness of which depends only on local conditions.This anisotropy is reproduced by the calculation, showing that small scales have little influence on it.

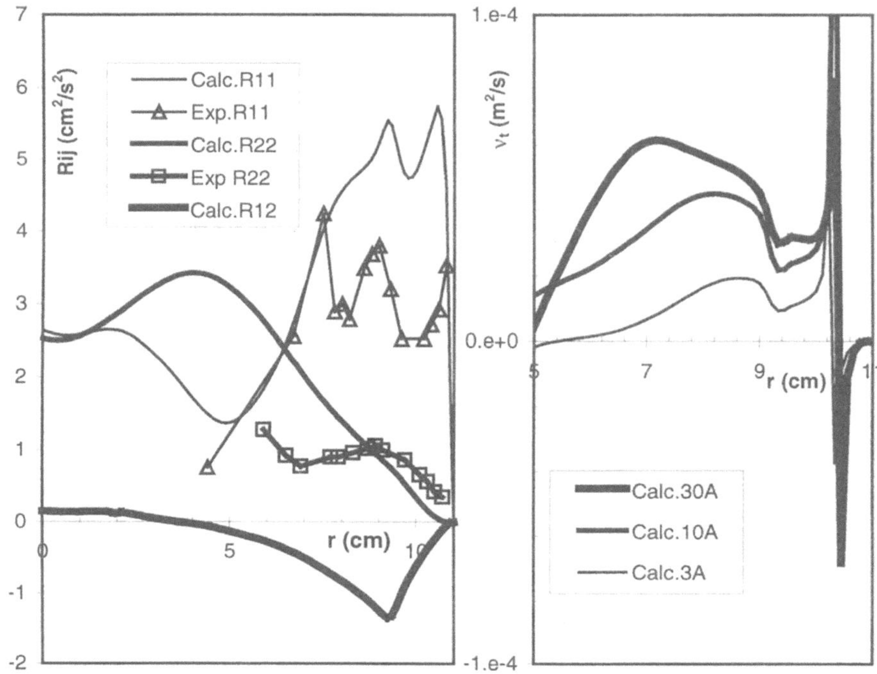

Figure 10 : Radial distribution of the Reynolds stress components at 10A (R11=$\overline{v'_\theta v'_\theta}$, R22=$\overline{v'_r v'_r}$, R12=$\overline{v'_\theta v'_r}$, left) and of the kinematic viscosity calculated at I=3,10,30A ($\nu_t = -\overline{v'_\theta v'_r}/(\partial V_\theta/\partial r)$, right)

The calculated Reynolds shear stress (divided by ρ) $\overline{v'_\theta v'_r}$ has also been plotted in *Figure 10*: it is opposite in sign to $\partial V_\theta/\partial r$ (as it would result from a turbulent viscosity model), in the turbulent layer above the electrode. In the wall layer, however, the sign of $\overline{v'_\theta v'_r}$ remains unchanged, whereas $\partial V_\theta/\partial r$ is negative : a flux of momentum opposite to the mean flow gradient exists in this zone, due to large eddies produced above the electrode. More precisely, this counter-gradient transport is related to detachements of the parallel layer, visible on *Figure 8* as two zones of negative vorticity (clockwise vorticity as in the parallel layer) passing near the wall before each large structure. From the Reynolds shear stress calculated results we can evaluate in the most turbulent zone (r=8 to 10cm) a turbulent kinematic viscosity $\nu_t \sim 5.10^{-5}$ m²/s for I=10 and 30A (~2.10^{-5} for I=3A), so that the ratio ν_t/ν is of the order of 100 : according to local dimensionnal analysis, this ratio should measure the characteristic Reynolds number of the eddies responsible of the turbulent mixing

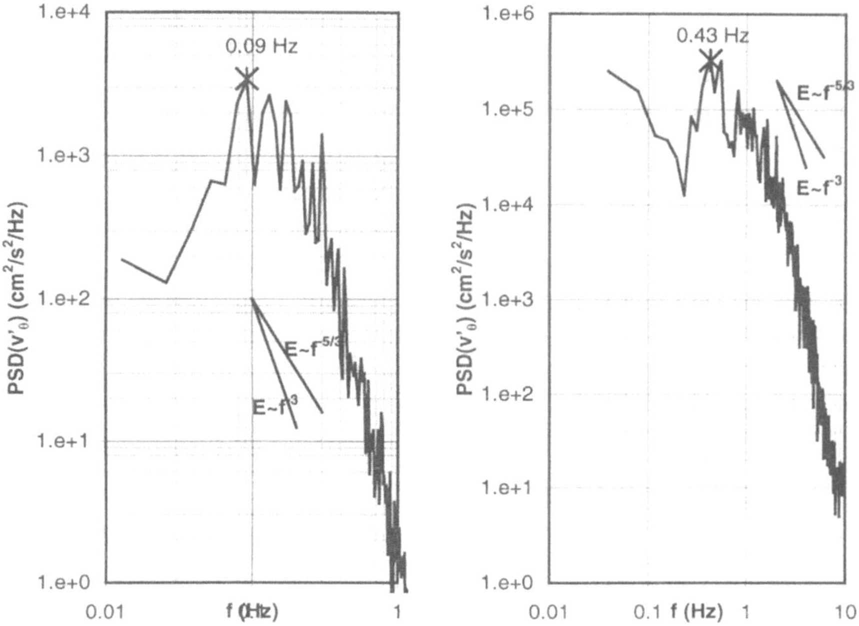

Figure 11 : Power Spectral Density (PSD) of the calculated velocity fluctuation v'_θ at r=9.1cm, for I=3A(left) and I=30A(right).

We can also derive a characteristic length l_t for large vortices from the power spectrum of v'_θ(*Figure 11*) and from the mean velocity V_θ, as was made for experiments : the frequency of the maximum is just half the value obtained in experiments, but of course, a peak also exists at the experimental frequency (there is in fact just a different balance between harmonic frequencies). Using the calculated mean velocity at r=9.1cm, we obtain l_t= 41;69;73cm for I=3;10;30A, so that the increase of size obtained in the experiments is also reproduced in the calculations. Of course, the real size of the main eddies is smaller because of their velocity, smaller than V_θ. We can also derive a characteristic velocity for those large vortices from the variance of the velocity ($v_t = v'^2_\theta$), so that their characteristic Reynolds number is of the order of $l_t.v_t/\nu$~41000;172000;438000 for I=3;10;30A : this very high value compared to ν_t/ν~100 indicates that the efficiency of the turbulent mixing is very low for such large and energetic vortices.

6. Conclusion

A mercury cell located in a vertical magnetic field is used to observe the singular properties of quasi-2D turbulent shear layers, with the help of numerical large eddy simulations for analysis. A number of striking results have been obtained, among which the most noticeable seem to be the following :

- The velocity field is dominated by a number of large coherent structures fed from the instability of the shear layer *via* a energy transfer towards the large scales.
- This kind of turbulence is, as expected, a good way to efficiently transport the momentum, but the intensity of this transport is not so high as could be expected from the characteristics of the large eddies and from the mean velocity profile.
- Detachements of the wall parallel layer have been noticed in the simulations, creating a counter-gradient transport in this zone.
- The energy spectrum exhibit peaks (at low wave numbers) corresponding to the large structures and an inertial range (at higher wave numbers) whose typical law is $k^{-5/3}$ or k^{-3} depending on the importance of the Hartmann damping compared with the friction in the shear layer parallel to the applied magnetic field.

References

Alemany A., Moreau R., Sulem P. and Frisch U. (1979) *J.de méca.Théor.et Appl.*, vol 18, pp.277-313

Brouillette E.C. and Lykoudis P.S. (1967) *Phys.Fluids*, vol 10, pp.995-1001

Gardner R.A. and Lykoudis P.S.(1971) *J.Fluid Mech.*, vol.47, pp.737-764

Kljukin A.A. and Kolesnikov Y.B. (1989) in *Liquid Metal Magnetohydrodynamics*, ed. Lielpeteris J. and Moreau R., Kluwer Acad. Pub., pp.449-454

Lehnert B.(1955) *Proc. Royal Soc. London*, A233, p.299

Lielausis O.A.(1975) *Atomic Energy Review*, vol.13, p.527

Sommeria J, Moreau R (1982) Why, how, and when, MHD turbulence becomes two-dimensional. *J. Fluid Mech.* **118**, 507-518

Sommeria J, Nguyen Duc J.M., Caperan P. (1989) in *Liquid Metal Magnetohydrodynamics*, ed. Lielpeteris J. and Moreau R., Kluwer Acad. Pub., pp.441-448

Moreau R (1990) *Magnetohydrodynamics*, Kluwer Academic Publishers, Dordrecht.

Fluent (1996) *Fluent UNS User's Guide, release 4.4.* Fluent Inc., Lebanon, New Hampshire

TRANSPORT OF MOMENTUM AND HEAT IN OSCILLATORY MHD FLOW

S. CUEVAS AND E. RAMOS
Centro de Investigación en Energía, UNAM
A.P. 34, Temixco, Mor. 62580, MEXICO
E-mail: scg@coatl.cie.unam.mx

Abstract. We analyze the laminar flow and heat transfer induced by a slightly curved wall oscillating in a viscous, electrically conducting fluid which is otherwise at rest, under the action of a transverse magnetic field. The electrically insulating wall as well as the fluid at infinity are maintained at different constant temperatures. The analysis is based on perturbative solutions and special attention is paid to the effect that the magnetic field has on the non-linear *streaming* contributions. The competing action of inertial transfer and magnetic suppression in the transport of vorticity in the boundary layer is analyzed. Diffusive heat transfer solutions including viscous and ohmic dissipation are found. The convective effect of the streaming flow in the boundary layer is also discussed.

1. Introduction

Oscillatory boundary layer flows have been widely studied in hydrodynamics [1]. However, only a few works have been devoted to analyze the consequences that magnetic fields produce in oscillatory flows of electrically conducting fluids near to solid objects. Although this is an interesting topic of basic research, particularly the understanding of nonlinear effects, the study of these kind of flows is mainly justified by two relevant practical applications. One is related to the possibility of transforming acoustic power into electrical power through a liquid-metal MHD thermoacoustic transducer [2]. These devices, which have been mainly developed at Los Alamos National Laboratory, utilize the thermoacoustic effect to set in oscillatory motion a liquid metal (sodium) inside a closed container immersed in an applied magnetic field, transverse to the flow. The second important application is the control of oscillatory flows in crystal growth processes with the Czochralski method, via an axial magnetic field [3]. This example is of particular interest since it has been demonstrated in model experiments and numerical simulation the existence of macroscopic oscillations of the

A. Alemany et al. (eds.),
Transfer Phenomena in Magnetohydrodynamic and Electroconducting Flows, 107–122.

flow and temperature in a Czochralski melt, which appear to be the cause of a defect of the crystals known as growth striations. It was found by numerical simulation of a gallium arsenide melt that the presence of an axial magnetic field was effective in suppressing the oscillatory flow. This would help to obtain more uniform crystals. In spite of the former applications, a detailed study of the mechanisms involved in the transport of heat and momentum in oscillatory boundary-layer flows with magnetic fields has not been previously done.

Preceding works concerning non-MHD oscillatory boundary layer flows deal with either oscillatory motions with a vanishing mean or flows presenting a steady (non-zero) mean. The latter area of research was opened by Lighthill [4] who established an approach to analyze the boundary layer response to fluctuations of the external flow about a steady mean flow. On the other hand, net transport of momentum and heat are known to exist in oscillatory flows with a zero mean near solid surfaces. The net transport of momentum, known as steady or acoustic streaming, is a second order effect originated by viscous Reynolds-like stresses in the unsteady boundary layer. The source of these stresses is the average effect of nonlinear convective interactions of the first harmonic (streamwise and normal) velocities in the Stokes layer. The phase difference in these velocities produces a non zero average over one cycle that results in a slight increase of the displacement amplitude of the fluid in one preferred direction. This leads to a pattern of steady vortices superimposed on the primary oscillatory flow. The persistence of the streaming motion beyond the inner layer is one of the characteristic features of this phenomenon. It has been pointed out that the convection or 'stirring' produced by the steady streaming could be effective in accelerating certain kinds of rate processes at liquid-solid interfaces; in particular, the steady streaming is able to enhance the heat transfer [5]. The present understanding of these phenomena stems from the pioneering work of Rayleigh [6] who offered the first theoretical explanation to the observations previously reported by Faraday [7]. Among the many relevant contributions that followed Rayleigh's work, stands out that of Schlichting, who used the boundary layer approximation to match the inner and outer solutions of the flow induced by an oscillating cylinder [8]. Stuart [9] proposed an important qualitative classification of the flow regimes in terms of the streaming Reynolds number. This dimensionless parameter is defined as $R_s = (U_\infty^*)^2/\omega\nu$, where U_∞^* is the velocity amplitude of the oscillating body, ω is the oscillation frequency and ν is the kinematic viscosity of the fluid. R_s appears to play a role analogous to the role of the ordinary Reynolds number for steady flow past a solid body. Hence, for $R_s \gg 1$ the outer streaming presents a boundary-layer structure with thickness $O(R_s^{-1/2}L)$, within which the steady velocity decays to zero, where L is the characteristic length of the flow. In turn, for flows with $R_s \ll 1$, the persistent streaming has a Stokes flow structure and the motion decays to zero over a much wider region. Substantial contributions to the understanding of this problem were made by Riley [10, 11, 12], who carried out a thorough study using the method of matched asymptotic expansions. A few works have explored the heat transfer in oscillatory flows with a vanishing mean. Net transport of heat by diffusive or convective means can arise in these systems at different orders of approximation. Richardson [13] discussed the streaming convection mechanism from a heated circular cylinder subjected to oscillations relative to the fluid in which it is immersed. Some of his results were criticized in a detailed study by Davidson [14] who used matched asymptotic expansions and a numerical method to obtain the heat transfer in the former system for small and large streaming Reynolds numbers. Gopinath and Mills [15] addressed the heat transfer analysis due to acoustic streaming from a sphere oscillating in a quiescent fluid. The same authors investigated the steady heat transfer due to streaming flow in a cylindrical duct with end walls maintained at different temperatures [16]. None of the former works have included viscous dissipation in their analysis. The effect of a transverse magnetic field on the boundary layer flow of an electrically conducting, incompressible fluid generated by an outer oscillatory irrotational flow with zero mean was recently explored by Cuevas and Ramos [17]. It was found that the action of a transverse magnetic field prevents the streaming motion from intruding the outer flow, unlike the hydrodynamic case where the streaming persist beyond the Stokes layer. In the present article, we extend the previous study to consider the effect of a transverse magnetic field on the dynamics and heat transfer of an oscillatory two dimensional flow, induced by the periodic motion of an electrically insulating solid wall. Assuming the low-magnetic Reynolds number approximation, the perturbation of the applied magnetic field is also determined. The emphasis is on how the transport mechanisms of momentum and heat are affected by the magnetic interaction; of particular interest is the steady transport caused by the streaming.

2. Formulation of the Problem

We consider an incompressible, electrically conducting, viscous fluid limited by an infinite electrically insulating wall under a uniform, transverse magnetic field. The wall is assumed slightly curved so that accelerations due to curvature as well as corrections in the y-coordinate due to the non-flatness of the wall are neglected. This approximation is frequently invoked in the analysis of streaming flows [9, 11]. The wall performs an oscillatory motion about a zero mean in the horizontal direction (x^*-direction) which can be expressed as the real part of $U^*(x^*,t^*) = U_o^*(x^*)e^{i\omega t^*}$, where t^* and ω are the time and the angular frequency of the oscillating wall, respectively, $*$ denoting a dimensional quantity. The x-dependence of U_o^* is due to the curvature of the wall in the streamwise direction. The applied magnetic field is expressed as $\mathbf{B_o}^* = B_o^*\hat{\mathbf{y}}$, where B_o^* is the strength of the field and $\hat{\mathbf{y}}$ is the unit vector in the y^*-direction, transverse to the wall. The wall and the fluid far from it, are maintained at different constant temperatures T_w^* and T_∞^*, respectively. The periodic motion of the wall acts as a source of vorticity and generates a primary oscillatory flow, namely, the Stokes layer, and a secondary flow originated by non-linear inertial interactions, which is composed by periodic and steady streaming contributions. Far from the wall, where the coordinate system is fixed, the fluid remains at rest. Under these conditions, an electrical current density in the z^*-direction is induced inside the boundary layer. Therefore, a perturbation to the applied magnetic field must exist. The current, in turn, interacts with the applied field originating a Lorentz force in the x^*-direction. The thermal behavior of the system is also altered by viscous and ohmic dissipation.

This problem has many independent length scales whose relative sizes have important implications in the physical nature of the flow as well as in the possible approximations used to model it. The ratio of the displacement amplitude of the fluid, U_∞^*/ω, to the characteristic length of the flow, L, gives the particularly important parameter $\epsilon = U_\infty^*/\omega L$. All the considered situations in this work correspond to the small amplitude of oscillation condition, $\epsilon \ll 1$, which ensures the absence of boundary-layer separation [9]. The squared ratio of L to the viscous diffusive thickness $(\nu/\omega)^{1/2}$, gives the second important dimensionless parameter, namely, the *oscillation* Reynolds number, $Re_\omega = \omega L^2/\nu$. Re_ω indicates the extent of penetration of viscous diffusive effects. From these two basic parameters, the ordinary Reynolds number for steady flow past a body can be obtained, that is, $Re = \epsilon Re_\omega = U_\infty^* L/\nu$. Similarly, the streaming Reynolds number is given by $R_s = \epsilon^2 Re_\omega = (U_\infty^*)^2/(\omega\nu)$. When heat is transferred, a further length scale related to the penetration depth of the thermal boundary layer, namely, $(\kappa/\omega)^{1/2}$, must be considered, where κ is the thermal diffusivity of the fluid. The squared ratio of the viscous and thermal diffusive lengths is the Prandtl number, $Pr = \nu/\kappa$. Another important dimensionless parameter is the magnetic Reynolds number $Rm = \mu\sigma U_\infty L$, where μ and σ are the magnetic permeability and electrical conductivity of the fluid, respectively. Here, we assume that Rm is much smaller than unity, which implies that the induced magnetic field is much smaller than the applied field.

In what follows, assuming the incompressibility condition and the low magnetic Reynolds number approximation, three uncoupled problems, namely, dynamic, magnetic and thermal, will be established.

2.1. DYNAMIC PROBLEM

In dimensionless terms, the incompressibility condition is given by the equation

$$\frac{\partial u}{\partial x} + \frac{\partial v}{\partial y} = 0, \tag{1}$$

where the velocity components in x- and y-directions, u and v, respectively, are normalized with U_∞^*. Likewise, the coordinates (x, y, z), are normalized by L which, for instance, may be a representative dimension of the oscillating boundary. Because of Eq. (1), the velocity components can be obtained as the real part of the following expressions

$$u = \frac{\partial \psi}{\partial y}, \qquad v = -\frac{\partial \psi}{\partial x}, \tag{2}$$

where ψ is the stream function. The equation satisfied by ψ can be derived by applying the curl operator to the momentum balance equation. In dimensionless terms, we get

$$\frac{\partial}{\partial t}(\nabla^2\psi) - \epsilon\frac{\partial(\psi, \nabla^2\psi)}{\partial(x,y)} = \frac{1}{Re_\omega}\nabla^4\psi + \frac{M^2}{Re_\omega}\left[\frac{\partial(j_z B_x)}{\partial x} + \frac{\partial(j_z B_y)}{\partial y}\right]. \tag{3}$$

where the time t, the magnetic field $\mathbf{B} = (B_x, B_y, 0)$ and the current density j_z are normalized by $1/\omega$, B_o^* and $\sigma U_w^* B_o^*$, respectively. $\mathbf{B}$ comprises both the applied field as well as the perturbation due to the oscillatory motion of the fluid. Evidently, Eq. (3) corresponds to the third component of the equation for the vorticity, $\nabla \times \mathbf{u} = (0, 0, -\nabla^2\psi)$, where $\mathbf{u} = \nabla \times (0, 0, \psi)$ is the velocity vector. The dimensionless parameter $M = B_o^* L\sqrt{\sigma/\rho\nu}$, is the Hartmann number, where ρ is the mass density of the fluid. The boundary conditions that must be satisfied by Eq. (3) are the non-slip condition on the wall and the vanishing of the velocity as the distance from the wall tends to infinity. In terms of the velocity vector, these conditions are

$$\left.\begin{array}{ll} \mathbf{u} = U_o(x)e^{it}\hat{\mathbf{x}}, & \text{on the wall,} \\ \mathbf{u} \to 0, & \text{as } y \to \infty. \end{array}\right\} \tag{4}$$

Further, Eq. (3) is coupled to the magnetic quantities through the curl of the Lorentz force term. In addition to Eqs. (3-4), the system must satisfy

$$\epsilon\nabla \times \mathbf{E} = -\frac{\partial \mathbf{B}}{\partial t}, \qquad \nabla \times \mathbf{B} = Rm\mathbf{j}, \tag{5}$$

$$\nabla \cdot \mathbf{B} = 0, \qquad \nabla \cdot \mathbf{j} = 0, \tag{6}$$

$$\mathbf{j} = \mathbf{E} + \mathbf{u} \times \mathbf{B}, \tag{7}$$

where $\mathbf{E} = (E_x, E_y, E_z)$ stands for the electric field which is normalized by $U_w^* B_o^*$. Eqs. (5)-(7) are the Faraday's law of induction, Ampère's law, magnetic Gauss law, conservation of current condition and Ohm's law, respectively. In order to uncouple the flow problem from the magnetic one, we assume that the low-magnetic Reynolds number approximation holds. Hence, we express the magnetic field as a perturbation expansion on Rm, namely,

$$\mathbf{B} = \hat{\mathbf{y}} + Rm\mathbf{b} + O(Rm^2) \tag{8}$$

where $\mathbf{b}$ is the perturbation to the applied transverse field. By introducing the expansion (8) on Faraday's law, we find that at $O(1)$, $\nabla \times \mathbf{E} = 0$, where we have to assume that Rm is $O(\epsilon^2)$ or smaller. This means that, to first approximation, the electric field is derivable from a potential, namely, $\mathbf{E} = -\nabla\phi$. Also, to this approximation the curl of the Lorentz force in Eq. (3) reduces to $\partial j_z/\partial y$. As the magnetic field is contained in the plane where the two-dimensional flow takes place, the flow in this region is not affected by the electric field (which is determined by the conditions on the boundaries at $z = \pm\infty$) since E_z is constant [18]; then the curl of the Lorentz force is simply $\partial^2\psi/\partial y^2$, and the flow and magnetic problems are uncoupled. Here, we first address the dynamic flow problem and return to the determination of the induced magnetic field in subsection 2.2.

For arbitrary values of Re_ω and M, we assume that ψ can be expressed as a perturbation expansion on the small parameter ϵ, namely,

$$\psi(x, y, t, \epsilon, Re_\omega, M) = \psi_0(x, y, t, Re_\omega, M) + \epsilon\psi_1(x, y, t, Re_\omega, M) + O(\epsilon^2), \tag{9}$$

where the subindexes 0 and 1 denote the first and second approximations, respectively. Since $\epsilon \ll 1$, as a first approximation we may neglect the convective non-linear terms in Eq. (3) and look for a solution of the resulting linear equation for ψ_0, which is proportional to $\cos t$. The equation for the second approximation will then include forcing terms proportional to $\cos^2 t = \frac{1}{2}(1 + \cos 2t)$. Hence, in addition to periodic terms of twice the original frequency, the nonlinear contribution introduces time-independent terms that are responsible of the steady streaming phenomenon. With these considerations, we decompose ψ_1 in the form

$$\psi_1(x, y, t, Re_\omega, M) = \psi_1^{(s)}(x, y, Re_\omega, M) + \psi_1^{(u)}(x, y, t, Re_\omega, M), \tag{10}$$

where the superscripts (s) and (u) stand for the steady and unsteady parts, respectively. If we introduce Eqs. (9-10) in Eq. (3) and equate coefficients of powers of ϵ, we get

$$\frac{\partial}{\partial t}(\nabla^2\psi_0) - \frac{M^2}{Re_\omega}\frac{\partial^2\psi_0}{\partial y^2} = \frac{1}{Re_\omega}\nabla^4\psi_0, \tag{11}$$

$$\frac{\partial}{\partial t}(\nabla^2\psi_1^{(u)}) - \frac{1}{Re_\omega}\nabla^4\psi_1^{(u)} - \frac{M^2}{Re_\omega}\frac{\partial^2\psi_1^{(u)}}{\partial y^2} = \left[\frac{\partial(\psi_0, \nabla^2\psi_0)}{\partial(x,y)}\right]^{(u)}, \tag{12}$$

$$-\frac{1}{Re_\omega}\nabla^4\psi_1^{(s)} - \frac{M^2}{Re_\omega}\frac{\partial^2\psi_1^{(s)}}{\partial y^2} = \left[\frac{\partial(\psi_0, \nabla^2\psi_0)}{\partial(x,y)}\right]^{(s)}. \tag{13}$$

Notice that the left-hand side of Eqs. (11-13) includes the Hartmann braking term that leads to the suppression of vorticity. Since we are interested in boundary-layer solutions, we introduce the approximation $\nabla^2\psi \approx \partial^2\psi/\partial y^2$, which can be formally obtained by rescaling the y-coordinate with the boundary layer thickness and assuming $Re_\omega \gg 1$ [11]. This ensures the existence of a Stokes shear-wave layer near the wall. The boundary conditions to be satisfied by Eq. (11) are

$$\left.\begin{aligned} \tfrac{\partial\psi_0}{\partial y} = U_o(x)e^{it}, \quad \psi_0 = 0, \quad &\text{at } y = 0, \\ \tfrac{\partial\psi_0}{\partial y} \to 0, \quad &\text{as } y \to \infty. \end{aligned}\right\} \tag{14}$$

Under this conditions, the first order stream function is

$$\psi_0(x,y,t) = \frac{U_o(x)}{\alpha + i\beta}(1 - e^{-(\alpha+i\beta)y})e^{it}, \tag{15}$$

where

$$\left.\begin{matrix}\alpha \\ \beta\end{matrix}\right\} = \left[\frac{\sqrt{M^4 + Re_\omega^2} \pm M^2}{2}\right]^{\frac{1}{2}}. \tag{16}$$

Solution (15) leads to a harmonically fluctuating vorticity, ω_{z0}, of the same frequency as the oscillating wall, namely,

$$\omega_{z0} = U_o(x)e^{-\alpha y}(\alpha\cos(\beta y - t) + \beta\sin(\beta y - t)).$$

From an order of magnitude analysis, it is found that the first order vorticity generated at the surface of the wall is confined to a thin boundary layer of thickness $\delta_0 \approx 1/\sqrt{Re_\omega + M^2}$, where δ_0 is normalized by L. Evidently, for $M \to 0$, the solution (15) reduces to the classical solution of the Stokes' second problem with $\delta_0 \approx 1/\sqrt{Re_\omega}$. On the other hand, notice that for $M \to \infty$ the first order vorticity is suppressed.

Using solution (15), we now turn to solve for the second approximation. Under the boundary-layer approximation, Eqs. (12-13) must satisfy the following boundary conditions

$$\left.\begin{array}{ll} \psi_1 = \frac{\partial \psi_1}{\partial y} = 0, & \text{at } y = 0, \\ \frac{\partial \psi_i}{\partial y} \to 0, & \text{as } y \to \infty. \end{array}\right\} \tag{17}$$

Hence, the periodic and steady solutions are

$$\begin{aligned} \psi_1^{(u)}(x,y,t,Re_\omega,M) &= U_o\frac{dU_o}{dx}e^{2it}\left\{\frac{i}{2\lambda_1}(e^{-\lambda_1 y}-1) - \frac{e^{-\alpha y}}{2\sqrt{M^4+Re_\omega^2}}\right. \\ &\quad \left.[(\beta+i\alpha)(\cos\beta y - 1) + (\alpha - i\beta)\sin\beta y]\right\}, \end{aligned} \tag{18}$$

$$\begin{aligned} \psi_1^{(s)}(x,y,Re_\omega,M) &= U_o\frac{dU_o}{dx}\frac{Re_\omega}{2\sqrt{M^4+Re_\omega^2}}\left\{\frac{\sqrt{M^4+Re_\omega^2}-M^2}{2\sqrt{M^4+Re_\omega^2}+M^2}\right. \\ &\quad \left[\frac{1}{M}(e^{-My}-1) - \frac{1}{2\alpha}(e^{-2\alpha y}-1)\right] + \frac{e^{-\alpha y}}{\sqrt{M^4+Re_\omega^2}} \\ &\quad \left[\left(\frac{M^2}{Re_\omega}\beta - \alpha\right)(\cos\beta y - 1) + \left(\frac{M^2}{Re_\omega}\alpha + \beta\right)\sin\beta y\right] \\ &\quad \left.+\frac{1}{M}(e^{-My}-1)\right\}, \end{aligned} \tag{19}$$

where $\lambda_1 = \sqrt{M^2 + 2iRe_\omega}$. Solutions (18-19) represent the fluid response to the nonlinear convective effects given by right-hand side terms in Eqs. (12-13). In particular, the convective term in Eq. (13) can be interpreted as a rotational *body force* that, acting as a source of vorticity, generates the steady flow in the boundary layer. It is interesting to notice that, differently from the hydrodynamic case, solution (19) leads to the vanishing of the steady streaming velocity, given by

$$\begin{aligned} u_1^{(s)}(x,y,Re_\omega,M) &= \epsilon U_o\frac{dU_o}{dx}\frac{Re_\omega}{2\sqrt{M^4+Re_\omega^2}}\left\{\frac{\sqrt{M^4+Re_\omega^2}-M^2}{2\sqrt{M^4+Re_\omega^2}+M^2}\right. \\ &\quad [e^{-2\alpha y} - e^{-My}] + e^{-\alpha y}(\cos\beta y - \frac{M^2}{Re_\omega}\sin\beta y) \\ &\quad \left.-e^{-My}\right\}, \end{aligned} \tag{20}$$

as $y \to \infty$. This implies that $u_1^{(s)}$, as well as the second order steady vorticity $\omega_{z1}^{(s)}$, are confined to the Stokes shear-wave layer. When $M = 0$, $u_1^{(s)}$ reduces to the hydrodynamic solution, namely,

$$u_1^{(s)} = \epsilon U_o \frac{dU_o}{dx}\left(-\frac{3}{4} + \frac{1}{4}e^{-\eta} + \frac{1}{2}e^{-\eta}\cos\eta\right), \tag{21}$$

where $\eta = (Re_\omega/2)^{1/2}y$. In this case, the streaming velocity does not vanish at the outer edge of the boundary layer but tends to the well known result $-(3/4)(\epsilon U_o dU_o/dx)$ [1]. This finite velocity may be considered as the inner boundary condition for an outer flow. Then, the steady motion in the Stokes layer generates a slow transfer of vorticity that eventually leaves the inner layer, producing a second order steady vorticity beyond this layer [9]. The transport of vorticity in the outer flow is basically a viscous diffusion process when $R_s \ll 1$, while convective transport is dominant for $R_s \gg 1$. In contrast, the magnetic field inhibits the transfer of vorticity to the outer flow, confining both the periodic and steady motions to the inner layer. In the limit case, as $M \to \infty$, $u_1^{(s)}$ and consequently $\omega_{z1}^{(s)}$, vanish. This behavior is clearly shown in Fig. 1 where the streaming velocity is plotted as a function of the y-coordinate for $Re_\omega = 10$ and M ranging from 0 to 10. As discussed in [17], where an oscillating flow with a fixed boundary was analyzed, the transport of vorticity in the boundary layer can be understood as a competition of the transfer of vorticity away from the boundary by inertial effects and the suppression of vorticity by the magnetic field. The dimensionless parameter that governs this phenomenon is the *oscillation* interaction parameter $N_\omega = M^2/Re_\omega = \sigma B_o^2/\rho\omega$ which can be interpreted as the ratio of the characteristic time associated to the inertial vorticity transport to the (Joule) time of the magnetic suppression mechanism [17]. From an order of magnitude analysis of the steady momentum balance equation, taking U_∞^* as the characteristic velocity, the thickness of the second order streaming boundary-layer can be estimated as $\delta_1^{(s)} = 1/(Re + M^2)^{1/2}$. If we compare the first and second order thicknesses, we get $\delta_0/\delta_1^{(s)} = \sqrt{(\epsilon + N_\omega)/(1 + N_\omega)}$. Then, for the non MHD case ($N_\omega = 0$), $\delta_0 = \sqrt{\epsilon}\delta_1^{(s)}$, while for $N_\omega \gg 1$, $\delta_0 = \delta_1^{(s)}$, since for strong magnetic field interaction both thicknesses tend to disappear. If the streaming velocity $U_s^* = (U_\infty^*)^2/\omega L$ is taken as the characteristic velocity for dimensional analysis, the estimated thickness of the streaming boundary layer is $\delta_1^{(s)} = 1/(R_s + M^2)^{1/2}$. This could be a good estimation when the magnetic interaction is very week. In fact, for $M = 0$, this thickness reduces to that of the outer hydrodynamic streaming flow [1].

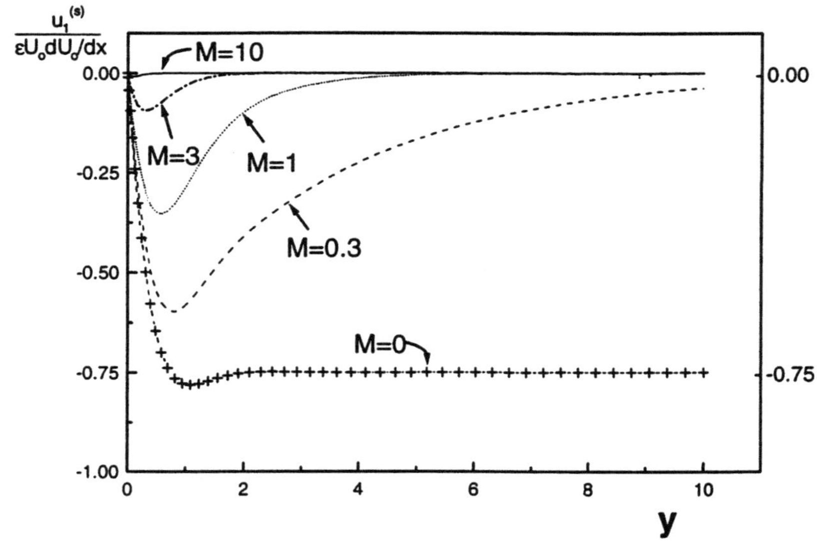

Figure 1. Steady streaming as a function of the y-coordinate for different Hartmann numbers and $Re_\omega = 10$.

2.2. MAGNETIC PROBLEM

Let us now address the magnetic problem. As it was previously noticed, to first approximation the curl of the electric field vanishes. For the sake of simplicity, we assume that perfect short-circuits are formed, so that the three electrostatic field components are zero everywhere [19]. Notice that it is only the first approximation of $\mathbf{E}$ which is assumed to vanish. As we mentioned before, the value of the electrostatic field does not affect the two-dimensional flow. Once (8) is introduced in Eqs. (5) and (7), the following equation for the perturbation field is found at $O(1)$,

$$\frac{\partial b_y}{\partial x} - \frac{\partial b_x}{\partial y} = u. \tag{22}$$

Because of the magnetic Gauss law, a magnetic flux function, $\chi(x, y, t)$, can be defined as

$$b_x = \frac{\partial \chi}{\partial y}, \qquad b_y = -\frac{\partial \chi}{\partial x}. \tag{23}$$

Therefore, the function χ satisfies the equation

$$\frac{\partial^2 \chi}{\partial x^2} + \frac{\partial^2 \chi}{\partial y^2} = -\frac{\partial \psi}{\partial y}. \tag{24}$$

Assuming the boundary layer approximation, Eq. (24) can be integrated once to give

$$b_x = \frac{\partial \chi}{\partial y} = -\psi(x, y, t, \epsilon, Re_\omega, M) + C_1,$$

but, since the wall is electrically insulated, $\partial\chi/\partial y = 0$ at $y = 0$, therefore, $C_1 = 0$. A second integration leads to

$$\chi = -\int \psi dy + C_2,$$

where C_2 can be determined from the condition $\chi = 0$ at $y = 0$. It is clear that, as well as the flow, the perturbed magnetic field consists of harmonic and time independent terms.

2.3. HEAT TRANSFER PROBLEM

The dimensionless heat transfer equation with viscous and ohmic dissipation reads

$$\frac{\partial \Theta}{\partial t} - \epsilon \frac{\partial(\psi, \Theta)}{\partial(x, y)} = \frac{1}{Pr Re_\omega} \nabla^2 \Theta + \frac{Ec}{Re_\omega} \Phi, \tag{25}$$

where $\Theta = (T^* - T^*_\infty)/(T^*_w - T^*_\infty)$ is the dimensionless fluid temperature, $Ec = U^{*2}_\infty / c_p (T^*_w - T^*_\infty)$ is the Eckert number and Φ is the dissipation function that comprises viscous and ohmic effects. Here, T^* and c_p are the dimensional temperature and heat capacity of the fluid, respectively. The boundary conditions to be satisfied by Eq. (25) are

$$\left.\begin{array}{ll} \Theta = 1 & \text{at } \ y = 0, \\ \Theta \to 0, & \text{as } \ y \to \infty. \end{array}\right\} \tag{26}$$

In addition, we assume that temperature dependence of the density and diffusivities, as well as natural convection effects, may be neglected. Similarly to the stream function, the temperature field is expressed as a perturbation expansion on the small parameter ϵ, namely,

$$\Theta = \Theta_0 + \epsilon \Theta_1 + \epsilon^2 \Theta_2 + O(\epsilon^3), \tag{27}$$

where, $\Theta_i = \Theta_i(x, y, t, Re_\omega, M, Pr, Ec)$. Since we are interested in electrically conducting fluids for which the Prandtl number is much smaller than unity, we assume that $Pr \ll 1$ and $Re_\omega \gg 1$, in such a way that $Pr Re_\omega = O(1)$; in addition, we take $Ec/Re_\omega = O(1)$. Hence, introducing expansions (9) and (27) in Eq. (25) and equating coefficients of powers of ϵ, we find that equations to $O(1)$ and $O(\epsilon)$ are, respectively,

$$\frac{\partial \Theta_0}{\partial t} - \frac{1}{PrRe_\omega}\nabla^2\Theta_0 = \frac{Ec}{Re_\omega}\Phi_0, \tag{28}$$

$$\frac{\partial \Theta_1}{\partial t} - \frac{1}{PrRe_\omega}\nabla^2\Theta_1 = \frac{\partial(\psi_0, \Theta_0)}{\partial(x,y)} + \frac{Ec}{Re_\omega}\Phi_1. \tag{29}$$

Notice that in the previous equations, the heat transfer is a purely diffusive process. In other words, convection heat transfer due to streaming flow is absent in this case. Since both viscous and ohmic dissipation involve squared terms of the velocity or its derivative, the right-hand side of equation (28) contains $\cos^2 t$ terms. Hence, in a similar way as in the second order flow calculation, terms proportional to $\cos 2t$ and steady-state terms will appear. On the other hand, Eq. (29) leads only to harmonic temperature contributions with frequencies of one and three times the original oscillation frequency; these are not considered here. For Eq. (28), we seek a solution of the form

$$\Theta_0 = \Theta_0^{(u)} + \Theta_0^{(s)}. \tag{30}$$

Introducing (30) in Eq. (28) and applying the boundary-layer approximation, we find that the unsteady contribution satisfies

$$\frac{\partial \Theta_0^{(u)}}{\partial t} - \frac{1}{PrRe_\omega}\frac{\partial^2 \Theta_0^{(u)}}{\partial y^2} = \frac{Ec}{Re_\omega}\left[\left(\frac{\partial^2 \psi_0}{\partial y^2}\right)^2 + M^2\left(\frac{\partial \psi_0}{\partial y}\right)^2\right]^{(u)}, \tag{31}$$

with boundary conditions $\Theta_0^{(u)} = 0$ at $y = 0$ and $\Theta_0^{(u)} \to 0$ as $y \to \infty$. The solution is

$$\begin{aligned}\Theta_0^{(u)} \quad = \quad & -\frac{PrEc\, U_o^2(x)(2M^2 + iRe_\omega)e^{2it}}{4(M^4 + Re_\omega) - PrRe_\omega(PrRe_\omega + 4iM^2)} \\ & [(M^2 - iRe_\omega(1 + Pr/2))(e^{-2\alpha y}\cos 2\beta y - e^{-\lambda y}) \\ & -(Re_\omega(1 + Pr/2) + iM^2)e^{-2\alpha y}\sin 2\beta y],\end{aligned} \tag{32}$$

where $\lambda_0 = \sqrt{PrRe_\omega}(1 + i)$. From an order of magnitude analysis, the thickness of the unsteady thermal boundary layer can be estimated as $\delta_{0T}^{(u)} = 1/\sqrt{Pr(Re_\omega + Ec(1 + M^2))}$. Since it is assumed that $Pr \ll 1$, the thermal boundary layer is much thicker than the dynamic layer δ_0. In turn, the steady contribution satisfies

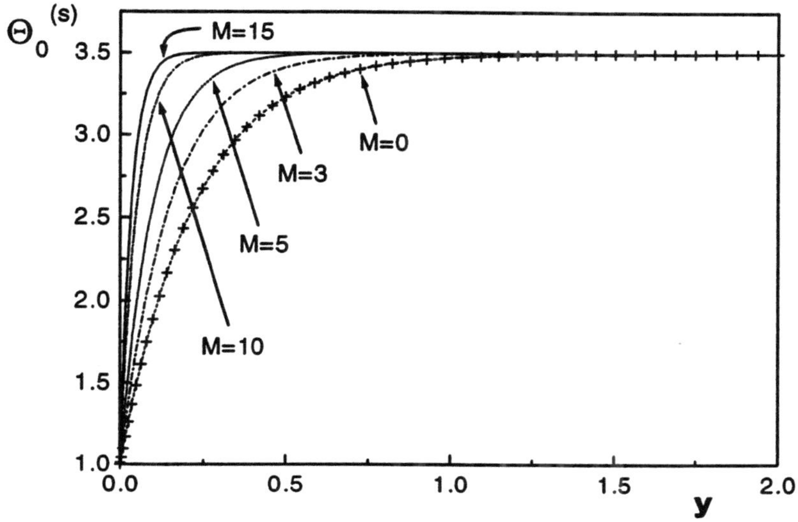

Figure 2. Steady temperature as a function of the y-coordinate for different Hartmann numbers. $Re_\omega = 10$, $Pr = 0.05$, $Ec = 100$.

$$-\frac{1}{PrRe_\omega}\frac{\partial^2\Theta_0^{(s)}}{\partial y^2} = \frac{Ec}{Re_\omega}\left[\left(\frac{\partial^2\psi_0}{\partial y^2}\right)^2 + M^2\left(\frac{\partial\psi_0}{\partial y}\right)^2\right]^{(s)}, \tag{33}$$

with boundary conditions $\Theta_0^{(s)} = 1$ at $y = 0$ and $\Theta_0^{(s)} \to 0$ as $y \to \infty$. However, it is not possible to satisfy both conditions simultaneously. Therefore, as in the hydrodynamic problem [1], we satisfy the condition at $y = 0$ and relax the condition at infinity. In this form, we get

$$\Theta_0^{(s)} = 1 - \frac{U_o^2(x)PrEc}{2}(e^{-2\alpha y} - 1). \tag{34}$$

In Fig. 2, solution (34) is shown for different Hartmann numbers. This steady solution indicates that a net heat transfer can take place in the boundary layer. Far from the wall, the steady temperature tends to the finite value $\lim_{y\to\infty}\Theta_0^{(s)} = 1 + U_o^2(x)PrEc/2$, which will be reached even in the non-MHD case. The fact that the temperature does not vanish as $y \to \infty$, implies the existence of a second thermal region within which the steady temperature adjusts from the previous value, taken at the outer edge of the inner layer, to zero. In the outer thermal region, the oscillatory flow vanishes, so that viscous and ohmic dissipation are practically absent and, therefore, the thermal behavior is governed by Laplace equation.

The impossibility of satisfying the thermal boundary condition at infinity in the previous case, may be related to the absence of convective transfer in the boundary layer. Once this transfer mechanism is included, a balance between convective and diffusive processes is expected to arise. In order to consider the action of the streaming flow in the heat transfer, the case $1/(PrRe_\omega) = O(\epsilon^2)$ must be considered. Accordingly, the heat transfer equation can be written as

$$\frac{\partial \Theta}{\partial t} - \epsilon \frac{\partial(\psi, \Theta)}{\partial(x,y)} = \frac{\epsilon^2}{PrR_s} \nabla^2 \Theta + \frac{\epsilon^2 Ec}{R_s} \Phi. \tag{35}$$

In fact, we assume that $Pr \ll 1$ and $R_s = \epsilon^2 Re_\omega \gg 1$, in such a way that $PrR_s = O(1)$; as well, we assume that $Ec/R_s = O(1)$. Substituting (9) and (27) in Eq. (35) and equating coefficients of powers of ϵ, we get

$$\frac{\partial \Theta_0}{\partial t} = 0, \tag{36}$$

$$\frac{\partial \Theta_1}{\partial t} - \frac{\partial(\psi_0, \Theta_0)}{\partial(x,y)} = 0, \tag{37}$$

$$\frac{\partial \Theta_2}{\partial t} - \frac{\partial(\psi_1, \Theta_0)}{\partial(x,y)} - \frac{\partial(\psi_0, \Theta_1)}{\partial(x,y)} = \frac{1}{PrR_s} \nabla^2 \Theta_0 + \frac{Ec}{R_s} \Phi_0. \tag{38}$$

Since Eq. (36) indicates that Θ_0 is a time-independent function and ψ_0 is a harmonic function of time, it is possible to integrate Eq. (37) once to obtain the form of Θ_1 as [14]

$$\Theta_1 = -i f_1 e^{it} + f_2, \tag{39}$$

where $f_i = f_i(x, y, \epsilon, Re_\omega, M, Pr, Ec)$. Using (39), we can separate the steady and time dependent parts of Eq. (38), to get the equation that satisfies $\Theta_0^{(s)}$, namely,

$$-\frac{\partial(\psi_1^{(s)}, \Theta_0^{(s)})}{\partial(x,y)} = \frac{1}{PrR_s} \nabla^2 \Theta_0^{(s)} + \frac{Ec}{R_s} \Phi_0, \tag{40}$$

subject to boundary conditions

$$\left. \begin{array}{ll} \Theta_0^{(s)} = 1, & \text{at } \ y = 0, \\ \Theta_0^{(s)} \to 0, & \text{as } \ y \to \infty. \end{array} \right\} \tag{41}$$

In Eq. (40), the heat transfer due to diffusion and convection by streaming flow are equally important. In fact, both mechanisms determine the mean first order heat transfer in the boundary layer. The solution of Eqs. (40-41) requires a numerical solution that will not be presented here.

3. Concluding Remarks

The transport of momentum and heat induced by a slightly curved wall oscillating in a quiescent conducting fluid under a transverse magnetic field was analyzed. Assuming oscillations of small amplitude, a perturbation method was used to calculate the dynamic response of the boundary layer. This comprises a primary flow oscillating with the same frequency as that of the wall and a secondary flow composed by a periodic contribution (oscillating with twice the original frequency) and the steady streaming motion. The net transfer of momentum associated to the streaming flow arises as a result of nonlinear convective interactions of the first harmonic velocities in the Stokes layer. The MHD streaming flow problem provides a neat example of the magnetic suppression of vorticity. In fact, the field prevents the transfer of vorticity away from the boundary, confining the motion to the inner layer. The parameter N_ω governs the competition between the inertial transfer of vorticity and the magnetic suppression mechanism. For a strong enough magnetic field, both the first order harmonic vorticity as well as the second order streaming vorticity can be completely annihilated. This result is consistent with numerical simulations that predict the suppression of an oscillatory motion in a Czochralski melt through an axial magnetic field [3]. Assuming perfect short-circuits, the perturbation of the applied magnetic field was determined using the approximation $Rm \ll 1$ and the electrically insulating wall condition. The perturbed field presents the same structure as the boundary layer flow. The transport of heat was analyzed for low-Prandtl number fluids considering that both diffusive and dissipative terms in the heat transfer equation are $O(1)$. This leads to a purely diffusive transport in which the convective action of the streaming flow is absent. As a consequence of dissipation, the perturbation solution leads to a first order temperature field which comprises a periodic and time-independent contributions. The net heat flux associated to the steady field can be increased by increasing the Eckert number and its effect on the wall heat transfer may be important. The steady temperature field does not disappear as the distance from the wall increases; consequently, in order to satisfy the boundary condition at infinity, an outer thermal region must be solved. In the motionless outer region the thermal behavior is governed by the Laplace equation. The impossibility of satisfying the condition at the edge of the thermal boundary layer may be related to the absence of convective transport. The inclusion of convection by streaming flow may provide a balance to the diffusive process in the boundary layer. This requires diffusive and dissipative terms to be $O(\epsilon^2)$ in the heat transfer equation.

Finally, although we have only considered streaming effects produced by vibrations of a solid curved wall, it is expected that this theory can also be useful when the oscillatory perturbation is of different origin, in particular, of acoustic origin. It appears that, in both cases, the frictional dissipation originated by the oscillating tangential relative velocity between the fluid and the solid wall, takes place in a Stokes boundary layer that presents the same structure [20].

Acknowledgment. This research was supported by DGAPA-UNAM and CONACYT under projects IN103797 and G0044E, respectively.

References

1. Telionis, D. P. (1981) *Unsteady Viscous Flows*, Springer-Verlag, New York.
2. Swift, G. W. (1988) A liquid-metal magnetohydrodynamic acoustic transducer, *J. Acoust. Soc. Am.*, **83**, 350-361.
3. Munakata, T. and Tanasawa, I. (1990) Onset of oscillatory flow in Czochralski growth melt and its suppression by the magnetic field, *J. Crystal Growth*, **106**, 566-576.
4. Lighthill, M. J. (1954) The response of laminar skin friction and heat transfer to fluctuations in the stream velocity, *Proc. Roy. Soc. A*, **224**, 1-23.
5. Gould, R. K. (1961) Effects of acoustic streaming on heat transfer, Ph.D. Thesis, Brown Univ., Providence, Rhode Island.
6. Rayleigh, Lord, (1884) On the circulations of air observed in Kundt's tubes and on some allied acoustical problems, *Phil. Trans. A*, **175**, 1. Rayleigh, Lord, (1945) *The Theory of Sound*, vol. 2, Dover, New York, 333-342.
7. Faraday, M. (1831) *Phil. Trans. Roy. Soc. London*, **121**, 229.
8. Schlichting, H. (1932) Berechnung ebner periodischer Grenzschichtstromungen, *Phys. Z.*, **33**, 327-335.
9. Stuart, J. T. (1966) Double boundary layers in oscillatory viscous flow, *J. Fluid Mech.*, **24**, 673-687.
10. Riley, N. (1965) Oscillating viscous flow, *Mathematika*, **12**, 161-175.
11. Riley, N. (1967) Oscillatory viscous flows. Review and extension, *J. Inst. Math. Appl.*, **3**, 419-434.
12. Riley, N. (1975) Unsteady boundary layers, *SIAM Review*, **17**, 274-29.
13. Richardson, P. D. (1967) Heat transfer from a circular cylinder by acoustic streaming, *J. Fluid Mech.*, **30**, 337-355.
14. Davidson, B. J. (1973) Heat transfer from a vibrating circular cylinder, *Int. J. Heat Mass Transfer*, **16**, 1703-1727.
15. Gopinath, A. and Mills, A. F. (1993) Convective heat transfer from a sphere due to acoustic streaming, *ASME J. Heat Transfer*, **115**, 332-341.
16. Gopinath, A. and Mills, A. F. (1994) Convective heat transfer due to acoustic streaming across the ends of a Kundt tube, *ASME J. Heat Transfer*, **116**, 47-53.
17. Cuevas, S and Ramos, E. (1997) Steady streaming in oscillatory viscous flow under a transverse magnetic field, *Phys. Fluids*, **9**, 1430-1434.
18. Moffatt, H. K. and Toomre, J. (1967) The annihilation of a two-dimensional jet by a transverse magnetic field, *J. Fluid Mech.*, **30**, 65-82.
19. Moreau, R. (1990) *Magnetohydrodynamics*, Kluwer Academic Publishers, Dordrecht.
20. Lighthill, J. (1978) Acoustic streaming, *J. Sound Vibration*, **61**, 391-418.

ROADS TO TURBULENCE FOR AN INTERNAL MHD BUOYANCY-DRIVEN FLOW DUE TO A HORIZONTAL TEMPERATURE GRADIENT

L. DAVOUST, R. MOREAU and R. BOLCATO
EPM-MADYLAM, ENSHMG BP 95, 1340 rue de la piscine,
Domaine universitaire, 38402 St Martin d'Hères, France

Transition to chaos for a magnetohydrodynamic flow driven by a horizontal temperature gradient is experimentally studied. In the core of a horizontal cylinder filled with mercury, the buoyancy-driven flow destabilises when decreasing the applied vertical magnetic field $\mathbf{B_o}$ (Hartmann number steps: δHa~0.021). For a moderate Grashof number ($Gr_{\Delta T}$~10^4-10^5), transition is characterised by time-dependent temperature oscillations whose features are made evident from spatial correlations on temperature measurements. The destabilisation of the flow gives simultaneously rise to two waves. One horizontal transversal wave is detected within the central cross-sections whereas the other is travelling along the axis of the enclosure. Since temperature and velocity fields are non-linearly coupled, these two waves, which arise from a sub-critical bifurcation, exhibit consistently the same frequency f_1. Decreasing more $\mathbf{B_o}$, a Hopf supercritical bifurcation yields a third and last axial standing wave (low frequency f_2) before the arising of chaos. For a larger Grashof number ($Gr_{\Delta T}>10^6$), transition is dramatically modified by the presence of stabilisation or oscillating windows whose occurence is dependent both on space and on the control parameter Ha.

1. Introduction

This paper aims at presenting recent experimental results on the transition to turbulence for a magnetohydrodynamic (MHD) flow, driven by a horizontal temperature gradient $\mathbf{G}$ and submitted to a permanent vertical magnetic field $\mathbf{B_o}$. This flow is investigated in the central region of a horizontal small aspect-ratio cylinder, $\varepsilon=R_o/L$, where R_o and L denote the radius and the length (figure 1). The temperatures T_1 and T_2 relate to the cold and the hot side end walls.

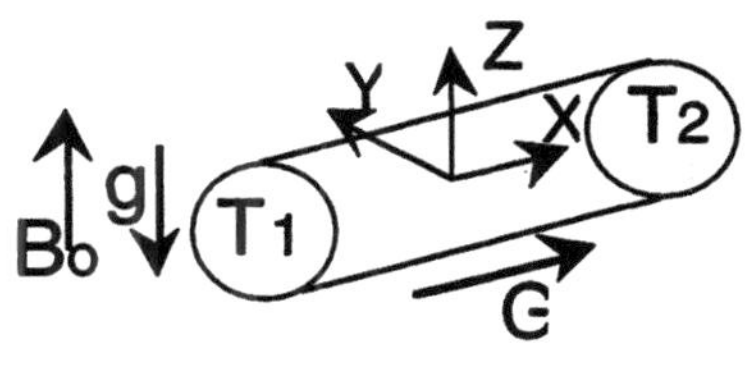

Figure 1. Configuration under study.

Such a study is relevant to the horizontal Bridgman crystal growth technics. It is now widely known that unsteady instabilities, which develop within the melt and vanish in a steady magnetic field (Hurle *et al*, 1974), are responsible for striations within crystals so

A. Alemany et al. (eds.),
Transfer Phenomena in Magnetohydrodynamic and Electroconducting Flows, 123–133.

built. Some previous theoretical, numerical and experimental developments (Cowley, 1995, Davoust *et al*, 1997) allow to consider the laminar core region of the flow to be known: it consists of a main recirculating loop - the so-called Hadley recirculation - on which are superimposed four additionnal 3-D vortices localised within the four quadrants of the middle cross-sections. The linear stability study of such an internal MHD buoyancy-driven flow was recently carried out in a parallelepipedic cavity (Bojarevics, 1995). Both steady 2-D modes distributed along the X-axis and 3-D oscillating modes distributed along the Y-axis were predicted. It was also found that only a Hartmann number of 10 is sufficient in stabilising the flow as $Gr_{\Delta T}$ is of the order of 10^4. The symbol $Gr_{\Delta T}$ designates the Grashof number built from the imposed temperature difference ΔT ($=T_2-T_1$) :

$$Gr_{\Delta T} = \frac{\beta|\mathbf{g}|\Delta T\, R_0^4}{L\,\upsilon^2},$$

where $\mathbf{g}$, β, ν stand for the gravity, the volumetric expansion coefficient and the molecular viscosity.

Up to now, only a few experimental investigations of liquid metals convective flows in H-G configurations have been performed, the symbol H-G referring to as **H**orizontal **G**radient. Let's quote Pratte's *et al* (1990) and Braunsfurth and Mullin's (1996) investigations for which a H-G parallelepipedic enclosure filled with gallium is involved. According to Pratte *et al* (1990), the aspect-ratios (width or height over length) modifies both the critical Grashof number $Gr^*_{\Delta T}$, above which time-dependent oscillations arise, and the nature itself of transition. Moreover, transition also occurs varying the Prandtl number (Braunsfurth and Mullin, 1996).

The whole of these experimental studies suffer somewhat of an ill-defined boundary condition: some additional Marangoni effects are more than likely to be expected where a thin layer of silicon oil is supposed to provide a thermally insulating condition on the upper bound of the enclosure. For the sake of comparison with numerics (Ben Hadid *et al*, 1997), care is here devoted to truly achieve *ad hoc* boundary conditions.

2. Experimental facilities

The investigation is performed on the MASCOT experiment (***MA**gnetic **S**tabilisation of **CO**nvection and **T**urbulence*) involving a horizontal cylindrical cell (radius 2cm, length 40cm) filled with mercury and submitted to a permanent vertical magnetic field $\mathbf{B_0}$. Two copper disks stand for the two side end walls whose temperatures T_1 (cold wall) and T_2 (hot wall) are controlled with a precision of ±0.01K. The cylindrical longitudinal wall, made of glass, is surrounded by, respectively, a thick copper wall and a thermally insulating coating. To investigate the stabilisation/destabilisation of the flow, a computer allows the temperature recordings by the way of a HP3852A data acquisition system and simultaneously, pilots the increase/decrease of $\mathbf{B_0}$, step by step (figure 2). Each step of the Hartmann number is such that $\delta Ha=0.021$.

The experimental cell is equipped with 55 thermocouples distributed on eleven half-circles at the longitudinal wall. In addition, a movable device, involving two independent radial ranks equipped with three thermocouples, is introduced within the liquid domain from the hot end. This allows the characterisation of the time-dependent oscillations since the two ranks are able to rotate and to translate from an independent manner. A close inspection is devoted to the ascertainment that neither electric nor mechanical perturbations of the flow are encountered.

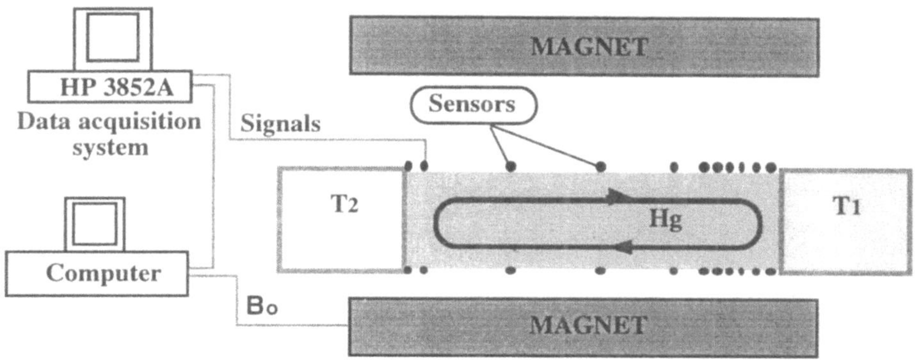

Figure 2. Control of the transition to turbulence and simultaneous data acquisitions.

3. Transition to turbulence: global features

As already mentionned before, the control parameter of the transition stand as the Hartmann number Ha. Whatever the value of the applied Grashof number $Gr_{\Delta T}$, destabilisation of the laminar core flow occurs when the Hartmann number just goes down to its critical value, $Ha<Ha^*_1$. Then a 3-D oscillating mode is made evident. Now, increasing Ha up to Ha^*_3 in order to re-stabilise the flow, an hysterisis pattern is found so long as the Grashof number remains smaller than $9\ 10^5$, suggesting thus a subcritical bifurcation ($Ha^*_3>Ha^*_1$). As far as this moderate level of convection is concerned, <u>global features</u> of the transition within the core and the end recirculating flows are reviewed elsewhere in a paper (Davoust *et al*, 1996).

For a higher value of the Grashof number, $Gr_{\Delta T}$ of the order of 10^6 and more, the key feature seems to be the fact that it's almost impossible to experimentally characterize the bifurcation. Curiously, as soon as the Hartmann number is larger than Ha^*_3, the stabilisation of the former 3-D oscillations is no longer achieved at a given axial position. For example, increasing a little bit more the Hartmann number from Ha^*_3 to a value Ha^*, one and indeed two oscillating windows (inside which time-dependent oscillations are standing), are made evident (figure 3b). Now, at this fixed value of the Hartmann number, when the sensors are moving within the vertical longitudinal mid-plane XZ to other axial positions, stabilisation and oscillating windows are successively observed. And since these windows stand as quite stable in time, any intermittent pattern has to be drawn away. Two attempts to interpret this remarkable phenomenum may be put forward. First, the non-negligible changes of physical characteristics such as thermal and kinematic diffusivities make also the Prandtl number Pr to be considered as a possible and aditionnal control parameter of the transition (Braunsfurth & Mullin, 1996). Second, taking account of the theoretical predictions by Bojarevics (1995) at least qualitatively, some steady 2-D convective cells distributed along the axis are predicted which may interact with the 3-D oscillating modes, experimentally observed and also predicted. Then, since the axial length of the experimental cell is much larger than its radius (axial anisotropy), the number of the 2-D cells may consistently increase as Ha decreases. More exactly, at the few X-locations of the eyes of these 2-D steady cells, the oscillating temperature signals related to the 3-D time-dependent rolls may become straight undetectable from the background electronic noise.

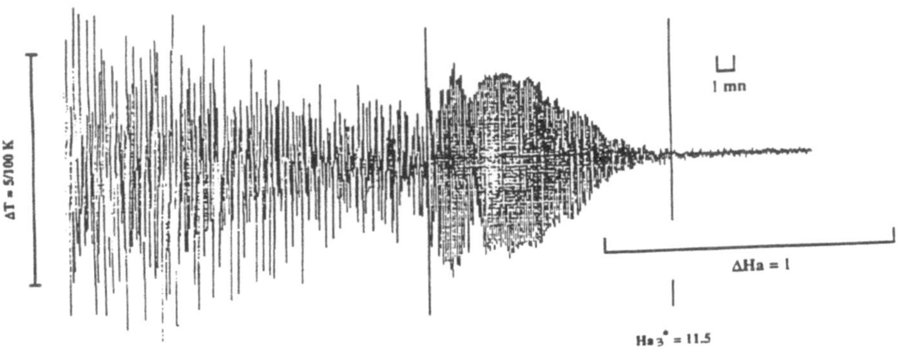

a. $Gr_{\Delta T} \sim 8.38\ 10^5$ (X=200 mm, $\alpha = 90°$)

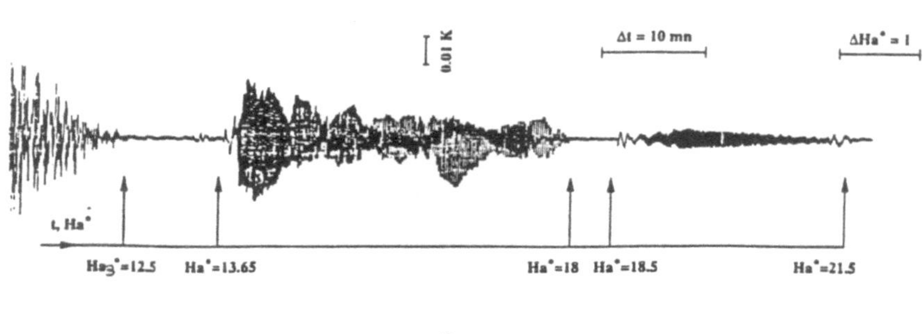

b. $Gr_{\Delta T} \sim 1.13\ 10^6$ (X=300 mm, $\alpha = 90°$)

Figure 3a-3b. Typical damping of the oscillating temperature signals.

The stabilisation and destabilisation levels of the time-dependent oscillations, are gathered on the overall stability diagram displayed on figure 4. The critical Hartmann number Ha^*_3 for which oscillations are first damped, is found to follow the scaling law $Ha^*_3 \sim \sqrt{Gr_{\Delta T}}$, in agreement with Hurle's *et al* (1974) investigations.

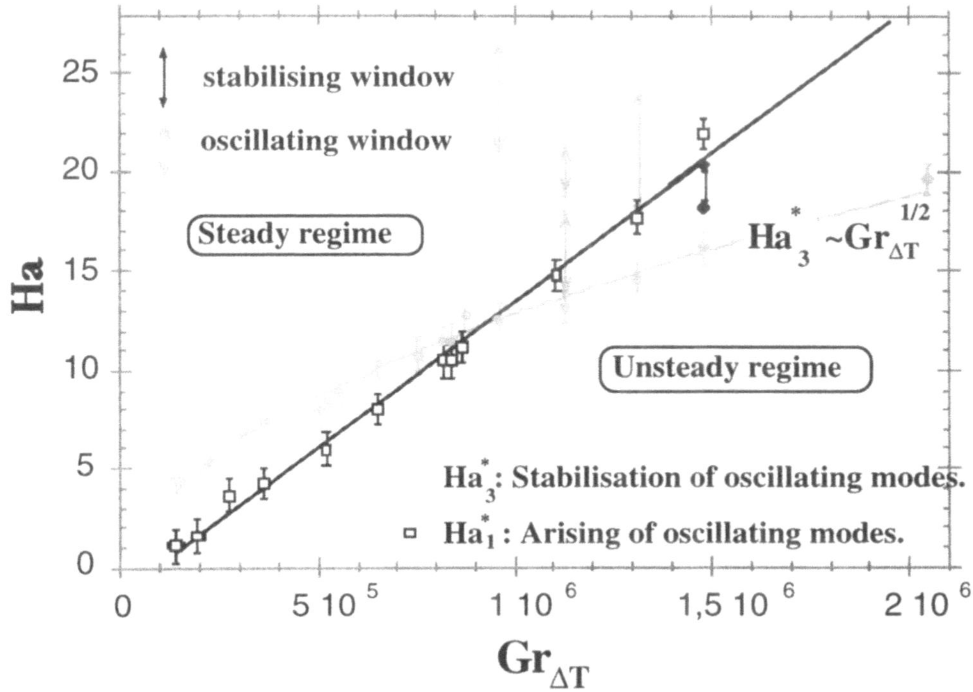

Figure 4. Overall stability diagram for the temperature oscillations (in the plane $Gr_{\Delta T}$-Ha)

4. Transition for a moderate Grashof number

At a moderate level of the Grashof number, $Gr_{\Delta T}=2.62\ 10^5$, any stabilisation or oscillating windows are ruled out. Whatever the location within the cell, the laminar flow is properly separated from turbulence by a single oscillating regime (figure 3a). This allows a close study of the transition to chaos. The investigation now focuses on the fully -developed core flow of the Hadley primary loop. All exhibited temperature signals are recorded from thermocouples whose respective location is reported on figure 5. For the sake of clearness, let's mention that the cold and hot side end walls are located, respectively, at X=0 and X=400 mm.

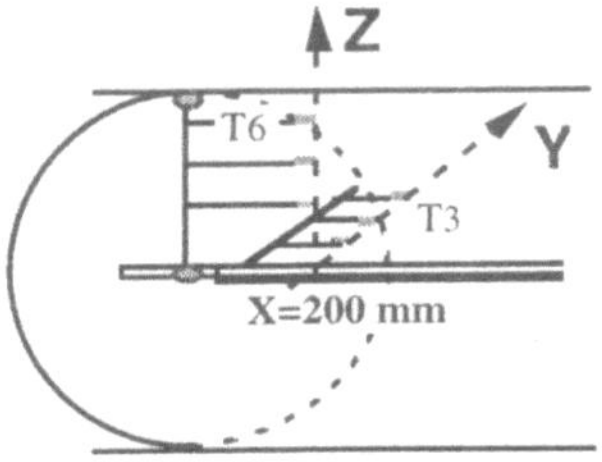

Figure 5. Sensors locations in the middle cross-section (X=200 mm).

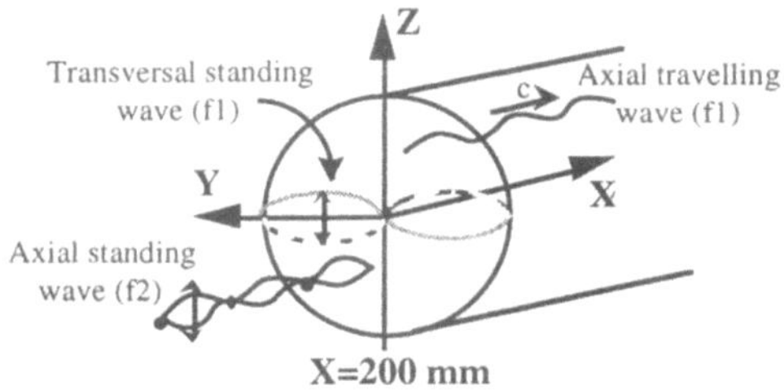

Figure 6. Nature of the three waves: travelling or standing, axial or transversal.

4.1. CHARACTERISATION IN TERMS OF WAVES

When Ha goes down to Ha^*_1, we observe a primary transition from a steady laminar to an oscillating regime: this bifurcation #I gives simultaneously rise to two waves of same frequency (f_1~44mHz). As the first one is travelling along the axis of the cell within the vertical longitudinal mid-plane XZ, the other is detected within the vertical transversal mid-plane YZ (figure 6). This last transversal wave is *a priori* identified as a standing wave since no delays are observed between the three temperature signals delivered from a same radial rank. And moreover, just beyond threshold, this wave is first characterised by a small amplitude (figure 7a, T_3-sensor).
Decreasing more Ha, because of a strong non-linear coupling between temperature and velocity fields, this amplitude grows more and more and the temperature oscillation measured by the T_6-sensor within the XZ mid-plane is enriched with harmonic foils. This classical phenomenum is also made evident on power spectral densities (PSD) where the fundamental peak at frequency f_1 becomes smaller than its first harmonic peak at frequency $2f_1$ (figure 7d, T_6 sensor).

Then, decreasing more Ha to Ha^*_2, a third axial standing wave of a much smaller frequency (f_2~8mHz) arises from a secondary bifurcation referred to as bifurcation #II (figures 5, 6, 8a & 9a). And then, as the level of spectral energy is growing ($Ha \ll Ha^*_2$), chaos progressively arises from the breakup of a quasi-periodic state (T^2-torus) based on the two former frequencies f_1 and f_2 (figures 9a-9c).

4.2. CHARACTERISATION IN TERMS OF HOPF AND SUB-CRITICAL BIFURCATIONS

Now, increasing the Hartmann number, chaos disappears and the subsequent oscillating temperature signals exhibit again the basic frequencies f_1 and f_2. As Ha goes beyond the critical value Ha^*_2, all the oscillating signals are again characterised by the single frequency f_1. Since there's no sign of an hysterisis pattern for the bifurcation #II, a supercritical behaviour is thus expected.

The fully-developed core of the Hadley flow is mainly controlled by the balance between the torques of buoyancy and Laplace forces. Such a balance leads to a modified Rayleigh number (Cowley, 1995),

$$Ra_G = {}^{Pr\, Gr_G}\!\Big/_{Ha^2},$$

where symbols Pr and Gr_G are referred to as the Prandtl and the Grashof number. This last Grashof number is built from the core axial temperature gradient G as :

$$Gr_G = \frac{\beta |g| G R_0^4}{\upsilon^2}.$$

As a matter of fact, Ra_G is found to be the control parameter of the bifurcation #II which is stated as a Hopf supercritical bifurcation (figure 8b). And the subsequent critical modified Rayleigh number denoted $Ra_G{}^{**}{}_1$ is found to be $Ra_G{}^{**}{}_1=766$. Increasing more the Hartmann number, the bifurcation #I behaves as subcritical since *e.g.*, the total stabilisation of the two other oscillating waves (of frequency f_1) occurs for a value $Ha^*{}_3$ of the Hartmann number higher than $Ha^*{}_1$. Both the 3-D oscillating mode present within the gap $Ha^*{}_2$-$Ha^*{}_1$ and the low frequency f_2 related to the axial standing wave have also been observed by direct numerical simulations (Ben Hadid *et al*, 1997).

As usual in very small or very large aspect-ratio systems (see *e.g.* Manneville, 1991, for the case of the R-B configuration), the nucleations/dislocations or merely the slow aperiodic motion of some quasi-steady structures - such as 2-D cells - drive to a weak time-modulation of the temperature signals and to a subsequent increasing level of intrinsic noise. Such a slow dynamical process induce a negative slope on PSD. This typically non-linear behaviour, also observed on PSD (figures 9a-9c) and phase portraits, may stem from some quasi-steady 2-D cells already present during transition to chaos.

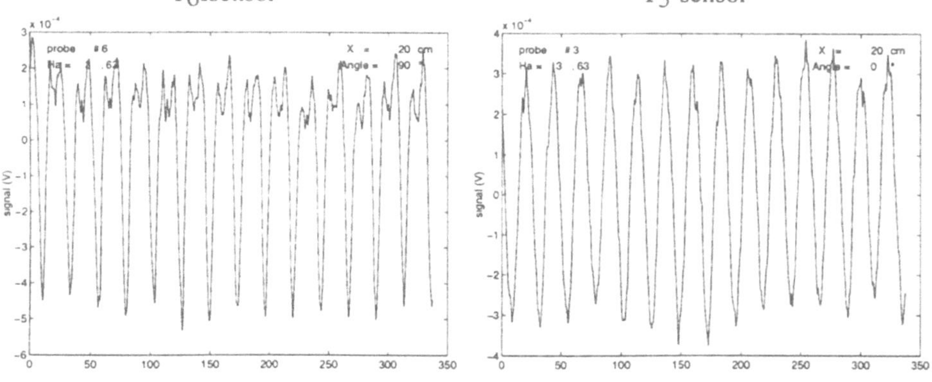

a. Ha = 3.63

T6-sensor T3-sensor

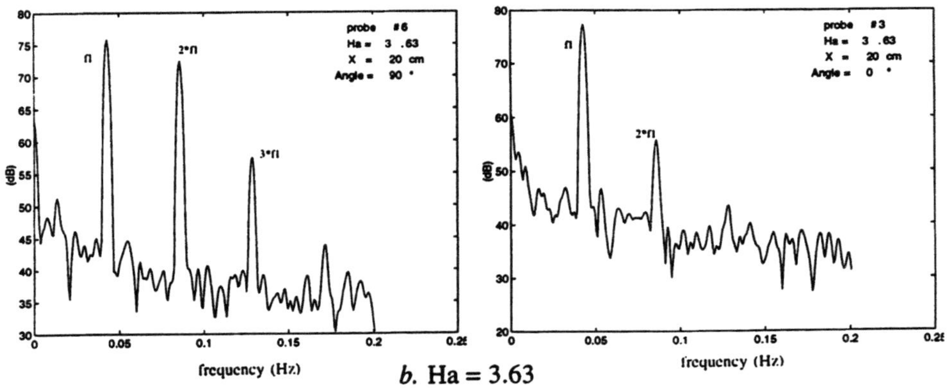

b. Ha = 3.63

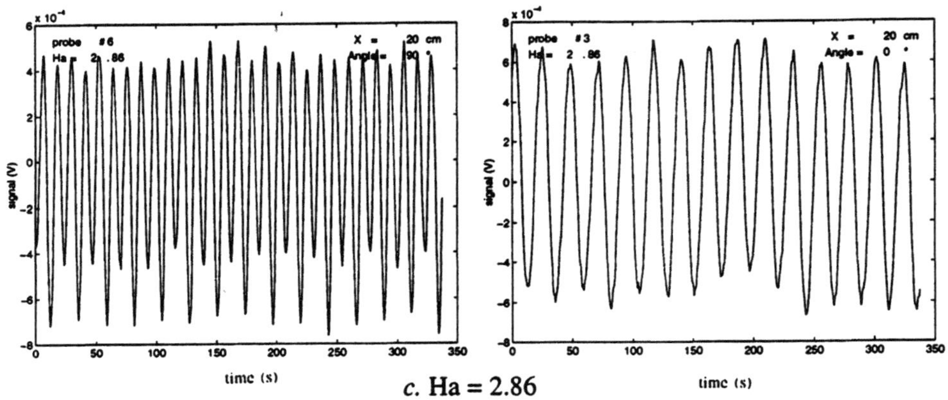

c. Ha = 2.86

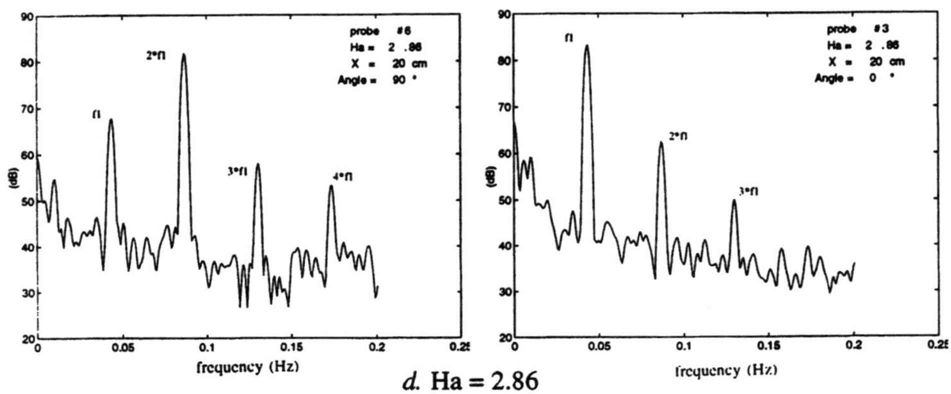

d. Ha = 2.86

Figure 7a-7d. Temperature signals and subsequent power spectral densities

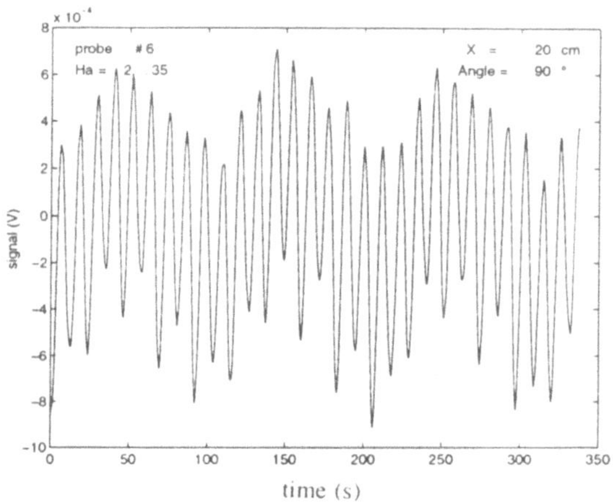

Figure 8a. Temperature signal at Ha=2.35 < Ha^*_2 < Ha^*_1 (T_6-sensor)

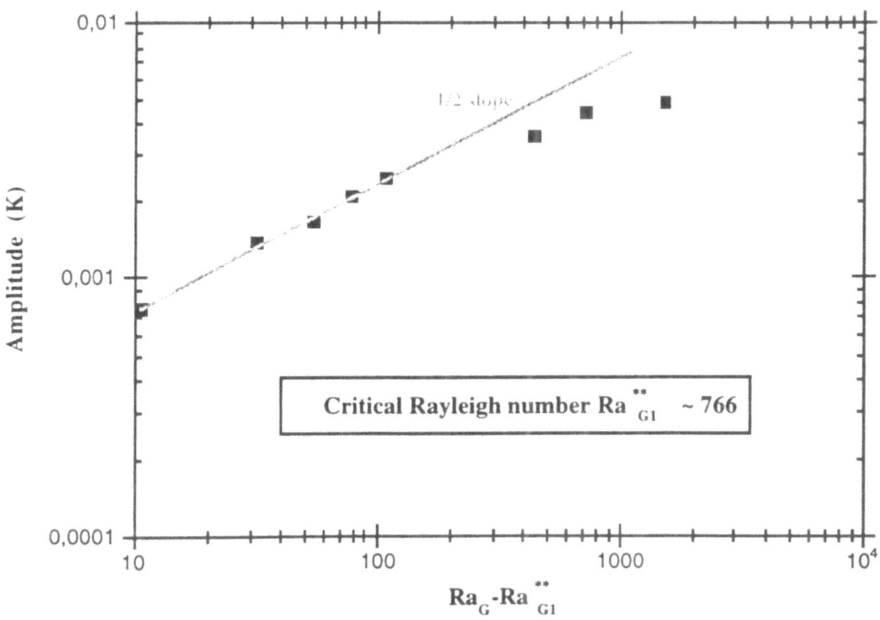

Figure 8b. Amplitude of the modulation due to the axial standing wave versus the difference (Ra_G-$Ra_G{}^{**}{}_1$)

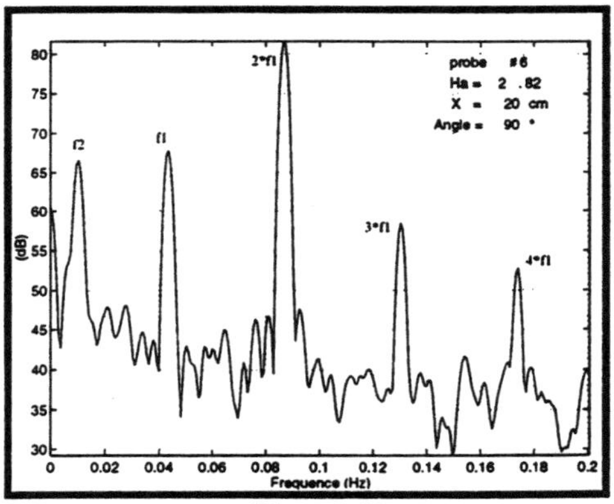

a. Ha = 2.82

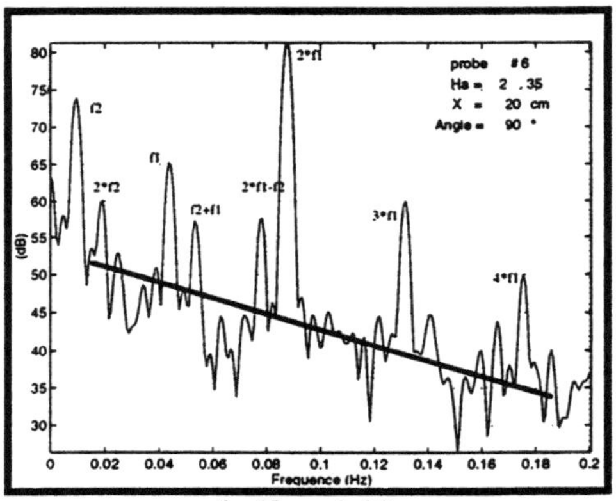

b. Ha = 2.35

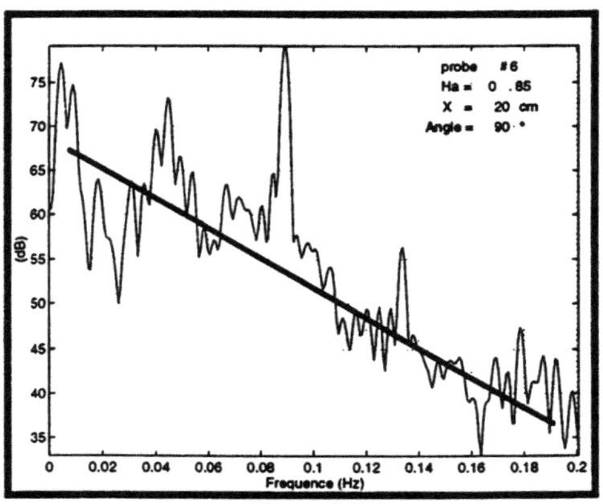

c. Ha = 0.85

Figure 9a-9c. Power spectral densities of the T_6-temperature signal during transition to chaos.

References

Ben Hadid, H., Henry, D. and Touihri, R. (1997) Unsteady three-dimensional buoyancy-driven convection in a circular cavity and its damping by magnetic field, *J. Cryst. Growth*, **180**, 433-441

Bojarevics, V. (1994) Buoyancy driven flow and its stability in a horizontal rectangular channel with an arbitrary oriented transversal magnetic field, *Magnetohydrodynamics* **31**, 333-343

Braunsfurth, M.G. and Mullin, T. (1996) An experimental study of oscillatory convection in liquid gallium, *J. of Fluid Mech.* **37**, 199-219

Cowley, M.D. (1995) The effect of a buoyancy-driven magnetohydrodynamic flow on the temperature distribution in a horizontal cylinder, *Magnetohydrodynamics*, **31**, 228-238

Davoust, L., Moreau, R., Cowley, M. D. and Bolcato, R. (1997) Buoyancy-driven convection with a uniform magnetic field, part II : experimental investigation, to be submitted

Davoust, L., Moreau, R., Neubrand, A.C. and Bolcato, R. (1996) Experimental results on MHD free convection in a horizontal cylinder, *Proceeding of the Eighth Beer-Sheva Int. Seminar on MHD Flows and Turbulence*, Jerusalem, Israel, to appear in Progress in AIAA

Hurle, D.T.J., Jakeman, E. and Johnson, C.P. (1974) , Convective temperature oscillations in molten gallium, *J. of Fluid Mech*, **64**, 565-576

Manneville, P. (1991) *Chaos et structures dissipatives*, Collection ALEA-SACLAY, pp. 114-117 and pp. 247-261

II - ELECTROCHEMICAL PROBLEMS WITH OR WITHOUT MAGNETIC FIELDS

A MODEL OF THE ANODE FROM THE CHLORATE CELL

PHIL BYRNE AND DANIEL SIMONSSON
The Faxén Laboratory
The Royal Institute of Technology
Applied Electrochemistry, Teknikringen 42
100 44 Stockholm, SWEDEN

ED FONTES AND DIDIER LUCOR
Eka Chemicals AB
R & D Department, Box 13 000
850 13 Sundsvall, SWEDEN

This article presents a mathematical model that describes the physio-chemical properties of a chlorate cell. The chlorate cell involves the electrochemical evolution of a gas that impinges a large velocity on the electrolyte flowing between two participating electrodes that are only 3 mm apart. The model is written to take into account the diffusion, migration and convection transport mechanisms, of the electrolyte species, and assumes that the hydrodynamic flow is the laminar development of fluid between two parallel plates. The transport, of each of the species, and kinetic data, for all the chemical and electrochemical reactions and bireactions, are also involved in the model. Properties of the species are worked out with consideration to the high concentration and temperature of an electrolyte. The most important result from the model is the development of a diffusion layer of species along the height of the anode. This diffusion layer creates a substantial concentration overpotential, for chloride oxidation, especially at the top of the anode surface. This gives a non-uniform current density distribution on the anode that worsens as flow velocity decreases. Moreover, the model shows that the major bireaction is the production of oxygen from the electrochemical decay of hypochlorite and not from water splitting.

1. Introduction

Electrochemical processes are not areas that have been commonly associated with mathematically modelling turbulent and laminar flows. Yet, because there exists an inherent correlation between flow of electrolyte and its species' mass transfer, to the system's electrodes, investigations of the fluid dynamics within such processes are increasing in popularity.

The chlorate system has been fairly well examined and a number of mathematical models and relationships have been found and formulated. Ibl and

A. Alemany et al. (eds.),
Transfer Phenomena in Magnetohydrodynamic and Electroconducting Flows, 137–152.

Landolt (1965 and 1968) used a previously postulated equation for the mass transfer coefficient at gas evolving electrodes (Venczel, 1965 and Ibl and Venczel, 1970) to develop an analytical model. They could then relate the rate of oxygen evolution to the mass transfer of reactants to the electrode. Beck (1969) wrote a macroscopic model separating the electrolysis and chemical reactors, found in most chlorate processes. Jaksic and co-workers (1969, 1974 and 1976) have probably worked most with modelling the chlorate's system, although their kinetics are based on graphite electrodes. They have incorporated both Ibl and Landolt's, and Beck's approaches to describe the industrial process rather concisely. Yet, none of the above models, or others that have been written (Rousar, 1969, Heal *et al.*, 1977 and Boxall and Kelsall, 1992) take into account the kinetics of the electrochemical reactions on DSA® electrodes, which is what modern chlorate processes use. In addition, none of the models consider the development of hydrodynamic conditions, to therefore include the migration and convective terms, along with their formulated diffusion terms, in the mass flux equation.

2. Definition of the problem

2.1 THE SYSTEM

The production of sodium chlorate ($NaClO_3$) occurs through the electrochemical oxidation of chloride ions and reduction of water, usually from a brine solution (NaCl), usually in a series of 'cellboxes' illustrated by the following figure:

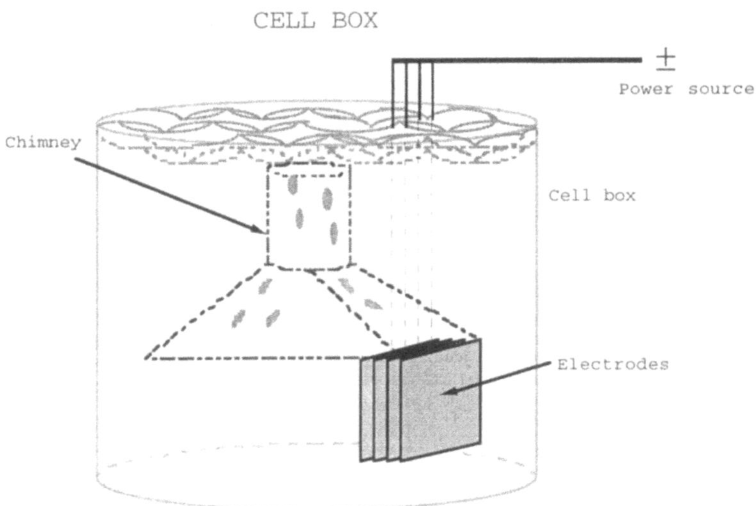

Figure 1. A schematic diagram of the Chlorate Process' cellbox.

Sodium chlorate is then used as the feedstock to paper bleach, chlorine dioxide (ClO_2), that is usually manufactured on-site of pulp and paper mills. Chlorine dioxide provides a brighter and stronger paper than other bleaching reagents, and does not pollute the surroundings as chlorine bleaching once did (Beal, 1994). Subsequently, sodium chlorate production has stabilised in world markets, as chlorine dioxide becomes accepted as a paper bleach, and was estimated at over 1.5 million tons in 1990 (Bradley *et al.*, 1990). Sodium chlorate is manufactured in large containers known as cell boxes. Up to and over one hundred electrodes are placed within each cell box, with a 3mm gap between each anode and cathode, and a 20cm gap between the electrodes and cell box bottom.

Hydrogen gas is produced on the cathode and 'drags' the electrolyte up the length of the electrodes with its buoyancy. This effect is known as 'gas-lift', and in many chlorate processes drives the plant-circulation and cellbox-recirculation of electrolyte. The upward flow is collected in a chimney above the electrodes, that can be up to several metres long, and provides the upward channel to the electrolyte recirculation existing in the cell box or in an external loop.

2.2 THE CHEMISTRY

The overall reaction is written as:

$$NaCl + 3H_2O \rightarrow NaClO_3 + 3H_2(g) \qquad (1)$$

The pioneering work behind the chemical and electrochemical formulations of the chlorate electrolyte process was done by Foerster (1924). His reaction scheme is summarised by the following chemical and electrochemical reactions.

Firstly, chloride ions migrate and diffuse through the diffusion layer to the anode surface in the electrochemical production of chlorine:

$$2Cl^- \rightarrow Cl_2(aq) + 2e^- \qquad (2)$$

The aqueous chlorine then undergoes an homogeneous hydrolysis reaction, to produce hypochlorous acid:

$$Cl_2(aq) + H_2O \underset{k_3^b}{\overset{k_3^f}{\rightleftarrows}} HOCl + Cl^- + H^+ \qquad (3)$$

Hypochlorous acid exists in a solution equilibrium to produce hypochlorite ions:

$$\mathrm{HOCl} \underset{k_4^b}{\overset{k_4^f}{\rightleftarrows}} \mathrm{OCl^-} + \mathrm{H^+} \tag{4}$$

Modern-day chlorate processes include a subsidiary reactor to the cell box, so as to carry out the formation of chlorate, with the following scheme:

$$2\mathrm{HOCl} + \mathrm{OCl^-} \underset{k_5^b}{\overset{k_5^f}{\rightleftarrows}} \mathrm{ClO_3^-} + 2\mathrm{H^+} + 2\mathrm{Cl^-} \tag{5}$$

In addition to the process' reactions are a number of bireactions, all of which contribute to the loss of efficiency and the detriment of the process and equipment. The first reaction is the decay of hypochlorite and is either homogenous or heterogeneous in nature:

$$2\mathrm{OCl^-} \underset{k_8^b}{\overset{k_8^f}{\rightleftarrows}} 2\mathrm{Cl^-} + \mathrm{O_2}(g) \tag{6}$$

Homogeneously it occurs in the bulk solution, whilst it has also been known to be catalysed by the anode, without the transfer of electrons. In a hot, concentrated electrolyte, involving DSA® anodes, the major bi-reaction was postulated (Kotowski and Busse, 1986) and later proven (Hardee and Mitchell, 1989) to be the following:

$$\mathrm{ClO^-} + \mathrm{H_2O} \rightarrow \mathrm{Cl^-} + 2\mathrm{H^+} + \mathrm{O_2}(g) + 2e^- \tag{7}$$

The final bireaction is the ever-present electrochemical splitting of water. In our conditions it has the following reaction scheme:

$$2\mathrm{H_2O} \rightarrow \mathrm{O_2}(g) + 2\mathrm{H_2O} + 4e^- \tag{8}$$

Hydrogen is produced on the cathode in an assumed 100% efficiency:

$$2\mathrm{H_2O} + 2e^- \rightarrow \mathrm{H_2}(g) + 2\mathrm{OH^-} \tag{9}$$

In reality, there exists efficiency losses on the cathode through bireactions occurring there. The major one of these reactions is the back-reduction of the hypochlorite ion to chlorine, yet the addition of dichromate to the system greatly reduces this reaction. It is assumed that the system functions as if dichromate was present, although it has been itself neglected as a participant in our model.

2.3 THE EQUATIONS

The mathematical model shall aim to present mass transfer rates to the anode, velocity and concentration profiles, and current and potential distribution. It is a numerical model that requires a few assumptions in order to simplify it. These are namely that water concentration is constant throughout the system, diffusivity and ionic mobility remain unchanged for the respective species and that the influence of the gas bubbles, on the conductivity of the electrolyte, is negligible.

The model also assumes that oxygen and hydrogen evolution do not have a turbulitary effect on the hydrodynamics of the system. In addition, it does not take into account the accelatory effects that an increase in void fraction, up the length of the cell, gives to the electrolyte, since this effect is small compared to the lift effect created by the chimney. Therefore we assume that the initial average velocity of the electrolyte is between 0.5 ms^{-1} and 1.5 ms^{-1}, and that this is maintained through its length.

Electrodes sit in the cell box and electrolyte is dragged up from beneath them. The electrolyte enters the gap with a velocity profile that is uniformly distributed over the width of the gap. Viscous friction, from the surfaces of the two electrodes, works in developing respective sub-viscous layers up the length of the cell. This eventually results in the development of a laminar profile across the gap. The velocity vector is obtained by solving the Navier-Stokes equation with the continuity equation (Schlichting, 1960). Using the appropriate boundary conditions, we obtain these vector values at every point, in the electrode gap, and can use them in the below mass transport equation (12).

The mathematical model's basic equation is a mass balance for each species over the gap between the electrodes, and is presented as the rate change of concentration. This takes into account the respective mass fluxes and chemical production/consumption that occurs by way of the chemically homogenous equations:

$$\frac{\partial c_i}{\partial t} = -\nabla \cdot \mathrm{N}_i + \mathrm{R}_i \qquad (10)$$

where R_i is production, c_i the concentration and $\mathbf{N}_i$ the mass flux of species i. At the situation of steady-state, concentration change is negligible such that:

$$\frac{\partial c_i}{\partial t} = 0 \qquad (11)$$

Mass flux shall be calculated from the below equation (Newman, 1991):

$$N_i = -D_i \nabla c_i - z_i u_i c_i F \nabla \Phi + \mathrm{v} c_i \tag{12}$$

where D_i is diffusivity, c_i is concentration, u_i is ionic mobility and z_i is the electrical charge of species i, F is Faraday's constant, Φ is the potential field and $\mathbf{v}$ the velocity vector. The above mass transfer equation can be described by the three terms on the right hand side as transfer due to concentration diffusion, electrical migration and convection respectively. The final condition within the domain of the gap is that of electroneutrality and is expressed by:

$$\sum_i z_i c_i = 0 \tag{13}$$

The boundaries are the anode, cathode and appropriate boundaries for the top and bottom of the investigated region. Mass transfer, from the anode for each of the species, is either non-existent or present through virtue of the electrochemical reactions; (2), (7) and (8), or the heterogeneous form of (6). The kinetics of these relationships are mathematically described by the following three Butler-Volmer equations:

$$i_2 = i_{0,2}\left[\left(\frac{c^s_{Cl}}{c^\infty_{Cl}}\right)^2 \exp\left\{\frac{2F}{RT}\left[\left(\Phi_m - \Phi_l\right) - \Delta\Phi^0\right]\right\} - \left(\frac{c^s_{Cl2}}{c^\infty_{Cl2}}\right)\right] \tag{14}$$

$$i_6 = i_{0,6}\left(\frac{c^s_{OCl}}{c^\infty_{OCL}}\right)^1 \exp\left\{\frac{0.5F}{RT}\left[\left(\Phi_m - \Phi_l\right) - \Delta\Phi^0\right]\right\} \tag{15}$$

$$i_8 = i_{0,8}\left(\frac{c^s_H}{c^\infty_H}\right)^{-1} \exp\left\{\frac{1.5F}{RT}\left[\left(\Phi_m - \Phi_l\right) - \Delta\Phi^0\right]\right\} \tag{16}$$

where i is current density, i_0 is exchange current density, c is concentration, F is Faraday's constant, R is the universal gas constant, T is absolute temperature, Φ is the actual potential and $\Delta\Phi^0$ is the standard potential. Both exchange current density and the standard potential are inherent properties to the investigated chemical reaction and system itself. They are presented in table 4. The superscripts ∞ and s relate to the regions of the bulk solution and the electrolyte at the surface of the anode, respectively. The subscript m denotes the electrical potential in the metal whilst the equivalent l denotes potential in the electrolyte.

2.4 THE INPUT PARAMETERS

Bulk concentrations at the bottom of the cell are input values to the model, and table 1 gives an approximate concentration for a chlorate process electrolyte. With the additional information that an electrolyte has a general pH of 6.5, we can work out the bulk concentrations to be used as the basis of our model.

TABLE 1: Basic concentration of a chlorate electrolyte.

SPECIES	CONCENTRATION (g/L)
NaCl	110
$NaClO_3$	600
NaOCl	3

The majority of the species' inlet concentrations are easily calculated, although a certain number of these have to take into account the conditions that can affect their respective equilibrium. The start concentration of the aqueated chlorine molecule is calculated from the Activation Energy and Van´t Hoff equations (Ashour *et al.*, 1996) and data taken from Dean (1992), under the assumption that the hydrolysis reaction (3) is in equilibrium at the inlet of the cell. The water equilibrium constant has been calculated from an equation taken from Olofsson (1975) whilst Dean (1992) gives a good approximation of oxygen's solubility. The equilibrium constant, between hypochlorous acid and the hypochlorite ion, was taken from Chao (1968) and recalculated using Van´t Hoff's equation. The start concentration of each species used in the equations is presented in table 2.

TABLE 2: Starting concentrations of the components in a chlorate process electrolyte.

LABEL	SPECIES	CONC. (mol/m^3)
c_1	Chloride ion	1.88×10^{3}
c_2	Chlorine gas	Not applicable
c_3	Aqueated chlorine molecule	2.83×10^{-1}
c_4	Hypochlorous acid	1.47×10^{1}
c_5	Hypochlorite ion	2.56×10^{1}
c_6	Hydrogen ion	3.16×10^{-4}
c_7	Hydroxide ion	4.79×10^{-3}
c_8	Chlorate ion	5.63×10^{3}
c_9	Aqueated oxygen molecule	5.80×10^{-3}
c_{10}	Sodium ion	7.54×10^{3}

The mass transfer equation (12) requires input values for diffusion coefficients and ionic mobilities. The more common ion diffusivities were taken from Newman (1991), whilst Chao (1968) had investigated HOCl, OCl^- and $Cl_2(aq)$, in systems similar to ours. Both authors present their data from fairly ideal systems, so that the values required recalculation to values that are more realistic to the concentrations and temperatures found in a chlorate process electrolyte. This was done through an approximating equation (Ashour *et al.*, 1996):

$$D_2 = D_1 \frac{\mu_1}{T_1} \frac{T_2}{\mu_2} \tag{17}$$

Mobility is found by using the Nernst-Einstein equation. The calculated diffusion coefficients are presented in table 3.

TABLE 3: Diffusivities of the components in an electrolyte.

LABEL	SPECIES	DIFFUSIVITY (m^2/s)
D_1	Chloride ion	1.71×10^{-9}
D_2	Chlorine gas	Not applicable
D_3	Aqueated chlorine molecule	1.83×10^{-9}
D_4	Hypochlorous acid	1.11×10^{-9}
D_5	Hypochlorite ion	1.07×10^{-9}
D_6	Hydrogen ion	7.86×10^{-9}
D_7	Hydroxide ion	4.43×10^{-9}
D_8	Chlorate ion	1.51×10^{-9}
D_9	Aqueated oxygen molecule	Not applicable
D_{10}	Sodium ion	1.11×10^{-9}

The production term, in equation (18) below, is directly related to the kinetics of the chemical reactions that occur in the chlorate electrolyte. These take place in the domain of the equation and have both forward and backward rates of reactions. If we have the following form of chemical reaction:

$$\mathrm{A} + \mathrm{B} \underset{k_i^b}{\overset{k_i^f}{\rightleftarrows}} \mathrm{C} + \mathrm{D} \tag{18}$$

then the rate of production of species C is expressed as:

$$\mathrm{R_C} = k_C^f c_\mathrm{A} c_\mathrm{B} - k_C^b c_c c_\mathrm{D} \tag{19}$$

where R is rate of production, c is concentration and k is rate of reaction. The superscripts, f and b, signify the forward and backward directions respectively. The rates of these reactions have been extremely hard to come by. A few of the reaction rates have been assumed, whilst the majority have to be estimated from more ideal conditions to the extremes encountered in an industrial concentrated electrolyte.

Despic *et al* (1972) used data gleaned from Spalding (1962) and found the forward rate, of the hydrolysis reaction (3), for a concentrated solution at 25°C. Using his graph, our ionic strength (I = 6.786 mol/kg) and the Arrehenius equation, we could recalculate it to conditions of a concentrated electrolyte at 70°C. The backward rate of reaction is is obtained by assuming equilibrium exists. This gives us the condition that production is zero and the possibility to calculate the backward rate of reaction.

No kinetic data of the dissociation of hypochlorous acid, in the equilibrium reaction (4), has been found in the literature, so the model assumes that this dissociation occurs in a similar fashion to water. Water's re-association constant was calculated by Moore (1986) and is thus assigned as the backward rate of reaction. The forward rate is found with the above proviso pertaining to equilibrium.

The water equilibrium reaction is added to give as accurate a value of pH as possible. The electrochemical reactions, and a few of the homogeneous ones, are quite sensitive to pH. Using the pK_W from Olofsson (1975) and the backward rate of reaction given by Moore we were able to present the kinetics. Since these above two reactions are so fast, it was difficult for the model to converge. We were therefore forced to moderate the respective reaction rates, keeping the equilibrium constant. Yet, due to the fact that these two reactions are basically in a state of continual equilibrium, the results remained unaffected by our moderate changes in these constants.

The most trustworthy reaction rates, in the production of chlorate, (5), were found by Peters (1993) in an undergraduate thesis project for The Swedish Royal Institute of Technology. His solution simulated an electrolyte rather well, even to the extent of temperature and containing sodium dichromate.

The homogeneous decay of hypochlorite is one of the bireactions. Lister (1956) had tested the rate of oxygen production, from hypochlorite decay, on a number of metals. The most appropriate of these metals was manganese and the ionic strength of the solution was 3.5 at 60°C. We recalculated his value using the Arrhenius equation and the equilibrium proviso to find the backward and forward rates of reaction.

The values for the forward and backward rates, in the homogenous reactions, and exchange current densities and equilibrium potentials, in the electrochemical reactions, are presented in the following table:

TABLE 4. Kinetic data for the chemical and electrochemical reactions.

EQN	k_i^f	k_i^b	i_0	E^0
3	$2.08\ s^{-1}$	$6.74 \cdot 10^{-2}\ m^6mol^{-2}s^{-1}$		
4	$7.70 \cdot 10^4\ s^{-1}$	$1.40 \cdot 10^8\ m^3mol^{-1}s^{-1}$		
5	$1.80 \cdot 10^{-8}\ m^6mol^{-2}s^{-1}$			
6	$4.49 \cdot 10^{-7} m^3mol^{-1}s^{-1}$	$1.43 \cdot 10^{-8}\ m^3mol^{-1}s^{-1}$		
H_2O	$2.12 \cdot 10^1\ s^{-1}$	$1.40 \cdot 10^8\ m^3mol^{-1}s^{-1}$		
14			$1.0 \cdot 10^1\ Am^{-2}$	1.36 V
15			$2.0 \cdot 10^{-13}\ Am^{-2}$	0.96 V
16			$1.0 \cdot 10^{-6}\ Am^{-2}$	1.23 V

3. Results and discussion

The model was run for three scenarios where the start velocity was the only variable. These velocities were: 0.5 ms^{-1}, 1.0 ms^{-1} and 1.5 ms^{-1}.
The following figure shows how the veloc ity profile is developed up the length of a singular cell, at a 1.0 ms^{-1} start velocity. The base of the cell is approximately 15 cm from the cell box bottom, whilst the cell itself has dimensions of 40 cm × 40 cm × 3 mm. The gap between the electrodes is 3 mm. The electrolyte enters the gap with a velocity profile that is uniformly distributed over the width of the gap. Viscous friction, from the surfaces of the two electrodes, works in developing respective sub-viscous layers up the length of the cell. This eventually results in the development of a laminar profile across the gap, shown in figure 2.

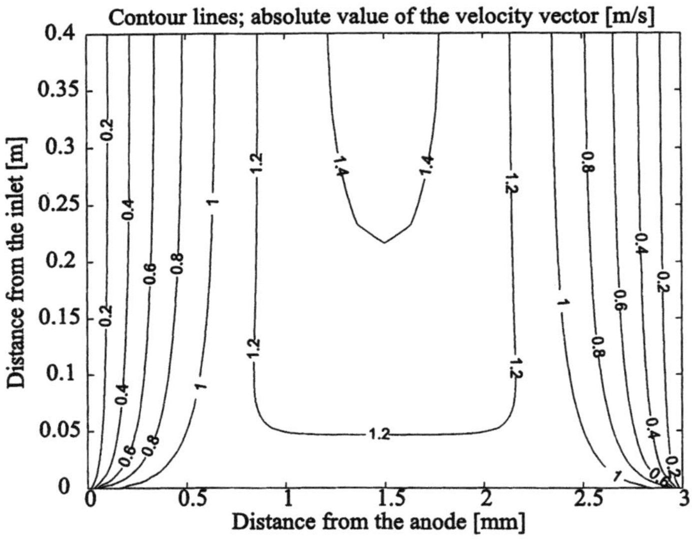

Figure 2. Absolute value of the velocity vector for a single cell with initial velocity of 1.0 ms^{-1}.

According to Schlichting (1960), velocity in the middle of the channel reaches a maximum of 1.5 ms^{-1}, when the initial velocity is 1.0 ms^{-1}. The above figure shows that the maximum velocity is 1.44 ms^{-1} in the channel's centre, such that the profile, theoretically, never fully develops in the cell.

The chloride ion profile (Figure 3) shows a large concentration gradient over quite a small distance.

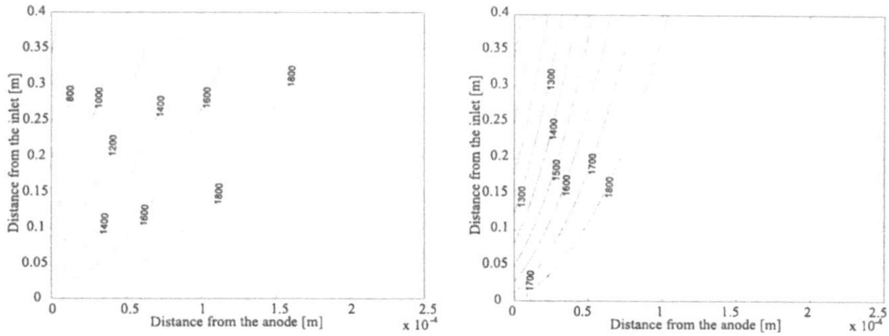

Figure 3. The chloride ion concentration profile, near the anode, for the start velocities of: a) 0.5 ms^{-1}, b) 1.5 ms^{-1}.

At the top of the cell, the horizontal concentration drop is as much as 1800 mol m^{-3} (105.3 gl^{-1}) to 600 mol m^{-3} (34.7 gl^{-1}) over 0.16 mm, at a 0.5 ms^{-1} start velocity. Additionally, the concentration difference up the length of the anode is of the same magnitude. This means that the top region of the anode has less than half the available number of chloride ions to electrolyse, than at the bottom, and therefore requires a greater effort (*i.e.* concentration overpotential) to achieve the same effect (*i.e.* chlorine production). In comparing the different velocities we see that this vertical concentration difference, and thus concentration overpotential, increases with decreasing velocity.

The initial current density was set at 2.6 kAm^{-2} and is the value at the bottom of the anode. This current density decreases in a geometric fashion up the length of the anode (Figure 4). The first reason for this is that the bottom region of the anode is continuously fed fresh feed electrolyte, which is being sucked up from underneath the cell. As the electrolyte proceeds up the height of the cell, more and more of the feed chloride is depleted from the domain close to the surface. The other reason for this decrease in current density, is that the development of a laminar profile brings about a growing sub-viscous layer, that hinders the transport of fresh chloride from the gap's centre to the surface.

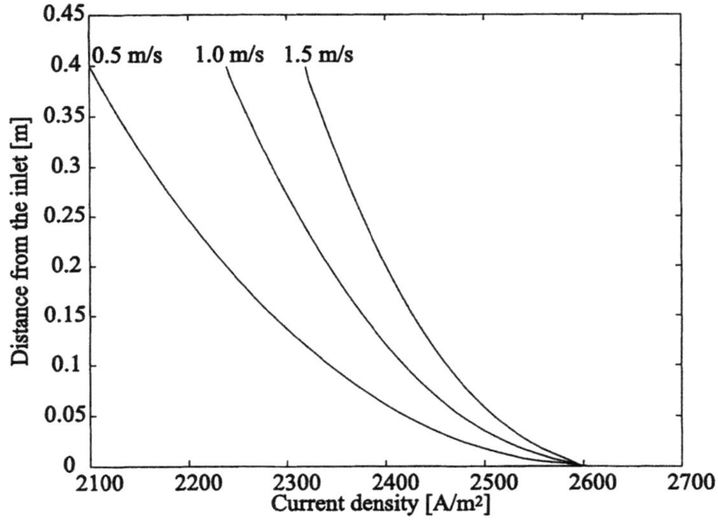

Figure 4. The total current density distribution, on the anode with a start current density of 2.6 kAm^{-2}, and for the start velocities of: 0.5 ms^{-1}, 1.0 ms^{-1} and 1.5 ms^{-1}.

The effect that velocity has on these profiles is evident from a comparison. The greater the start velocity, the less varied is the current density distribution. Integration under the graph run at 1.5 ms^{-1} gives an average current density of 2.41 kAm^{-2}, whilst that at 0.5 ms^{-1} has an average of 2.26 kAm^{-2}. These values are equivalent to 6.12 $kgday^{-1}$ and 5.76 $kgday^{-1}$ production of sodium chlorate respectively; if we assume that the process can be run at 100% efficiency. The reason for these variations is fairly evident as the convective effects are greater, at larger start velocities, whilst the alternative lower velocities allow a more pronounced, and hindering, velocity profile to be developed.

The below diagram (Figure 5) shows that the current density, for the electrochemical decay of the hypochlorite ion, increases up the length of the anode. It also shows that there exists a local minimum in all three circumstances.

Initially, there is quite a high concentration of hypochlorite in the electrolyte, dragged up from underneath the cell. In addition to this, is the lack of a hydrodynamic hindrance, from the existence of a viscous layer. The consequence of these two factors is an initial high current density for the decay of hypochlorite. This current density decreases as, firstly, the hypochlorite ion concentration falls and, secondly, the amount of produced hydrogen ions increases. The resulting minima has the significance that the local oxygen production, and consequently anode corrosion, is lowest at this point.

The hydrolysis reaction's production of hypochlorous acid starts to become prevalent, after the minima, along with a falling concentration of chloride ions. This results in a steady increase of hypochlorite decay up the length of the anode. Conversely, as the diffusion coefficients of the chloride ion and aqueated chlorine molecule are larger than those for the hypochlorite ion and hypochlorous acid molecule (see Table 3), the viscous layer's presence would have a diminishing effect on hypochlorite decay. Yet, this effect is far less than those given by the overall hypochlorous acid production and falling reaction availability of the chloride ion.

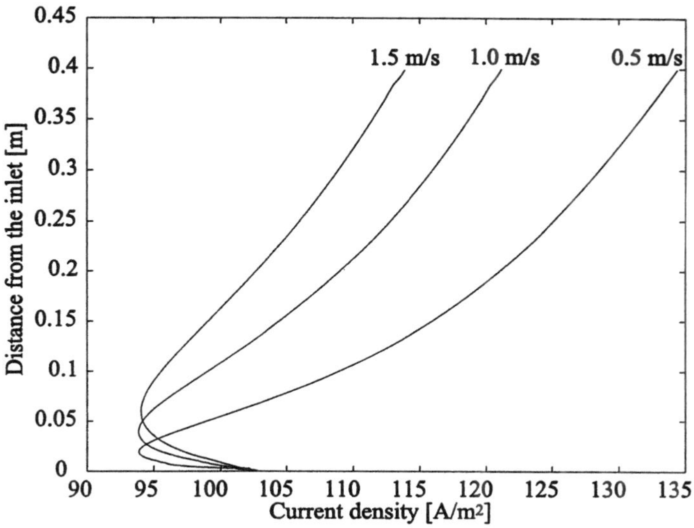

Figure 5. The current density distribution for the electrochemical decay of hypochlorite, on the anode for the start velocities of: 0.5 ms^{-1}. 1.0 ms^{-1}. 1.5 ms^{-1}.

The positive effect that an increased velocity has over the total current density distribution (Figure 4) is also evident from the above diagram (Figure 5). Current density distribution, at the faster initial velocities, is much less varied than at the slower velocities. Furthermore, integration under the graph run at 1.5 ms^{-1} gives an average current density of 103.6 Am^{-2}, whilst that at 0.5 ms^{-1} has an average of 119.2 Am^{-2}. These values are equivalent to 181.4 $lday^{-1}$ and 208.8 $lday^{-1}$ production of oxygen gas, respectively.

An interesting set of results is given by the below diagram (Figure 6). Figure 5 showed that the current due to hypochlorite decay is at a maximum at the top of the anode. Yet, figure 6 shows that the same ion's concentration is at a minimum at this same point. Both the hydrolysis and electrochemical hypochlorite decay reactions produce hydrogen ions, which in turn hold the hypochlorite ion concentration down, through virtue of the equilibrium reaction. This points to the conclusion that the equilibrium reaction is very fast, so as to keep the electrode fed with these ions. An interesting point to note is that the hypochlorite concentration is 5-6 $molm^{-3}$ (0.38-0.45 gl^{-1}) at the top of the anode.

A reasonable assumption is that this concentration would decrease to virtually zero, the limiting current density being reached, if the anode was extended.

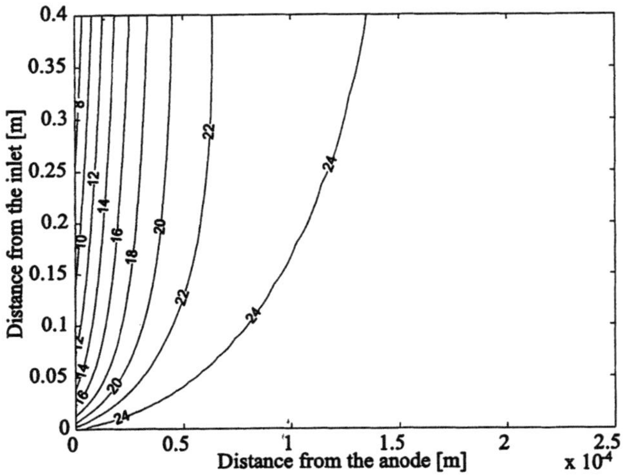

Figure 6. The hypochlorite ion concentration profile, near the anode, for the start velocity of 1.0 ms^{-1}.

Figure 7 shows the insignificance that water splitting has on the process.

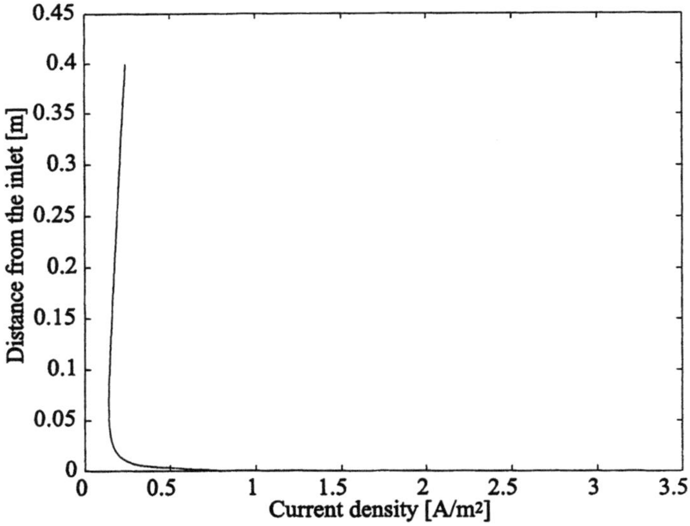

Figure 7. The current density distribution for the electrochemical decay of water (water splitting), on the anode, for the start velocity of 1.0 ms^{-1}.

The equivalent graph (Figure 5) has current density, for oxygen production from hypochlorite decay, ranging from 94-121 Am^{-2}. Water splitting's contribution is over two orders of magnitude smaller, pointing to the conclusion that it is not at all significant, as a participating reaction.

4. Conclusion

The presented model gives the transport relationships, with respect to the fluid dynamic variations occurring up the length of a cell, of the species found in a chlorate process' electrolyte. The analytical model investigated three different inlet velocities; 0.5 ms^{-1}, 1.0 ms^{-1} and 1.5 ms^{-1}, and the profile was assumed to be laminar in nature. The results show that the chloride reaction's concentration overpotential increases with decreasing velocity, up the length of the cell. This in turn brings about a more varied current distribution at these lower velocities. The model also shows that oxygen evolution, shown mostly to be due to the electrochemical decay of the hypochlorite ion, also varies along the anode's length. Its rate of production reaches a minima, before increasing to the top of the cell. Once again it was seen that the distribution of current, to the oxygen evolution reaction, is far more varied at the lower velocities.

References

ASHOUR, S. S., RINKER, E. B. and SANDELL, O. C. Absorption of chlorine into aqueous bicarbonate solutions and aqueous hydroxide solutions, *AIChE Journal.* **42** (1996) 671.

BEAL, M. G. *Modern Chlor-alkali Technology. **Vol 6,** Chap1.* The Royal Society of Chemistry, Cambridge, 1994.

BECK, T. R. A contribution to the theory of electrolytic chlorate formation, *Jnl. Electrochem. Soc.* **116** (1969), 1038.

BOXALL, C. and KELSALL, G. H. Hypochlorite electrogeneration. II: Thermodynamics and kinetic model of the anode reaction layer, *ICHEME Symposium series* nr **127** (1992) 59.

BRADLEY, F., LEDER, A. and SASANO, T. *CEH Product Review.* Chemical Economics Handbook, SRI International, 1990.

CHAO, M. S. The diffusion coefficient of hypochlorite, hypochlorous acid and chlorine in aqueous media by chronophotography, *Jnl. Electrochem. Soc.* **115** (1968), 1172.

DEAN, J.A. *Lange's Handbook of Chemistry. **14th. Ed.*** M^c^Graw-Hill (1992). 6-69.

DESPIC, A. R. JAKSIC, M. M. and NIKOLIC, B. Z. The of kinetic and hydrodynamic factors on current efficiency in the chlorate cell process, *Jnl. Appl. Electrochem.* **2** (1972), 337.

FOERSTER, F. The electrolysis of hypochlorite solutions, *Trans. Am. Electrochem. Soc.* **46** (1924), 23.

HARDEE, K. L. and MITCHELL, L. K. The influence of electrolyte parameters on the percent oxygen evolved from a chlorate cell, *Jnl. Electrochem. Soc.* **136** (1989), 3314.

HEAL, G. R., KUHN, A. T. and LARTEY, R. B. A parametric study and computer-based simulation of an undivided sodium hypochlorite electrolyzer, *Jnl. Electrochem. Soc.* **124** (1977), 1690.

IBL, N. and LANDOLT, D. On the mechanism of anodic chlorate formation in dilute NaCl solutions, *Jnl. Electrochem. Soc.* **115** (1968), 713.

IBL, N. and VENCZEL, J. Unterscung des stofftransports an gasentwickelnden elektroden, *Met. Oberfläche.* **24** (1970), 365.

JAKSIC, M. M., NIKOLIC, B. Z., CSONKA, I. M. and DJORDJEVIC, A. B. Studies on the chlorate cell process. IV: Effect of neutral salts on conversion of available chlorine to chlorate, *Jnl. Electrochem. Soc.* **116** (1969), 684.

JAKSIC, M. M., DESPIC, A. R., CSONKA, I. M. and NIKOLIC, B. Z. Studies on the chlorate cell process. V: Theory and practice of a modified technology for electrolytic chlorate production, *Jnl. Electrochem. Soc.* **116** (1969), 1316.

JAKSIC, M. M. Mutual effect of current density, pH, temperature, and hydrodynamic factors on current efficiency in the chlorate cell process, *Jnl. Electrochem. Soc.* **121** (1974), 70.

JAKSIC, M. M. Mass transfer and optimisation of Faradaic yields in a chlorate cell process. *Electrochim. Acta* **21** (1976), 1127.

KOTOWSKI, S. and BÜSSE, B. *Modern Chlor-alkali Technology.* ***Vol 3, Chapl.*** Ellis Horwood Ltd, Chichester (1986)

LANDOLT, D. and IBL, N. On the mechanism of anodic chlorate formation in concentrated NaCl solutions, *Electrochim. Acta* **15** (1970), 1165.

LISTER, M. W. Decomposition of sodium hypochlorite: the catalysed reaction, *Can. Jnl. of Chem.* **34** (1956), 479.

MOORE, W. J. *Basic Physical Chemistry.* ***1st. Ed.,*** Prentice-Hall Int. Inc. (1986) 370.

NEWMAN, J. S. *Electrochemical Systems. 2nd Ed.* Prentice-Hall Inc., Englewood Cliffs, New Jersey, 1991.

OLOFFSON, G. and HEPLER, L. G. Thermodynamics of ionization of water over wide ranges of temperature and pressure, *J. Soln. Chem.* **4** (1975) 127.

PETERS, G. Undergraduate thesis Kungl Tekniska Högskolan, Stockholm, Avdelningen för Tillämpad Elektrokemi (1993).

ROUSAR, I. Calculation of current density distribution and terminal voltage for bipolar electrolyzers; application to chlorate cells, *Jnl. Electrochem. Soc.* **116** (1969), 676.

SCHLICHTING, H. *Boundary Layer Theory. 4th. Ed.* McGraw-Hill, 1960.

SPALDING, C. W. Reaction kinetics in the absorption of chlorine into aqueous media, *A. I. Ch. E. Jnl.* **8** (1962) 685.

VENCZEL, J. *PhD Thesis* ETH, Zurich, Prom nr **3673**, (1965).

SODIUM CHLORATE ELECTROSYNTHESIS CELL UNDER NATURAL CONVECTION : SIMULATION OF THE TRANSIENT AND STEADY STATE WORKING BEHAVIOUR

OZIL Patrick, AUROUSSEAU Marc
Laboratoire d'Electrochimie et de Physico-chimie des Matériaux et des Interfaces, U.M.R. INPG-CNRS 5631 associée à l'UJF, ENSEEG, BP 75, 38402 St Martin d'Hères (France)

AND

MITU Stelian
University Politehnica of Bucarest, Department of Chemical Engineering 1 Polizu street, 78126 Bucarest (Roumanie)

A model based on mass, voltage and mechanical energy balances is proposed for predicting the behaviour of an electrochemical reactor under natural convection due to gas evolution. It takes into account the slipping between gas and liquid phases. The friction pressure drops related to the two-phase flow are evaluated by using an empirical relation as proposed by Lockhart and Martinelli while the changes in electrolyte velocity and direction in the cell are estimated by adding equivalent fittings. The gap between electrodes is described as a stack of elementary stirred-tank electrochemical reactors in which gas evolution leads to a gas fraction varying with height. The theoretical model for hydrodynamics is successfully validated from experiments and then the complete model is applied to predict the transient behaviour of a chlorate electrosynthesis cell. Practical conclusions are obtained from the model : the cell productivity decreases when the electrode length increases and an optimal value is found out for the distance between electrodes.

1. Introduction

The practical purpose of this work is the simulation and the optimisation of a sodium chlorate electrosynthesis cell. This process is a complex one because it is based on both electrochemical and chemical reactions. It requires suitable solutions in order to increase mass and heat transfers by improving hydrodynamic conditions. Such complexity explains why very different technologies were developed by a lot of manufacturers (Canoxy-Squamish, Solvay, Chemwest, DeNora, Huron, Kemanord, Krebs, Pennwalt, Ugine Kuhlmann, OCC, Oulu, Eltchec...) interested in the same

A. Alemany et al. (eds.),
Transfer Phenomena in Magnetohydrodynamic and Electroconducting Flows, 153–167.

profitable industrial operation (see e.g. Viswanathan and Tilak, 1984). More generally, this paper presents how to predict the theoretical behaviour of an electrochemical reactor in which an internal recycle occurs because of a gas-lift effect resulting from gas evolution at electrodes. The two-phase flow of the gas-liquid electrolyte within the cell has to be taken into account very accurately to model and design such cells.

2. Electrochemical and chemical reaction scheme

The reaction scheme generally accepted nowadays (Ibl and Vogt, 1981, Beck and Brännland, 1972, Jaksic, Nikolic *et al.*, 1969) for describing the sodium chlorate electrosynthesis does not differ from the one initially proposed by Foerster and Müller about 90 years ago (see Hine, 1988).

At the anode, a main electrochemical reaction occurs :

$$2Cl^- \rightarrow Cl_2 + 2e^- \qquad \text{(a1)}$$

with some possible parasitic reactions :

$$6ClO^- + 3H_2O \rightarrow 2ClO_3^- + 4Cl^- + 6H^+ + 3/2\,O_2 + 6e^- \qquad \text{(a2)}$$

or

$$2H_2O \rightarrow 4H^+ + O_2 + 4e^- \qquad \text{(a3)}$$

$$4OH^- \rightarrow O_2 + 2H_2O + 4e^- \qquad \text{(a4)}$$

$$ClO_3^- + H_2O \rightarrow ClO_4^- + 2H^+ + 2e^- \qquad \text{(a5)}$$

At the cathode, the main electrochemical reaction is :

$$2H_2O + 2e^- \rightarrow H_2 + 2OH^- \qquad \text{(c1)}$$

while two parasitic reactions remain possible :

$$ClO^- + H_2O + 2e^- \rightarrow 2OH^- + Cl^- \qquad \text{(c2)}$$

$$ClO_3^- + 3H_2O + 6e^- \rightarrow 6OH^- + Cl^- \qquad \text{(c3)}$$

Moreover several main chemical reactions take place in the electrolyte :

$$Cl_2 + H_2O \xrightarrow{K1} HClO + H^+ + Cl^- \qquad \text{(e1)}$$

$$HClO \xleftrightarrow{Ke} ClO^- + H^+ \qquad \text{(e2)}$$

or

$$HClO + OH^- \longleftrightarrow ClO^- + H_2O$$

and

$$2HClO + ClO^- \xrightarrow{K3} ClO_3^- + 2H^+ + 2Cl^- \qquad \text{(e3)}$$

or

$$2HClO + ClO^- + 2OH^- \longrightarrow ClO_3^- + 2H_2O + 2Cl^-$$

Let us note that the following reactions proposed by Lister, 1952:

$$2HClO \longrightarrow HCl + H^+ + ClO_2^- \qquad \text{(e4)}$$

$$HClO + ClO_2^- \longrightarrow Cl^- + H^+ + ClO_3^- \qquad \text{(e5)}$$

in place of reaction (e3) as suggested by Foerster were invalidated by Tilak *et al.*, 1981. Two other parasitic reactions in solution can be considered :

$$ClO^- \longrightarrow Cl^- + 1/2\,O_2 \qquad \text{(e6)}$$

$$Cl_2\big)_{dissolved} \longrightarrow Cl_2\big)_{gas} \tag{e7}$$

Generally sodium dichromate is added within the electrolyte. Several roles are attributed to this additive (see e.g. Tilak *et al*, 1981) : it suppresses the corrosion of steel cathodes, it acts as a buffer keeping the electrolyte pH within the range 5-7 which is optimal for chlorate production kinetics as shown by Jaksic, Despic *et al.*, 1969. Moreover the formation of a CrO_2 film on the electrode prevents parasitic cathodic reactions (c2) and (c3), thus reducing the losses of hypochlorite and chlorate ions and inhibits oxygen evolution at the anode due to the reaction (a3).
On the other hand, parasitic reactions (e6) and (e7) play an insignificant role under industrial conditions (see Ibl and Vogt, 1981) because firstly reaction (e6) is catalysed by elements such as Co, Ni or Cu but not by Fe or Mg and secondly the concentration of dissolved chlorine remains low.
A lot of research has been devoted to chlorine hydrolysis which leads to antagonistic conclusions about the zone where this reaction (e1) occurs and depending on the operating conditions. Some authors such as Jaksic, Nikolic *et al.*, 1969, 1975, or Wanngard, 1990, determined that the reaction is complete within the anodic diffusion layer while others concluded that the hydrolysis also occurs in the bulk (see Jaksic *et al.*, 1987, Hammar and Wranglen, 1964). Scott, 1988, showed from modelling that the reaction will be complete within the diffusion layer if the kinetic rate constant is higher than 1 s^{-1} while it can range in practice from 0.17 to 11.0 s^{-1}. The discordance of literature led us to estimate the thickness of the reaction and diffusion layers in order to conclude where chlorine hydrolysis should take place under our conditions. The thickness m of the reaction layer is expressed from the chlorine diffusivity D_{Cl_2} and the kinetic rate constant k for hydrolysis as :

$$m = \sqrt{D_{Cl_2}/k} \tag{1}$$

On the one hand, chlorine diffusivity was evaluated by Despic *et al.*, 1972, as 6.7 10^{-10} m^2s^{-1} at 25°C and for a ionic strength ranging from 5 to 6. On the other hand, the kinetic rate constant $k_{25°C,\infty}$ at 25°C for an infinite dilution was estimated as 21 s^{-1} by Spalding, 1962, and as 13.5 s^{-1} by Brian *et al.*,1966. Then Jaksic obtained $k_{25°C}$=0.17 s^{-1} for this constant at 25°C and for a ionic strength ranging from 5 to 6, by assuming that the influence of the ionic strength for chloride solutions was the same as the one shown by Spalding, 1962, for sulphate solutions. If applying this rough estimation method to our working conditions (60°C, ionic strength between 5 and 6), we deduce : $k_{25°C}$=0.37 s^{-1}. Considering this last value, equation (1) leads to a reaction layer thickness of 43 µm. Nevertheless, it seems careful to consider this value as an upper limit and rather to propose for m a probable range from 2 to 50 µm.
The estimation of the diffusion layer thickness δ will be presented and discussed in more detail further in this paper. It leads to propose 50-80 µm as the range of δ values. In conclusion, the hypothesis of a complete chlorine hydrolysis seems to be right and so we will admit here that reaction (e1) only occurs within the anodic diffusion layer. When considering industrial operating conditions at high temperature with

concentrated brine, the kinetic scheme can be simplified and most authors agree to propose a scheme reduced to the only following reactions :
- four electrochemical reactions : anodic reactions (a1), (a2), cathodic ones (c1), (c2)
- three major chemical reactions occurring in solution : reaction (e1) within the anodic diffusion layer and reactions (e2), (e3) within the electrolyte bulk.

Let us consider now the steps limiting the rate of electrochemical reactions at the anode and the cathode.
At the anode, the hypothesis of a mass-transfer control for the hypochlorite reduction (a2) is sustained by many authors as Beck and Brännland, 1972 (40°C, 5.6 A dm^{-2}) or Hammar and Wranglen, 1964 (6.6 - 212 A dm^{-2}) for graphite anodes, Heal *et al.*,1977, for platinated titanium anodes (25°C, 0.025-0.40 A dm^{-2}), Janssen and Barendrecht, 1985, for platinum anodes (25°C, only for current densities higher than 15 A dm^{-2}).
At the cathode, convection-diffusion is generally considered again as the limiting step for reaction (c2) for example by Hammar and Wranglen, 1964, for stainless steel cathodes or Heal *et al.*, 1977, for platinated titanium anodes.

As suggested by literature, we will assume here that reactions (a2) and (c2) are controlled by the diffusion mass transfer of ClO^- ions and so the corresponding current densities can be respectively written as :

$$i_{a2} = \frac{F\, D_{ClO^-}}{\delta_a} C^{ClO^-} \qquad i_{c2} = \frac{2F\, D_{ClO^-}}{\delta_c} C^{ClO^-} \tag{2}$$

where D_{ClO^-} is the diffusion coefficient of hypochlorite, C^{ClO^-} its concentration, δ_a and δ_c the thickness of the anodic and cathodic diffusion layers.
Moreover the total current densities i_a at anode and i_c at cathode verify :

$$i_a S_a = i_c S_c \qquad i_a = i_{a1} + i_{a2} \qquad i_c = i_{c1} + i_{c2} \tag{3}$$

S_a, S_c being the electrode surface areas and i_{a1}, i_{a2}, i_{c1}, i_{c2} the current densities related to the corresponding electrochemical reactions.

3. Model of the cell

Modelling the transient behaviour of the electrolyser implies to simultaneously solve the differential mass, mechanical energy and voltage balances all along the cell at each time. Its complexity comes from the two-phase characteristic of the electrolyte and the numerous reactions occurring both at electrodes and in the bulk.

3.1. HYDRODYNAMICS

The description of hydrodynamics within the electrolyser requires a special attention due to the cell geometry and to the two-phase flow. Therefore the gap between electrodes is considered as a stack of elementary stirred-tank electrochemical reactors in which gas evolution leads to a gas fraction ε varying with height and implies an internal recycle of a two-phase electrolyte under the gas-lift action (Figure 1).

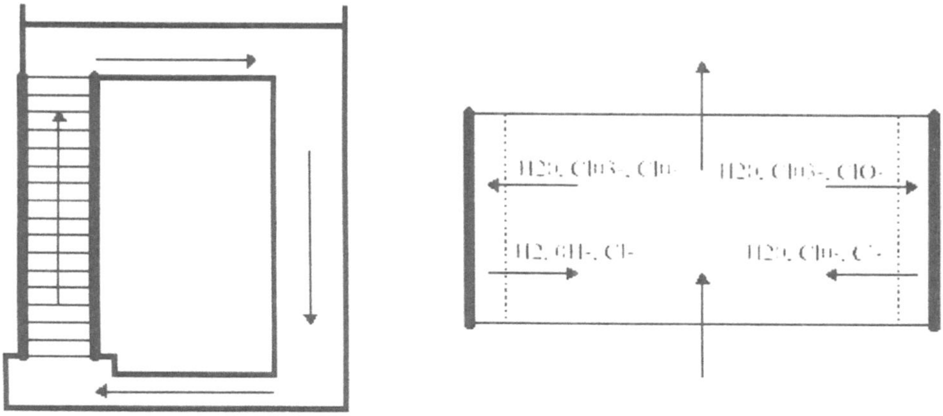

Figure 1. Geometrical model of the electrolyser and elementary cell within the gap

In most cases, the bubble diameter for electrogenerated gases is about 5 10^{-5} m corresponding to a bubble velocity U_∞ in a liquid at rest below 2 10^{-3} m s^{-1} and so negligible in front of the internal recycle velocity $U_0 \cong 0.1$ m s^{-1}. Consequently one can assume that there is no slipping between phases and the homogeneous model can be used considering the two-phase electrolyte as a pseudo one-phase fluid having mean physical properties. The gas fraction is then defined from the liquid and gas flowrates Q_{liq}, Q_{gas} as :

$$\varepsilon = Q_{gas} \big/ \left(Q_{liq} + Q_{gas}\right) \tag{4}$$

In the present case for chlorate electrosynthesis, measurements from photos showed us that the bubble diameter for evolving hydrogen was about 2 10^{-4} m. The velocity U_R of the gas phase related to the liquid phase can be estimated close to 0.02 $m.s^{-1}$ from the following relation proposed by Vogt, 1981 :

$$U_R = U_\infty [1-\varepsilon]^4 \tag{5}$$

valid for $\varepsilon \leq 0.3$ as an approximation of the expression provided by Brauer and Thiele, 1973, under the classical general form used by Bhaga and Weber, 1972. Such velocity cannot be neglected in front of the internal recycle velocity which ranges here from 0.07 to 0.20 $m.s^{-1}$.

Let us note $<X>$ the mean of a variable X for a cross-section A perpendicular to the flow :

$$<X> = \iint_A X dA \tag{6}$$

At a given height, the ratio between the gas flowrate and the total flowrate can be expressed from the velocities U_L, U_G of the liquid and gas phases as :

$$\frac{Q_{gas}}{Q_{liq}+Q_{gas}}=\frac{<\varepsilon U_G>}{<(1-\varepsilon)U_L+\varepsilon U_G>} \tag{7}$$

i.e. by introducing velocity U_R (=U_G-U_L) :

$$\frac{Q_{gas}}{Q_{liq}+Q_{gas}}=\frac{<\varepsilon(U_L+U_R)>}{<U_L+\varepsilon U_R>} \tag{8}$$

If assuming that the velocities and the gas fraction are constant in each elementary stirred-tank reactor, the gas fraction within the gap can be evaluated from :

$$\varepsilon=Q_{gas}\Big/\left(Q_{liq}+Q_{gas}+U_{\infty}[1-\varepsilon]^5 A\right) \tag{9}$$

A being the cross section area of the gap (see Roy, 1992).
So the homogeneous model is no more suitable and the electrolyte must be considered as a two-phase fluid. Then hydrodynamics are described by using simplified equations at a given height deduced from the more general formulation of Dehaye, 1981, and Boure, 1987:

- mass balances

$$\frac{d}{dh}\left[<(1-\varepsilon)\rho_L U_L>\right]=0 \qquad \frac{d}{dh}\left[\rho_G<\varepsilon U_G>\right]=\dot{m}_G \tag{10}$$

where $\dot{m}_G$ is the gas production term, ρ_L, ρ_G the liquid and gas densities.

- momentum balance

$$\frac{d}{dh}\left[<(1-\varepsilon)\rho_L U_L^2+\varepsilon\rho_G U_G^2>\right]+\left[(1-\varepsilon)\rho_L+\varepsilon\rho_G\right]g+\frac{dP_{2\varphi}}{dh}+\frac{dP_{Acceler}}{dh}+\frac{dP}{dh}=0 \tag{11}$$

Equation (11) takes into account kinetic, gravity, and friction terms. Its integration from the bottom (h=0) to the top (h=H) of the cell, both for the upstream and the downstream, finally leads to :

$$\rho_L g\left[H-\int_0^H \varepsilon(h)dh\right]+\rho_L U_0^2(1-\varepsilon_0)\left(\frac{1-\varepsilon_0}{1-\varepsilon_H}-1\right)+\Delta P_{2\varphi}^{upstream}=\rho_L g(1-\varepsilon_0)H+\Delta P_{2\varphi}^{downstream} \tag{12}$$

U_0 being the velocity of the electrolyte internal recycle at the gap lowest part.
The main problem is now the estimation of the friction pressure drops $\Delta P_{2\varphi}$ related to the two-phase flow. Here an empirical correlation based on the theory developed by Lockhart-Martinelli, 1949, is used rather than the equation proposed by Friedel, 1979, to predict with a suitable accuracy these friction pressure drops:

$$\Phi^2=\Delta P_{2\varphi}/\Delta P_L=1+C/\chi+1/\chi^2 \quad \text{with : } \chi^2=\Delta P_L/\Delta P_G \tag{13}$$

This relation requires first to evaluate the friction pressure drops $\Delta P_L, \Delta P_G$ corresponding to the flow of each phase considered as alone in the same system. This is done by using the classical Fanning's equation and by adding equivalent fittings which allow to take into account the changes in electrolyte velocity and direction in the cell.

Then a problem remains about the estimation of constant C which strongly acts on the predicted behaviour of the cell as previously shown for the simulation of a perchlorate cell by Ozil *et al.*, 1994. This large effect is illustrated by figure 2 which shows how the C value influences the theoretical profile of the gas fraction within the gap between electrodes in a perchlorate cell.

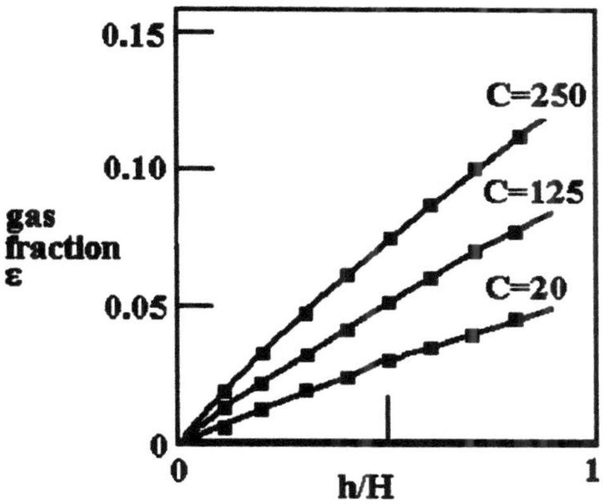

Figure 2. Influence of parameter C on the gas fraction profile within the gap in a perchlorate cell.

For a water-air flow, this constant ranges from 5 to 20 as a function of the flow regime for each phase depending on the Reynolds number to be compared to its critical value $Re_{critical} = 1000$ (Table 1)

liquid phase	gas phase	C
laminar	laminar	5
turbulent	laminar	10
laminar	turbulent	12
turbulent	turbulent	20

Table 1. Determination of the constant C [from Chisholm, 1969 a]

On the other hand Chisholm, 1967 b, proposed for other systems the following relation based on the phase densities ρ_L, ρ_G:

$$C = (\rho_L/\rho_G)^{0.5} + (\rho_G/\rho_L)^{0.5} \tag{14}$$

which leads in the present case to higher C values about 100-200.

In order to obtain a right C value, specific experimental studies were performed on a modified chlorate pilot plant specially designed for this purpose (Figure 3). Complementary investigations were done with a brine used for bromate electrosynthesis. All these studies allow to verify the experimental validity of equation (8) with a value of C close to 125 (Figure 4).

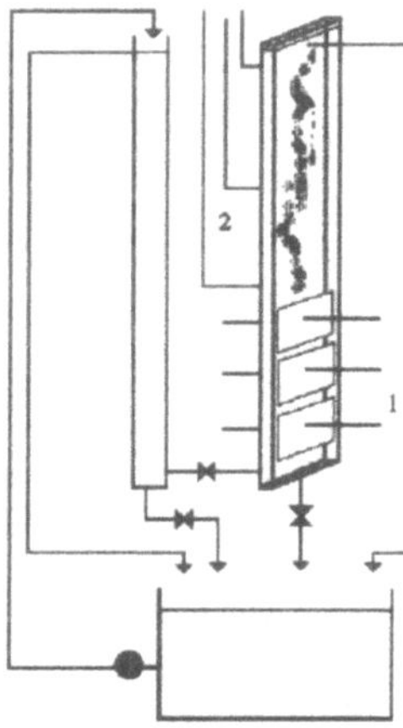

Figure 3. Pilot cell for hydrodynamic studies
1 electrodes - 2 manometers - 3 liquid flowrate measurement

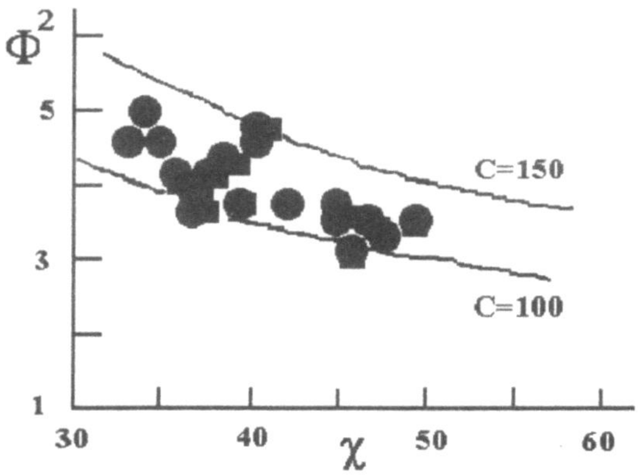

Figure 4. Experimental correlation $\Phi^2 = f(\chi)$

The theoretical model for hydrodynamics was independently tested with this pilot cell and it appeared to provide a suitable predicted velocity for the two-phase electrolyte recycle as shown on figure 5.

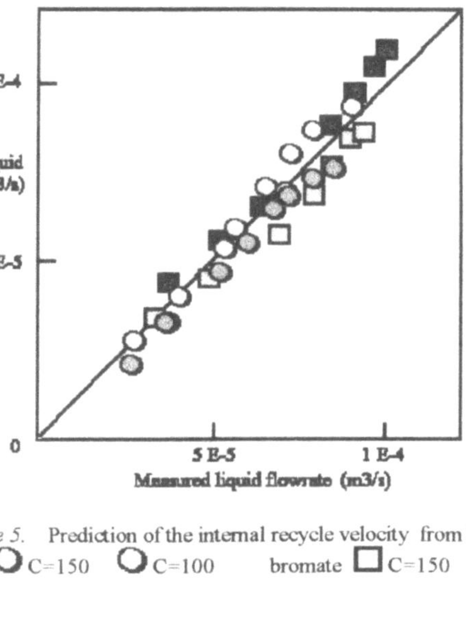

Figure 5. Prediction of the internal recycle velocity from the model
(chlorate ○ C=150 ○ C=100 bromate □ C=150 ■ C=100)

3.2. EVALUATION OF THE DIFFUSION LAYER THICKNESS

The thickness of the diffusion layers depend not only on the electrolyte flow but also on the local gas evolution (respectively hydrogen at cathode and oxygen at anode). Several authors such as Vogt, 1981, and Manju Sharma *et al.*, 1988, chose to use a global mass transfer coefficient to characterise the anode of a chlorate cell. Here we prefer to consider two separate effects to be added as suggested by Beck, 1969 : forced convection due to electrolyte flowing without gas evolution and free convection due to gas evolution without electrolyte flowing. Then the thickness of the diffusion layer δ is evaluated from the thickness δ_{flow} and δ_{gas} of the corresponding diffusion layers :

$$\delta^{-1} = \delta_{flow}^{-1} + \delta_{gas}^{-1} \tag{15}$$

Here the operating conditions correspond to a laminar flow in the cell (electrolyte velocity u). It can be characterised by the following classical relation valid for a flat plate of length L_e and where ν is the kinematic viscosity of the electrolyte (see e.g. Pickett, 1979) :

$$\delta_{flow} = 1.5\left[\frac{\nu}{D_{ClO^-}}\right]^{-1/3}\left[\frac{\nu L_e}{u}\right]^{1/2} \tag{16}$$

The effect of gas evolution on mass transfer depends on many factors such as the nature of the evolving gas, the electrolyte pH and temperature and the current density. Most of research is devoted to oxygen and hydrogen evolution (see e.g. Janssen and Hoogland, 1970, Janssen, 1978) and it allows to conclude that the thickness of the diffusion layer without electrolyte flow is correlated to the current density of the evolution process as :

$$\delta_{gas} = K i_{gas}^{-n} \tag{17}$$

The value of exponent n depends on the range of current density because bubble coalescence can occur when this electrical parameter is large enough, thus modifying the diffusion layer (Table 2).

	Electrolyte	Current density (A m^{-2})	n
O_2 evolution	acid	70 - 1000	0.28
	acid	> 1000	0.57
	basic	< 300	0.33
	basic	> 300	0.87
H_2 evolution	acid or basic	< 400	0.50
	acid	> 5000	0.62
	basic	> 5000	0.36

Table 2. Value of exponent n vs. gas nature and current density [from Janssen and Hoogland, 1973]

For the present simulation the following correlations were used :

$$\delta_{O_2} = 35\ 10^{-3}\ i_{O_2}^{-0.33} \qquad \delta_{H_2} = 66\ 10^{-3}\ i_{H_2}^{-0.5} \tag{18}$$

for a current expressed in mA cm^{-2} and a layer thickness in cm. The n values can be compared to those used by Heal *et al.*, 1977, for modelling an hypochlorite cell ($n_{O_2} = 0.33$, $n_{H_2} = 0.5$). The K values were deduced from these works with a temperature correction using the apparent activation energies determined by Janssen, 1978.

3.3. VOLTAGE BALANCE

The voltage balance is written for each elementary cell j as :

$$E = E_{\min} + \eta_a + |\eta_c| + RI \tag{19}$$

including anodic η_a and cathodic η_c overvoltages as well as ohmic drop RI in addition to the minimal potential E_{min}. Here the overvoltages are evaluated from experimental Tafel equations and the ohmic drop which depends both on current density i_j and gas fraction ε_j is expressed as :

$$RI = r_{2\varphi_j} e\ i_j = r_{liq}\left[1 - \varepsilon_j\right]^{-1.5} e\ i_j \tag{20}$$

by using the Bruggeman equation (Bruggeman, 1935) which is mainly recommended as being reliable for predicting the resistivity $r_{2\varphi}$ of the gas-liquid electrolyte (see Kreysa and Kuhn, 1985).

3.4. MASS BALANCES

The bulk of each elementary cell j within the gap exchanges mass vertically with the bulks of the two adjacent cells and horizontally with the corresponding parts of the electrode diffusion layers (figure 1). The parts of these layers included in cell j are supposed to exchange mass only with the bulk of cell j. Then the mass balance related to cell j (volume AΔH) between times t and t+dt can be written for any species s as :

$$U_0 A\left[C_{j-1}^s - C_j^s\right] + N_s - A\Delta H\left[1 - \varepsilon_j\right]\frac{dC_j^s}{dt} = 0 \tag{21}$$

N_s being the production or consumption rate of species s within the cell j by electrochemical reactions at anode and cathode and by chemical reactions in the bulk.

3.5. NUMERICAL SOLVING

Twenty elementary cells at least are necessary in order to appropriately describe the behaviour within the gap. The numerical model requires to solve the coupled mass, voltage and mechanical energy balances for each elementary cell at each time. It is based on an iterative scheme (Figure 6) presenting at each time three levels as previously used for the simulation of a perchlorate cell at steady-state by Ozil *et al.*, 1988 :

- iteration on the current density distribution as a function of height within the gap,
- iteration on the gas fraction of hydrogen evolving at cathode,
- iteration on the velocity of the electrolyte internal recycle.

The order of the three iteration loops has a great importance to insure convergence of computations. Initial profiles of ClO^- concentration, internal recycle velocity and current density all along the cell are arbitrarily chosen for starting computations. Then iterative calculations are performed until the consistency between these profiles is reached at each time. Finally the transient behaviour of the electrolyser can be analysed from the evolution of the main parameters i.e. current density, recycle velocity, species concentrations and gas fraction as a function of time and of location in the cell.

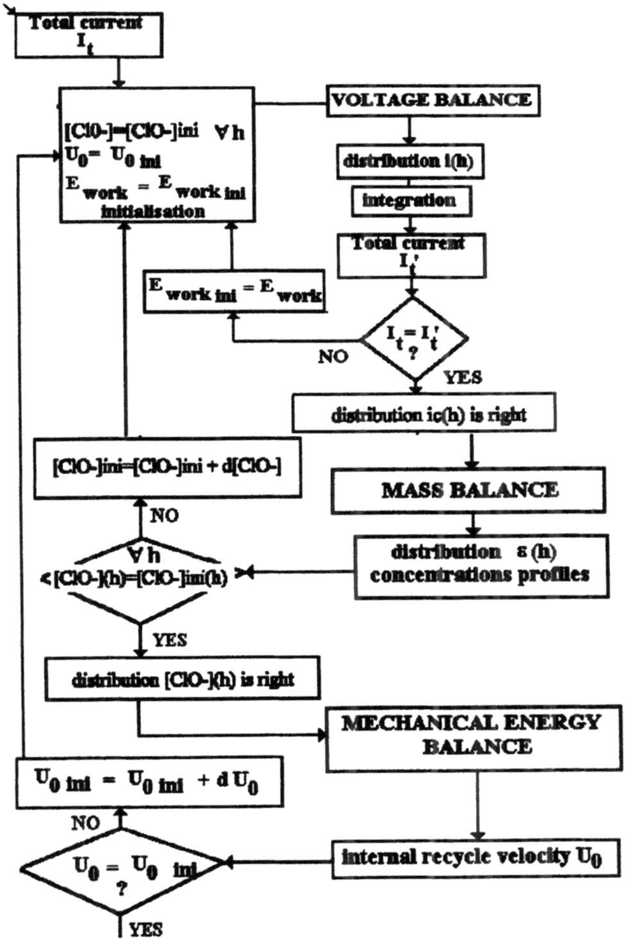

Figure 6. Simplified organigram of the cell model

Figure 7 clearly shows the influence of hydrodynamics on the working of the electrosynthesis. When the length of electrodes increases, the gas fraction increases and can reach high values in the upper part of the gap between electrodes. The presence of gas becomes a significant obstacle against insuring a uniform distribution of current.

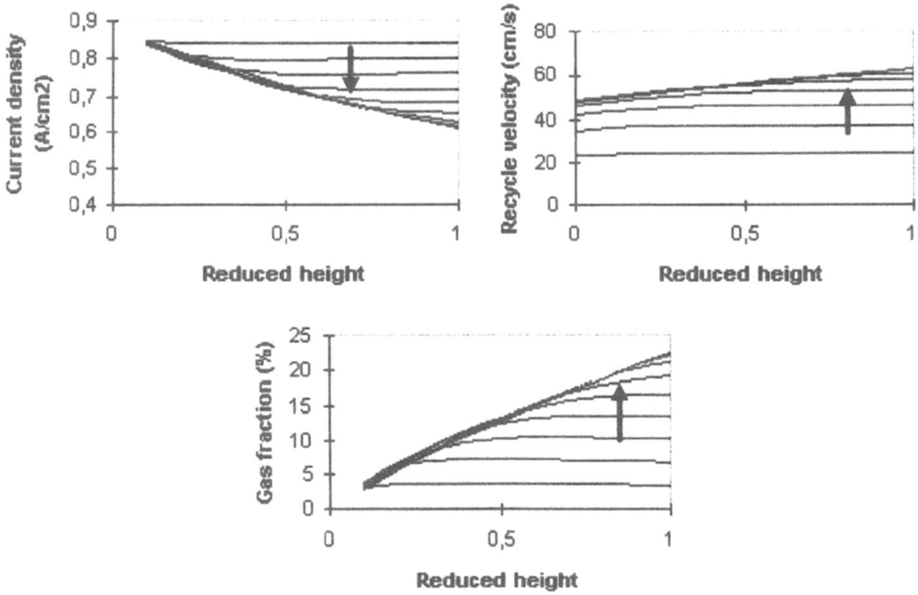

Figure 7. Typical results from simulation : profiles of current density, recycle velocity and gas fraction within the gap between electrodes vs. time from starting to steady-state (↑)

In such conditions the electrosynthesis efficiency decreases because the expected improvement of mass transfer due to gas-lift is not benefit enough to compensate for the poor flow of current within the gap (Figures 8 and 9).

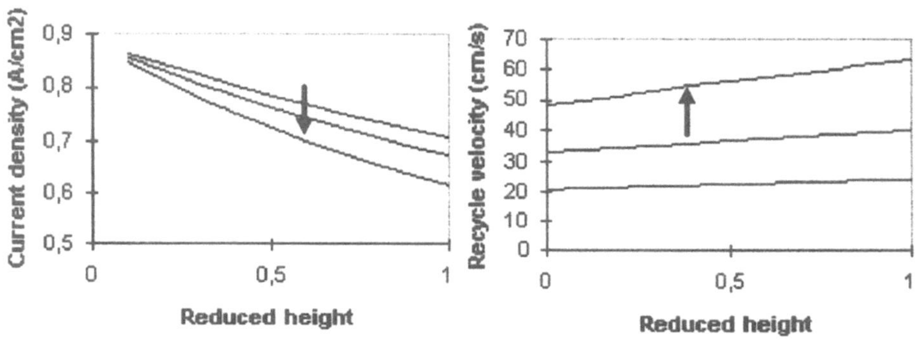

Figure 8. Influence of the electrode length on the current and velocity profiles (increasing length ↑)

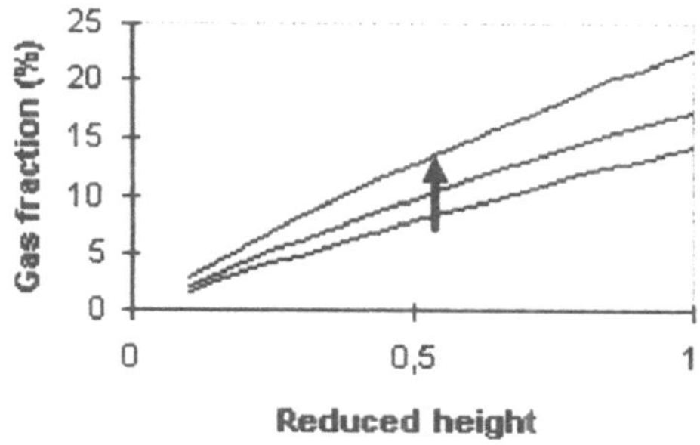

Figure 9. Influence of the electrode length on the gas fraction profile (increasing length ↑)

Thus the cell productivity defined as the chlorate production for a 1 m^2 electrode surface area per second decreases when the electrode length increases (Figure 10).

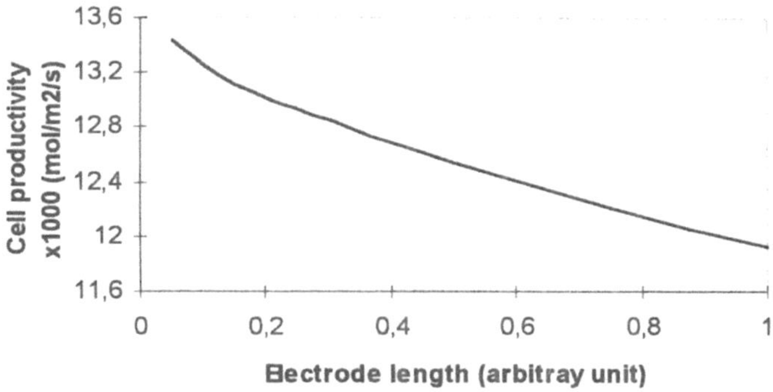

Figure 10. Cell productivity vs. electrode length

Moreover simple physical considerations show that an optimal value should exist for the distance between electrodes.

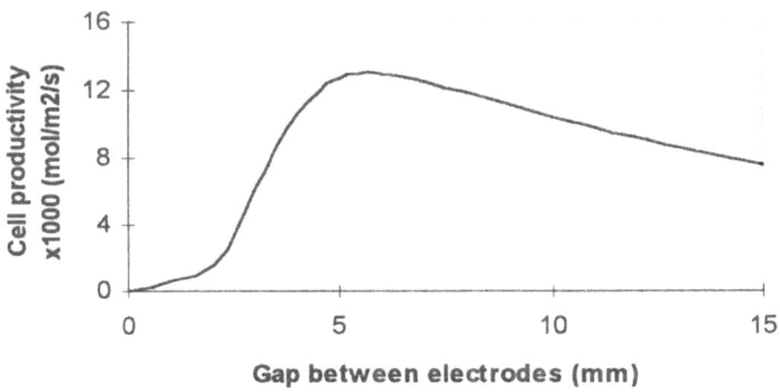

Figure 11. Cell productivity vs. gap between electrodes

On the one hand, when the gap is decreased, the gas fraction within the gap increases leading to an increased recycle velocity which improves the mass transfer. On the other hand, the flow of the current simultaneously becomes more difficult because of gas thus limiting the chlorate production. It is clear that an optimal gap should result from these opposite trends. In fact modelling points out an optimal value for the gap (Figure 11).

Conclusions

Modelling appeared to be an efficient way to study the behaviour of a chlorate electrosynthesis cell. More generally, this study highlights the large influence of hydrodynamics for electrochemical cells wherein gas-lift occurs and the necessity of an accurate knowledge of the electrolyte two-phase flow in such conditions.

References

BHAGA, D. and WEBER, M.E (1972) Holdup in vertical two and three phase flow, *Can. J. Chem. Eng.*, **50**, 323-336.

BECK, T.R. and BRÄNNLAND, R. (1972) Analysis of a Chlorate Cell System with Separate Reactor, *J. Electrochem. Soc.* **119 (3)**, 320-325.

BECK, T.R. (1969) A contribution to the theory of electrolytic chlorate formation, *J. Electrochem. Soc.* **116**, 1038-1041.

BOURE, J. (1987) Two-phase flow models : the closure issue, *Multiphase. Sci. Technol.*, **3**, 3-30

BRAUER H. and THIELE, H. (1973), Bewegung von partikel schwärmen , *Chem. Ing. Techn.*, 45, 902-912

BRIAN, P.L.T., VIVIAN, J.E. and HABIB, A.G. (1966) The effect of temperature on the rate of absorption of chlorine into water, *Chem. Eng. Science* **21**, 551-558.

BRUGGEMAN, D.A.G. (1935) The calculation of various physical constants of heterogeneous substances. I. The dielectric constants and conductivities of mixtures composed of isotropic substances, *Ann. Physik* **24**, 636-664.

CHISHOLM, D. (1967) A theoretical basis for the Lockhart-Martinelli correlation for two-phase flow, *Int. J. Heat Mass. Transfer* **10**, 1767-1778.

CHISHOLM, D. (1967) Pressure gradients during the flow of incompressible two-phase mixtures through pipes, venturis and orifice plates, *British Chem. Eng.* **12 (9)**, 1368-1371.

DELHAYE, J.M. (1981) Local instantaneous equations, in *Thermohydraulics of two-phase flow systems for industrial design and nuclear engineering*, Delahaye, J.M., Giot, M., Riethmuller, M.L., Ed., Mc Graw Hill, New York,

DESPIC, A.R., JAKSIC, M.M. and NIKOLIC, B.Z. (1972) The effect of kinetic and hydrodynamic factors on current efficiency in the chlorate cell process, *J. Applied Electrochem.* **2**, 337-343.

FRIEDEL, L. (1979) Improved friction pressure drop correlations for horizontal and vertical two phase pipe flow, *European Two Phase Flow Group Meeting* **paper E2**, Ispra, 5-8 June.

HAMMAR, L. and WRANGLEN, G. (1964) Cathodic and anodic efficiency losses in chlorate electrolysis, *Electrochimica Acta* **9**, 1-16.

HEAL, G.R., KUHN, A.T. and LARTEY, R.B. (1977) A Parametric Study and Computer-Based Simulation of an Undivided Sodium Hypochlorite Electrolyzer, *J. Electrochem. Soc.* **124 (11)**, 1690-1697.

HINE, F. (1988) *Electrode process and electrochemical engineering*, Plenum Press, New York.

IBL, N. and VOGT, H. (1981) Inorganic synthesis, in J.O.M. Bockris, B.E. Conway and R.E. White, *Comprehensive treatise of electrochemistry, Vol. 2 : Electrochemical processes*, Plenum Press, New York, pp. 167.

JAKSIC, M.M., NIKOLIC, B.Z., CSONKA, I.M. and DJORDJEVIC, A.B. (1969) Studies on Chlorate Cell Process : IV. Effect of Neutral Salts on Conversion of Available Chlorine to Chlorate, *J. Electrochem. Soc.* **116 (5)**, 684-687.

JAKSIC, M.M., DESPIC, A.R., CSONKA, I.M. and NIKOLIC, B.Z. (1969) Studies on Chlorate Cell Process, *J. Electrochem. Soc.* **116 (9)**, 1316-1322.

JAKSIC, M.M., NIKOLIC, B.Z. and SPASOJEVIC, M.D. (1975) Les cellules à chlorate à circulation d'électrolyte simple, *Chem. Techn.* **27** **(9)**, 534-538.

JAKSIC, M.M., LACNJEVAC, C., CEZNER, V. and ROUSAR, I. (1987) Hydrodynamic optimization of energy in the chlorate cell process and sea water electrolysis, *Meeting of the Electrochemical Society, Honolulu, Hawaii* **88** **(2)**, 112-126.

JANSSEN, L.J.J. and BARENDRECHT, E. (1985) Mechanism of mass transfer of indicator ions to an oxygen-evolving and a hydrogen-evolving electrode in alkaline solution, *Electrochimica Acta* **30** **(5)**, 683-694.

JANSSEN, L.J.J. and HOOGLAND, J.G. (1970) The effect of electrolytically evolved gas bubbles on the thickness of the diffusion layer, *Electrochimica Acta* **15**, 1013-1023.

JANSSEN, L.J.J. (1978) Mass transfer at gas evolving electrodes, *Electrochimica Acta* **23**, 81-86.

JANSSEN, L.J.J. and HOOGLAND, J.G. (1973) The effect of electrolytically evolved gas bubbles on the thickness of the diffusion layer II, *Electrochimica Acta* **18**, 543-550.

KREYSA, G. and KUHN, M. (1985), Modelling of gas evolving cells. I. The gas voidage problem, *J. Applied Electrochem.* **15**, 517-526

LISTER, M.W. (1952) La décomposition de l'acide hypochloreux, *Canadian J. Chem.* **30**, 879-889.

LOCKHART, R.W. and MARTINELLI, R.C. (1949) Proposed correlation of data for isothermal two-phase, two-component flow in pipes, *Chem. Eng. Progress* **45** **(1)**, 39-48.

MANJU SHARMA, USHA SHANBHAG and SADANANDA RAO, B.K. (1988) Circulation rate on the pseudo mass transfer of hypochlorite in a chlorate cell in the absence of dichromate, *Bull. Electrochem.* **4** **(3)**, 259-261.

OZIL, P., DELACHAUME, J.C., CAIRE, J.P. and MILLET, J.C. (1988) Modélisation du foctionnement continu d'un électrolyseur avec dégagement gazeux, *Récents Progrès en Génie des Procédés* **6** **(2)**, 341-346.

OZIL, P., CAIRE, J.P. and DELACHAUME, J.C. (1994) Modelling of an electrochemical cell under gas-lift conditions for sodium perchlorate electrosynthesis, *Annual meeting of Electrochemistry 94, Edinburg, july 5-7.*

PICKETT, D.J. (1979), *Electrochemical reactor design*, Elsevier, Amsterdam

ROY, T. (1992) Etude hydrodynamique d'une cellule d'électrolyse soumise à l'effet de gas-lift, Thesis, INPG, Grenoble

SPALDING, C.W. (1962) Reaction kinetics in the absorption of chlorine into aqueous media, *AIChE Journal* **8**, 685-689.

SCOTT, K. (1988) An analysis of electrochemical reactions followed by series or complex chemical reactions, *J. Electroanal. Chem.* **243**, 1-19.

TILAK, B.V., VISWANATHAN, K. and RADER, C.G. (1981) On the Mechanism of Sodium Chlorate Formation, *J. Electrochem. Soc.* **128** **(6)**, 1228-1232.

VISWANATHAN, K. and TILAK, B.V. (1984) Chemical, Electrochemical, and Technological Aspects of Sodium Chlorate Manufacture, *J. Electrochem. Soc.* **131** **(7)**, 1551-1559.

VOGT, H. (1981) A hydrodynamic model for the ohmic interelectrode resistance of cells with vertical gas evolving electrodes, *Electrochim. Acta* **26**, 1311-1317.

WANNGARD, J. (1990), Oxygen formation in chlorate electrolysis, communication at the Joint meeting of Dechema and Society of Chemical Industry : Electrochemical cell design and optimisation procedures, Bad Soden.

MHD AND MICRO-MHD EFFECTS IN ELECTROCHEMICAL SYSTEMS

Ryoichi Aogaki, Atsushi Tadano and Kenji Shinohara

National Research Laboratory for Magnetic Science
Japan Science and Technology Corporation
1-1-56, Shibashimo, Kawaguchi, Saitama, 333, Japan

Department of Product Design, Polytechnic University
4-1-1, Hashimoto-dai, Sagamihara, 229, Japan

Abstract

Using a new type of electrode, MHDE, copper electrodeposition from acidic copper sulfate solution was examined.

Resultantly, remarkable hysteresis effect of the magnetic field on reaction rate emerged, accompanied with drastic decrease of the reaction activity.

To clarify this phenomenon, another experiment, *i.e.*, copper dissolution in nitric acid was carried out. The results obtained exhibited the quite curious facts that in concentrated nitric acid, the dissolution process is almost blocked by magnetic field, whereas dilute nitric acid, on the contrary, makes acceleration of the dissolution.

All these results revealed the existence of micro-scopic MHD flow induced by the micro-scale electrochemical process and the magnetic field, which affects the unstable growth of the concentration fluctuation of copper ion to restrain or enhance the reaction.

1. Introduction

According to recent development of oxide-superconducting materials and high performance cryocooler, a new type of superconducting magnet, *i.e.*, liquid helium-free superconducting magnet is being produced, which easily allows us to utilize high magnetic field of the order of 10 T [1]. At the same time, much attention has been focused on various phenomena in high magnetic field.

For example, even on paramagnetic or diamagnetic materials, such a high magnetic field can exert the magnetic force comparable in magnitude to the gravity force on the earth. Beaugnon and Tournier demonstrated that the levitation of water and organic substances in high magnetic field of 27 T is possible [2]. So-called Moses and reversed Moses effects are another good examples, where suppression and elevation of liquid surfaces according to diamagnetic and paramagnetic susceptibilities take place [3,4]. Magnetic ordering effects such as the alignment of red blood cells [5] and polymer

A. Alemany et al. (eds.),
Transfer Phenomena in Magnetohydrodynamic and Electroconducting Flows, 169–179.

molecules [6] were also observed.

All these effects are characterized by the thermodynamic equilibrium. On the other hand, nonequilibrium phenomena far from the equilibrium have been also known. The MHD effect in electrochemistry is one of the theoretically well-established phenomena, resulting from the solution flow induced by the Lorentz force which is generated by the interaction between the electrolytic current and external magnetic field. Resultant enhancement of mass transfer process (*i.e.* diffusion current) can be observed. Aogaki et al. [7-9] derived a theoretical equation of the diffusion current in the case of a pair of electrodes embedded on the inner walls of a rectangular channel with two open ends, where electrolytic current flows between the electrodes. Figures 1.A. and 1.B. illustrate the schematic configuration of the electrode cell which we call the magneto-hydrodynamic electrode (MHDE) and its cross section, respectively.

The equation of the diffusion current density is written as

$$i = H^{*}(C_0 - C_s)^{4/3} B^{1/3} \tag{1}$$

where C_0 and C_s are bulk and surface concentrations of active species, respectively, B is external magnetic flux density, and H^* is a constant written as

$$H^{*} = 0.753(nFD)^{4/3}(\nu/D)^{4/9}\nu^{2/3}\rho^{-1/3}\gamma^{1/3}(\sqrt{X_2} - \sqrt{X_1})^{4/3}(X_2 - X_1)^{-1} \tag{2}$$

where n electron number participating reaction, F Faraday constant, D diffusion coefficient, ν dynamic viscosity, ρ density and γ is cell constant. As shown in Fig. 1.B., X_1 and X_2 are the coordinates of the electrode edges at the higher and lower parts of the stream, respectively.

Then, they deduced another equation to examine the electrochemical reaction in magnetic field [10, 11]. That is,

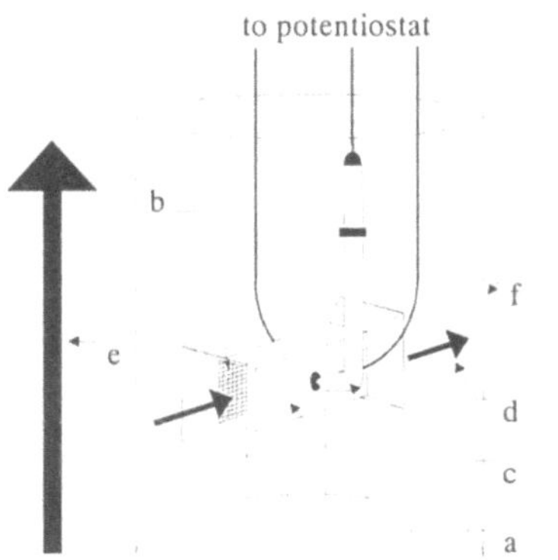

Figure 1.A. MHDE configuration.
a; working electrode, b; counter electrode, c; Luggin capillary, d; MHD flow, e; magnetic flux density, f; vessel.

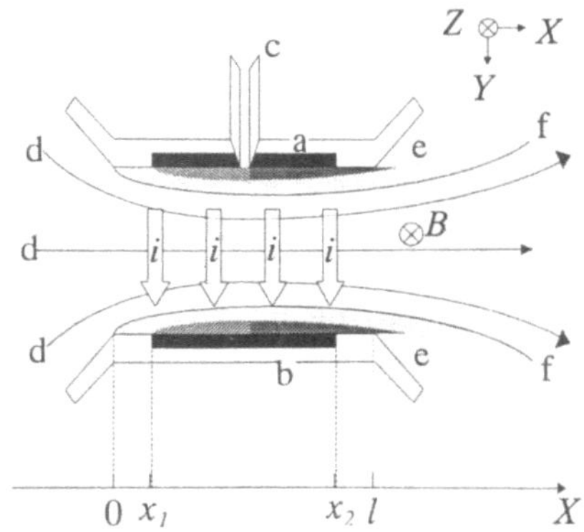

Figure 1.B. Cross section of MHD flow.
a; working electrode, b; counter electrode, c; Luggin capillary, d; stream line, e; diffusion layer, f; boundary layer, i; current density, *B*; magnetic flux density.

$$1/i - 1/i_k = 1/\left\{H^{*4/3} C_0 (Bi)^{1/4}\right\} \quad (3)$$

where i is observed current density, and i_k is reaction current density corresponding to reaction rate in chemical reaction.

Applying a constant potential to the working electrode, we can observe the current response by sweeping the magnetic flux density, so that the locus obtained from the plot of $1/i$ against $1/(Bi)^{1/4}$ may converge to a straight line expected from Eq. (3). We call this new method the cyclic magnetammetry (CM) to analyze the magnetic field effect in electrochemical reaction.

In the present paper, using CM, the MHD effect on concentration fluctuation of active species in copper electrodeposition is firstly examined, in which drastic suppression of the reaction takes place. Secondly, a new type of the MHD effect, *i.e.*, the micro-MHD effect is discussed with regard to copper chemical dissolution. In this case, the acceleration of the reaction together with the suppression are observed. Finally, the mechanism of the micro-MHD effect is clarified with CM mainly from the electrochemical viewpoint.

2. Experimental

The magnet used in experiments was a liquid helium-free superconducting magnet (Sumitomo Heavy Industries Ltd., Model HF10-100VHT-1), which can generate the magnetic flux densities up to 10 T, having a 10 cm bore diameter. For the purpose of rapid sweep of the magnetic flux density, a water-cooled resistance magnet at Institute for Material Science, Tohoku University was employed, which can sweep the magnetic field from 0 T to 15 T at the highest rate of 15T min^{-1}.

As mentioned above, Fig.1A shows the schematic configuration of the MHDE. The magnetic field is applied vertically to the electrolytic current flowing between the working and counter electrodes. The potential of the working electrode was controlled by an Ag/AgCl electrode through a Luggin capillary. The electrode was immersed in a vessel filled with test solution and whole vessel was settled inside the magnet bore. The flow rate of the solution was regulated by the magnetic flux density, so that the reaction rate (*i.e.*, reaction current) was controlled by the external magnetic field. Platinum and copper electrodes with 1× cm^2 area were used corresponding to the following redox and copper reactions, respectively. The test solutions were composed of: 10 mmol dm^{-3} $K_3[Fe(CN)_6]$ +10 mmol dm^{-3} $K_4[Fe(CN)_6]$ with 1 mol dm^{-3} KCl for the redox reaction, 0.3 mol dm^{-3} $CuSO_4$ + 0.5 mol dm^{-3} H_2SO_4 for the copper electrodeposition, and 3 mol dm^{-3} HNO_3 and 0.005 mol dm^{-3} HNO_3 + 2 mol dm^{-3} KNO_3 for the copper chemical dissolution. The temperature was kept at 11±1℃. Prior to experiments, all the solutions were deaerated by argon bubbling.

3. Results and discussion

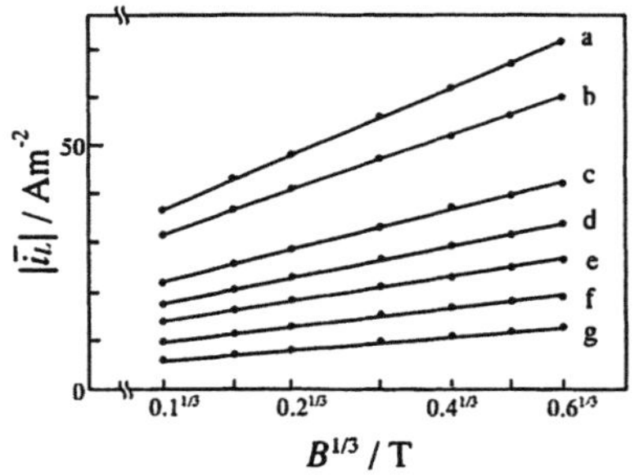

Figure 2. Enhancement of diffusion current by magnetic flux density. Diffusion current was measured for copper electrodeposition in copper sulfate solution with 0.5 mol dm^{-3} sulfuric acid [8].
a; $C_0 = 7.6\times10$ mol m^{-3}, b; $C_0 = 6.8\times10$ mol m^{-3},
c; $C_0 = 5.2\times10$ mol m^{-3}, d; $C_0 = 4.4\times10$ mol m^{-3},
e; $C_0 = 3.6\times10$ mol m^{-3}, f; $C_0 = 2.8\times10$ mol m^{-3},
g; $C_0 = 2.0\times10$ mol m^{-3}.

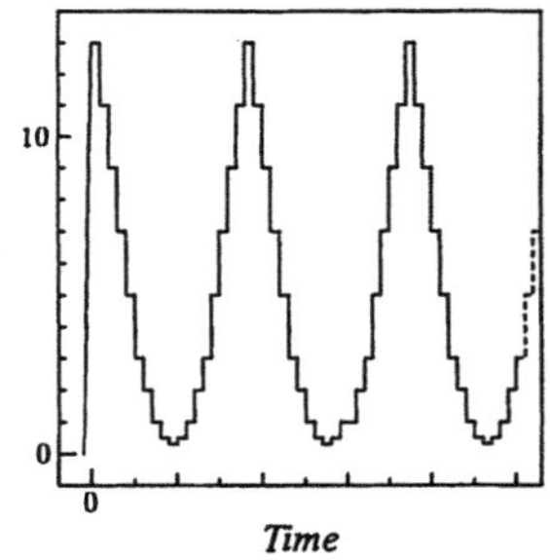

Figure 3. Steady sweep mode of magnetic flux density for CM.

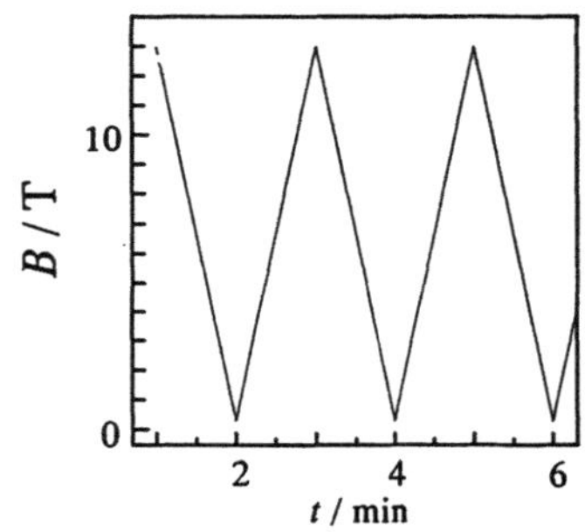

Figure 4. Nonsteady sweep mode of magnetic flux density for CM.

3.1 Application of the MHD effect to CM

Figure 2 shows the relationship between the limiting diffusion current and the magnetic flux density in copper electrodeposition. As shown in the figure, according to Eq. (1), the diffusion current increases in proportion to 1/3 order of the magnetic flux density. This means the acceleration of mass transport process of cupric ions by the MHD effect.

As the concentration of copper sulfate increases, the total reaction is gradually controlled by the electron transfer process (reaction) instead of the mass transfer process (diffusion). According to Eq. (3), we examined the copper electrodeposition in the mixed controlling state of diffusion and reaction. To make the plot corresponding to Eq. (3), two different sweep modes of the magnetic flux density were taken, which are shown in Figs. 3 and 4, *i.e.*, the steady mode and nonsteady mode. Figure 3 corresponds to the case when in various constant magnetic fields, electrochemical reaction proceeds, whereas in Fig. 4, current response is measured by sweeping the magnetic flux density.

First of all, for verifying the validity of Eq. (3), as shown in Figs. 5 and 6, the CM for the ferricyanide/ferrocyanide redox reaction was performed in these two sweep modes. Figure 5 is the result of the steady sweep mode, so that a straight line predicted by Eq. (3) is obtained. In Fig. 6, with decreasing rate, all the loci converge to the same straight line as obtained in Fig. 5 [11].

3.2 Copper electrodeposition

After above preparation, we carried out, as shown in Figs. 7 and 8, the CM's for the copper electrodeposition under mixed control of diffusion and reaction. In steady mode

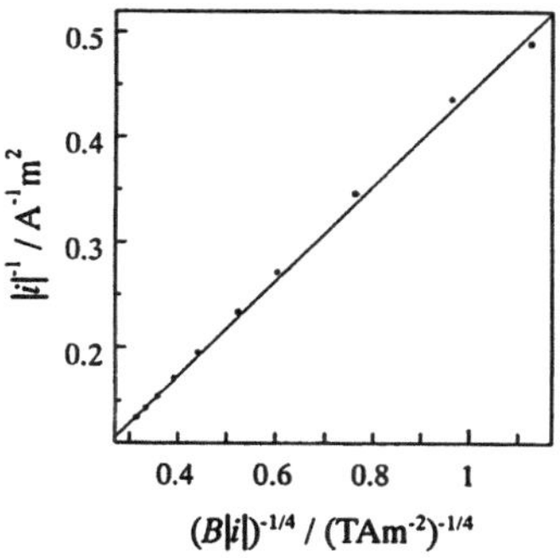

Figure 5. Locus of CM in steady sweep mode for ferricyanide reduction in ferricyanide/ferrocyanide system. $[K_3Fe(CN)_6]$; 10 mol m^{-3}, $K_4Fe(CN)_6]$; 10 mol m^{-3}, [KCl]; 1000 mol m^{-3}, $T = 283 \pm 1$ K, and overpotential is -150 mV.

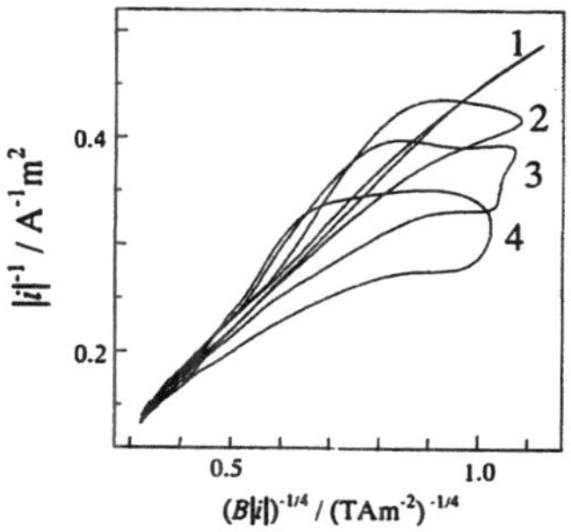

Figure 6. Loci of CM in nonsteady sweep mode for ferricyanide reduction in ferricyanide/ferrocyanide system. Experimental conditions except for sweep rate are the same as Fig. 5. 1; steady mode, 2; 1 T min^{-1}, 3; 3 T min^{-1}, 4; 7.5 T min^{-1}.

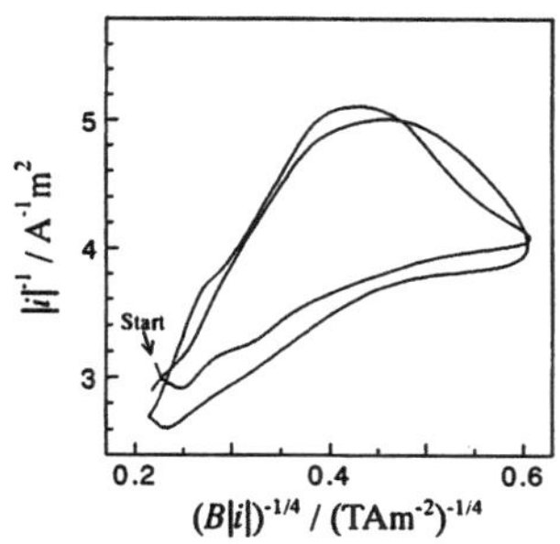

Figure 7. Locus of CM in steady sweep mode for copper electrodeposition in 300 mol m^{-3} copper sulfate with 500 mol m^{-3} sulfuric acid. $T = 283 \pm 1$ K and overpotential is -110 mV.

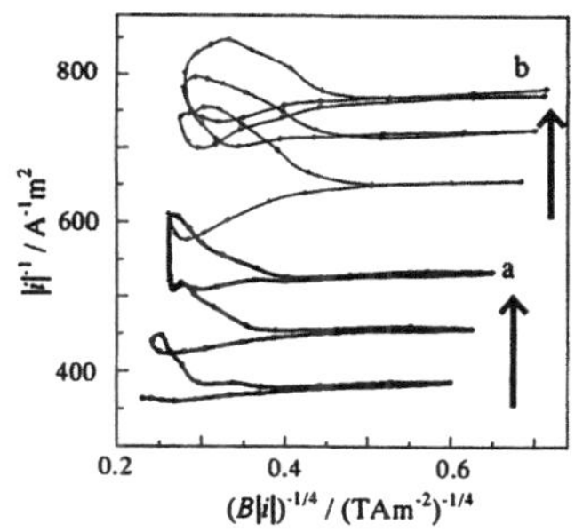

Figure 8. Loci of CM in nonsteady sweep mode for copper electrodeposition. All the experimental conditions are the same as Fig. 7. a; 3 T min^{-1}, b; 1 T min^{-1}.

shown in Fig. 7, the locus does not draw a straight line but indicates a large hysteresis, which is different from Fig. 5. For the nonsteady CM in Fig. 8, instead of the convergence to a straight line, the current decrease after every cycle makes the locus shifted upward because of the reciprocal plot of the current in the ordinate. This means that the reaction activity of the copper electrodeposition is lowered by the MHD effect.

Generally, in the reaction such as copper deposition, it is known that the heterogeneous concentration distribution of metal ions is formed in the neighborhood of the electrode surface. Extracting the component of spatial variation, we call it "the nonequilibrium fluctuation" since such distribution develops with the nonequilibrium processes such as reaction and diffusion. The sign of the fluctuation is defined negative because of the depletion of metal ions for deposition. Figure 9 schematically illustrates the formation process of the fluctuation. As shown in the figure, the concentration fluctuation autocatalytically develops with the reaction [12,13].

In the MHDE system, as shown in Fig. 10, the solution flow (the MHD flow) which

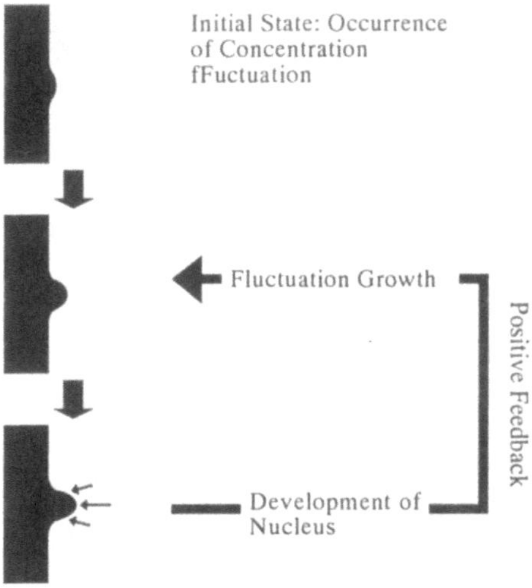

Figure 9. **Unstable growth of concentration fluctuation accompanied with electrodeposition.**

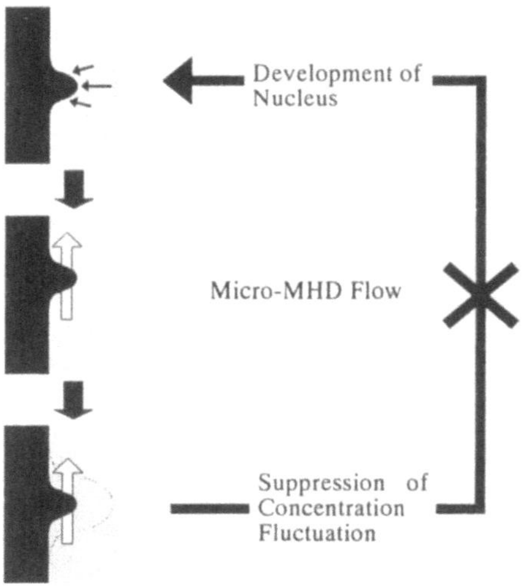

Figure 10. **Suppression of unstable growth of concentration fluctuation by MHD flow.**

simultaneously occurs inside the channel removes the fluctuation away from the electrode surface. This movement destroys the positive feedback cycle between the concentration fluctuation and the copper nucleus, and finally reduces the deposition rate. Such the negative catalysis effect by the MHD effect is assured by the difference of the morphology of the copper deposits in the absence and presence of high magnetic fields. As shown in Fig. 11, the surface deposited in reduced activity under high magnetic field looks like a flat surface formed in the presence of inhibitors.

3.3 Micro-MHD effect in copper chemical dissolution

The MHDE amplifies the depression of the fluctuation through a feedback process between the reaction and the solution flow. Thus, from the decrease of the electrolytic current shown in Fig. 8, we can easily determine whether the reaction is controlled by the unstable growth of the fluctuation or not.

In electrodeposition, since the electrode itself participates the reaction, unstable growth of the concentration fluctuation becomes important because the reaction is essentially heterogeneous along the surface. In such a system, locally heterogeneous current, *i.e.*, the current fluctuation also develops, and in the magnetic field, yields localized Lorentz forces which produce special convection cells of the solution in the vicinity of the electrode surface. These convection cells are supposed to have much smaller sizes in comparison with the usual MHD flow. Therefore, we especially call the effect "the micro-MHD effect". Figure 12 shows one of the examples of the extraordinary growth of copper crystal on the copper MHDE in copper sulfate solution; a sharp drill-like crystal appears, which is thought to reflect the

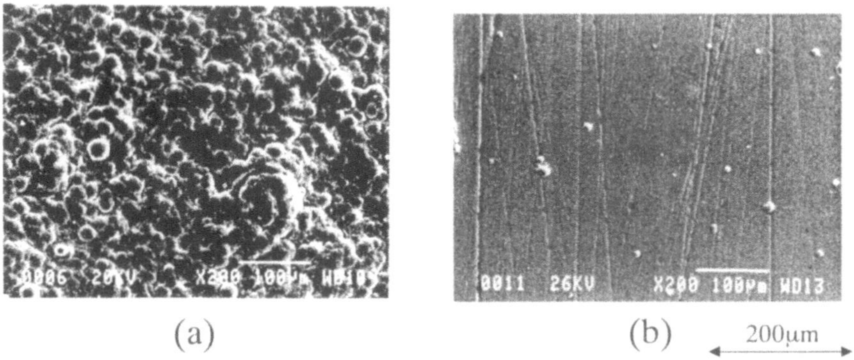

Figure 11. SEM photos of deposited copper surfaces in the absence (a) and presence (b) of magnetic field. Solution composition and temperature are the same as Fig. 7. Magnetic field were swept between 0.3 T and 13 T in the rate of 0.5 T min^{-1}. Deposition time is about 60 min. Applied overpotential is -110 mV [11].

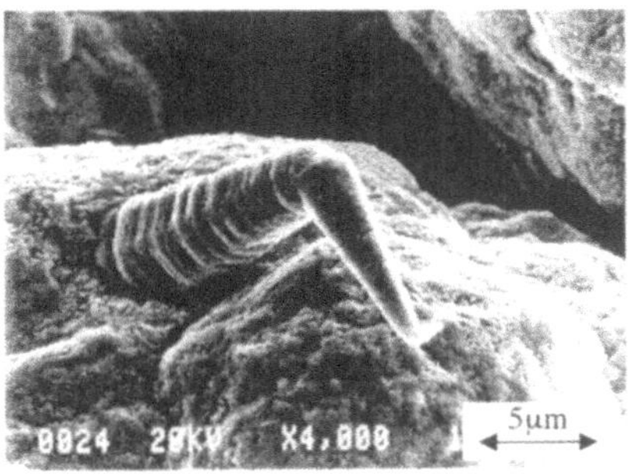

Figure 12. Extraordinary copper crystal growth on copper MHDE in 300 mol m^{-3} copper sulfate with 1000 mol m^{-3} sulfuric acid.
$T = 283 \pm 1$ K and overpotential is -90 mV.

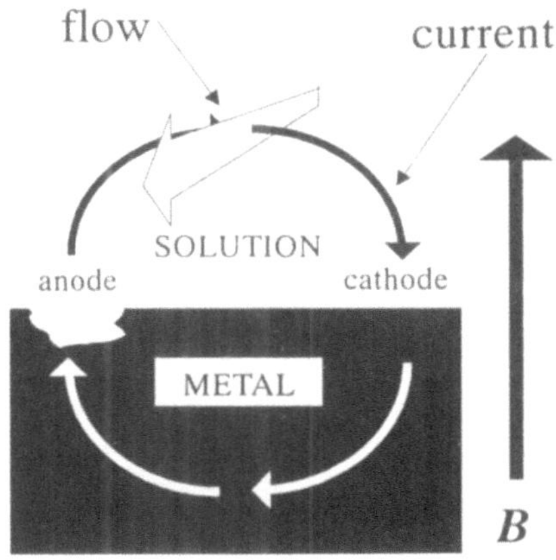

Figure 13. Micro-MHD flow and local cell

special convection mode of the micro-MHD flow.

To clarify the micro-MHD effect, it is the most convenient to pick up the case where these is no net electrolytic current which leads to the usual macroscopic MHD flow.

Copper chemical dissolution in nitric acid is one of the good examples, where no net current flows, and the reaction processes are expressed as follows; as shown in Fig.13, the reaction proceeds between the copper-dissolution sites (oxidation sites) and the nitric-acid decomposition sites (reduction sites) which separately exist on the copper surface, forming so-called electrochemical local cells.

According to the anion-adsorption model[14-16], at the oxidation site, the dissolution proceeds by the specific adsorption of nitrate ions on copper surface as follows,

$$Cu(ad) + 2NO_3^-(aq) \rightarrow Cu(NO_3)_2(ad) + 2e \tag{4}$$

$$Cu(NO_3)_2(ad) \rightarrow Cu^{2+}(aq) + 2NO_3^-(aq) \tag{5}$$

At the reduction site, nitric acid is complementarily decomposed in the following [17],

$$NO_3^-(aq) + 3H^+(aq) + 2e \rightarrow HNO_2(aq) + H_2O \tag{6}$$

where "ad" means the adsorption, and "aq" means the hydration.

In this cell, since a complete electric circuit is formed, we can not observe any net current, *i.e.*, the total oxidation current and reduction current are balanced with each other, and the rest potential is kept at the corrosion potential. As shown in Fig.13, local current flows not only in the metal phase, but also in the solution phase, so that the localized Lorentz force induced by the current and the external magnetic field makes a micro-MHD flow, which disturbs the concentration fluctuation yielded by dissolution, restricting the reaction in the similar way to Fig. 10. Figure 14 schematically elucidates the suppression of unstable growth of concentration fluctuation by the micro-MHD flow. Different from the deposition, reflecting the enrichment of dissolved copper ions, the fluctuation takes positive value.

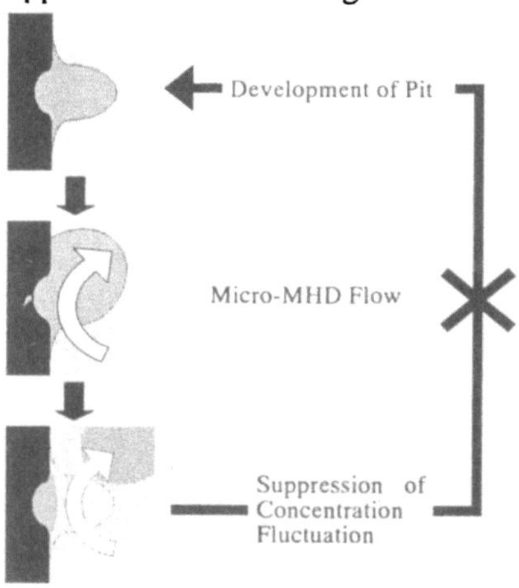

Figure 14. Suppression of unstable growth of concentration fluctuation by micro-MHD flow.

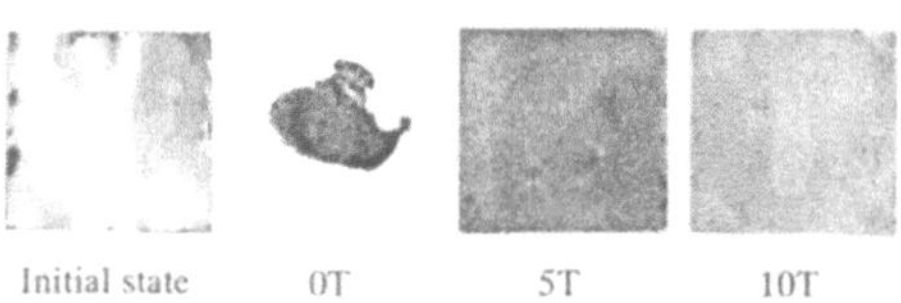

Figure 15. Photos of copper sheets (2.5×2.5 cm^2 in area and 0.1 mm thick) after 20 min immersion in 3 mol dm^{-3} nitric acid at the magnetic flux densities of 0 T, 5 T and 10 T. Temperature is 11±1 ºC.

Actually, a copper sheet (2.5×2.5 cm^2 in area and 0.1mm thick) dipped in 3 mol dm^{-3} nitric acid for 20 min was, as shown in Fig.15, drastically attacked in the absence of magnetic field. It was however effectively protected with increasing magnetic flux density [18]. Figure 16 shows the dependence of the dissolution rate on the magnetic flux density calculated from the weight-loss measurement [18]. At 10 T, the corrosion rate becomes 1/4 times smaller than that at 0 T.

Above discussion may hold when anodic dissolution controls the whole reaction; with the copper MHDE, taking the CM of anodic current in the same solution as Figs. 15 and 16, we can see, as shown in Fig. 17, a remarkable upward shift of the locus in the similar way to Fig. 8. This means that the anodic dissolution is restricted by the magnetic field; in this case, since the reaction is in the

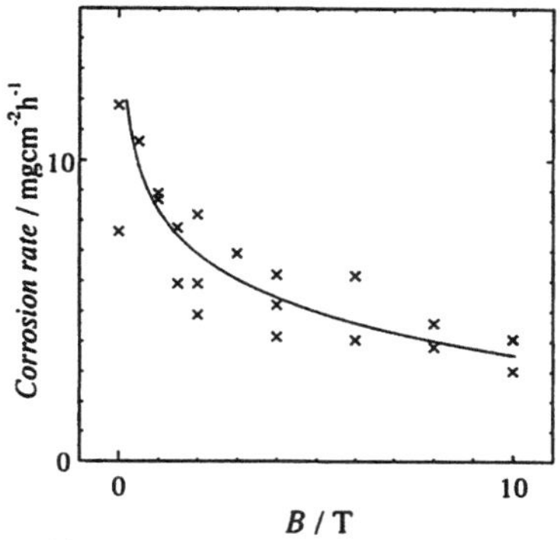

Figure 16. Decrease of dissolution rate of copper in 3 mol dm^{-3} nitric acid with magnetic flux density. Temperature is 11 ± 1 ºC.

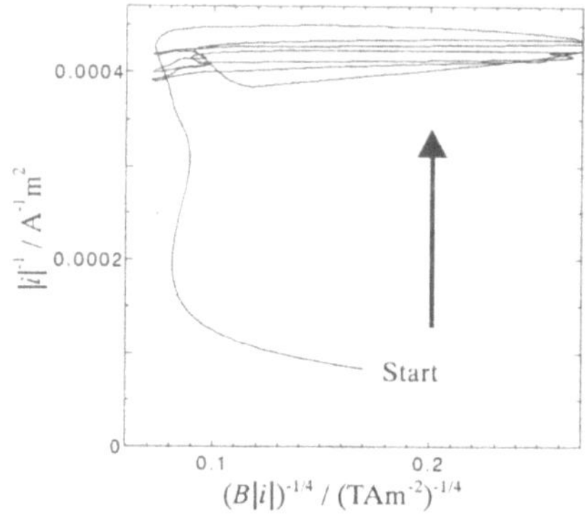

Figure 17. Locus of CM in nonsteady sweep mode for anodic reaction of copper chemical dissolution in 3 mol dm^{-3} nitric acid. Applied overpotential is 50 mV and Temperature is 11 ± 1 ºC.

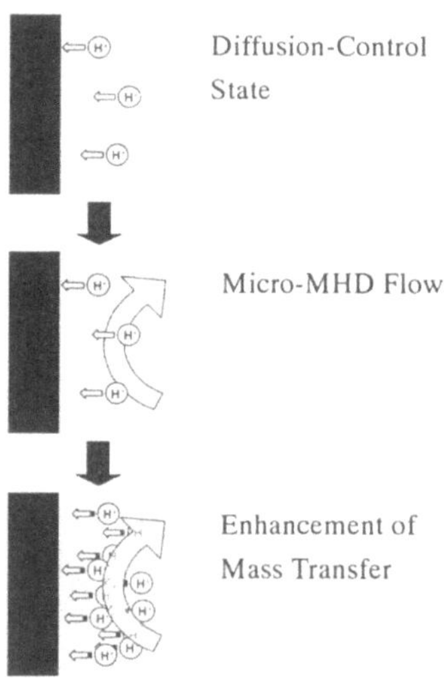

Figure 18. Acceleration of proton transfer by micro-MHD flow.

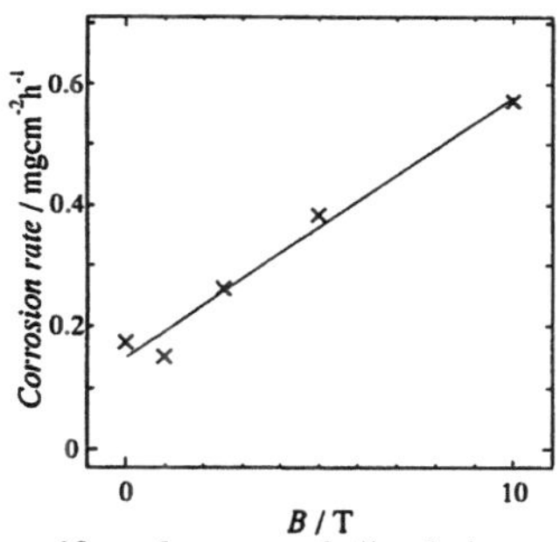

Figure 19. Increase of dissolution rate of copper with magnetic flux density. Solution is 0.005 mol dm^{-3} nitric acid + 2 mol dm^{-3} potassium nitrate. Temperature is 11 ± 1 ºC.

mixed control of anodic and cathodic reactions, the cathodic current also decreases [18].

If the cathodic reaction is the rate-determining step, we can predict a quite different result. The cathodic reaction is expressed by the reduction of nitric acid in Eq. (6). In the presence of small amount of proton and large amount of nitrate salt, it can be said that the mass transfer of proton controls the whole reaction processes. Though we can also expect the concentration fluctuation of proton, as shown in Figs. 5 and 6, just like ferricyanide/ferrocyanide system, there is no special growth mechanism of the fluctuation. Therefore, the micro-MHD flow simply assists the mass transfer of proton in the same way as Fig. 2. In Fig. 18, the promotion of proton transfer by the micro-MHD flow is schematically exhibited.

In Fig. 19, the copper dissolution rate measured in the solution containing quite small amount of proton is presented, which increases almost in proportion to the magnetic

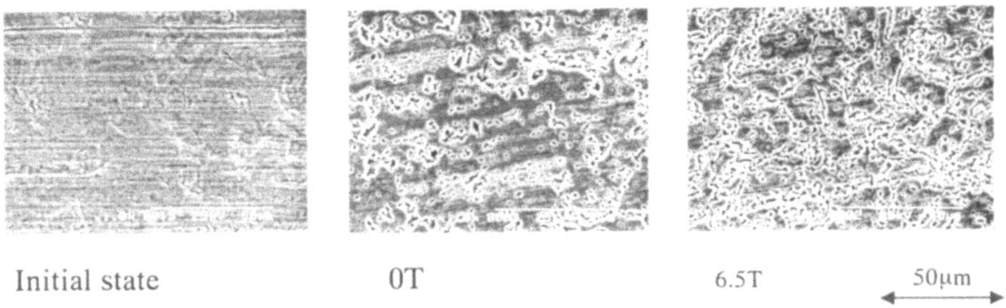

Figure 20. SEM photos of dissolved copper surface after immersion of 72 hr. Experimental conditions are the same as Fig. 19.

flux density. Figure 20 shows the SEM (scanning microscope) photos of the dissolved surface. In zero magnetic field, lots of pits emerge on the copper surface, whereas in high magnetic field, such pits are deepened and connected with each other, forming a kind of network. This suggests that the micro-MHD effect also gives some special effects to the pit formation in the similar way to Fig. 12.

To assure the role of cathodic reaction in the enhancement of dissolution, the CM of the cathodic current with the copper MHDE was taken in the same solution as Figs. 19 and 20. As shown in Fig. 21, the locus shifts, contrary to Fig. 17, downward, which means that the mass transfer of proton is enhanced by the micro-MHD effect to increase the reduction rate of nitric acid.

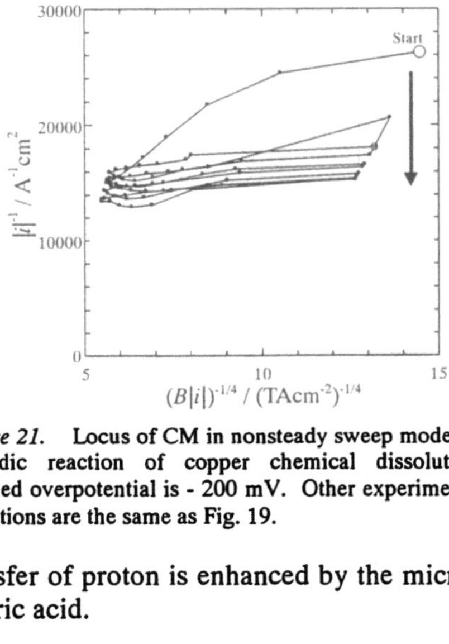

Figure 21. Locus of CM in nonsteady sweep mode for cathodic reaction of copper chemical dissolution. Applied overpotential is - 200 mV. Other experimental conditions are the same as Fig. 19.

4. Conclusions

In order to analyze the magnetic field effect on electrochemical reaction, the MHD effect and the cyclic magnetammetry (CM) were firstly examined. The experiments were performed for the ferricyanide/ferrocyanide redox reaction and the copper electrdeposition in copper sulfate solution.

Consequently, it was concluded that in a simple redox system, the locus of CM converges to a straight line obtained in steady state. On the other hand, when the unstable growth of the concentration fluctuation controls the whole processes of the reaction such as the copper electrodeposition, intense suppression of the fluctuation by the MHD flow gives the large hysteresis or non-periodic response to the locus without any convergence to a straight line. In this case, local current distribution, *i.e.*., current fluctuation becomes important, which generates a localized Lorentz force in magnetic field, and finally, local convection cells are induced. Since these cells have only small sizes similar to that of the fluctuation, we can easily discriminate the small convection cells from the usual MHD flow. Therefore, we call this flow and effect the micro-MHD flow and the micro-MHD effect, respectively.

To clarify the micro-MHD effect, copper dissolution in nitric acid was examined, where there is no net current, but localized current flows from the anodic site to the cathodic site in the solution phase, forming a local cell. If the anodic dissolution controls the whole reaction, resultant Lorentz force moves the solution to disturb the growth of the concentration fluctuation. Drastic reduction of dissolution rate was surely observed with increasing magnetic flux density. For reinforcing these discussion, we carried out the CM of anodic current in the similar way to the copper electrodeposition, the large hysteresis and non-periodic response characteristic of the micro-MHD effect emerged in the loci of the CM's.

In the case where the cathodic reaction is rate-determining, on the contrary, the micro-MHD flow enhances the proton transfer, accelerating the dissolution. Applying a certain overpotential to the working electrode of MHDE, we carried out the CM in the same solution; As expected above, a large increase of cathodic current was observed.

Acknowledgement

The authors wish to express their appreciation to High Field Laboratory for Superconducting Materials, Institute for Material Research, Tohoku University for granting the use of the water-cooled resistance magnet.

References

1. J. Sakuraba, M. Ishihara, S. Yasuhara, K. Jikihara, K. Watazawa, T. Hasebe, C.-K. Chong, Y. Yamada and K. Watanabe, *IEICE TRANS. ELECTRON.*, E78-C (1995) 535.
2. E. Beaugnon and R. Tournier, *J. Phys. III France* **1**(1991) 1423.
3. S. Ueno and M. Iwasaka, *J. Appl. Phys.* *75*(1994) 7177.
4. N. Hirota, T. Honma, H. Sugawara, K. Kitazawa, M. Iwasaka, S. Ueno, H. Yokoi, Y. Kakudate, S. Fujiwara and M. Kawamura, *Jpn. J. Appl. Phys.* **34** (1995)L991-L993.
5. A. Yamagishi, Handai Kyojiba, **7** (1993) 43.
6. H. Sata, M. Yamato and E. Ito, *Rep.Prog. Polym. Phys. Jpn.* **36** (1993)199.
7. R. Aogaki, K Fueki and T. Mukaibo, *Denki Kagaku* **43** (1975) 504.
8. R. Aogaki, K. Fueki and T. Mukaibo, *Denki Kagaku* **43** (1975) 509.
9. R. Aogaki, K. Fueki and T. Mukaibo, *Denki Kagaku* **44** (1976) 89.
10. S. Yamanaka, R. Aogaki, M. Yamato, E. Ito and I. Mogi, *Sci. Rep. RITU* A-vol.1, **38** (1993)399.
11. A. Tadano and R. Aogaki, E. Ito, S. Morisaki, and I. Mogi, *submitted to J. Electrochem. Soc.*.
12. R. Aogaki, *J. Chem. Phys.* **103** (1995) 8616.
13. A. Tadano and R. Aogaki, *J. Chem. Phys.* **106** (1997) 6126.
14. R. Aogaki, *J. Electrochem. Soc.* **142** (1995) 2954.
15. M. Asanuma and R. Aogaki, *J. Chem. Phys.* **106** (1997) 9930.
16. N. Sato, "*Electrode Chemistry*", Vol. 2, P. 365, Nittetsu-Gijutsu Jyoho Center, Tokyo (1994).
17. N. Sato, "*Electrode Chemistry*", Vol. 2, P. 359, Nittetsu-Gijutsu Jyoho Center, Tokyo (1994).
18. K. Shinohara and R. Aogaki, *in the course of submission.*

ANALYSIS OF MHD EFFECTS ON ELECTROCHEMICAL PROCESSES. EXPERIMENTAL AND THEORETICAL APPROACH OF THE INTERFACIAL PHENOMENA

CHOPART J.-P., DEVOS O., AABOUBI O., MERIENNE E.,
and OLIVIER A.

EP120 - CNRS, UFR Sciences
BP 1039 - 51687 Reims Cedex 2- France
e-mail : jp.chopart@univ-reims.fr

Magnetically induced effects on electrolytic currents have been studied for various electrochemical systems. For all investigated cases, significant modification was found to be strictly relevant to transport process and magnetic effects were proved to be MHD (*i.e.* MagnetoHydroDynamic) phenomena. This holds for a reversible system (ferri-ferrocyanide) as well as for non reversible systems such as the electrocrystallization of Zn, or Ni from a pure or adulterated Watts bath. Assuming that the well-known magnetic effects on electrolytic currents in the bulk of solutions keep valid in the close vicinity of the electrode, the MHD phenomena are explained by taking into account the space charge of the diffuse layer. Theoretical expressions which depend on bulk composition and hydrodynamic conditions can thus be obtained. A comparison with experimental results under stationary or dynamic conditions was undertaken. For the latter case, two methods were employed : either classical electrochemical impedance spectroscopy (EIS) or MHD transfer function analysis. For both cases, a good agreement was found between expected physical values and calculated ones. Moreover, new results on electrocrystallization of Ni and Zn have been found.

1. Introduction

The analysis of magnetic field effects on electric currents in electrolytic solutions has been performed for many years from experiments carried out either on bulk solutions or on the electrodes themselves. Reviews [1,2] on this subject emphasized the difficulties of understanding these phenomena which

A. Alemany et al. (eds.),
Transfer Phenomena in Magnetohydrodynamic and Electroconducting Flows, 181–198.

involve many physical concepts. Focus has to be applied on the conclusion of an overview by Tacken and Janssen [2] where they wrote : *"Mass transfer effects (with magnetic field) are well understood and practically relevant, yet difficult to model ... Studies of electron transfer and, particularly, corrosion in magnetic fields have produced new, and sometimes controversial, insights."*.

Until now, modifications of hydrodynamic flows by magnetic fields are generally analyzed as resulting from the Lorentz force which acts whenever a magnetic field is applied on an electric current. However, no calculation involving such a force has been able to obtain the exact magnitude of the convective effect in any case. Moreover, more recent results [3] have proved that preliminar explanations were unsuitable and that a full understanding demanded deeper insight.

As noted by Tacken and Janssen, this lack of comprehensive knowledge explains the difficulties to model such magnetohydrodynamic (*i.e.* MHD) behaviour and many erroneous conclusions about magnetic field effects on electrochemical systems which were falsely assumed unrelevant to transport phenomena [4]. Furthermore, an analysis of the literature devoted to MHD effects on convective phenomena shows an amazing disregard for magnetic field effects on the electric properties of the electrolyte itself despite evidence which had been highlighted before [5 and refs 1-51 in ref 1].

2. Magnetoelectrolytic effects in solution

In an electrolytic solution wherein a *dc* current flows, by superimposition of a magnetic field in one direction perpendicular to the current flow, transversal effects in the third direction are created ; indeed, temperature, concentration and potential gradients (hence electric fields) arise. But the most important fact which has to be emphasized is the fundamental difference between the classical Hall effect in metals and the one in electrolytic solutions : no charge separation is created in the bulk of the solution.

Let $\vec{O}x, \vec{O}y, \vec{O}z$ be direct orthogonal vectors, with an electric field $\vec{E}_x$ and a current flow $\vec{J}_x$ along the Ox axis. The superimposition of a magnetic field $\vec{B}_z$ along Oz creates an electric field $\vec{E}_y$ along Oy whose amplitude can be expressed as [5] :

$$E_y = \frac{\sum_k u_k \Gamma_k z_k^2 C_k}{\sum_k z_k^2 C_k} B E_x \tag{1}$$

where k represents ionic species of charge z_k, mobility u_k and concentration C_k. Γ_k is a characteristic ionic coefficient which may be written as :

$$\Gamma_k^{-1} = \mathrm{erfc}(P) + \frac{4P}{\sqrt{\pi}} e^{-P^2}\left(\frac{1}{2} + \frac{P^2}{3}\right) \tag{2}$$

with $P^2 = W_k/k_B T$. T is the absolute temperature, k_B the Boltzmann constant and W_k the intrinsic kinetic energy of the species k [5]. The concentration gradients $\partial C/\partial y$ which appear for each ionic species k are :

$$\frac{\partial C_k}{\partial y} = z_k \frac{F}{RT} C_k \left[E_y - u_k \Gamma_k B_z E_x\right] \tag{3}$$

When concentration gradients exist in the x-direction, as is the case in the diffusion layer, the tangential electric field is :

$$E_y = \frac{B_z}{\sum_k z_k^2 C_k}\left[E_x \sum_k u_k \Gamma_k z_k^2 C_k - \frac{RT}{F}\sum_k u_k \Gamma_k z_k \frac{\partial C_k}{\partial x}\right] \tag{4}$$

In the bulk of the electrolytic solution, with B = 1 T and E_x (due to the ohmic drop) equal to a few volts per meter, E_y is only a few mV/m [6]. In the diffusion layer, due to the second term of equation 4, E_y may be higher : generally the ratio of E_y to E_x is about 0.01 percent. Temperature gradients are very small [7], so as most often electrolytic solutions can be regarded as isothermal.

3. Magnetoelectrokinetic effect

By calculations similar to those of Olivier [6], we can get the E_y value in the close vicinity of the electrode, keeping in mind that in the present case, the electrolytic composition is imposed by the electrochemical reaction.

For a reversible redox system (*e.g.* hexacyanoferrate III/II) with a Pt working electrode, a supporting electrolyte and under cathodic limiting current conditions, if the magnetic field is parallel to the electrode plane surface, E_y is expressed as :

$$E_y = \frac{BE_x}{2n}\left[(n + z_{Ox}) u_{Ox} \Gamma_{Ox} - z_{Red} u_{Red} \Gamma_{Red}\right] \tag{5}$$

where n is the total number of electrons involved in the electrochemical reaction.

Another example is the electrodeposition of copper from cupric sulfate in a sulfuric acid medium, in which case the tangential field is :

$$E_y = u_{Cu^{2+}} \Gamma_{Cu^{2+}} B E_x \tag{6}$$

In the electrochemical double layer there is a volumic charge ρ, and therefore the electric field which exists there is very large. The volumic charge and the electric field are correlated and obey the Poisson's rule [8]. If linear condition can be supposed (that can be done for a planar electrode), the normal electric field E_X can be reached by integration of :

$$\rho(x) = \varepsilon\varepsilon_o \frac{dE_x}{dx} \tag{7}$$

where ε is the solution dielectric constant and $\varepsilon_o = 8{,}85.10^{-12}\ C^2N^{-1}m^{-2}$. Due to the non electroneutrality in the electric double layer, E_X can attain $10^{10}\ Vm^{-1}$. Therefore, if E_y is a few millivolts per meter in the bulk, it is largely higher in the close vicinity of the electrode where exists a volumic charge. Because of this charge and of the tangential field, a force arises by the so-called electrokinetic effect. This force F_y is expressed as :

$$F_y = \rho(x) E_y \tag{8}$$

Due to that force, a fluid motion is created in the vicinity of the electrode and developped in the bulk of the solution by viscid phenomena. The tangential component of the velocity v_y has to be determined by the resolution of the Navier-Stokes equation. Therefore, if gravitational effects are negligible and η represents the solution viscosity :

$$F_y = \eta \frac{\partial^2 v_y}{\partial x^2} \tag{9}$$

The fluid motion induces concentration profiles which can be calculated if the convective diffusion equation (10) is resolved [9]. With the same linear condition as before, this convective diffusion equation can be simplified under the form :

$$D \frac{\partial^2 C_k}{\partial x^2} = v_y \frac{\partial C_k}{\partial y} \tag{10}$$

where D is the diffusion coefficient of the electroactive species k which react at the electrode. Finally, the current density can be obtained from the local flux :

$$j(y) = -D\left(\frac{\partial C_k}{\partial x}\right)_{x=0} \tag{11}$$

and the total current by an integration over the whole electrode.

The complexity of the involved mathematical problem prevents from obtaining an exact solution in every case, only simplified expressions can be obtained. It must be emphasized that this drastic difficulty explains the very great number of experimental results in literature [1,2] which seem to be in disaccordance one another. In fact, to solve these equations and write the final current, the geometry of the cell as well as the electrolytic composition of the solution have to be taken into account.

With a supporting electrolyte and a reversible redox system, under the assumption that the repartition of non electroactive ionic species is a Boltzmann one and for limiting current polarization, the volumic charge ρ, the electric field E_X, and therefore the tangential electric field E_y, the electrokinetic force and thus the tangential fluid velocity v_y can be calculated. In this simplified case v_y is :

$$v_y(x) = A(x_c)\, B\left\{1 - \exp\left[-s\,(x - x_c)\right]\right\} \tag{12}$$

where x_c is the shear plane abscissa, (v_y ($x \leq x_c$) = 0), $A(x_c)$ and s are characteristic coefficients which depend on the concentrations of electrolytic species. Because of the s value, v_y reaches $A(x_c)\times B$ for very small x values and decreases by the viscous effect to reach a constant value which is imposed by the natural convection when no other convection exists. With this velocity expression, and experimental conditions of refs 10-14, the limiting current for a d-diameter circular electrode is calculated as :

$$I_{lim} = 1{,}36\, nFD^{2/3}\, C^*\left[A(x_c)\, s\, B\right]^{1/3} d^{5/3} \tag{13}$$

where C^* is the bulk concentration of the reactive species.

In the above experiments, the hydrodynamic conditions are those of a boundary layer type flow as in [10]. If hydrodynamic conditions are changed into those of a completely developed flow [11], the calculated limiting current on a disk electrode is :

$$I_{lim} = KnFC^*\left[DA(x_c)B\right]^{1/2} d^{3/2} \tag{14}$$

where K is an empirical constant which depends on the cell geometry. The calculation of K is outside the context of this paper.

Such relations between I_{lim} and fundamental parameters of the magnetohydrodynamic phenomenon well agree with many published results [1,2,12]. Owing to the numerous simplifications which were made either for experimental apparatus or theoretical calculations, it must be emphasized that both relations (13) and (14) are not valid in every cases. Therefore, it is not surprising to find experimental results with other relationships between limiting currents and MHD parameters [3,13].

4. Experimental results

4.1 MASS-TRANSPORT ANALYSIS

4.1.1. *Reversible Redox System*

Experiments were carried out with a horizontal Pt disk as working electrode in an equimolar $K_3Fe(CN)_6$ and $K_4Fe(CN)_6$ solution with 0 to 2 M KCl as supporting electrolyte [13-16]. For high level magnetic inductions, anodic and cathodic limiting currents obey equation 13. With solutions of Table 1, under potentiostatic control, all limiting currents on a horizontal electrode follow the equation (13) with a determined average shear abscissa x_C equal to 3.4 nanometers.

To confirm these stationary investigations, dynamic analyses were performed either by classical *a.c.* impedance (Fig.1) or by a new transfer function (*i.e.* MHD transfer function) [14], (Fig.2). Dynamic results were in perfect agreement with theoretical derivatives of stationary equations, either with a cell immersed in a classical electromagnet or with a new home-made apparatus [16]. Actually, because magneto-convective effects are generated in the electrochemical double-layer, the whole cell has not to be in the gap of an electromagnet and little permanent magnets near the working electrode can be used alternatively (Fig.3). Results are unchanged, but more experimental possibilities are then available [17]. Previous works [13-18] have supposed that the velocity gradient $\alpha = \left(\partial v_y / \partial x\right)$ was constant, in which case the integration of the convective diffusion equation leads to :

$$I_{lim} = 0.678\, n\, F\, C^* \, D^{2/3}\, \alpha^{1/3}\, d^{1/3} \qquad (15)$$

When the supporting electrolyte concentration is kept constant, the relationship

$$\alpha = k_e B C^* \qquad (16)$$

was experimentally established in numerous cases where k_e is an empirical constant deduced from the limiting current I_{lim}. In fact, the derivation of equation (12) gives a gradient $\alpha(x)$, but to explain the empirical rule (16) it has to be noticed that the product $A(x_c)\times s$, which is the right term to be taken into account, is rather proportional to C^*.

TABLE 1. s and $A(x_c)$ values for different electrolytic compositions. Supporting electrolyte = KCl, Ox = $K_3Fe(CN)_6$, Red = $K_4Fe(CN)_6$.

C_{KCl} mol/L	$C_{Ox} = C_{Red}$ mol/L	s $10^9 m^{-1}$	$A(x_c)$ anodic 10^{-10} $mT^{-1}s^{-1}$	$A(x_c)$ cathodic 10^{-10} $mT^{-1}s^{-1}$
0.5	0.10	3.5	10.6	10.6
0.5	0.25	4.4	15.0	12.5
1	0.01	3.9	0.7	0.5
1	0.05	4.1	4.6	4.5
1	0.10	4.4	8.6	7.2
1	0.25	5.1	11.0	9.0
2	0.10	5.8	3.9	4.8

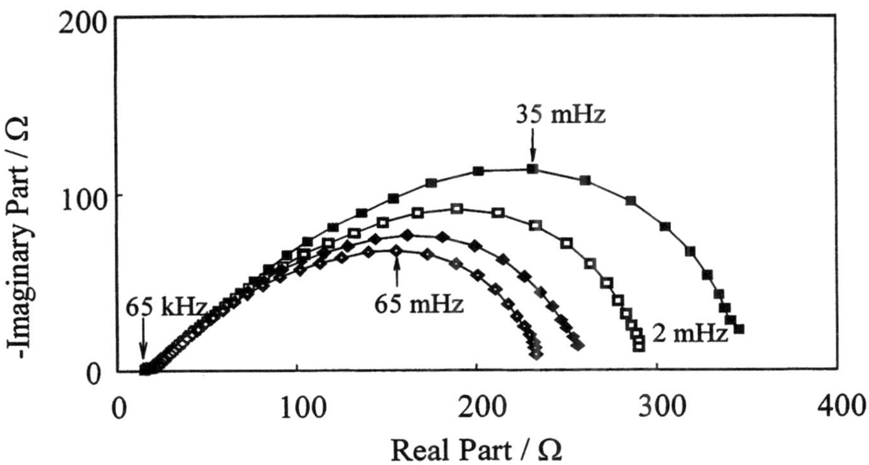

Figure 1. Electrochemical impedance plotted in a Nyquist plane at 3/4 of the anodic limiting current for various magnetic intensities. 0.2 M $K_4Fe(CN)_6$ and $K_3Fe(CN)_6$ in 1M KCl. (▪) B = 125 mT, (□) B = 177 mT, (♦) B = 244 mT, (◇) B = 322 mT.

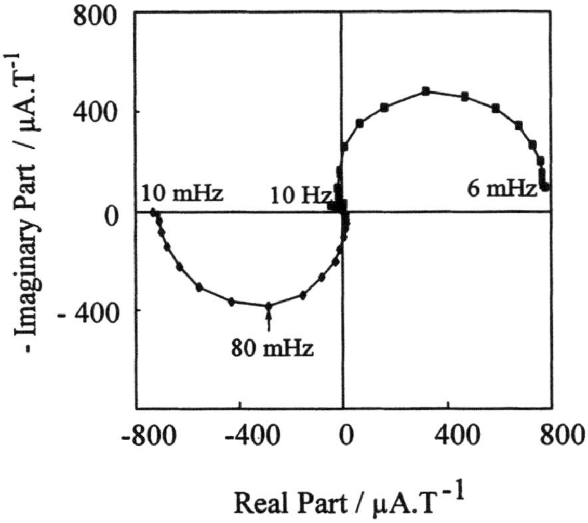

Figure 2. MHD transfer function $(\partial I/\partial B)_V$ plotted in a Nyquist plane. 0,15M $K_3Fe(CN)_6$ and $K_4Fe(CN)_6$ in 1M KCl at the limiting anodic. [(■) B = 303 mT] and cathodic [(♦) B = 136 mT] currents.

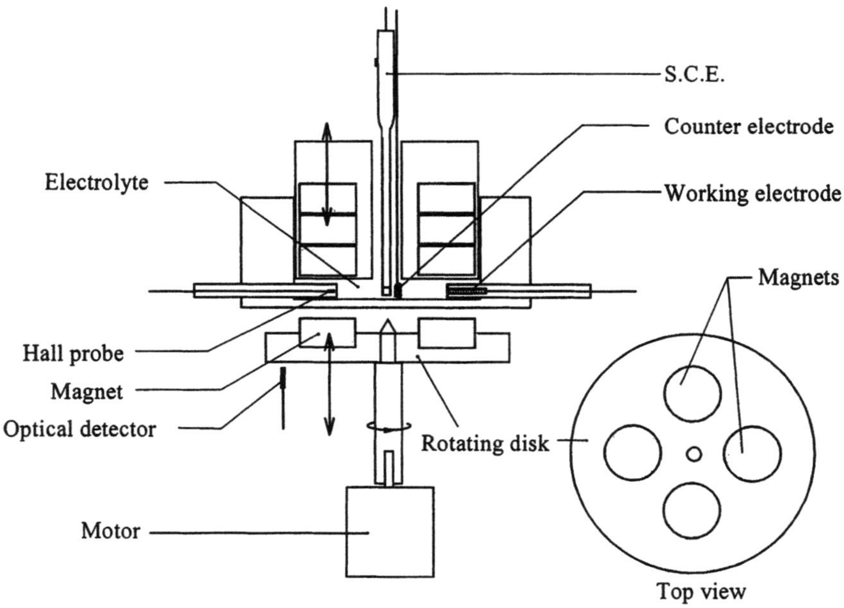

Figure 3. Schematic representation of the home-made experimental device [16].

4.1.2. *Zinc electrocrystallization*

One of the most important features of MHD investigations in electrochemistry is the very good accuracy between theoretical equations and experimental results for a convection regime weaker than that given by a classical convective apparatus (*e.g.* the rotating disk electrode). The experimental study of zinc electrocrystallization in basic media [18] is a piece of evidence for MHD possibilities. Under mass-transport control, the cathodic limiting current obeys equation (13) (Fig.4). The potentiostatic MHD transfer function $(\partial I/\partial B)_V$ behaves accordingly to the theoretical predictions, both regarding the low frequency limit and the reducibility by a characteristic adimensional frequency σ (Fig. 5). Here σ is defined by [14,16] :

$$\sigma = \omega\left(\frac{d^2}{\alpha D}\right)^{1/3} \tag{17}$$

where ω is the pulsation.

Two remarkable points have to be stressed.

The first one is the possibility to determine the diffusion coefficient of the reactive ion with a great accuracy. Indeed, under limiting current conditions, the experimental MHD transfer function modulus can be compared with the theoretical one which is calculated with an arbitrary diffusion coefficient value (Fig. 6). A good agreement is found with a value equal to 5.3 × 10^{-10} m^2s^{-1}, which is very close to that obtained from the Levich's expression in the case of a rotating disk electrode [17].

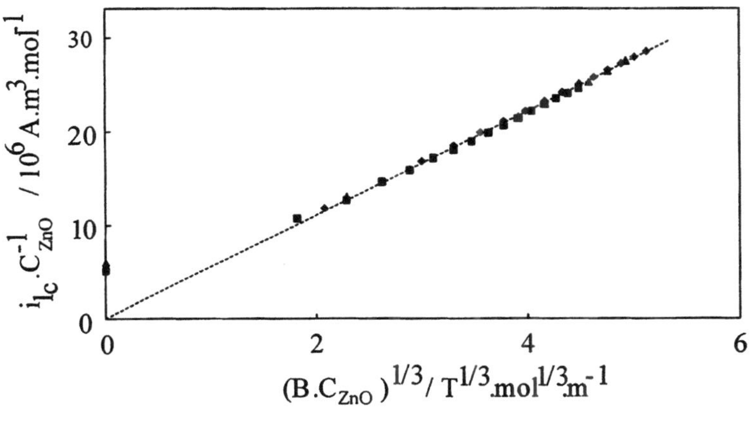

Figure 4. Ratio of cathodic limiting current to ZnO concentration *vs* $(B.C_{ZnO})^{1/3}$. (■) [ZnO] = 0.1 M, (◆) [ZnO] = 0.15 M, (▲) [ZnO] = 0.2 M. [KOH] = 7 M.

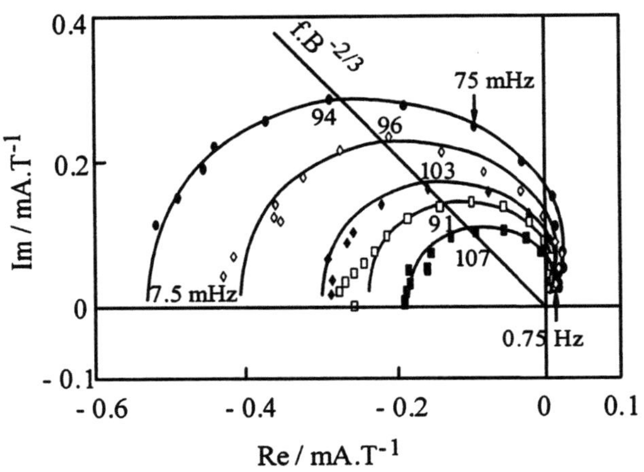

Figure 5. MHD transfer function $(\partial I/\partial B)_V$ plotted in a Nyquist plane at the limiting current for the $Zn(OH)_4^{2-}$ / Zn system and various magnetic fields. [ZnO] = 0.2 M, [KOH] = 7 M. (•) B = 0.3 T, (◊) B = 0.45 T, (♦) B = 0.6 T, (□) B = 0.75 T, (■) B = 0.9 T.

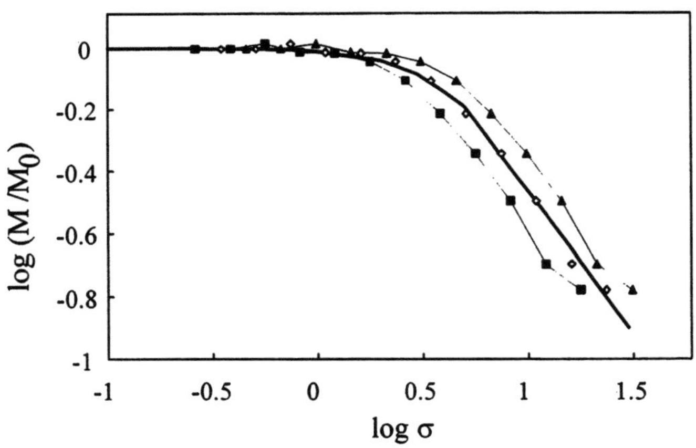

Figure 6. Normalized modulus of experimental and theoretical MHD transfer functions $(\partial I/\partial B)_V$ for the $Zn(OH)_4^{2-}$ / Zn system at the limiting current in reduced frequency. [KOH] = 7 M, [ZnO] = 0.2 M, B = 0.6 T. Bold solid line : theoretical MHD transfer function. (■) $D_{Zn(II)} = 4 \times 10^{-10} m^2.s^{-1}$, (◊) $D_{Zn(II)} = 5.3 \times 10^{-10} m^2.s^{-1}$, (▲) $D_{Zn(II)} = 7 \times 10^{-10} m^2.s^{-1}$.

The second point is the opportunity to gain a deeper insight in Zn electrocrystallization, because a response characteristic of two diffusional species is highlighted by classical electrochemical impedance at low current density (Fig.7). This response is perfectly analyzed by MHD theoretical expressions of the electrochemical impedance [18] and proves the existence of a diffusional control of zinc electrocrystallization by OH^- and Zn^{II} species.

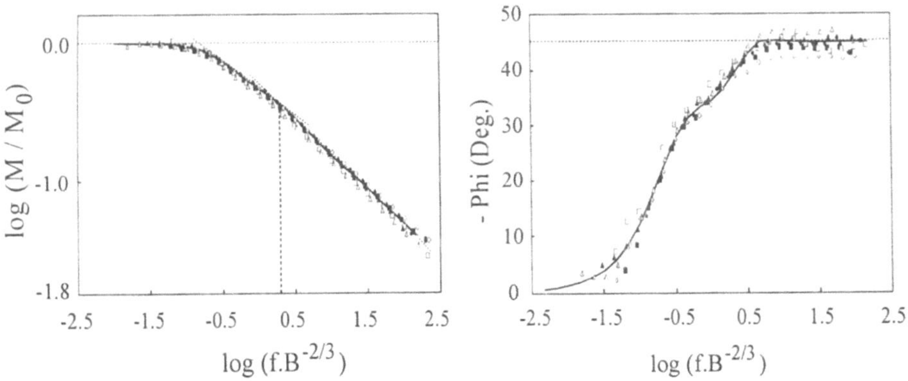

Figure 7. Experimental and calculated electrochemical impedances dV/dI for the zincate/zinc system plotted in a Bode plane as reduced frequency at 3/4 of the limiting current for various magnetic fields. [ZnO] = 0.2 M, [KOH] = 7 M. Solid line : calculated electrochemical impedance. (□) B = 0.15T, (■) B = 0.3 T, (◇) B = 0.45 T, (◆) B = 0.6 T, (Δ) B = 0.75 T, (▲) B = 0.9 T.

4.2 NICKEL ELECTROCRYSTALLIZATION

Nickel electrodeposition was performed in a Watts solution ($NiCl_2.6H_2O$ 35 g/L, $NiSO_4.7\ H_2O$ 300g/L, H_3BO_3 40g/L) with a pH equal to 4.5. In some experiments, small amount (0.01 mol/L) of 2-butyne-1,4-diol (BD) was added. This organic additive is well known to react on the nickel cathode under mass-transport control [19].

When a magnetic induction is applied parallel to the surface of the electrode, in the presence of BD in solution, the cathodic current decreases (Fig.8). The electrochemical impedance is drastically changed (Fig.9) due to the magnetically induced modifications of the time constants characteristic of the elementary steps. The potentiostatic MHD transfer function $(\partial I/\partial B)_V$ was analyzed [17] and through the low frequency limit study, a linear relationship between current and $B^{-1/3}$ determined. The good agreement between experimental and calculated current for magnetic field over 0.2 tesla is thus confirmed (Fig.8).

That cathodic current decrease as B increases results from rather complex modifications undergone by the catholyte due to the degradation process of BD [20]. This acetylenic diol, whose concentration increases due to the MHD convection, is readily desorbed from the cathodic surface whereupon it undergoes a prompt catalytic hydrogenation. The desorption liberates four neighbouring cathodic sites which are then available for the competitive reduction of Ni^{2+} and H^+ ions [20]. Since the reduction of H^+ is a mass-controlled process [21], the MHD convection favours the H^+ reduction to the detriment of that of Ni^{2+} ions. This explains why the interfacial pH is higher in the BD-adulterated electrolytes, thus favouring the precipitation of $Ni(OH)_2$. Actually, the long-term inhibiting species of Ni electrodeposition, at the end of the degradation process of BD, is the strongly-adsorbed Ni hydroxide [20].

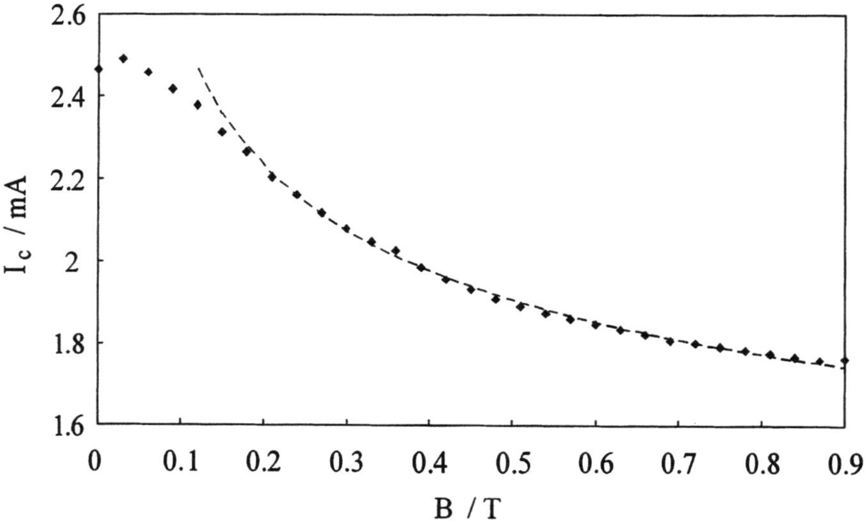

Figure 8. Ni electrodeposition current vs. magnetic field. Watts bath containing 10 mM 2-butyne-1,4-diol, V = - 930 mV/SCE. Dashed line : calculated current.

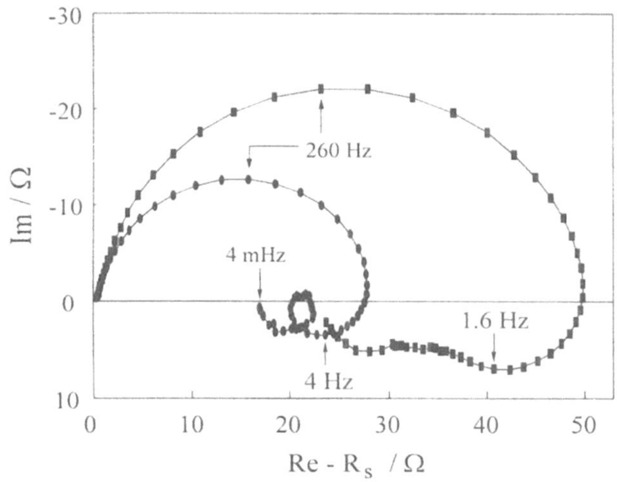

Figure 9. Electrochemical impedances (dV/dI) plotted in a Nyquist plane for Ni electrodeposition. Watts bath with 10 mM 2-butyne-1,4-diol, T = 30°C, V = - 930 mV/SCE. (•) B = 0, (▮) B = 0.9 T.

When the Watts solution contains no BD, no detectable effect on steady-state current occurs at the time of the superimposition of a magnetic induction. Nevertheless, by impedance analysis, a weak modification of the inductive loop can be detected (Fig.10) which can readily be interpreted as the enhancement of the hydrogen electrode reaction.

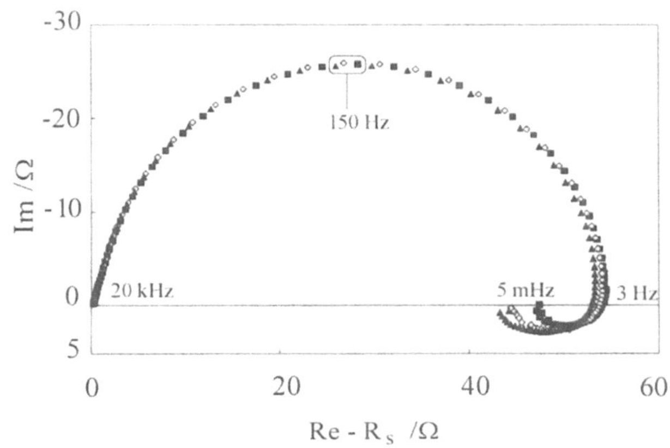

Figure 10. Electrochemical impedances in Nyquist plane for nickel electrodeposition. Watts bath, T = 30°C, E = - 0.84 V / SCE.
(▮) B = 0, (◇) B = 0.3 T, (▲) B = 0.6 T.

This hardly detectable effect becomes obvious through a MHD transfer function analysis (Fig.11). The same above mentioned relationship between current and magnetic field can be obtained from the low frequency limit as was the case with BD in solution. Once more, magnetic field effects essentially consist of an enhancement of the mass transport processes.

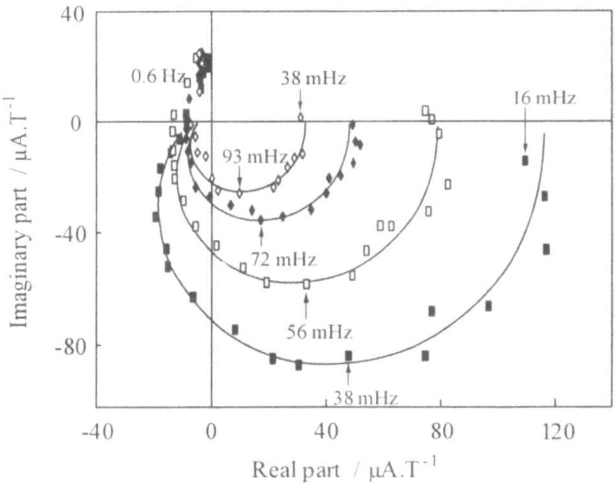

Figure 11. MHD transfer function $(\partial I/\partial B)_V$ for nickel electrodeposition. Watts bath, T = 30°C, V = - 880 mV/SCE. (▪) B = 0.3 T, (□) B = 0.45 T, (♦) B = 0.6 T, (◊) B = 0.75 T.

Figure 12. SEM image of H_2 bubble traces on a stationary nickel deposit prepared at -1 V/SCE from a Watts bath containing 5 mM 2-butyne-1,4-diol, in presence of a magnetic field (B = 0.6 T).

Structural investigations were also performed [4], both by SEM and X-Ray diffraction diagrams, which once more put the convective MHD effects in a prominent place : in Fig.12, the induced convective flow is noticeable by the perturbation of the deposit microrelief in the vicinity of H_2 bubbles when BD is added. The modifications of the X- Ray diffractograms (Fig.13) evidence the modifications of nickel deposit texture by magnetic field superimposition. These texture modifications depend on mass-transport processes *via* the inhibition by various species involved in the BD reaction.

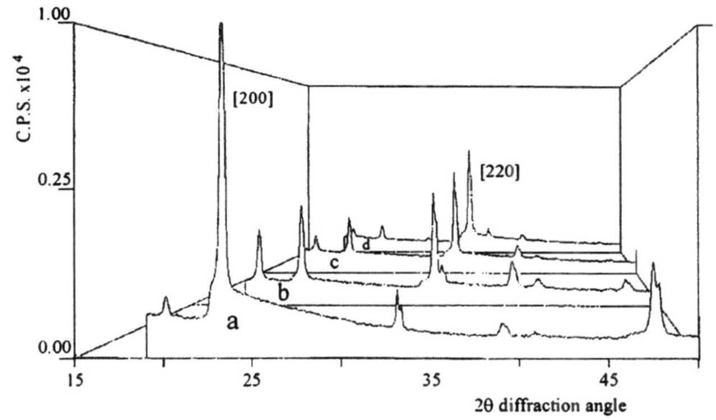

Figure 13. X-Ray diffractograms of nickel electrodeposits prepared at -1 V/SCE from a Watts solution containing 5.10^{-3} M 2-butyne-1,4-diol for various intensities of the superimposed magnetic field. (Mo-Kα X Ray source, vertical axis in a square root scale). a- B = 0, very strong <100> texture, b- B = 0.3 T, mixture of <100> and <110> textures, c- B = 0.6 T, medium <110> texture and d- B = 0.9 T, strong <110> texture.

4.3 TRANSFER KINETICS

The high frequency impedance analysis of the whole set of now available results [15,17,22], whatever the system studied, reveals that magnetic fields weaker than one tesla exert no detectable effect on the charge tranfer kinetics, namely regarding α and β transfer coefficients. This has been confirmed by radiotracer studies of the exchange current [23] and throws a conclusive light upon the controversial results quoted in ref 2.

5. Conclusion

The use of a 3-electrode cell with a perfectly controlled working electrode allowed us to show unambiguously that, if the magnetic field exerts no effect on the charge transfer, it conversely exerts a prominent influence on the mass transport process. Hence it emerges as an essential parameter for both the analysis and the understanding of steady-state as well as dynamic electrochemical processes. It has been proved necessary to take into account the intervention of the diffuse layer and of its charge density $\neq 0$ to enable us to interpret the origin of the magnetic-induced convection, thus permitting to correctly interpret the whole set of results concerning $K_3Fe(CN)_6/K_4Fe(CN)_6$, Zn and Ni systems. Moreover, the structural investigation implementation in the latter case has proved fruitful for the study of interfacial processes.

6. References

1. FAHIDY, T.Z. (1983) Magnetoelectrolysis, *J. Appl. Electrochem.*, **13**, 553.
2. TACKEN, R.A. and JANSSEN, L.J.J. (1995) Applications of magneto-electrolysis, *J. Appl. Electrochem.*, **25**, 1.
3. LEE, J., RAGSDALE, S.R., GAO, X., and WHITE, H.S. (1997) Magnetic field control of the potential distribution and current at microdisk electrodes, *J. Electroanal. Chem.*, **422**, 169.
4. DEVOS, O., OLIVIER, A., CHOPART, J.-P., AABOUBI, O., and MAURIN, G. (1998) Magnetic field effects on nickel electrodeposition, *J. Electrochem. Soc.*, **145**, 401.
5. TRONEL-PEYROZ, E.and OLIVIER, A. (1982) Applications of the Boltzmann equation to the study of electrolytic solution in the presence of electric and magnetic fields, *Phys. Chem. Hydrodyn.*, **3**, 251.
6. OLIVIER, A. (1979) Contribution à l'étude des effets magnéto-électriques en solution aqueuse, *Thesis*, Reims, France.
7. TRONEL-PEYROZ, E. (1978) Etude des effets de température galvanomagnétiques transverses dans les solutions aqueuses électrolytiques, *Thesis*, Reims, France.
8. BARD, A.J. and FAULKNER, L.R. *Electrochimie. Principes, méthodes et applications*, Masson, Paris, 1983.

9. LEVICH, V.G., *Physicochemical hydrodynamics*, Prentice-Hall Inc, Englewood Cliffs, N.J., (1962).

10. AOGAKI, R., FUEKI, K., and MUKAIBO, T. (1975) Application of magnetohydrodynamic effect to the analysis of electrochemical reactions -2. diffusion process in MHD forced flow of electrolyte solutions, *Denki Kagaku*, **43**, 509.

11. AOGAKI, R., FUEKI, K., and MUKAIBO, T. (1976) Diffusion process in viscous flow of electrolyte solution in magnetohydrodynamic pump electrodes, *Denki Kagaku*, **44**, 89.

12. NGOBOUM, G., ALEMANY, A., OLIVAS, P., and BARK, F.H. (1997) Mass transfer in electrochemical processes, 3rd International Conference on Transfer Phenomena in MHD and Electroconducting Flows, Abstract p.185, Aussois, France 22-26 Sept.

13. OLIVIER, A., CHOPART, J.-P., DOUGLADE, J., and GABRIELLI, C. (1987) Investigation of magnetic effects on mass-transport at the electrode : electrolyte interface by impedance technique, *J. Electroanal. Chem.*, **227**, 275.

14. AABOUBI, O., CHOPART, J.-P., DOUGLADE, J., OLIVIER, A., GABRIELLI, C., and TRIBOLLET, B. (1990) Magnetic field effect on mass transport, *J. Electrochem. Soc.*, **137**, 1796.

15. AABOUBI, O. (1991) Etude de l'influence du champ magnétique sur les phénomènes de transport de masse en électrochimie : impédance magnétohydrodynamique, *Thesis*, Reims, France.

16. DEVOS, O., AABOUBI, O., CHOPART, J.-P., MERIENNE, E., OLIVIER, A., GABRIELLI, C., and TRIBOLLET, B. (1997) A new experimental device for magnetoelectrochemical (M.E.C.) transfer function measurements, *Polish J. Chem.*, **71**, 1160.

17. DEVOS, O. (1997) Contribution à l'étude de l'électrodéposition métallique sous champ magnétique : analyses stationnaires et dynamiques, *Thesis*, Reims, France.

18. DEVOS, O., AABOUBI, O., CHOPART, J.-P., GABRIELLI, C., MERIENNE, E., OLIVIER, A., and TRIBOLLET, B. (1997) Magnetic field effect on the zinc electrodeposition, The 1997 joint International Meeting, Abstract 445, Paris, France 31 Aug.-5 Sept.

19. WIART, R. (1968) Contribution à l'étude cinétique de l'électrocristallisation du nickel. Influence de quelques inhibiteurs organiques, *Oberfläche-Surface*, **9**, 213, 241, 275.

20. AMBLARD, J., EPELBOIN, I., FROMENT, M., and MAURIN, G. (1979) Inhibition and nickel electrocrystallization, *J. Applied Electrochem.*, **9**, 233.

21. DORSCH, R.K. (1969) *J. Electroanal. Chem.*, **21**, 495.

22. CHOPART, J.-P., DOUGLADE, J., FRICOTEAUX, P., and OLIVIER, A. (1991) Electrodeposition and electrodissolution of copper with a magnetic field : dynamic and stationary investigations, *Electrochim. Acta*, **36**, 459.

23. FRICOTEAUX, P., OLIVIER, A., and DELMAS, R. (1992), Study of the exchange current at the Cu^{2+}-Cu interface by radiotracer : magnetic field effect, *J. Electrochem. Soc.*, **139**, 1096.

ENHANCEMENT OF ELECTROLYTIC MASS TRANSFER AROUND CYLINDERS BY EXPOSURE TO SWITCHING MAGNETIC FIELDS

SHIGERU MORI, MIKIO KUMITA and MASASHI TAKEUCHI*
Dept. of Chemistry & Chemical Engineering, Kanazawa University
2-Chome, Kodatsuno, Kanazawa 920, JAPAN
** M. Takeuchi is now with INOAC Corporation, Ltd., Anjo 446, JAPAN.*

Abstract. The effect of alternating magnetic fields, from 20 to 200 mHz and up to 250 mT in the horizontal direction with a symmetrical or asymmetrical rectangular waveform, on the mass transfer rate during diffusion-controlled electrolysis was experimentally investigated for a $K_3Fe(CN)_6$ - $K_4Fe(CN)_6$ system. The experiments were performed using inclined cylindrical Pt cathodes on a plane parallel to the magnetic field direction.

The instantaneous rates of the mass transfer under the periodic steady state are shown to be increased by the magnetic field exposure. The instantaneous data for the mass transfer enhancement obtained were studied to clarify its characteristic behavior by applying Fourier analysis. The mean enhancement of the mass transfer is roughly less than that for the static magnetic field case. A non-dimensional regression equation was determined for the mean mass transfer data by introducing a new parameter, the "frequency factor", and the magneto-diffusion factor.

1. Introduction

The mass transfer step is known to be rate-controlling in electrolytic treatment. Magnetic fields have the possibility of increasing or controlling the mass transfer rate by the MHD effects depending on their exposed directions and intensities and will become an important industrial tool in the future. This is suggested by the fact that electrochemical reactions are enhanced or suppressed by exposure to magnetic fields (Iwakura *et al.*, 1984). Research on electrolytic mass transfer in a magnetic field has been concentrated on vertical flat plates or rotating discs under a static magnetic field

A. Alemany et al. (eds.),
Transfer Phenomena in Magnetohydrodynamic and Electroconducting Flows, 199–212.

because of their simple conditions. Furthermore, obtaining a suitable correlation equation for the mass transfer data enhanced by the MHD effect has been difficult (Fahidy, 1990). This is due to such complicate interactions between relevant transport phenomena as schematically shown in Figure 1, even under conditions without forced flow. More recently, experimental investigations were carried out on the magnetoelectrolytic mass transfer around spherical and inclined cylindrical cathodes in horizontal or vertical magnetic fields (Mori *et al.*, 1994a, 1994b). The effective correlation of the obtained mass transfer coefficient was performed by introducing the "magneto-diffusion factor".

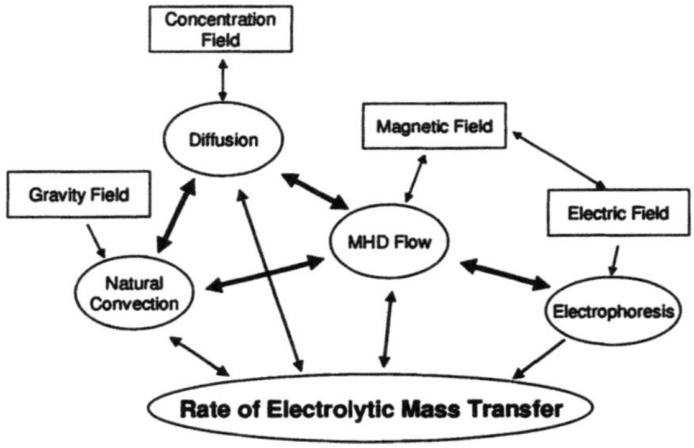

Figure 1. Schematic drawing of interactions between relevant transport phenomena.

On the other hand, the influence of the time variation of an imposed magnetic field on the electrolytic mass transfer has seldom been treated to date. The dynamics of magnetoelectrolysis was studied with an impedance technique employing a magnetic field with sinusoidal perturbation (Aaboubi *et al.*, 1990). Concerning the enhancement of the electrolytic mass transfer, the pulsation effect of the imposed magnetic flux for inclined cylindrical cathodes was experimentally studied and the time-averaged mass transfer coefficient was successfully correlated by taking into account the pulsation duty cycle (Mori *et al.*, 1996).

In the present article, the effect of the imposed switching magnetic field on the mass transfer rate around inclined cylindrical cathodes is dealt with to clarify the characteristics of the magnetoelectrolysis. The obtained instantaneous mass transfer enhancement is examined in detail by applying Fourier analysis, and the time-averaged mass transfer coefficient is correlated in order to establish a dimensionless regression equation by introducing a new parameter, the “frequency factor”.

2. Experimental

Using the cathodic reduction of $Fe(CN)_6^{3-}$ ion with a supporting electrolyte, experiments were conducted with a cylinder and the mass transfer rate was measured by the limiting current technique in order to investigate the directional switching effect of the imposed magnetic fields. The experimental setup, configuration of the electrolytic cell, shown respectively in Figures 2 and 3, and procedures were essentially similar to those in a previous study (Mori *et al.*, 1996).

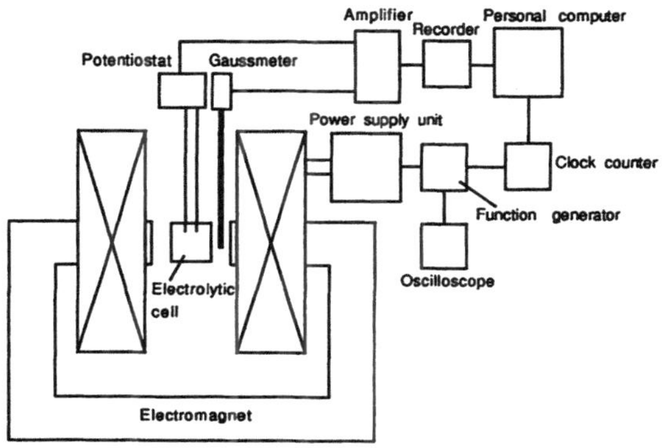

Figure 2. Schematic diagram of experimental apparatus.

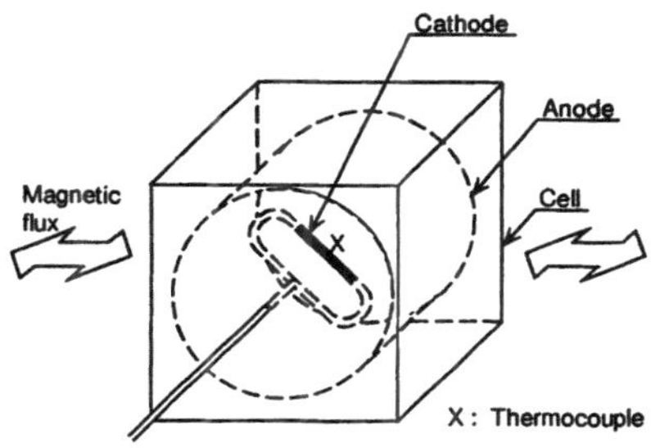

Figure 3. Details of electrolytic cell.

The anode of a short circular Pt-duct, 45 mm diameter × 30 mm long, was placed so as to surround the cathode in the center of the electrolytic cell made of acrylic resin plate. The cylindrical Pt cathode, 10.3 mm long and 2.0 mm in diameter, was located so that the cathode axis intersects that of the anode at right angles. The inclination angle of the cathode axis was specified to be $\theta = 0$, $\pi/6$ or $\pi/3$ rad from the horizon on a plane parallel to the horizontal magnetic field, switching its direction periodically with a specified frequency and magnetic flux density. The magnetic field was generated by supplying an electric switching current with a symmetrical or asymmetrical rectangular waveform (time ratio: 50% positive-to-50 % negative or 80 %-to-20 %, referred to as the Symmetry case $D_Y = 0.5$ or the Asymmetry case $D_Y = 0.8$) to an electromagnet from a power supply unit (Tamaoki Electronics Co., Ltd., TE-duo-1200K) controlled by a function generator (Iwatsu Electric Co., Ltd., FG-330). The electrolytic current was supplied by a potentiostat/galvanostat (Hokuto Denko Corp., Ltd., Model HA301) and the cell voltage was specified at 0.3 V so as to provide a diffusion-controlled limiting current, which was confirmed by preliminary experiments under any specified flux density of a static magnetic field.

The electrolyte was prepared as 0.15 M equimolar aqueous solution of $K_3Fe(CN)_6$ and $K_4Fe(CN)_6$ with a supporting electrolyte of 1.0 M NaOH, using analytical grade reagents. The solution was preliminarily treated sufficiently by O_2 gas bubbling to eliminate the influence of dissolved oxygen. The $Fe(CN)_6^{3-}$ concentration was determined exactly from the absorbance measured at 420 nm by a spectrophotometer (Shimazu Corp., Ltd., Model UV-160A) before each experiment.

To eliminate thermal natural convection caused by heat generation in the magnet coil, both the coil and cell were cooled by an electric fan and the room temperature was adjusted to 23.0 ± 0.2 ºC by an air conditioning system of the inverter type. The temperature of the solution was monitored by a calibrated C-C thermocouple.

After the solution charged into the cell was confirmed to be stationary at constant temperature, the specified cell voltage was applied. First, a limiting current in natural convection was recorded by an autograph recorder when the steady state was attained. Next, the switching magnetic field was imposed under specified conditions. The steady periodic changes in the limiting current and the magnetic flux density as outputs from the potentiostat and a gaussmeter (F. W. Bell Co., Ltd., Model 9200 with a Hall-effect sensor, Model STG92-0404) were simultaneously recorded during one period of magnetic field oscillation using a personal computer. The data sampling was performed, synchronized with the signal from the function generator, through an amplifier (Yokogawa Electric Works, Co., Ltd., DDC-3131) and a multi-channel A/D converter (Micro-Science Co., Ltd., DAS-1298XPC 12BIT/16CH). Finally, the magnetic field was turned off and the resulting limiting current was reconfirmed to be the same as that of the initial condition.

The experiment was repeated for different magnetic field conditions: flux densities, up to B = 250 mT at an interval of 50 mT; frequencies, f = 20, 80, 140 and 200 mHz, and for each inclination angle of the cathode.

The following well-known relation gives the mass transfer coefficient k from the observed limiting current, I.

$$k = I/(zAF\Delta C) \tag{1}$$

The instantaneous enhancement of the mass transfer rate is given as a ratio of the rate in the magnetic field to that with no magnetic field:

$$k/k_0 = I/I_0 \tag{2}$$

3. Results and Discussion

The mass transfer coefficient obtained at different inclination angles from the preliminary experiments under the conditions without a magnetic field was compared with the following empirical correlation:

$$\mathrm{Sh}_0 = 0.53\mathrm{Ra}^{1/4} \tag{3}$$

which was proposed for the free convective mass transfer around a horizontal cylinder (Schütz, 1963) and later extended to inclined-cylinder cases by modifying the characteristic length in Sh and Ra from the cylinder diameter d to $(d/\cos\theta)$ (Nippon Kikai Gakkai, ed., 1965). The present data were confirmed to be in good agreement with Eq.(3), showing that the effect of the geometrical configuration of the present experimental setup was negligible, at least for simple natural convection.

3.1. INSTANTANEOUS ENHANCEMENT

Figure 4 shows typical time variations of the applied magnetic flux density for the Symmetry and Asymmetry cases and the resulting instantaneous enhancement of the mass transfer, k/k_0, over one cycle under the conditions of f = 80 mHz, B = 250 mT and $\theta = 0$ rad.

The waveforms of the applied magnetic fields are known to be almost regularly rectangular except that a small spike appears when the magnetic field direction is switched.

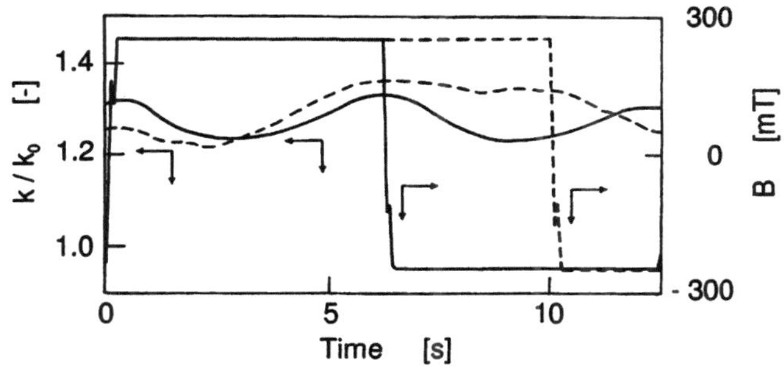

Figure 4. Changes in imposed magnetic fields and the resultant enhancement of mass transfer coefficient during one cycle (θ =0 rad, B =250 mT, f = 80 mHz; solid line: Symmetry case; broken line: Asymmetry case).

For the Symmetry case, the resultant time course of the enhancement of the mass transfer shows a harmonic oscillation of two cycles during one cycle of the imposed magnetic fields. Such a regular oscillation in the enhancement does not always appear in the experimental results for the other frequencies in the Symmetry case. Especially, for the Asymmetry case, the behavior does not show such a regularity like that with the applied magnetic field.

To investigate the above secondary periodic behavior of the mass transfer in detail, the instantaneous enhancement data were treated by applying Fourier analysis by taking into account the first twenty harmonic components. Examples of the obtained power spectra for f = 140 mHz and $\theta = 0$ rad are shown in Figure 5 for several values of the magnetic flux density. In the Symmetry case, the intensity of the second component is

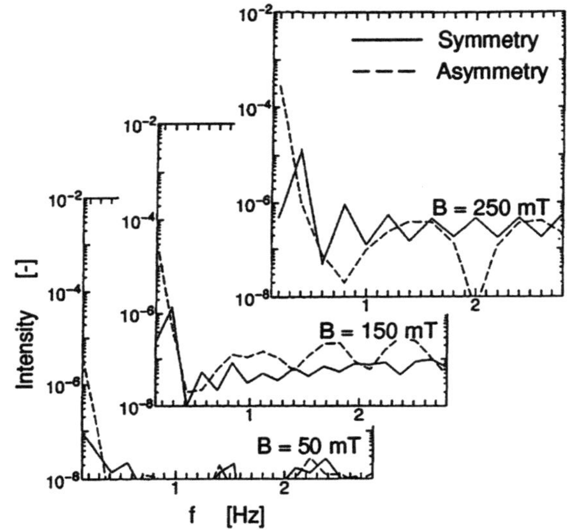

Figure 5. Power spectra for different magnetic flux densities (θ =0 rad, f = 140 mHz).

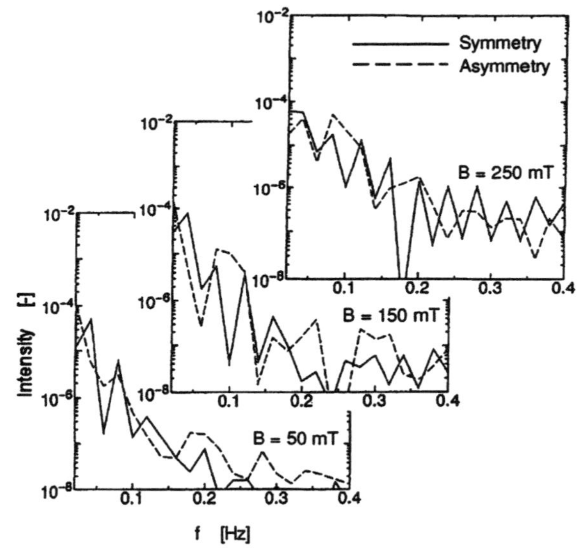

Figure 6. Power spectra for different magnetic flux densities (θ =0 rad, f = 20 mHz).

predominant for the major runs as pointed out previously and the harmonic with the frequency of integer times the second-component's frequency shows more intense peaks in the spectrum. On the contrary, in the Asymmetry case, the fundamental component is predominant and several broad peaks appear in the spectrum. Figure 6 shows the power spectra at θ = 0 rad for a lower frequency case, f = 20 mHz. Although the spectra for the Symmetry case show almost the same regular behavior as those f or f = 140 mHz, those for the Asymmetry case tend to show diminished regularity of the broad peaks.

3.2. TIME-AVERAGED ENHANCEMENT

Mean enhancement data are convenient for engineering purposes; therefore, the instantaneous enhancement of the mass transfer was integrated over the entire cycle and averaged. The averaged enhancement of mass transfer thus obtained, k^*/k_0, for θ=0 rad is plotted in Figure 7 versus B with parameters f and D_Y, compared with the data for the static magnetic field. As has been pointed out in previous articles concerning static or pulsating magnetic fields (Fahidy, 1990; Mohanta & Fahidy, 1974; Mori *et al.*, 1994a, 1994b, 1996), the averaged enhancement generally increases non-linearly with the flux density of the imposed switching magnetic field. However, the averaged enhancement for the Symmetry case decreases with an increase in the frequency and while that for the Asymmetry case is not appreciably influenced in the high B region by the frequency. As a result, application of switching magnetic fields is found not to be very effective when compared with that of the static ones, especially for the Symmetry case in the high f region.

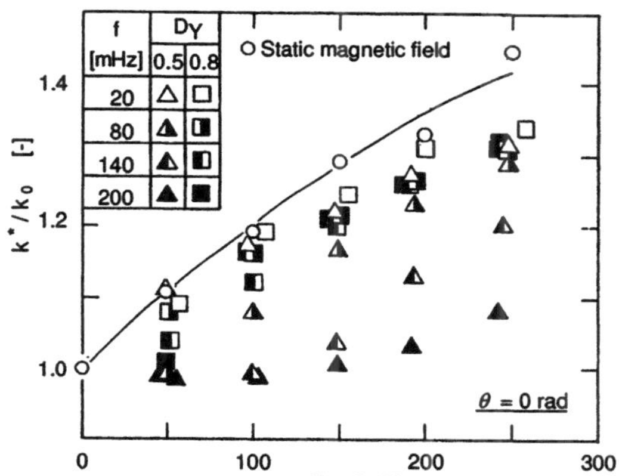

Figure 7. Averaged enhancement of mass transfer coefficient at θ =0 rad versus magnetic flux density.

In Figures 8 and 9, the results for $\theta = \pi/6$ and $\pi/3$ rad are shown, respectively. Although the dependence of the averaged enhancement on each parameter is seen to be generally very similar to that in Figure 7, the enhancement in the low B region decreases with an increase in the inclination angle. This means that the natural convection effect is predominant in the low B region.

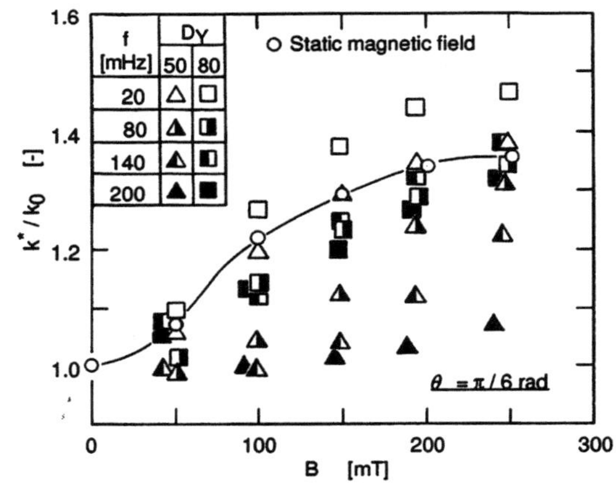

Figure 8. Averaged enhancement of mass transfer coefficient at $\theta = \pi/6$ rad versus magnetic flux density.

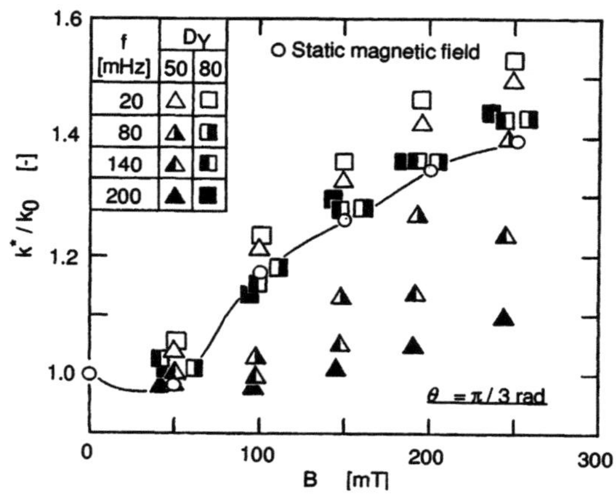

Figure 9. Averaged enhancement of mass transfer coefficient at $\theta = \pi/3$ rad versus magnetic flux density.

3.2. CORRELATION OF AVERAGED EXPERIMENTAL DATA

Normalizing relevant governing equations is a useful technique for developing proper parameters for establishing a dimensionless regression equation. Under the MHD approximation and some additional assumptions for simplicity, we can derive the following fundamental equations in dimensionless form, using reference physical quantities:

For diffusion

$$\frac{DC_i^*}{Dt^*} = \frac{1}{\mathrm{Sc}_i}\Delta^* C_i^* - \frac{z_i v_i \mathrm{F} d E_0}{\nu}(\nabla^* C_i^*) E^* - z_i v_i \mathrm{F} B_0 (\nabla^* C_i^*)(u^* \times B^*)$$
$$- z_i v_i \mathrm{F} B_0 C_i^* (\nabla^* \times u^*) B^* + z_i v_i \mathrm{F} B_0 C_i^* u^* (\nabla^* \times u^*) \tag{4}$$

For momentum

$$\frac{Du^*}{Dt^*} = -\nabla^* P^* + \Delta^* u^* - \mathrm{Gr} K - \frac{d^3 \sigma E_0 B_0}{\rho \nu^2}(u^* \times B^*) - \mathrm{Ha}^2 (u^* \times B^*) \times B^*$$
$$+ \sum_i \frac{M_{\mathrm{D}i}}{\mathrm{Sc}_i}(B^* \times \nabla^* C_i^*) \tag{5}$$

For magnetic field

$$\frac{DB^*}{Dt^*} = (B^* \nabla^*) u^* + \frac{1}{\mathrm{Pr_m}} \Delta^* B^* \tag{6}$$

with dimensionless variables:

$$u^* \equiv du/\nu\ , \qquad t^* \equiv \nu t/d^2\ , \qquad P^* \equiv d^3 P/(\rho_0 \nu^2),$$
$$\nabla^* \equiv d\nabla\ , \qquad B^* \equiv B/B_0\ , \qquad C_i^* \equiv C_i/C_0\ , \qquad E^* \equiv E/E_0$$

Thus we obtained important parameters as follows under the diffusion-controlled conditions:

$$\mathrm{Gr} \equiv \frac{g \zeta d^3 \Delta C}{\nu^2}, \qquad \mathrm{M}_{\mathrm{D}i} \equiv \frac{z_i \mathrm{F} B_0 \Delta C_i d^2}{\rho \nu}, \qquad \mathrm{Sc}_i \equiv \frac{\nu}{D_i}$$

and the magnetic Prandtl number, $\mathrm{Pr_m}$, is neglected in the data correlation in this study.

The following Hartmann number and dimensionless parameters appeared in the second to fifth terms on the right-hand side of Eq. (4) and the fourth term on the right side of Eq. (5) should be additionally taken into account when the electrophoresis has substantial effect.

$$\mathrm{Ha} \equiv dB_0 \{\sigma/(\rho \nu)\}^{1/2}$$

Based on the above discussion, first we correlated well the mass transfer coefficients for inclined cathode cylinders in static magnetic fields by introducing the following magneto-diffusion factor, Eq. (7), where the inclination angle was taken into account (Mori *et al.*, 1994a).

$$M_D = zFB\Delta C(d/\cos\theta)^2/(\rho\nu) \tag{7}$$

From the viewpoint of the amount of energy consumed in the mass transfer enhancement, there is no essential difference between exposure to static magnetic fields and that to switching ones. Therefore, the correlation of the averaged mass transfer coefficients in the present cases is attempted in the same manner as that for the static magnetic field. The Sherwood number is plotted versus $M_D Ra^{1/2}/Sc$ in Figure 10 with the frequency f as a parameter, compared with the regression equation, Eq. (8) (Mori *et al.*, 1994a), given for the case with the horizontal static magnetic field.

$$Sh = 2.72\left\{\frac{M_D Ra^{1/2}}{Sc}\right\}^{0.244} \tag{8}$$

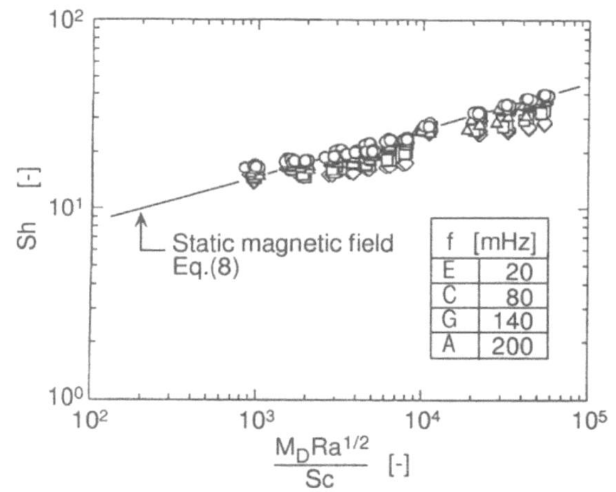

Figure 10. Correlation of mean mass transfer coefficient in the same manner as that for a static magnetic field

Clustering of the experimental data around Eq. (8) is found to be fairly good, however, a systematic shift in data points is observed in the graph in accordance with a change in the frequency. The flow condition in a case where higher-order harmonics prevail in a power spectrum might be substantially different from that in the other cases where the fundamental or the second component predominates over the spectrum. Consequently, data points with a weight of the most intensive component being less than 20 % were excluded from Figure 10 to obtain Figure 11. Thus the remaining data were then found to be in good agreement with Eq. (8) for the static magnetic field case.

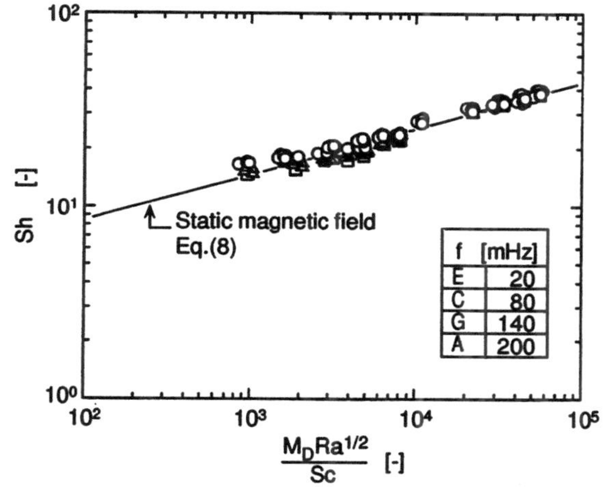

Figure 11. Correlation of mean mass transfer coefficient in the same manner as that for a static magnetic field (data with less fluctuation).

In the above data correlation, however, no switching characteristics of the imposed magnetic field are taken into consideration resulting in only the restricted data being correlated well. Therefore, another suitable regression equation will be studied now to correlate all the data by properly taking into account the contribution of the frequency, which is a characteristic factor of the periodic switching magnetic field. When MHD flow is induced by imposing a magnetic field in one direction, the flow development would depend on the imposed time. In the switching magnetic field case, the time imposed in one direction is considered to be substantially D_Y/f within one oscillation period. Accordingly, we introduced a new dimensionless parameter, the "frequency factor", defined as follows;

$$F_D \equiv D_Y/(f\tau) \tag{9}$$

where τ is the relaxation time of the magnetic field effect on the mass transfer, measured by preliminary experiments. Finally, the effect of the magnetic fields is assumed to be expressed by the product of $F_D{}^n$ with $M_D Ra^{1/2}/Sc$. We determined the exponent of F_D as n = 0.35 by trial-and-error. Thus all of the experimental mass transfer data are re-plotted in Figure 12 against $F_D{}^{0.35} M_D Ra^{1/2}/Sc$ with the frequency as a parameter, without any restriction. The present data are shown to be clustering very well around the following regression equation with a correlation coefficient of 0.906, regardless of any variation in frequency.

$$\mathrm{Sh}=3.61\left\{F_{\mathrm{D}}{}^{0.35}\left(\frac{M_{\mathrm{D}}\mathrm{Ra}^{1/2}}{\mathrm{Sc}}\right)\right\}^{0.221} \tag{10}$$

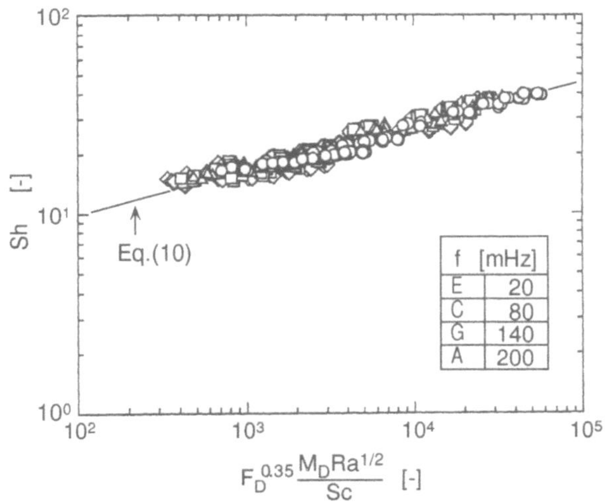

Figure 12. Correlation of mean mass transfer coefficient by introducing a "frequency factor".

4. Conclusions

The effect of a switching magnetic field in the horizontal direction with a symmetric or asymmetric rectangular waveform on electrolytic mass transfer was experimentally studied for a $K_3Fe(CN)_6$ - $K_4Fe(CN)_6$ system with supporting electrolyte using an inclined cylindrical cathode on a plane parallel to the magnetic field direction.

The instantaneous mass transfer rate under the periodic steady state was enhanced by applying the magnetic fields. Utilizing Fourier analysis, the instantaneous enhancement data were examined to elucidate the characteristic behavior depending on the waveform, frequency and flux density of the imposed magnetic fields.

Roughly speaking, the mean mass transfer enhancement for the alternating magnetic fields is comparatively less than that for the static magnetic field. By introducing a new parameter, the "frequency factor", and the magneto-diffusion factor, the averaged mass transfer coefficient correlated very well with a new non-dimensional regression equation, Eq. (10). When the perturbation of the instantaneous data is comparatively small, the mean mass transfer coefficient data can be represented fairly well by the dimensionless regression equation for the static magnetic field case in spite of the absence of a term related to frequency in the equation.

Acknowledgments

The authors are grateful to Miss N. Kajita and Mr. T. Okamoto for their assistance in the experiments. The electromagnet employed in the experiments was provided by the Laboratory of Magnetic Field Control and Applications, Kanazawa University.

Nomenclature

A	=	surface area of cathode	[m^2]
B	=	magnetic flux density	[T]
C_i	=	concentration of ion species i	[mol/m^3]
ΔC	=	concentration difference between bulk solution and interface	[mol/m^3]
d	=	diameter of cylindrical cathode	[m]
D	=	diffusion coefficient	[m^2/s]
D_Y	=	time ratio of switching magnetic field	[-]
E	=	intensity of electric field	[V/m]
f	=	frequency	[Hz]
F	=	Faraday constant	[C/eq.]
F_D	=	frequency factor, defined by Eq. (9)	[-]
g	=	gravity acceleration	[m/s^2]
Gr	=	Grashof number, defined as $g\,\zeta(d/\cos\theta)^3/\nu^2$	[-]
Ha	=	Hartmann number, defined as $d\,B_0\{\sigma/(\rho\,\nu)\}^{1/2}$	[-]
I	=	limiting current	[A]
k	=	instantaneous mass transfer coefficient	[m/s]
K	=	vertical unit vector	[-]
k*	=	mass transfer coefficient averaged over one cycle	[m/s]
M_D	=	magneto-diffusion factor, defined as $z\mathrm{F}B_0\Delta C\,(d/\cos\theta)^2/(\rho\,\nu)$	[-]
P	=	pressure	[Pa]
$\mathrm{Pr_m}$	=	magnetic Prandtl number, defined as ν/ν_m	[-]
Ra	=	Rayleigh number, defined as GrSc	[-]
Sc	=	Schmidt number, defined as ν/D	[-]
Sh	=	Sherwood number, defined as $k^*(d/\cos\theta)/D$	[-]
t	=	time	[s]
u	=	flow velocity	[m/s]
v_i	=	mobility of ion species I	[mol/(N s)]
z	=	valency of ion	[eq./mol]
ζ	=	densification factor	[m^3/mol]
μ	=	magnetic permeability	[H/m]
θ	=	inclination angle of cathode	[rad]
ν	=	kinematic viscosity	[m^2/s]
ν_m	=	magnetic viscosity	[m^2/s]
ρ	=	density	[kg/m^3]
σ	=	electrical conductivity	[S/m]
τ	=	relaxation time	[s]

<Subscripts and Superscripts>

0	=	without magnetic field; or specified reference quantity
*	=	averaged over one cycle; or normalized quantity
i	=	ion species i

References

Aaboubi, O., J.P. Chopart, A. Olivier *et al.* (1990) Magnetic field effects on mass transport, *J. Electrochem. Soc.* **137**, 1796-1804.

Fahidy, T. Z. (1990) The statistical indeterminacy of the magnetic field effect on electrolytic mass transfer, *Electrochim. Acta* **35**, 929-932.

Iwakura, C., T. Edamoto & H. Tamura (1984) Effect of a relatively weak magnetic field on electrochemical reaction, *Denki Kagaku* **52**, 596-601.

Mohanta, S. & T. Z. Fahidy (1974) Mass transfer in a magnetoelectrolytic flow cell, *Electrochim. Acta* **19**, 835-840.

Mori, S., K. Satoh & A. Tanimoto (1994a) Electrolytic mass transfer around inclined cylinders in static magnetic fields, *Electrochim. Acta* **39**, 2789-2794.

Mori, S., K. Satoh & A. Tanimoto (1994b) Enhancement of electrolytic mass ransfer around spheres by applying static magnetic fields, *J. Chem. Eng. Japan* **27**, 803-807.

Mori, S., M. Kumita, M. Takeuchi & A. Tanimoto (1996) Effects of pulsating magnetic fields on electrolytic mass transfer around cylindrical cathodes, *J. Chem. Eng Japan* **29**, 229-234.

Nippon Kikai Gakkai, ed. (1965) *Dennetsu Kogaku Shiryo*, p.36, Tokyo, Japan.

Schütz, G. (1963) Natural convection mass-transfer measurements on spheres and horizontal cylinders by an electrochemical method, *Int. J. Heat Mass Transfer* **6**, 873-879.

LAMINAR DEVELOPING MASS TRANSFER IN ANNULUS WITH POWER LAW-FLUIDS

O. OULD-DRIS, A. SALEM
Laboratoire de Mécanique des fluides, Institut de Physique, USTHB, B.P.32 El Alia, 16111, Alger, Algérie

J. LEGRAND
Laboratoire de Génie des Procédés, UPRES-EA 1152 - Université de Nantes, IUT de Saint-Nazaire, B.P. 406, 44602 Saint Nazaire Cedex, France

and

C. NOUAR
LEMTA - URA CNRS No 875, 2 avenue de la Forêt du Haye, B.P. 160, 54504 Vandoeuvre Cedex, France

Abstract : The aim of the present work is to investigate the mass transfer evolution in developing flow of Newtonian and non-Newtonian electrolyte solution flowing through an annulus with radius ratio a = 0.70. The non-Newtonian solution can be described in terms of power-law model. The rheological behaviour of the electrolyte was experimentally determined. Using the electrochemical method, the mass transfer coefficients were determined for different electrode lengths L= 0.1; 0.3 and 0.5 m, always located just downstream of the cell entrance, taking into account both developments of the hydrodynamic and concentration boundary layers. The results corresponding to Newtonian and non-Newtonian power law fluid flows were presented and were compared to the literature data.

A. Alemany et al. (eds.),
Transfer Phenomena in Magnetohydrodynamic and Electroconducting Flows, 213–227.

1. Introduction

The study of developing flow of non-Newtonian fluids in the entrance region of the annulus is important because of its application in many polymer industrial processes. Experimental investigations on non-Newtonian developing annular flow are mostly devoted to pressure drops and velocity profiles determination. For inelastic liquids the pressure drop per unit length in the entrance section is higher than that in the fully developed region. Then, the entrance section, although not very long, may contribute a significant influence to the overall pressure drop and has to be taken into account in design (Bhattacharyya and Tiu, 1974). In the experimental works relating to the annular entry flow three different geometries of the entry are usual, namely, square entrance, rounded entrance, and a square entrance with the inner tube extending to the upstream side of the contraction. When a fluid enters an annulus in laminar flow through an abrupt contraction, it undergoes a change in the flow pattern from an initial condition close to a flat velocity profile at the inlet, to a fully developed condition downstream. In the numerical or theoretical works, a perfectly flat entry profile is generally assumed. In order to avoid the singularity due to the discontinuity in the velocity at the entrance, the plug flow profile can be replaced by a boundary layer polynomial with continuous derivatives (Gaidos and Darhy, 1988). The case of convective heat transfer has been studied for different types of non-Newtonian fluids. Information available on the simultaneous development of momentum and thermal boundary layers is greatly limited. Moreover, the evolution of the rheological properties with temperature is not always taken into account (Nouar *et al.,* 1994). This evolution constitutes a major difference between heat and mass convective problems in non-Newtonian fluids. However, few information dealing with mass transfer in developed or developing annular non-Newtonian flow is available in the literature. Mass transfer in fully developed laminar and turbulent annular flow has been analyzed by several workers (Ross and Wragg, 1965 ; Turitto, 1975 ; Remorino *et al.*, 1979), but the case of developing annular flows with developing concentration boundary layers has received less attention (Qi and Savinell, 1991 ; Ould-Rouis *et al.,* 1995). Further, except for the work of Remorino *et al.* (1979), all reported studies are limited to Newtonian fluids. Using the electrochemical method, Remorino *et al.* (1979) have studied the mass transfer in annuli for non-Newtonian fluids. They have adapted the Lévêque solution for the case of mass transfer to the inner wall of an annulus in fully developed flow of non-Newtonian fluid obeing to the power-law model. They have shown that experimental data for both Newtonian and non-Newtonian fluids (n = 0.77 to 1) can be successfully represented by :

$$Sh = 1.614\left[Re_n . Sc_n . de / L . \Gamma(a,n)\right]^{1/3} \tag{1}$$

Re_n and Sc_n are the Reynolds and Schmidt numbers respectively, modified for power-law fluids:

$$Re_n = \frac{de^n . \rho . Um^{2-n}}{8^{n-1} . K . (3n+1)^n / (4n)^n} \tag{2}$$

and

$$Sc_n = \frac{K . (3n+1)^n / (4n)^n . (8 Um / de)^{n-1}}{D . \rho} \tag{3}$$

where de = 2.(R2 - R1) is the equivalent diameter, K and n are respectively the consistency parameter and the flow behaviour index, Um is the mean velocity and D the diffusivity. The parameter $\Gamma(a,n)$ is fonction of both aspect ratio, $a = R1/R2$, and flow behaviour index n ; it is expressed by :

$$\Gamma(a,n) = \frac{(1-a^2) . a^2 . \left[\left(\frac{\lambda(a,n)}{a}\right)^2 - 1 \right]^{1/n} (1-a)}{4 . \Omega(a,n)} \tag{4}$$

where $\lambda(a,n)$ is the dimensionless radius for maximum velocity and $\Omega(a,n)$ is the dimensionless flow rate. Their values were obtained numerically by Fredrickson and Bird (1958).

The present paper investigates the simultaneous development of diffusional and momentum boundary layers in annular laminar power-law fluid flow. The overall mass transfer coefficients are determined by using an electrochemical method. The mass transfer evolution is studied downstream a calming section constituted by an annular packed bed, which allows to obtain a rather uniform axial velocity profile with an over-velocity region near the walls, where the porosity tends to unity. The mass transfer results are analyzed according to the Levêque solution in annulus.

2. Experimental

2.1. DESCRIPTION OF THE APPARATUS

The experimental device is schematically shown in figure 1. The annulus test section was made of two coaxial cylinders, 600 mm long with 38 mm and 54 mm respectively for internal and external diameters.

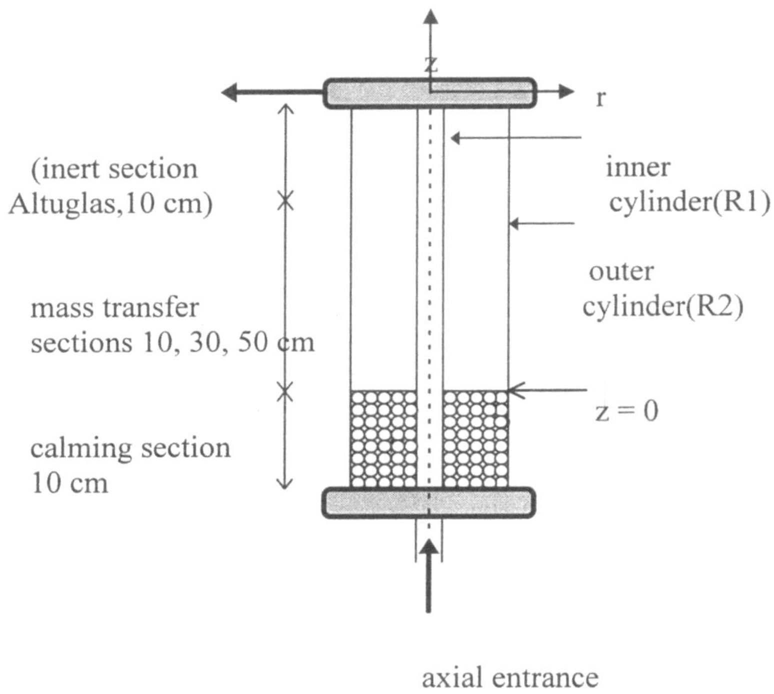

Figure 1 : Description of the annular cell.

Each wall is composed by an interchangeable nickel cylinder of 100 ; 300 and 500 mm. Between the test section and the outlet, an altuglas extension of 100 mm long was added to eliminate the end effects. Before the entrance in the annular test section, the fluid flowed through annular packed bed of spheres of 2 mm in diameter (figure 1).
The features of the hydraulic circuit are given in figure 2. The electrolyte solution, which was maintained at constant temperature of 25°C, flowed from the tank through a centrifugal pump to the flowmeters which allowed the measurement of the volumic flow rate. Then, the electrolyte

axially entered the annulus and flowed out to the tank through a tangential outlet (figure 2). We have used a centrifugal pump by verifying that the rheological behaviour of the solution do not change during the experiments.

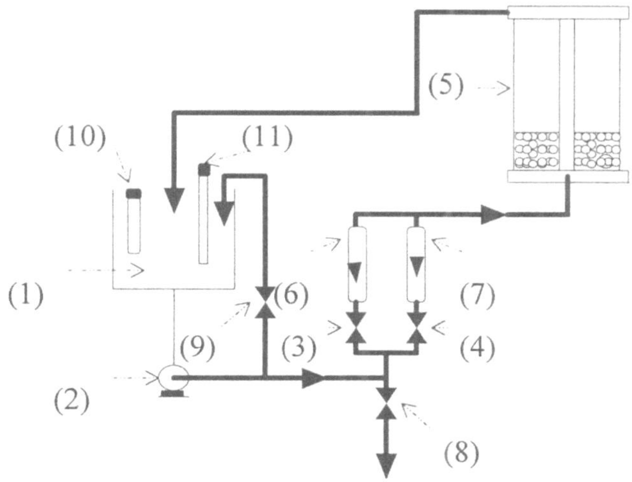

Figure 2 : A general view of the experimental apparatus (1) tank, (2) centrifugal pump, (3) and (4) flow adjustment gates, (5) electrochemical cell, (6) and (7), flowmeters, (8) drain pipe, (9) by-pass, (10) temperature regulation,(11) nitrogen injection.

2.2. ELECTROLYTE

Measurements of the overall mass transfer coefficients were performed using a polarographic method. Two different electrolytes were used in the present study. The Newtonian electrolyte was an aqueous 0.5 M NaOH solution supporting a mixture of $2\ 10^{-3}$ M potassium ferricyanide, $5\ 10^{-2}$ M potassium ferrocyanide. At 25°C, the physical properties of this solution are : a density ρ of 1028 kg/m^3, a kinematic viscosity ν of 8.72×10^{-7} m^2 .s^{-1} and a diffusion coefficient D of ferricyanide ions in the mixture of 6.95×10^{-10} m^2. s^{-1} . Thus, the Schmidt number $Sc = \nu / D$ is equal to 1255. The Newtonian solution is used to test the equipment and our experimental conditions. The non-Newtonian solutions contained 6 or 8 kg m^{-3} carboxymethylcellylose (CMC) sodium salt (BDH high viscosity), 1 mol m^{-3} potassium ferricyanide $K_3\,Fe(CN)_6$, 10 mol m^{-3} potassium ferrocyanide $K_4Fe(CN)_6$, 10^2 mol m^{-3} sodium carbonate, sodium hydrogenocarbonate 10^2 mol m^{-3}. Its density was $\rho = 1020$ kg m^{-3} at 25°C.

The behaviour of this polymer solution can be described in terms of the Ostwald de Weale law model $\tau = K.\dot{\gamma}^n$. The behaviour index n and the consistency parameter K were determined from shear stress τ, against shear rate $\dot{\gamma}$, data obtained in a rotational rheometer using a double gap concentric cylinder. The rheological equations at 25°C, may be written as follows:

$$\tau = 0.09.\dot{\gamma}^{0.77} \quad \text{for} \quad 6\ \text{kg.m}^{-3}\ \text{CMC solution}$$
$$\tau = 0.31.\dot{\gamma}^{0.67} \quad \text{for} \quad 8\ \text{kg.m}^{-3}\ \text{CMC solution} \qquad (5)$$

The rheological equations are obtained for $23\text{s}^{-1} \le \dot{\gamma} \le 837\text{s}^{-1}$; the value $\dot{\gamma}\ = 837\,\text{s}^{-1}$ corresponds to the largest estimated value of shear rate reached in mass transfer experiments. As shown in figure 3, the rheological behaviour of the non-Newtonian electrolyte is constant during the experiments.

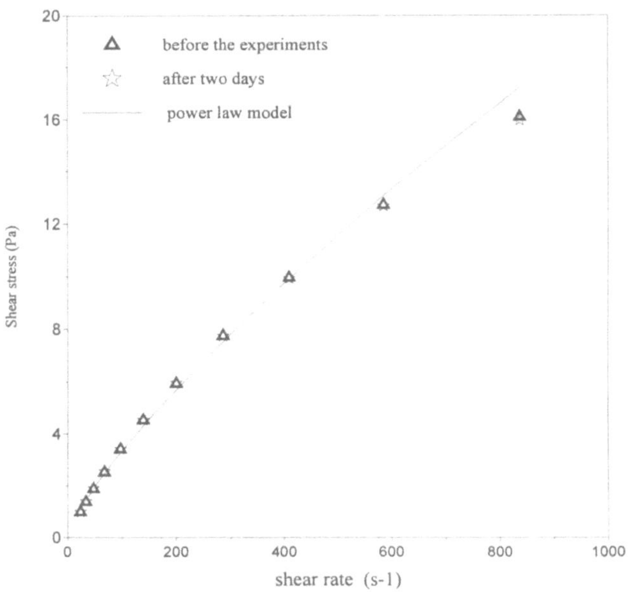

Figure 3 : Rheological behaviour of the electrolyte solution (6 kg.m^{-3} CMC).

The diffusivity D was determined with the rotating disk electrode by using the theoretical equation giving the average molar flux at the disk $J = I/(F.\pi.R^2)$, which was derived analytically by Grief and Paterson (1973) :

$$J = \frac{N(n)}{0.893} C_0 D^{2/3} \left(\frac{K}{\rho}\right)^{-1/3(n+1)} R^{(1-n)/3(1+n)} \omega^{1/(n+1)} \tag{6}$$

where C_0 the concentration of ferricyanide ions in the solution , F the Faraday number, and :

$$N(n) = \left[12.a.\left(\frac{1+n}{7+5n}\right)^2\right]^{1/3} \tag{7}$$

On the other hand, the electrochemical method with rotating platinum disk electrode of radius $R = 10^{-3}$ m allowed to obtain the limiting current from current - potential curves corresponding to ferricyanide ions reduction at the disk. The relation obtained between the limiting current and the rotation speed ω is expressed for $6 kg.m^{-3}$ CMC solution by :

$$I = 5.49 x 10^{-7}.\omega^{-0.56} \qquad \text{with} \qquad 50 s^{-1} \leq \omega \leq 524 s^{-1} \tag{8}$$

The equations (6), (7), and (8) lead to 0.77 for index behaviour, *i.e.* the same value than the one determined with the rheometer. Then, the diffusion coefficient D is equal to $5.81 \times 10^{-10}\ m^2 .s^{-1}$. For $8 kg.m^{-3}$ CMC solution, the diffusion coefficient is obtained by the same way and equal to $5.65 \times 10^{-10}\ m^2 s^{-1}$.

2.3. ELECTROCHEMICAL METHOD

The electrochemical determination of the mass transfer coefficients between the electrolyte and the active sections of both inner and outer walls of the annulus was performed using an electrochemical method, based on the cathodic reduction of ferricyanide ions. The electrochemical reaction studied was:

$$Fe(CN)_6^{3-} + e^- \rightarrow Fe(CN)_6^{4-} \tag{9}$$

The overall mass transfer coefficients k was obtained by means of diffusion controlled conditions on the cathode using the following relation :

$$k = \frac{I}{n_e . F . A . C_0} \tag{10}$$

where I is the measured limiting current, n_e the number of the electrons involved in the electrochemical process ($n_e = 1$), F the Faraday constant, C_0 the potassium ferricyanide concentration and A the cathode surface area. The main interest of this electrochemical method is that the mass transfer surface remains unaltered during the experiments. The mass transfer experiments were performed in a cylindrical annulus with the aspect ratio of the inner radius to the outer one $a = R1/R2 = 0.70$. The mass transfer evolution were determined for different interchangeable electrode lengths L = 10, 30, and 50 cm, always located just downstream of the cell entrance. These nickel electrodes were placed at the entrance of the cell, then hydrodynamic and diffusional boundary layers were simultaneuously developed.

3. Results

The mass transfer results were expressed by the mean Sherwood number $Sh = k.de/D$. For Newtonian solution we have represented the mass transfer results in figure 4 by the evolution of the ratio $Sh/Sc^{1/3}$ against the product $Re.\phi(a).de/L$, where $Re = de.Um/\nu$ is the Reynolds number for Newtonian fluid and $\phi(a)$, the annulus geometrical factor expressed (Ross and Wragg, 1965) by :

$$\phi(a) = \frac{1-a}{a} \left(\frac{0.5 - \frac{a^2}{1-a^2} \ln(1/a)}{(\frac{1+a^2}{1-a^2}) \ln(1/a) - a} \right) \tag{11}$$

We have compared our experimental results with correlations experimentally established by Qi and Savinell (1991) for developing flow. It is shown that our results are in a good agreement with the authors for low Reynolds numbers with a little deviation probably due to a different aspect ratio, a, and the experimental conditions.

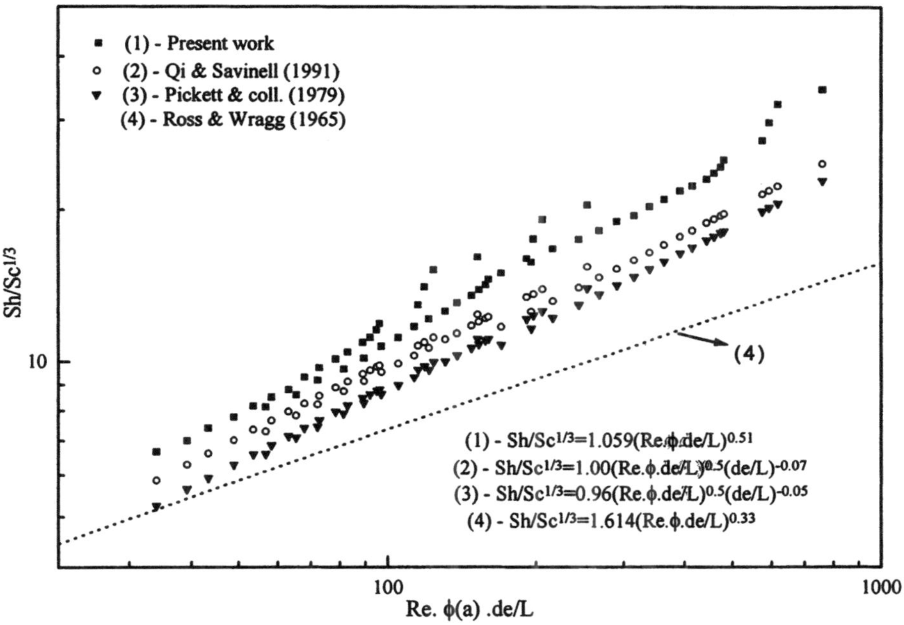

Figure 4 : Mass transfer results for Newtonian fluid in an annulus.

For non-Newtonian solution , we have studied the influence of the electrode length and the evolution of mass transfer in both walls of the annulus (figure 5). As for Newtonian fluid, it is shown that the shorter the cathode is, the higher the mass transfer rate will be. The mass transfer on the inner wall remains more important to that along the outer wall. This difference is related to the boundary layers evolution. The inner boundary layer is always thinner to the outer one (Ould-Rouis *et al.*, 1995).

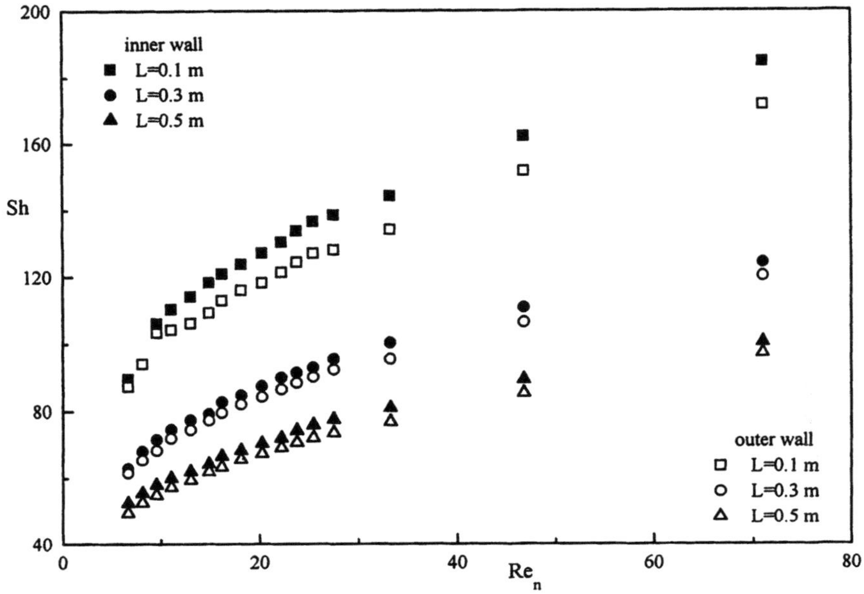

Figure 5 : Relationship of Sh to Re_n for various electrode lengths for power-law fluid in both walls of an annulus (6 $kg.m^{-3}$ CMC solution).

The comparison between the evolution of mass transfer for two non-newtonian solutions (n = 0.67 and 0.77) is given in figure 6. For a given Reynolds number, the mass transfer coefficients increase with a decrease in index behaviour, which is related to an increase of Schmidt number.

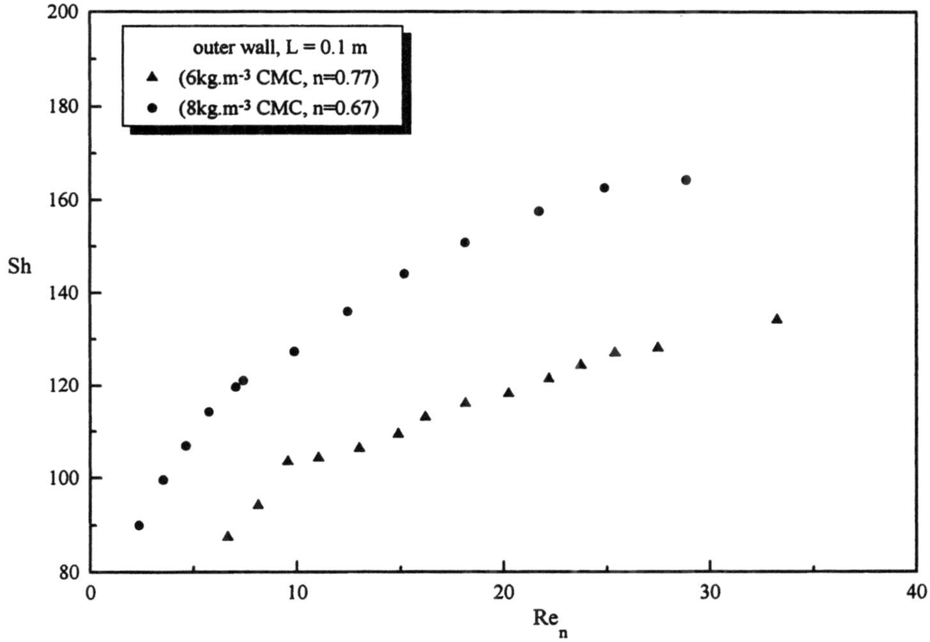

Figure 6 : Mass transfer comparison for power law fluid flows.

In figure 7, we have represented our experimental results for developing axial annular flow for power-law fluids in terms of $Sh/Sc^{1/3}$ against the expression $Re_n.\Gamma(a,n).de/L$, where the different terms are previously defined. The comparison of our results in developing annular flow with the data obtained in fully developed flow (eq. 1) shows that the laminar flow regime is pratically fully developed as soon as the fluid enters in the cell for power-law fluid. The entrance length for flow of inelastic power-law fluids increases with increasing flow behaviour index (Tiu and Bhattacharyya, 1974 ; Mehrotra and Patience, 1990). By assuming that the entrance length, Le, for non-Newtonian fluids could be determined from equation established for Newtonian fluids by replacing Re by Re_n, we have used the following expression :

$$L_e = 2.68 \times 10^{-2} (R2 - R1) Re_n \tag{12}$$

This equation was obtained by Ould-Rouis *et al.* (1995) in laminar annular Newtonian fluid flow with entrance conditions similar of the ones studied in the present study.
The experimental domain of Re_n investigated in our work is comprised between 5 and 70 (figure 5). Then, the entrance lengths (eq. 12) vary between 0.1 and 1.5 cm, thus they are negligeable with respect to the lengths of the mass transfer section, L = 10 to 50 cm.

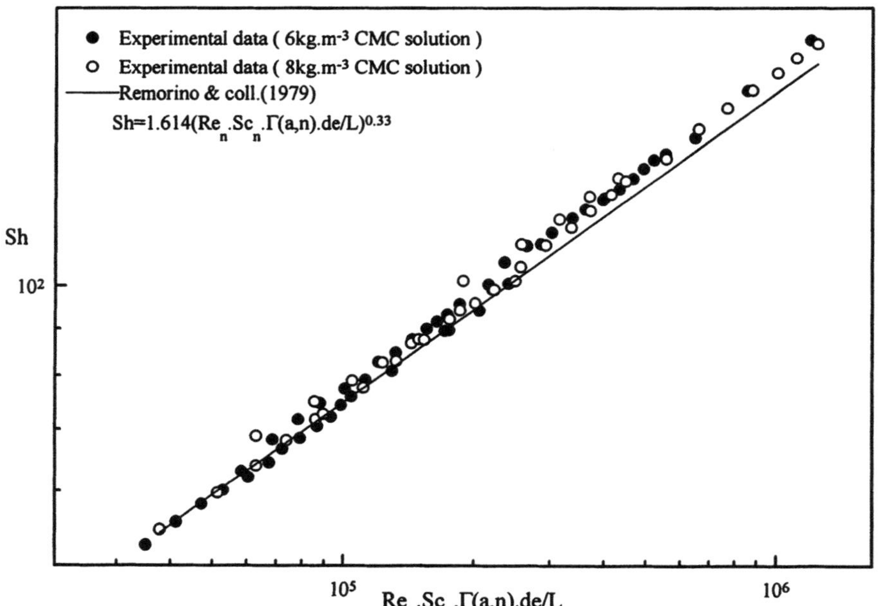

Figure 7 : Mass transfer results for power-law fluid flow in an annulus.

The comparison is made (figure 8) between our experimental results for power-law fluid flow, which is pratically fully developed at the entrance of the cell, with the fully developed Newtonian fluid flow represented by the relation (13), obtained by Ross and Wragg (1965) :

$$Sh / Sc^{1/3} = 1.614[Re.\phi(a).de / L]^{0.33} \tag{13}$$

This equation is equivalent to equation (1) for Newtonian fluid.

Figure 8 shows the same evolution, which confirms that experimental data for Newtonian and non-Newtonian power-law fluids can be successfully represented by the equation (1).

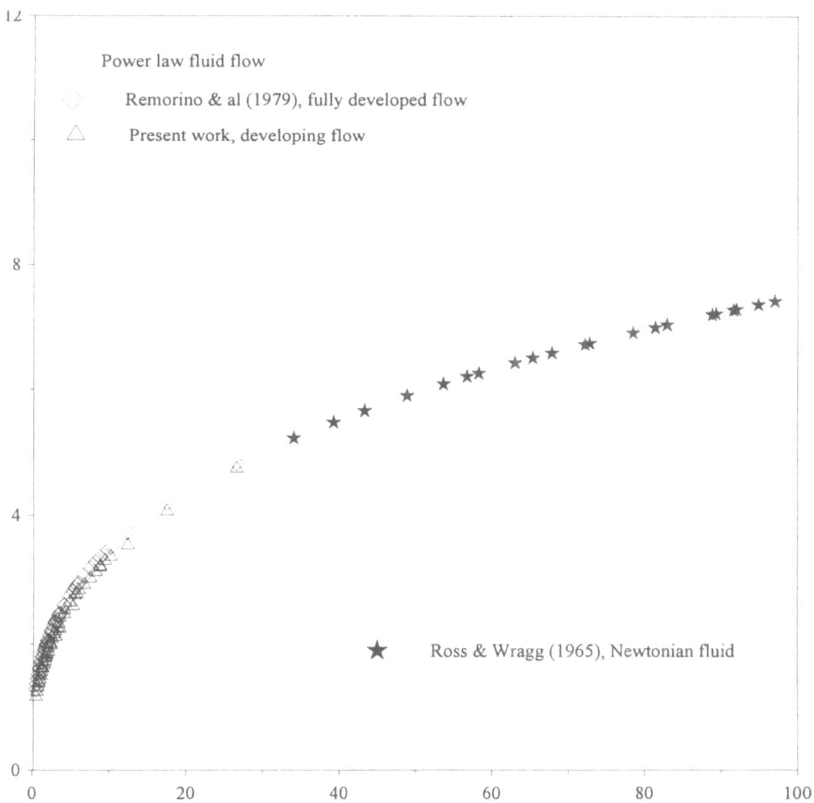

Figure 8 : $Sh / Sc_n^{1/3} vs (Re_n.\Gamma(a,n).de / L)$ Comparison of experimental data obtained for Newtonian and power-law fluids flow

4. Conclusion

The electrochemical method was carried out to study the mass transfer in the entrance region of annular cell for non-Newtonian power-law fluid laminar flow through an annulus. Two types of electrolyte were used : a Newtonian solution in order to test the validity of the polarographic method and two non-Newtonian solution obeing to the power-law model. The laminar flow regime of an Ostwald fluid axially flowing in an annulus was pratically fully developed at the entrance of the cell, due to the large apparent viscosity of non-Newtonian fluid leading to low generalized Reynolds number. This results was confirmed by calculating the hydrodynamic entrance length which can be considered as negligeable in our experimental conditions. Then, a same correlation may correlate mass transfer data for both Newtonian and non-Newtonian fluids flow. The correlation corresponds to the Lévêque solution for the case of developing mass transfer in annular flow for fully developed laminar flow regime.

Acknowledgements

The authors would like to thank the Algerian and French Governements for their financial support under the auspices of Comité Mixte d'Evaluation et de la Prospective de la Coopération Interuniversitaire franco-algérienne. Accord-Programme 95 MDU 333.

References

Bhattacharyya, S. and Tiu, C. (1974) Developing pressure profiles for non-Newtonian flow in an annular duct, *AIChE Journal* **20**, 154-158.

Fredrickson, A.G and Bird, R.B. (1958) Non-Newtonian flow in annuli, *Ind. Eng. Chem.* **50**, 347-352.

Gaidos, R.E. and Darby, R. (1988) Numerical simulation and change in type in the developing flow of a nonlinear viscoelastic fluid, *J. Non-Newtonian Fluid Mech.* **29**, 59-79.

Grief, R. and Paterson, J. (1973) Mass transfer to a rotating disk in a non-Newtonian fluid, *Physics of Fluids* **16** 1816-1823.

Mehrotra, A.K. and Patience, G.S. (1990) Unified entry length for Newtonian and power-law fluids in laminar pipe flow, Can. J. Chem. Engng **68**, 529-533.

Ould-Rouis, M., Salem, A., Legrand, J. and Nouar, C. (1995) Etude numérique et expérimentale des transferts de matière et de quantité de mouvement dans un écoulement annulaire laminaire non établi, *Int. J. Heat Mass Transfer* **38**, 953-967.

Nouar, C., Devienne, R. and Lebouché, M. (1994) Convection thermique pour un fluide de Herschel- Bulkley dans la région d'entrée d'une conduite, *Int. J. Heat Mass Transfer* **37**, 1-12.

Pickett, D.,J. (1979) *Electrochemical Reactor Design*, Elsevier, Amsterdam, 121-170.

Qi, J. and Savinell, R.F. (1991) Mass transfer in a laminar-flow parallel plate electrolytic cell with simultaneous development of velocity boundary layers, *J. Appl. Electrochem* **20**, 885-892.
Remorino, M.R., Tonini, R.D. and Bohm, U. (1979) Mass transfer to the inner wall of an annulus with non-Newtonian fluids, *A.I.Ch.E. Journal* **25**, 368-370.
Ross, T.K. and Wragg, A.A. (1965) Electrochemical mass transfer studies annuli, *Electrochimica Acta* **10**, 1093-1106.
Tiu, C. and Bhattacharyya, S. (1974) Developing and fully developed profiles for inelastic power-law fluids in an annulus, *AIChE Journal* **20**, 1140-1144.
Turitto, V.T. (1975) Mass transfer in annuli under conditions of laminar flow, *Chem. Engng Sci.* **30**, 503-509.

STUDY OF NEAR WALL HYDRODYNAMICS AND MASS TRANSFER UNDER MAGNETIC FIELD INFLUENCE

S. MARTEMIANOV[1,2], A. SVIRIDOV[2,3]

1 - Laboratoire d'Etude Thermique, UMR CNRS 6608
ESIP - 40 avenue du Recteur Pineau 86022 , Poitiers Cedex, France

2 - Frumkin Institute of Electrochemistry RAN
31, Leninsky prospekt, Moscow 117071 Russia

3 - Moscow State University of Aviation Technology
27, Petrovka, Moscow, Russia

Abstract

The influence of magnetic field on near wall hydrodynamics and mass transfer was studied by means of two experimental methods. The first one is based on direct mass transfer measurements by means of radio-active isotopes. The second method is electrochemical flow measurements. The experiments show that non-stationary magnetic field is more efficient for intensification of the near wall mass transfer than the stationary one. Also, non-uniform in space magnetic field is more efficient for increasing of wall shear stress value and thus for intensification of mass transfer processes than a uniform one of the same intensity. This effect can be explained by a resonance brake action of inertial and electromagnetic forces on the mean flow. As a result, near wall liquid is significantly accelerated.

1. Introduction

This paper aims at presenting some experimental results on magnetic field influence on near wall hydrodynamics and mass transfer. It is well-known that by means of magnetic field which is perpendicular to a mean flow direction it is possible to significantly increase wall shear stress and by this

A. Alemany et al. (eds.),
Transfer Phenomena in Magnetohydrodynamic and Electroconducting Flows, 229–240.

way to intensify near wall mass transfer. This effect can be important in some electrochemical technologies, as a total rate of fast electrochemical reactions is controlled by mass transfer of electrochemical species to an electrode surface. Processes of hydrodynamical clearning of pipelines and different nature conduits give another example when intensification of near wall mass transfer is very important.

For stationary flows of a simple geometry the magnetic field influence can be taken into account by means either analytical models (Hartmann layer) or numerical simulations. For practical application it is important to study the effect of non-uniform in space and non-stationary magnetic field on near wall hydrodynamics and mass transfer. In order to achieve this goal two experimental methods are used in the present study.

The first method is based on direct mass transfer measurements by means of radio-active isotopes. In this method a contamination with some known amount of radio-active isotope Fe-59 is brought on the inner surface of a tube. By measuring of a radiation rate on the flow outlet it is possible to estimate the rate of particle's removal from the surface.

The secon method is electrochemical flow measurements. Electrochemical method is one of the most efficient instruments for hydrodynamical and mass transfer studies in electrolyte flows [1]. This method is founded on measurements of the limiting diffusion current of a fast electrochemical reaction which takes place on a surface of a flash monted with a flow electrode. Some electroactive species should be dissolved in the electrolyte in order to provide the electrochemicl reaction. These species does't influence on the flow hydrodynamics and are used only as a marker. Electrochemical method was successfully used for measurements in gas-liquid flows [2], for turbulent studies [1 - 4], for measurements in swirling decaying flows [5] and in impinging jet flows [6], for mass transfer measurements at a stirred liquid-liquid interfaces [7, 8], in film flows [9] and in non-newtonian flows [10, 11]. There are two modifications of electrochemical diagnostics of liquid flows : amplitude method and delay-time method.

2. Amplitude method in electrochemical flow measurements

This method is widely practised in laboratory conditions when reseacher makes his own choice of electrolyte solution and marker species. The amplitude method makes use of electrochemical current amplitude measurements in order to obtain information about wall shear stress and mass transfer coefficient. For detection of local wall shear stress value a microelectrode is embedded into a surface of a flow which is inert from electrical point of view. Non-segmented microelectrode gives an information about scalar value of the wall shear stress. Both mean and pulsation characteristics of wall shear stress can be measured.

Two-segmented microelectrode allows one to measure mean and fluctuating characteristics of two components of the wall shear stress. Three-segmented electrodiffusion probes [12, 13] give a possibility to obtain an information about normal velocity component in steady-state regime. The systeme of three working electrodes with different electrode's length in a flow direction is suitable to obtain an information about fluctuations of the normal velocity component [4].

Theoretical background of electrochemical flow measurements is well developped and transfer characteristics of electrodiffusion probes are calculated for different electrode's configurations [1 - 4, 10, 12, 13]. From the other hand, manufacturing of electrodiffusion probes is rather difficult problem. Indeed, the characteristics of these probes are strictly dependent on the working electrode geometry. An important step was made recently with the use of microtechnologies (photolithography). These technologies allow one to realize electrochemical probes fulfilling as close as possible the theoretical requirements [14].

For mass transfer studies electrochemical method makes use of relatively large working electrodes [5, 7, 8, 16]. Diffusion layer near a surface of a large electrode is fully developped and the electrochemical current through this electrode is directly related to the mass transfer coefficient. In order to obtain a local value of the mass transfer coefficient a small microelectrode is embedded into a large electrode and both electrodes are polarized by the same potential. Interpretation of mass transfer measurements in turbulent flow regime is rather difficult problem. An important step in interpretation of turbulent mass transfer measurements was made after developping of a linear statistical theory of turbulent diffusion layer [15, 16] and non-linear modification of this theory [17, 18].

3. Delay-time method in electrochemical flow measurements

In order to spread the field of applicability of electrochemical method on industrial and natural electrolytes and in particular to adapt this method to sea water conditions the delay-time modification of electrochemical diagnostics was developped. The delay-time sensors use a pair of working electrodes as a sensitive element. One of these electrodes generates some electroactive species which are detected downstream the flow by the second electrode. The diffusion of these species is controlled by the near wall hydrodynamics. If the upstream electrode generates impulses of electroactive species, these impulses are detected on the downstream electrode with some delay-time. This delay-time is related to local wall shear stress value and can be predicted for laminar and turbulent flow regimes [16]. Delay-time method was used for wall shear stress measurements in some natural and industrial electrolytes [16, 19].

4. Influence of non-stationary magnetic field on near wall mass transfer

Herein some experimental results deal with the influence of non-stationary magnetic field on near wall mass transfer are presented. These results were obtained by direct mass transfer measurements by means of radio-active isotopes. A contamination (steel filings) with some known amount of radio-active isotope Fe-59 was brought on the inner surface of a tube. The rate of particle's removal from the surface was estimated by measuring of a radiation rate on the flow outlet. The ratio N of initial radiation rate of the contamitation and the radiation rate on the flow outlet gives the efficiency of particle's revomal from the surface.

Experiments were provided with 6% aqueous solution of NaCl in a tube of the diameter ∅ = 20 mm. The length of working section was 300 mm. The mean flow rate was 200 mm^3/s. The mean flow was modified by mechanical pulsator which created the flow rate fluctuations with the amplitude 120 mm^3/s and the frequency f = 15 Hz.

Non-stationary magnetic field was created by inductors of the length 400 mm and the width of 60 mm. The intensity of the magnetic field was 24000 A/m, the frequency of the magnetic impulses $\mathbf{f_m}$ was varied from 6 Hz to 60 Hz, and their duration was equal to 5% of the period of the flow fluctuations. The magnetic impulses were shifted versus mechanical pulsations of the flow on the phase angle φ . The direction of the magnetic field was perpendicular to the mean flow direction.

The Fig. 1 gives the dependence of the efficiency N of particle's removal from the wall versus the phase angle φ. The most efficient particle's removal is achived when magnetic impulses overtake the mechanical pulsations on the phase angle $\varphi = \pi/4$. This fact can be explained by the following model. The efficiency of partical's removal is related to the maximum wall shear stress value. It is known, that in the pulsating flow the liquid which moves near the wall, overtakes the mean flow pulsations due to the influence of inertial forces. For relatively high frequency range ($f \geq 10$ Hz) this phase shift is about $\pi/4$. If magnetic impulses overtake the mean flow pulsations on the phase angle $\varphi = \pi/4$, electromagnetic force significantly increase the braking effect of the inertial forces on the mean flow. At the moments when the wall shear stress value achives its maximum value, we obtain some type of resonance effect due to common influence of inertial and electromagnetic forces.

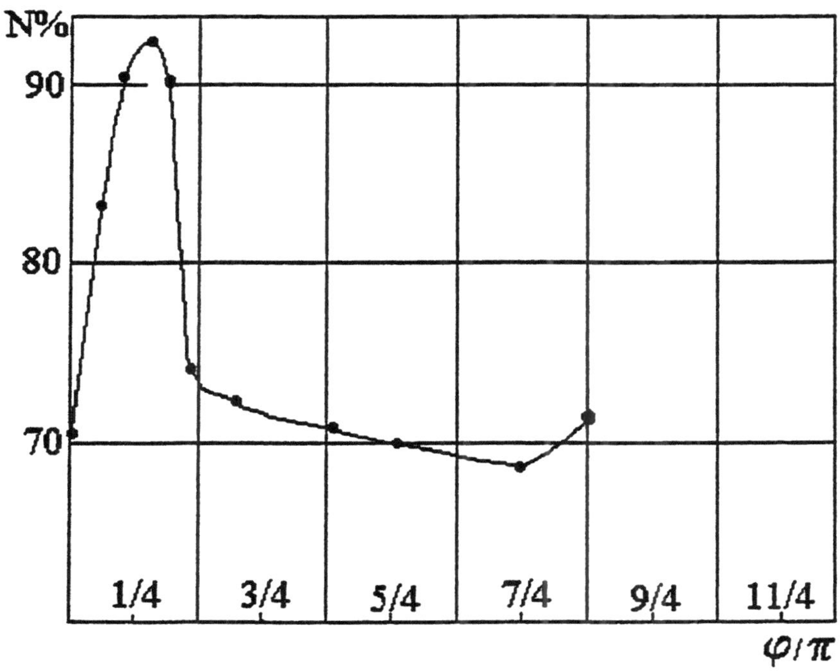

Fig. 1. Efficiency N of particle's removal from the wall. The magnetic impulses are shifted versus mechanical pulsations of the flow on the phase angle φ.

Fig. 2 shows the dependence of the efficiency N of particle's removal versus the ratio f/f_m, where f is the frequency of the mechanical pulsations and f_m is the frequency of the magnetic impulses. In all the experiments the magnetic impulses overtake the mean flow pulsations on $\pi/4$, and their duration was equal to 5% of the period of the flow pulsations. For comparison on the Fig. 2 is presented the efficiency of particle's removal by means of stationary magnetic field (B) and without magnetic field influence (C). The experiments show that the most effect can be achived when the frequency of the magnetic impulses is equal to the frequency of the flow pulsations. This fact is in a good agreement with the above mentioned resonance effect (common brake action of inertial and electromagnetic forces).

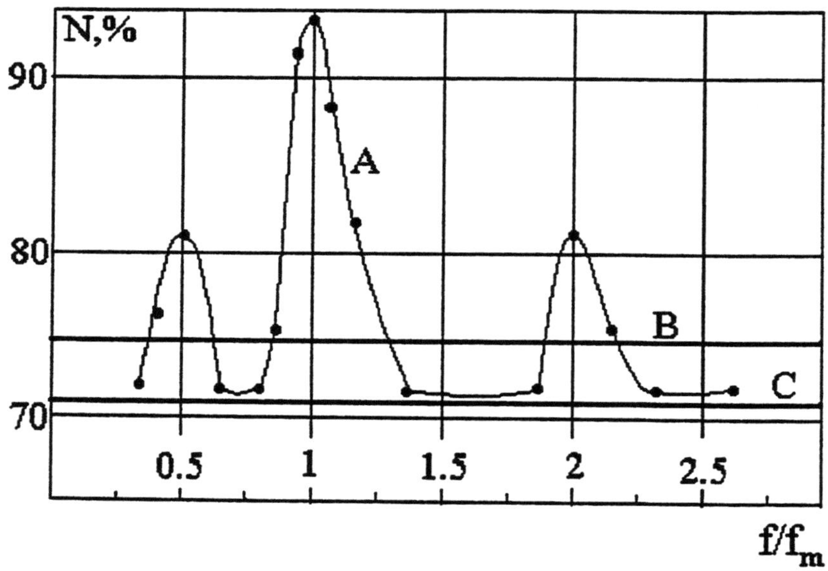

Fig. 2. Efficiency N of particle's removal from the wall versus the frequency ratio. Here f is the frequency of the mechanical pulsations and f_m is the frequency of the magnetic impulses.

5. Influence of non-uniform in space magnetic field on near wall hydrodynamics

It is well-known that perpendicular to the mean flow magnetic field significantly influences on near wall hydrodynamics. This influence leads to development of a Hartmann layer in the vicinty of solid surfaces and to an increasing of wall shear stress value. As a result, mass transfer in the magnetic field is perturbated as well. In order to predict magnetoelectrolytic mass transport a rigorous magnetohydrodynamic equations should be solved. In practice, however, simplified models and empirical relations are used for simple flow configurations [20 - 23]. For more complex cases as for example for non-uniform in space magnetic field experimental information is very scanty.

Herein we shall present some qualitative results concerning the influence of non-uniform in space magnetic field on near wall hydrodynamics. These results where obtained by using electrochemical technique (amplitude method). Electrolyte was an aqueous solution of NaCl with a small amout of electroactive species which was used as a marker in order to provide wall shear stress measurements. Non-segmented microelectrode (ϕ = 0.4 mm) was placed on the outlet of the working section and was used for the mean wall shear stress measurements. The current amplitude of the measuring electrochemical current **i** is directly related to the local wall shear stress value on the outlet of the working section. The influence of the magnetic field manifests itself in increasing of wall shear stress value and this effect can be detected by means of electrochemical current measurements. The most important influence of the magnetic field on the wall shear stress is manifested by the maximum electrochemical current value.

In our experiments magnetic field was provided by means of three electromagnets which were either connected together or separated, see Fig.3.

The experiments were made in steel tubes of different diameter **D** (mm) : 10, 14, 16, 18, 20, 21, 22, 24, 26, 30. The space period of the magnetic field along the working section **H** was a constant. The ratio **H/D** was varied between 2-6. The length of the electromagnets was **l** = 3 cm and and the distance between the electromagnets **L** was varied. The intensity of the magnetic field was about 42500 A/m.

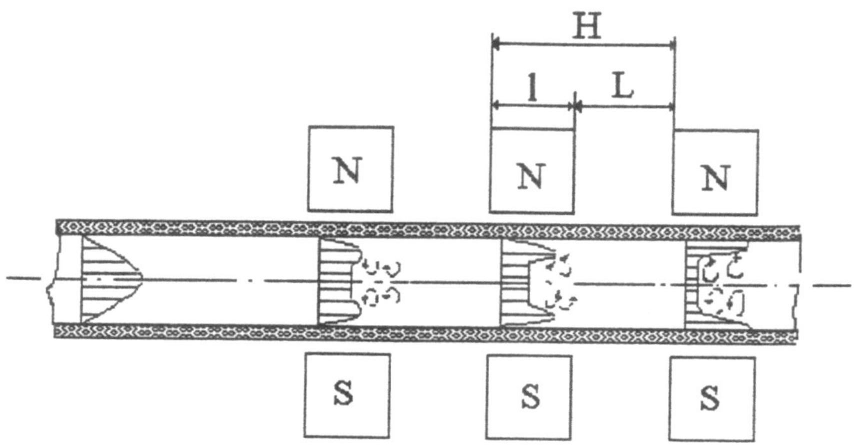

Fig. 3. Scheme of experimental installation

The influence of the magnetic field on the wall shear stress value is presented on Fig. 4. Here α is the angle between the direction of the magnetic field and the mean flow direction. The curve 1 corresponds to the case when the magnetic field is absent, the curve 2 corresponds to the case of a uniform magnetic field and the curve 3 corresponds to the case **l = L = 1.5 D.** As it can be expected, the influence of the magnetic field is the most important when this field is perpendicular to the flow direction. It can be also noted that the influence of non-uniform magnetic field is more important than the uniform one.

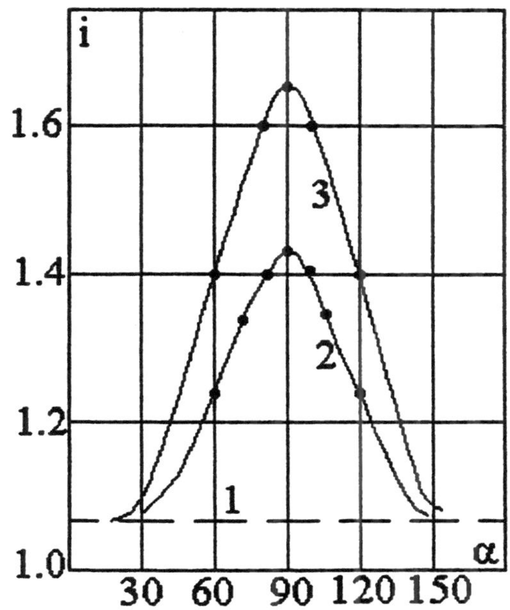

Fig. 4. Dependence of the wall shear stress (proportional to i) from the angle α between the direction of the magnetic field and the mean flow direction.

The influence of the **L/ l** ratio on the wall shear stress value is presented on Fig. 5. The different curves correspond to different electrolyte concentration which was varied from 1% (curve 1) to 6% (curve 4). We can note that the influence of the magnetic field is the most pronouned when the distance between electromagnets is equal to their length.

The influence of **H/D** ratio on the wall shear stress is presented on Fig. 6 for two different intensities of the magnetic field. We can see that the maximum effect can be obtained with the ratio **H/D** = 2.9 -3.3.

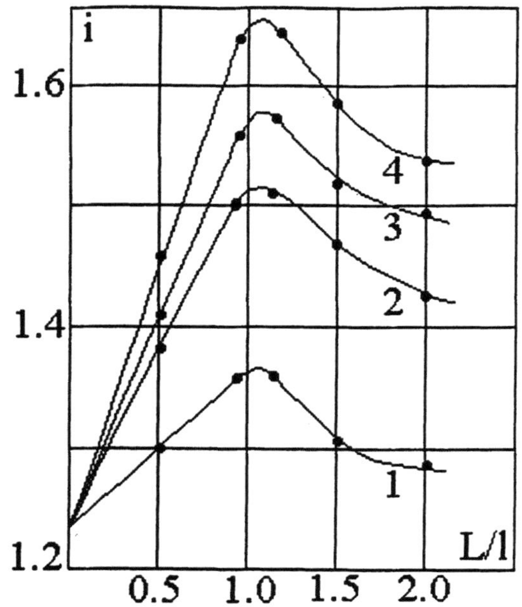

Fig. 5. Dependence of the wall shear stress (proportional to i) from the **L/l** ratio.

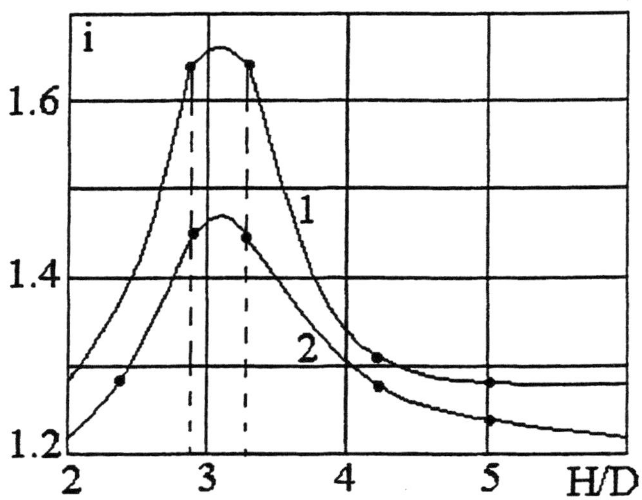

Fig. 6. Dependence of the wall shear stress (proportional to i) from the **H/D** ratio.

6. Discussion

Mass transfer and wall shear stress measurements in electrolyte flows were provided by means of electrochemical technique and by means of radio-active isotopes. These experiments show that non-stationary magnetic field is more efficient for intensification of the near wall mass transfer than the stationary one. This effect can be explained by a resonance brake action of inertial and electromagnetic forces on the mean flow. As a result, near wall liquid is significantly accelerated. The same effect (acceleration of near wall liquid) was obtained by means of non-uniform in space magnetic field. It was shown, that non-uniform in space magnetic field is more efficient for increasing of wall shear stress value and thus for intensification of mass transfer processes than a uniform one of the same intensity.

7. Acknowledgment

This work was supported in part by Russian Fundamental Science Foundataion (RFFI) by the grant No. 960332168a.

8. References

1. Mitchell J., Hanratty T.J. // J. Fluid Mech., **26** (1966) 199.
2. Nakoryakov V.E., Burdukov A.P., Kashinsky O.N., Geshev P.I. / Electrodiffusional Method for Studying Local Structure of Turbulent Flows, Novosibirsk, « Institute of Thermophysics » (1986).
3. Deslouis C., Huet F., Gil O ., Tribollet B . // Exp. In Fluids, **16** (1993) 97.
4. Martemyanov S.A. , Petrovsky N.V. , Grafov B.M. // J. Appl. Electrochemistry, **21** (1991) 1099.
5. Legentilhomme P., Aouabed H ., Legrand J. // Chem. Eng. J., **52** (1993) 137.
6. Baleras F ., Deslouis C., Tribollet B ., Sobolik V. // J. Applied Electrochemistry, **24** (1994) 676
7. Fautrelle Y., Debray F., Dalard F // Proceedings of the 4 Workshop on « Electrochemical Flow Measurements », Lahnstein - Germany (1996).
8. Cognet G., Martemyanov S.A., Paskal G, Sobolik V. // Comptes-Rendus de l'Academie des Sciences Paris, **320**, Série II b, (1995) 505 .
9. Alekseenko S.V., Nakoryakov V.E., Pokusaev B.G. / Wave Flow of Liquid Films, Novosibirsk : Siberian publishing fiem All-Russian Inc. « Nauka » (1992).

10. Pokryvailo N.A., Wein O., Kovalevskaia N.D. / Electrodiffusion Diagnostics in Suspension and Polymer Solutions, Minsk : " Nauka i Tekhnika Inc. " (1988).
11. Sobolik V., Martemyanov S. , Cognet G. // J. Appl. Electrochemistry, **24** (1994) 632.
12. Wein O., Sobolik V. // Collect. Czech Chem. Comm., **52** (1987) 2169.
13. Wein O., Sobolik V. // Collect. Czech Chem. Comm., **54** (1989) 3043.
14. Deslouis C. // Proceedings of the 4 Workshop on " Electrochemical Flow Measurements ", Lahnstein - Germany (1996).
15. Vorotyntsev M.A., Martemyanov S.A., Grafov B.M.//J. Electroanal. Chem., 179 (1984) 1.
16. Grafov B.M., Martemyanov S.A., Nekrasov L.N. / Turbulent Diffusion Layer, Moscow, " Nauka Inc. " (1990).
17. Skurygin E. F., Vorotyntsev M.A., Martemyanov S.A. // J. Electroanal. Chemistry , 259 (1989) 285.
18. Martemyanov S.A, Skurygin E. F., Grafov B.M. // Electrokhimiya, **32** (1996) 1411.
19. Senuta A.S., Martemyanov S.A, Kosmukhambetov A.R. // Electrokhimiya, **32** (1996), 708.
20. Fahidy T.Z. // J. Applied Electrochemistry, **13** (1983) 553.
21. Fahidy T.Z. // Electrochim. Acta, **18** (1973) 607.
22. Aaboubi O., Chopart J.P., Douglade J., Olivier A., Gabrielli C., Tribollet B. // J. Electrochem. Soc., **137** (1990) 1796.
23. Tacken R.A., Janssen L.J. J. // Journal of Applied Electrochemistry, **25** (1995) 1.

MOTIONS AND MASS TRANSFER IN A MERCURY CORELESS INDUCTION FURNACE

Y.FAUTRELLE**, F.DEBRAY* and J.ETAY**

** LCMI - CNRS 25 Avenue des Martyrs 38000 Grenoble FRANCE*
*** EPM-MADYLAM INPG/ENSHMG BP 95 38402 St Martin d'Hères Cedex FRANCE*

We deal with electromagnetically driven flows and their influence on the mass transfer through an interface between a mercury pool and a covering water layer. This experiment is devoted to simulate a refining ladle. The liquid metal pool is located in the single phase coil which may be supplied with alternating electric current with various frequencies. According to the coil current frequency, various types of flow regime may be obtained, namely : bulk flows, surface wave motions. Firstly, we review some experimental results concerning the characteristics of the turbulent flow. Secondly, we analyze the effect of various types of surface flow on the mass transfer coefficient by means of an electrochemical method, which is used to measure the instantaneous mass transfer across the liquid metal interface.

1. Introduction

Coreless induction furnaces are widely used in ladle metallurgy to heat and achieve metallurgical operations such as alloying or metal refining. Among them, mass transfers through the liquid metal interface is one of the key points in such processes. We may quote for example the desulfurization process for steel refining. The main aim consists in enhancing the efficiency of the metal-slag transfer reaction. Contactless, single-phase electromagnetic stirring is an elegant solution to this problem. However, its efficiency remains questionable when compared with other stirring devices as gas bubbling. Thus, many investigations have been performed in order to determine the ability of electromagnetically driven flows to stir and to promote refining reactions.

The present paper reviews some experimental results concerning the characteristics of the turbulent flow field in a single phase coreless induction furnace when the frequency of the applied magnetic field varies. We also shall give some indications about the ability of various types of electromagnetically driven motions to promote mass transfers through a liquid-liquid interface. Those results are extracted from experiments performed with mercury [1] to [8].

A. Alemany et al. (eds.),
Transfer Phenomena in Magnetohydrodynamic and Electroconducting Flows, 241–249.

2. Experimental set-up

The measurements have been achieved by means of a mercury model (density $\rho = 13.6\ 10^3\ kg.m^{-3}$, electrical conductivity $\sigma = 10^6 (\Omega.m)^{-1}$ and magnetic permeability $\mu = \mu_o = 4\pi\ 10^{-7}\ SI$) (cf. Figure *1*). It consists of a *178 mm*-diameter *2a* and *124 mm*-height mercury pool which may be covered by an aqueous electrolyte layer. The liquid metal pool is located in a solenoidal coil supplied with alternating electric currents, whose frequency *f* may be varied from *1* to *4700 Hz*.

According to the frequency range, the present experiments allow us to generate various types of motions in the mercury , namely :
- for $1Hz < f < 10Hz$: the motion consists of free surface waves whose wavelength depends on the current frequency (the so-called surface stirring case),
- for $f > 20\ Hz$: a bulk flow is generated, the free surface being almost flat (bulk stirring case),
- for $10\ Hz < f < 20\ Hz$, the motion consists of a bulk flow with superimposed surface waves whose amplitudes are weak.

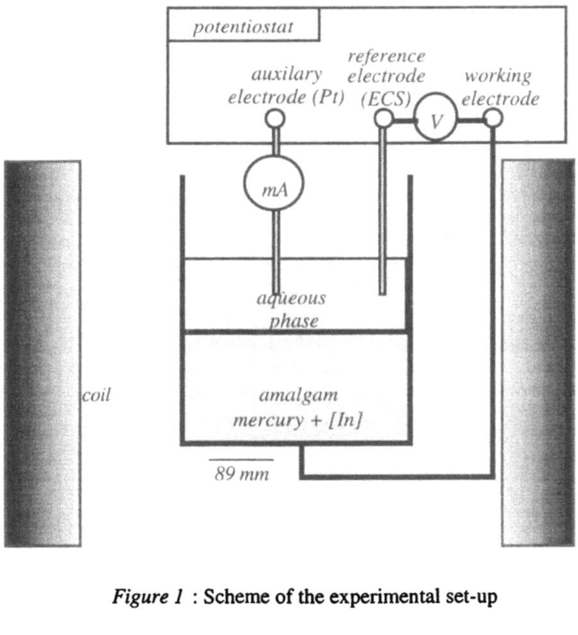

Figure 1 : Scheme of the experimental set-up

To measure the mass transfers, we use an electrochemical method which has been developed by Debray and others [5]. The instantaneous mass transfer coefficient *k(t)* is extracted from the measurement of the Faraday current *i* in the following way :

$$k(t) = \frac{nF}{S[In]_b} i(t)$$

where n, F, S, $[In]_b$ are respectively the valence of the Indium ion ($n = 3$), the Faraday constant, the surface of the mercury pool, the indium concentration in the amalgam.
The overall mass transfer coefficient K is obtained by integrating $k(t)$ with respect to time divided by the period of integration.

3. Bulk stirring case

We, firstly, analyze the case where the mercury flow consists of a bulk electromagnetic stirring. In the used frequency range ($f = 14\ Hz - 4700\ Hz$), the mercury free surface is almost flat. The flow is usually turbulent, but it may be split into a mean part and turbulent fluctuations.
In the present geometry, the mean flow configuration generally consists of two counter-rotating vortices in a meridian plane. It is possible to relate the electrical parameters, namely frequency $\omega = 2\pi f$ and magnetic field strength B_0, to the liquid metal velocity U :
$U/U_A = f(R_\omega, \textit{geometry of the furnace})$ where $R_\omega = \mu\sigma\omega a^2$ and $U_A = B_0/\sqrt{\mu\rho}$
The evolution of the ratio U/U_A with respect to the frequency f, or equivalently the shield parameter R_ω, is illustrated by Figure 2 below.

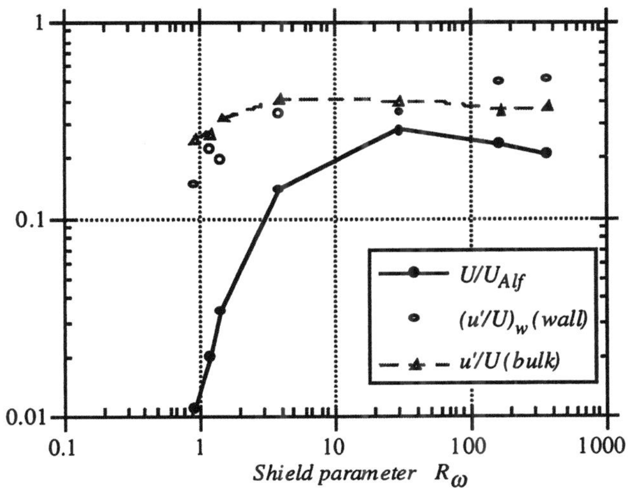

Figure 2 : Evaluations of the typical mean velocity U, the bulk turbulence intensity u'/U and the turbulent intensity $(u'/U)_w$ near the wall with respect to the shield parameter $R_\omega = \mu\sigma\omega a^2$.
The mean velocity values are normalized by the Alfvén speed U_A. The typical velocity U is obtained from the average value of the velocity along the pool axis

It is found that the characteristic mean velocity U admits a maximum value when R_ω is of order of 30. In the large R_ω limit, the decay of U is quite slow :

$$U/U_A \propto R_\omega^{-1/4}$$

This can be simply explained using the balance between the mean part of the electromagnetic forces in the electromagnetic skin depth $\delta = \left(\frac{2}{\mu\sigma\omega} \right)^{1/2}$ and inertia [1] or [2] i.e. $\sigma\omega B_0^2 a\ \delta \approx \rho U^2$

For small values of R_ω, U decays very rapidly. That result cannot be explained by simple theoretical estimates stemming from the equilibrium between inertia and the mean part of the electromagnetic forces [1]. Note that in the low R_ω limit, the oscillatory part of the Lorentz forces becomes predominant and may drive additional fluctuations.

The evolution of the turbulent fluctuations with R_ω are shown in Figure *2* also. It is observed that the turbulence level quantified by the turbulent intensity u'/U, is generally high. The bulk turbulent intensity u'/U admits a maximum value approximately for $R_\omega = 3.9$, whilst the wall turbulent intensity $(u'/U)_W$ increases significantly with R_ω. The latter result is likely related to the fact that the skin effect leads to an increase of the mean velocity shear, which is responsible for an associated increase of turbulent fluctuations. For low $R\omega$ limit, the bulk turbulent intensity decreases more slowly than the mean velocity. This tendency confirms the predominance of the oscillatory part of the Lorentz forces.

4. Surface stirring case

In the present section, we consider the case where the liquid metal motion consists of surface waves. The frequencies f of the coil current range approximately from $1\ Hz$ to $10\ Hz$. Various mode patterns appear according to this magnetic field frequency. Those modes, whose eigenfrequency is of order of the coil current frequency, are excited by the oscillating part of the Lorentz forces thanks to resonant effects. Three main cases are identified [6] :

(i) regime I : the flow configuration consists of forced axisymmetric surface waves whose oscillating frequency is equal to $2f$; those waves are directly forced by the oscillatory part of the Lorentz forces;

(ii) regime II : the flow pattern consists of subharmonic asymmetric waves whose frequency is f; the leading mechanism is very similar to the so-called parametric resonance;

(iii) regime III : it consists of asymmetric surface waves resulting from resonant mode-combination whose frequencies f_1, f_2 are such that $2f = f_1 + f_2$.

Photographs *1*, *2*, and *3* represent these regimes. The photograph *1* presents a top view of the surface of the mercury exhibiting concentric waves. They oscillate at twice the magnetic field frequency and illustrate regime I. The photograph *2* presents the same view in the regime II. A non axisymmetrical mode arises exhibiting clearly *3* maxima in the azimuthal direction. The corresponding power spectrum, i.e. the Fourier Transform of the transfer electrical current i, on figure *3* indicates that the dominant frequency is the magnetic field frequency. The two lower peaks in this spectrum are first and second harmonics of this frequency. The photograph *3* presents waves related to the so-called regime III. It is observed that the surface pattern contains two dominant non-axisymmetric types of modes having respectively *4* and *6*

Photo 1 - Top view of the mercury free surface
magnetic field frequency *3.06 Hz* , I_{coil} = *52 A*
Regime 1 : directly forced regime
The waves oscillate at *6.12 Hz*

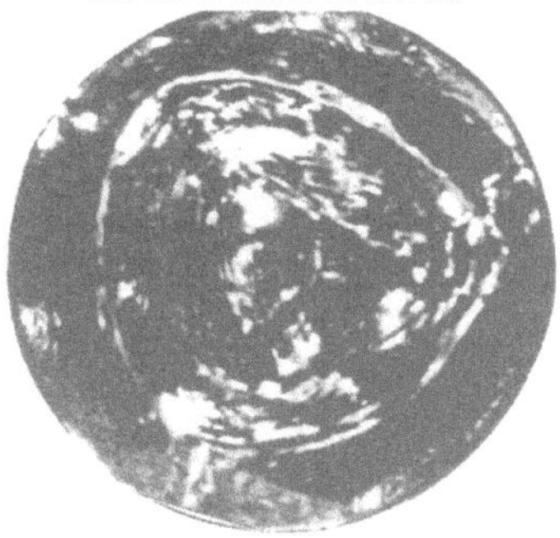

Photo 2 - Top view of the mercury free surface
(mode*3,1*)
magnetic field frequency *3.46 Hz*, I_{coil} = *105 A*
Regime II : subharmonic transition

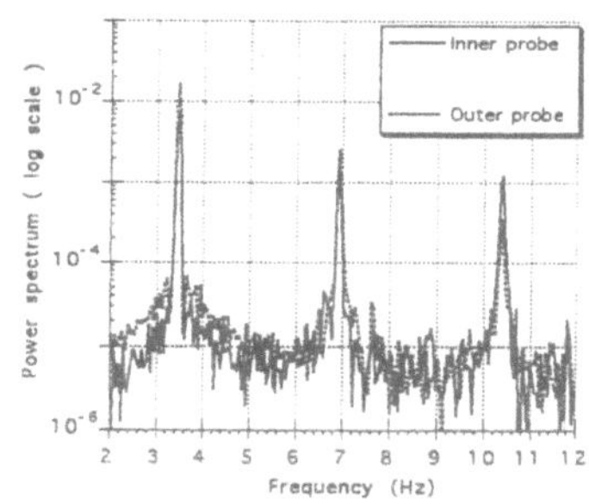

Figure 3 - Power spectrum corresponding to photo *2*
The dominant frequency is *3.46 Hz*.
Two lower peaks are first and second harmonics of this frequency.

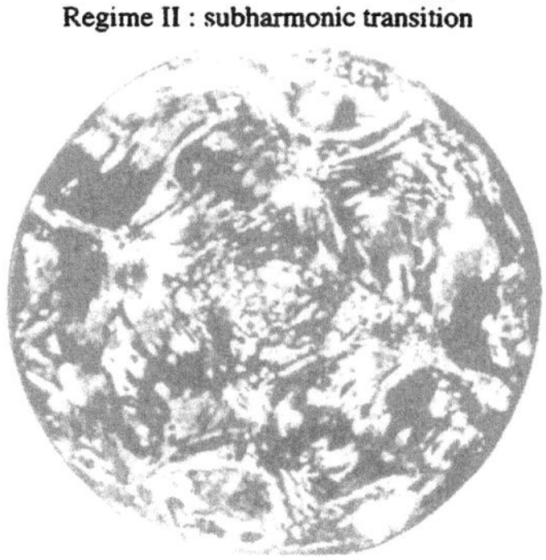

Photo 3 - Top view of the mercury free surface
magnetic field frequency *4.58 Hz*, I_{coil} = *75 A*
Regime III : mode combination transition
The surface pattern contains two dominant non-axisymmetric type of modes having m=*4* or 6

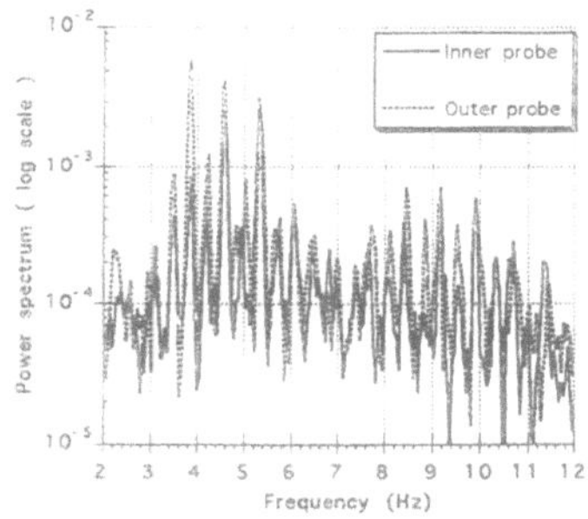

Figure 4 - Power spectrum corresponding to photo *3*
Note three dominant peaks whose frequencies are respectively
3.83Hz, *4.58 Hz* and *5.33 Hz*

maxima in the azimuthal direction. On figure *4*, the power spectrum relative to these waves exhibits three dominant peaks whose frequencies are $\omega_1 = 3.83\ Hz$, $\omega_2 = 4.58\ Hz$ and $\omega_3 = 5.33\ Hz$. Then, the eigenfrequency of the mode(*6,1*) theoretically is *4.69 Hz*. It is likely that the peak corresponding to *4.58 Hz* is a trace of the subharmonic instability of the mode (*6,1*). The eigenmodes whose frequencies are closer to ω_1 and ω_2 are respectively (*4,1*) and (*4,2*). It is worth to notice that $\omega_1 + \omega_2$=twice the magnetic field frequency. It is why the regime is called mode combination regime.

It has been shown that both regimes II and III are the result of a kind of parametric-type electromagnetic instability [7]. It is of interest to determine the dependence between the amplitude h and the magnetic field strength B_0. We gathered the various data extracted from [4] and [5] in Figure *5* That figure shows the evolution of the wave maximum amplitude normalized by the pool radius as a function of the interaction parameter $N = \sigma B_0^2 / \rho\omega$ which involves the magnetic field strength and inertia effects. When the mercury surface is free, the wave amplitude h is approximately proportional to N in the range considered here. This latter behaviour is identical to what we find in the linear regime [4]. Indeed, we recall that in linear regime where inertia is neglected, the electromagnetic forces $\sigma\omega a B_0^2$ are balanced by the local acceleration term $\rho h \omega^2$, then it follows $h/a \propto N$.

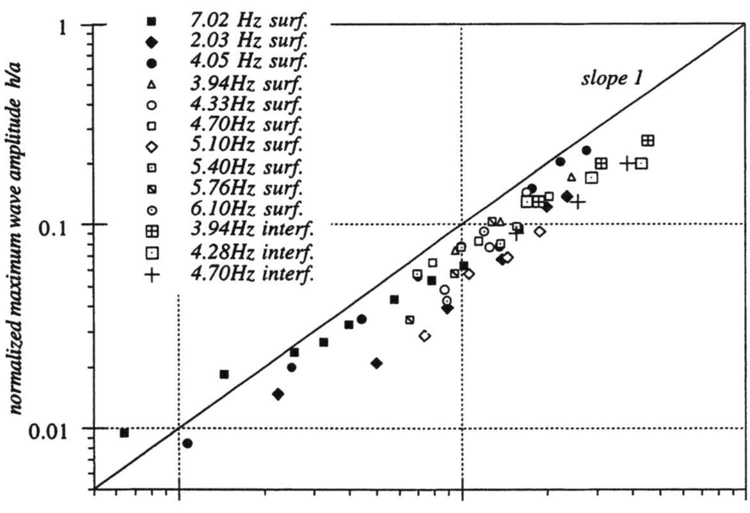

Figure 5 : Evolution of the wave amplitude h as a function of the interaction parameter N; the black and white points are respectively extracted from [4] and [8]; "interf." and "surf." correspond to measurements achieved respectively with and without a covering water layer

5. Interface mass transfers

When there is no current in the coil, it has been demonstrated [9] that a chemical stratification exists. A chemical Rayleigh number characteristic of the destabilization of the surface layer was found. Experiments with continuous magnetic field were performed which prove the stabilizing effect of such a field.

When there is a current I in the coil, the mass transfer coefficient K has been measured for various current frequencies [8]. The results are shown in Figure *6* . The selected frequencies correspond to modes which are generated by resonance (usually type II-regime).

For large coil current values, we observe that : (i) the mass transfer coefficient K evolves approximately like I^2, (ii) the value of K seems to become independent of the wave frequency what is suggested by the constriction of the data as the intensity increases.

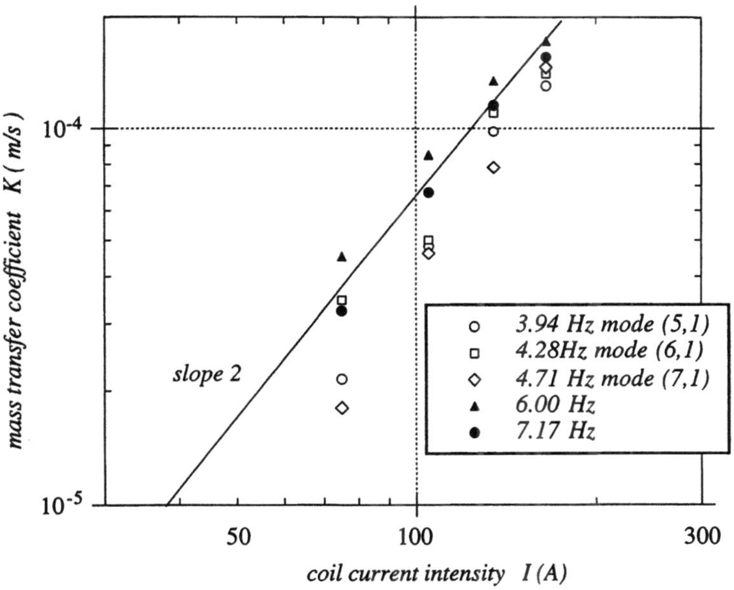

Figure 6 : Evolution of the mean mass transfer coefficient K as a function of the coil current intensity I *[4]* the mode pattern is indicated for some frequencies when possible

The interpretation of those results are not obvious. Some clues about the transfer mechanism may be obtained by regarding the dependence between K and h.
Let us write the instantaneous mass transfer coefficient K as

$$K \propto \sqrt{\frac{D}{t_c}}$$

where D is the molecular chemical diffusion of the Indium in mercury ($D = 1{,}9.10^{-9}$ $m^2.s^{-1}$) and t_c , so called the renewal time, defined as $t_c = l/u$ where l is a length scale characteristic of the transfer and u the characteristic velocity.

In the present case $u = \omega h$. For l, we choose the capillary length which has two definitions :

$$l = l_1 = \frac{\gamma}{\rho u^2} \quad \text{or} \quad l = l_2 = \left(\frac{\gamma}{\rho g}\right)^{1/2}$$

An order of magnitude of these lengths is obtained using $\gamma = 0.38 N.m$, $\rho = 13.6\ 10^3 kg.m^{-3}$, $\omega = 8\pi\, rd.s^{-1}$, $h = 10^{-2} m$, $l_1 = 0.57 mm$ and $l_2 = 1.9 mm$
Introducing l_1 in the previous law leads to the Levich-Davies correlation [10] :

$$K \propto \left(\frac{\rho \omega^3 h^3 D}{\gamma}\right)^{1/2}$$

The expression of l_2 leads to $K \propto \left(\frac{\rho g}{\gamma}\right)^{1/4} (D\omega h)^{1/2}$.

Now, thanks to simultaneous wave amplitude and transfer measurements, it was possible to determine the dependence between the mass transfer coefficient K and the maximum wave amplitude h or, more precisely, the "free surface velocity" ωh. Experimental data are shown on Figure 7. In the range considered here, we note that the mass transfer coefficient evolution is not contradictory with the Levich-Davies correlation which predicts a scaling such as $K \propto (\omega h)^{3/2}$. That result is fully true for f = *3.94* and *4.70 Hz*. Note however that the numerical coefficient involved in the Levich-Davies estimate is clearly irrelevant in the present configuration. That discrepancy could be related to the fact that the Levich-Davies correlation involves the turbulent velocity fluctuation, and the use of the surface velocity may not be presently appropriate. Clearly, the scale l_2, which leads to $K \propto (\omega h)^{1/2}$ is not relevant here.

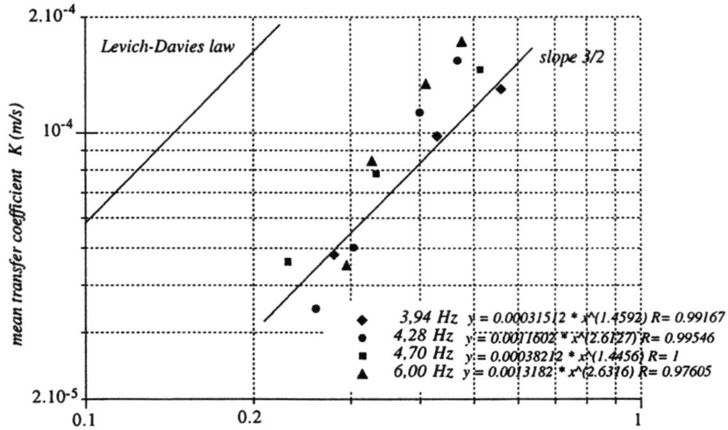

Figure 7 : Evolution of the mean mass transfer coefficient K as a function of the free surface velocity defined from the maximum wave amplitude h and its pulsation ω; the data are extracted from Debray and others [7].

From figure *5* we note that :

$$\frac{h}{a} \propto \frac{\sigma B_0^2}{\rho\omega} \propto \frac{I^2}{\omega} \quad \text{or} \quad \omega h \propto I^2$$

Introducing this value in the previous K expression leads to $K \propto I^3$. This latter result is inconsistent with the previous (i) comment on figure *6* stating $K \propto I^2$. Explanation of this clue is to be found.

6. Conclusion

Many investigations have been achieved as for the coreless induction furnace, and many results have been obtained. However, there still remains two phenomena which are not yet fully understood. The first one concerns the effect of the alternating part of the Lorentz forces on the bulk flow as well as on the turbulence, especially in the low frequency limit ($R_\omega \approx O(1)$). Preliminary results seem to indicate that the corresponding fluctuations exert an additional damping effect on the mean velocities. The second open question concerns the mass transfer through the liquid-liquid interface. The present measurements indicate that the Levich-Davies law is unable to fit with all the data. Moreover in the case where the mass transfer evolves as predicted by this law, the lengthscale characteristic of the transfer proposed by Levich leads to a result inconsistent with all our data.

References

1. E. Taberlet and Y. Fautrelle, "Turbulent stirring in a experimental induction furnace", *J. Fluid Mech.*, 159 (1985), 409-431.
2. D.J. Moore and J.C.R. Hunt, "Electromagnetic stirring in the coreless induction furnace", Proc 3rd Beer-Sheva Symposium on MHD-Flows and Turbulence, Progress in Astronautics and Aeronautics vol 84 Ed A. Branover and A. Yakhot, (1983) 359-373
3. N. El Kaddah, J. Szekely, E. Taberlet and Y. Fautrelle, "Turbulent recirculating flow in induction furnaces: a comparison of measurements with predictions over a range of operating conditions", *Met. Trans. B*, 17 (1986), 687-694.
4. J. M. Galpin and Y. Fautrelle, "Liquid metal flows induced by low frequency alternating fields", *J. Fluid Mech.*, 239 (1992), 383-408.
5. F. Debray, Y. Fautrelle and F. Dalard, "Mass Transfer Measurement as a tool to study interfacial instabilities", *Experiments in Fluids*, 19 (1995), 353-358.
6. F. Debray, I. Lagarde and Y. Fautrelle, "Electromagnetically driven parametric instability in an annular layer", *Applied Scientific Research*, 51 (1993), 31-36.
7. J. M. Galpin, A. Sneyd and Y. Fautrelle, " Parametric instability in low frequency magnetic stirring", *J. Fluid Mech.*, 239 (1992), 409-427.
8. F. Debray, Y. Fautrelle, M. Burty and J. M. Galpin, "Surface waves and mass transfer induced by a low frequency electromagnetic field", Proc. Int. Symposium on Electromagnetic Processing of Materials EPM '94, Oct. 25-28, 1994, Nagoya (Japan), ed. by The Iron and Steel Institute of Japan (1994), 29-34.
9. F. Debray, L. Davoust, Y.Fautrelle and J. Etay, "Mesure et contrôle par électrochimie d'une convection d'origine solutale à une interface entre deux liquides en présence ou non d'un champ magnétique", *Int. J. Heat and Mass Transfer,* 40 (1997), n°8, 1985-1989.
10. J. T. Davies, 1972, "Turbulence Phenomena", Academic Press (1972).

SEA WATER MHD : ELECTROLYSIS AND GAS PRODUCTION IN FLOW

P. BOISSONNEAU, J-P. THIBAULT
LEGI
BP 53X
38041 GRENOBLE Cedex, FRANCE
E-Mail: Jean-Paul.Thibault@hmg.inpg.fr

The work presented is principally experimental, it concerns mainly the coupling between sea water electrolysis and hydrodynamics (in both ways). The first series of measurements presented serve as a reference background to our study. It is realised in the conventional working conditions of an electrochemical cell containing an agitated solution of NaCl 35g/l. The working electrode is a rotating disc of Platinum. Using polarography and chronovoltametry in the range of tension and current studied, we observe two anodic reactions are in competition : production of chlorine or di-oxygen. The second series of measurements is much more relevant to sea water MHD propulsion. A real flow(maximum 10 m/s) of a solution of NaCl 35g/l, is imposed in a semi-transparency test section (4 cm x 4 cm cross section, 1m length). The working electrodes are rectangular plates of Platinum coated Titanium (Pt/Ti) placed as the two horizontal walls. The electrochemical measurements using the same method as in the first series, demonstrate that the wall region hydrodynamics of a TBL (Turbulent Boundary Layer) is strongly promoting the mass transfer and thus completely change the anodic limit current. The hydrodynamic measurements are based on the combined use of a PDPA (Phase Doppler Particle Analyser) and flow visualisation. They demonstrate that the electrolysis micro-bubbles does perturb the TBL and that they can be considered as passive tracers of the flow. Most of the experimental results are explained using basic theoretical concept as for instance : the hydrodynamic boundary layer and the diffusion layer, combined to balance principle as for instance : Faraday law (for the total amount of electrolysis gas produced).

Nomenclature

C local concentration of the active species
Co concentration of the active species to the electrode surface
Cs concentration of the active species in the bulk flow
D diffusion coefficient of the active species
E reaction potential
Eo standard reaction potential

A. Alemany et al. (eds.),
Transfer Phenomena in Magnetohydrodynamic and Electroconducting Flows, 251–266.

F Faraday constant
i current
J current density
jlim limiting diffusion current
n number of electrons
N mass flux
Uflow flow velocity
u ionic mobility
Umax maximum velocity
Vgas total gas volume
Vtest measurement volume (liquid + gas)
x longitudinal co-ordinate
y transversal co-ordinate
y^+ local boundary layer transversal position

Greek symbols

α volume gas fraction
α_{max} maximum gas fraction
δ hydrodynamic boundary layer thickness
δd diffusion boundary layer thickness
δv viscous layer thickness
μ dynamic viscosity
ν kinematic viscosity
ρ density
τ wall shear stress
τ_{app} apparent shear stress
τ^t turbulent shear stress
Φm bubbles mean diameter

Dimensionless numbers

Pe Peclet number
Re Reynolds number
Rm magnetic Reynolds number
Sc Schmidt number

1. Introduction

Recent applications of MagnetoHydroDynamics (MHD) in sea water are divided in two families The first one, MHD propulsion which is closely connected to the large scale development of superconducting electromagnet [1]. As a matter of fact several studies [2,3,4,5] demonstrate the need of magnet having very high magnetic induction (10 T) applied to a large flow volume (hundreds of m^3). Meanwhile applied DC currents (10 -

1000 A/m^2) must be moderate for minimising the Joule losses. An important part of the works is the seek of electrodes materials cheap enough and tough to sea water corrosion [5]. The second application of MHD in sea water, concerns electromagnetic flow control (EMFC). Previous work [6] showed drag reductions and delayed transition to turbulent flow, using electromagnetic forces generated by alternate electrodes and magnetic elements wall-flush distributed on a flat plate. Concerning turbulent boundary layer control, various possibilities have been tested : streamwise (parallel to the mean flow) body forces applied for controlling separation [7] or transverse (wall normal) forces applied for generating or redistributing vorticity in the boundary layer [8] and consequently producing a reduction of turbulent intensity as well as skin friction.

In any case, the current is brought by electrodes into sea water. Even if sea water is an electrolyte, it can be considered, for the bulk flow, as a poorly conducting fluid (4 - 5 S/m). Unfortunately, the application at the electrode, of DC or low frequency AC currents creates a sea water electrolysis which generates gas bubbles. Sea water electrolysis is well known for industrial application (i.e. chlorine production) which needs high currents (of the order of 10000 A/m^2. On the contrary, few works have been realised having working conditions compatible with the one of sea water MHD [2,3,11] i.e. moderate currents (less than 1000 A/m^2) applied to a high velocity sea water flow (of the order of 10 m/s).

The works presented here is aimed at extending the knowledge and comprehension of the couplings between electrochemistry and hydrodynamics. The experimental results concern electrochemical reactions in a flowing NaCl (35 g/l) solution. Two aspects of the coupling are approached. First, electrochemical measurements allow to compare standard electrochemistry (using a conventional cell) to the one of a real flow (using the salt water facility of our laboratory). In this part we observe the effect of the flow conditions on the electrochemical reactions. The working conditions considered are corresponding to a MHD thruster (i.e. flow velocity in the range : 0,5 to 10 m/s and a current density in the range : 0 to 1000 A/m^2). The results with a real flow are compared to those obtained in standard electrochemical conditions in cell. In addition various electrodes materials have been tested. Second, we analyse the possibility of electrolysis bubbles to affect on the flow (i.e. boundary layer). Two complementary methods are used and measurements are realised in various axial and transversal positions inside the test section of the facility. On one hand, Laser Doppler based measurements permit to determine flow velocity, turbulence rate and bubbles sizes. On the other hand, quantitative flow visualisation, using high magnification optics, allow a statistical count of bubbles in a measurement volume, which determine the local gas concentration in the turbulent boundary layer developing on the electrode.

2. Influence of the flow conditions on electrochemical reactions

2.1. ELECTROLYSIS OF A NaCl (35 g/l) SOLUTION IN CELL

As a basis of comparison for our studies let us first of all present some results on sea water electrolysis in standard conditions. The measurements are realised in an electrolysis cell containing an agitated (35 g/l) solution [Fig. 1]. The applied current density ranging from 10 to 1000 A/m^2. The working electrode (anode or cathode) is a Platinum Rotating Disk Electrode (RDE). One can observe three major reactions in sea water involving the following reacting species : H^+, Cl^-, H_2O.

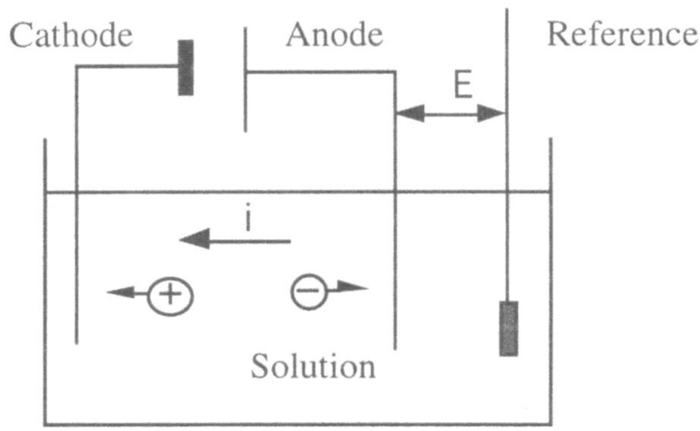

Figure 1 : Standard electrolysis cell with 3 electrodes

In the following : Eo is the standard potential corresponding to an equilibrium (zero current) between the electrode surface and the electrolyte. Electrochemical reactions appears when an electric current is imposed into the cell, between anode and cathode. The reaction potential E is given by the voltage drop between an active electrode (anode or cathode) and a reference one which is in equilibrium into the cell. The reference electrode can be either a Normal Hydrogen Electrode (NHE) or a Saturated Calomel Electrode (SCE) or a Silver/Silver chloride electrode (Ag/AgCl).

The following reactions are observed :

-A single cathodic reduction :

$$2H^+ + 2e^- \text{-->} H_2 \qquad (Eo = 0\ V/NHE),$$

Note that in the range of the imposed current density, only the H^+. The Na^+ reduction cannot be observed because its reaction needs higher potential (Eo = -2,71 V/NHE) than the one used to achieve to desired working conditions.

- Two anodic oxidations :

(1) Cl^- oxidation :

$$2Cl^- \text{-->} Cl_2 + 2\ e^- \qquad (Eo = 1{,}36\ V/NHE)$$

(2) H_2O oxidation :

$$2H_2O \text{-->} O2 + 4H^+ + 4e^- \qquad (Eo = 1{,}23\ V/NHE)$$

The two oxidations observed occur depending on the operation mode. The following measurements present the key elements of this competition. The diagram in figure 2 gives a typical cyclic evolution of the current density versus the reaction potential referred to an SCE (Saturated Calomel Electrode).

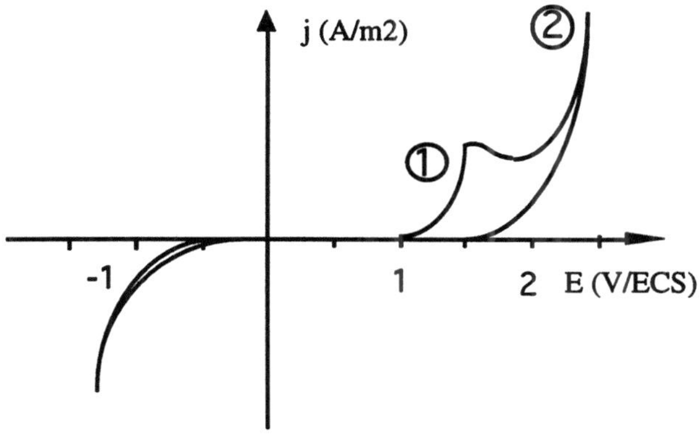

Figure 2 : Cyclical current versus potential diagram in a NaCl 35 g/l solution.

Regarding the cathodic part (negative part of the curve) only one pick, corresponding to the H^+ species reduction, is show. Contrarily on the anodic part (positive part of the curve), two picks succeed one another. The first one corresponds to the Cl^- species oxidation (oxidation (1)), and the second one to the H_2O species (oxidation (2)). Oxidation (1) appears first because of kinetic reasons. Contrary oxidation (2) is more stable than oxidation (1). This is based on thermodynamic reasons which show that the standard potential (zero current) of oxidation (2) is lower than the one of oxidation (1). In fact, when an electric current is imposed into the electrolyte, the first species to react are those which have the lower diffusion coefficient (D $[Cl^-]$ < D $[H_2O]$). Oxidation (2) appears once the Cl^- species are almost entirely consummated at the electrode vicinity. This analysis is well confirmed by figure 3 which represents the evolution of the potential reaction versus time in response to a constant DC current in a standard electrolysis cell.

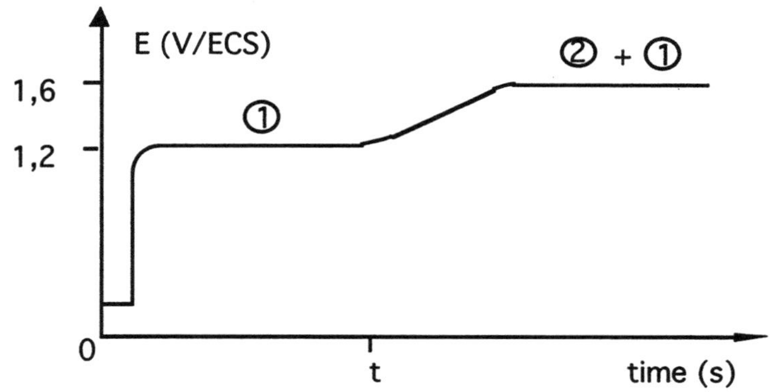

Figure 3 : Potential versus time diagram in a NaCl 35 g/l solution.

2.2. ELECTROLYSIS OF A NaCl (35 g/l) SOLUTION IN FLOW

2.2.1. *Experimental set-up*

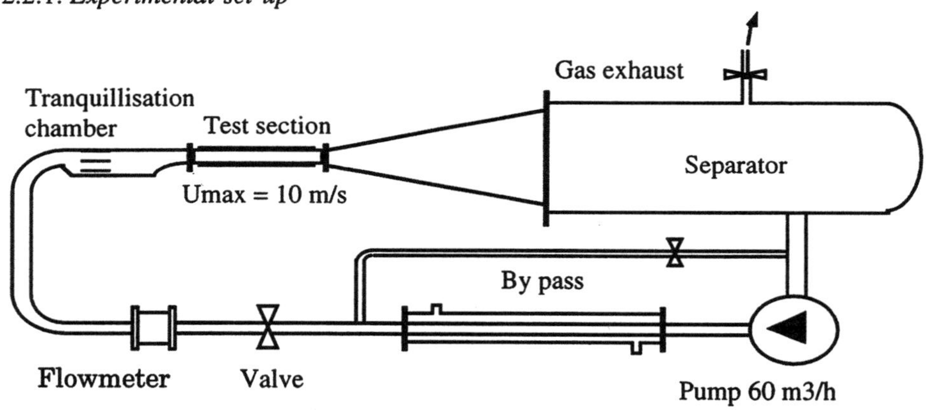

Figure 4 : Sea water electrolysis facility.

The results presented in the previous section were obtained in standard electrochemical conditions. In order to demonstrate the role of the flow conditions necessary to sea water MHD applications, measurements have also to be realised with a flowing electrolyte. We use for this the facility shown figure 4. It is composed of an hydraulic circuit made of a non corrodable material, with a 60 m^3/h pump which imposes the electrolyte flow convection in the test section. The flow rate measured by an electromagnetic flow meter is adjusted by a valve. The velocity can be adjusted in the range 0 - 10 m/s. The temperature is maintained constant by a heat exchanger. The flow is "prepared" upstream the test section by a tranquillisation chamber which is composed of a hole net and honeycomb. This makes the flow uniform and laminar upstream a convergent part which initiate the boundary layer. Note that these elements fix the entrance conditions of the test section, equivalent to the leading edge of a flat plate and are in good similitude with the one of the entrance of a sea water MHD thruster.

Figure 5 : Sea water MHD facility test section.

The test section [Fig. 5] is one meter long with a 4 cm x 4 cm cross section. It is made of altuglass which permits visualisation and anemometer measurements through the vertical walls. On the horizontal walls are embedded eight segmented working electrodes, with a 12 cm x 4 cm area and a 1 cm thickness. These electrodes are made of Titanium coated with a film of about a 5 μm thickness Platinum, which allow that they act (on the electrochemical point of view) like massive platinum electrodes meanwhile they have a reduced coast. These electrodes are supplied by pairs (on both side of the flow) to measure the reaction evolution. Four axial positions along the developing boundary layer can be used depending on the pair of electrode. In the centre of each working electrode an Ag/AgCl reference electrode can be placed. It gives a local reference potential equivalent to a SCE (Eo = 0,22 V/NHE). Downstream the test section, is situated a gas separator, 2.5 m long. First a divergent part slows down the flow and a cylinder part, 0.3 m diameter, allows electrolysis bubbles to be decanted. The pressure inside the facility can be adjusted from 1 to 3 bars. The control parameters are temperature, flow velocity, pressure. The pH of the solution is tested periodically. Electrical power is supplied by a stabilised DC current generator (40 V, 30 A) controlled by a PC. The reaction potential are registered by a Keithley DAS 1602 acquisition board. The voltage drop between a given working electrode and its reference electrode corresponds to the electrolysis over potential. The voltage drop between opposite side electrodes corresponds to the working potential which comprises the two electrodes over-potential and the Ohmic voltage drop in the bulk flow.

2.2.2. *Measurements*

The differences between the standard electrochemical measurements, given in I.1, and flowing electrolysis measurements are firstly due to experimental conditions. To quantify these differences, we calculate the Reynolds and Peclet numbers which characterise hydrodynamic and mass transfer conditions in each case.

For the standard conditions in cell, a rotating disk working electrode (RDE) is used. The flow and mass transfer regimes correspond to : Re = 500, and Pe = 5.10^5. As regard to the flowing electrolysis facility, the size of the electrodes used in the facility is much larger than the RDE used in the electrolysis cell. The current density range (0 - 1000 A/m^2), is imposed by the stabilised generator described before. Consequently even in the conditions of a very moderate flow velocity (U = 0,5 m/s)), the Reynolds and Peclet numbers are much larger tin the real flow than in the cell (i.e. Re = 5.10^4, Pe = 5.10^7). This means that the electro-active species transport to the electrode surface is more important in the case of the convected electrolyte inside the electrolysis facility. This approach of the problem, using non-dimensional parameters, shows that hydrodynamic behaviour is really different in cell and in flow. The table 1 gives a statement of these differences.

	Electrolysis cell	Electrolysis facility
Electrode material	platinum	platinum coated titanium
Activate area	3.10^{-6} m^2	48.10^{-4} m^2
Electrical generator	Potentiostat, imax = 2 A	DC Generator
Current density range	1 - 600 A/m^2	1 - 1000 A/m^2
Flowing	Rotating Disk Electrode 50 < Re < 500	Turbulent Boundary Layer $10^4 < Re < 10^8$
Results	Two reactions Cl_2 (1,2 V/NHE) O_2 (1,6 V/NHE)	One reaction (1,2 V/(Ag/AgCl))

Table 1 : Comparison between experimental processing in cell and in flow.

2.2.3. *Results*

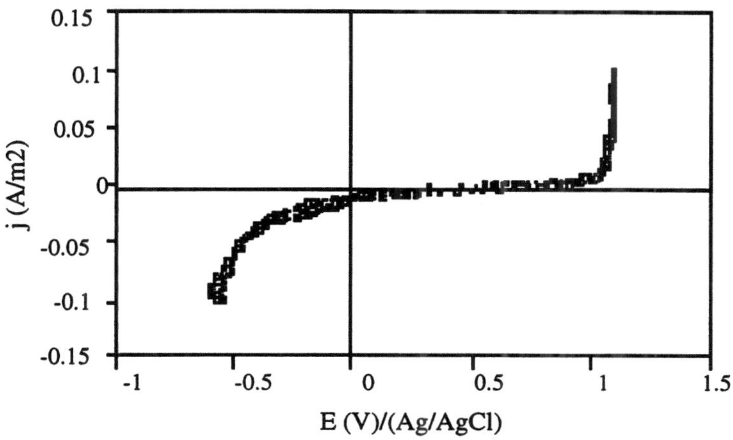

Figure 6 : Cyclical current versus potential diagram in flow, Uflow = 2 m/s.

The curves in figure 6 give the cyclic evolution of current versus potential for an electrolyte flow : U = 2 m/s. Note that the results are the same for flow velocities ranging from 0,5 to 10 m/s. In each case only one oxidation (Cl^- species, E = 1,2 V / Ag/AgCl) appears on the positive (anodic) part of the voltamperogram. This reaction is stable in time, as shown on figure 7 which represents chronopotentiometric measurements, obtained by imposing a constant current density during an imposed period time. Measurements have been realised for five hours and no transition to the oxidation 2 (H_2O species oxidation) have been observed. This means that oxidation 1 is stable in time, contrarily to what we observed in standard electrolysis results cell.

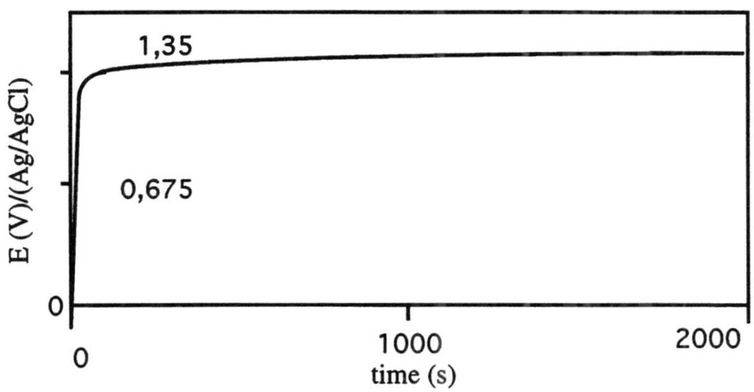

Figure 7 : Potential versus time diagram in flow, Uflow = 2 m/s.

2.2.4. *Discussion*

We now plan to explain why anodic electrolysis of a NaCl (35 g/l) solution is so much different when measurements are realised in cell or in flow. In fact, electro-active species (EAS) transfer is different when the electrolyte is a strongly convective flow. The EAS transfer mechanism can be analysed as following. In any case, the reaction involved at the electrode surface leads to the impoverishment of the EAS. The EAS concentration evolve from Cs in the bulk flow to Co at the electrode surface, [Fig. 8]. This evolution takes place in a zone, very close to the electrode surface, where appears a concentration gradient. This zone is called the diffusion layer δ_d [12].

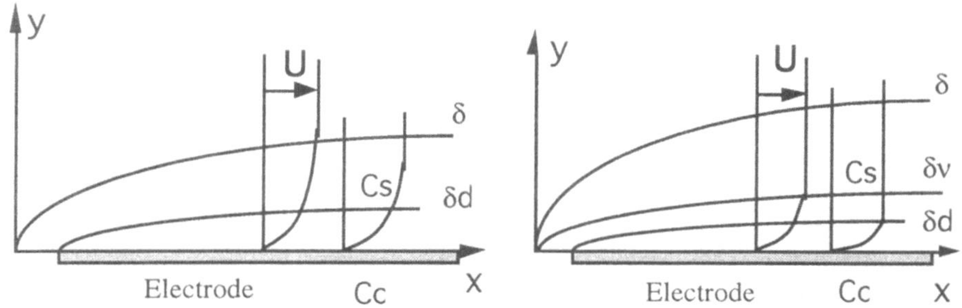

a- Laminar diffusion layer. b- Turbulent diffusion layer.

Figure 8 : Diffusion layer in laminar (a) and turbulent flow (b).

When the electrolyte is flowing the diffusion layer is confined into the hydrodynamic boundary layer. This thickness : δ_d is given by the following relation [13] :

$$\frac{\delta}{\delta_d} = Sc^{1/3}$$

Where Sc is the Schmidt number, ratio between the cinematic viscosity ν and molecular D diffusivity : $Sc = \nu \,/\, D$. Sc is generally high for the electrolytes (Sc ~ 1000). In the diffusion layer we can neglect species migration and convection. Thus the current passing through this layer is entirely controlled by a diffusion mechanism. Consequently the mass flux N is fixed by the concentration gradient into the diffusion layer as following :

$$N = - D \,\nabla C$$

Where C is the local concentration value. As shown on figure 8, two cases are possible depending on whether the flow is laminar or turbulent. In the first case, the diffusion layer is confined into the boundary layer as describe before, $\delta_d = \delta \,.Sc^{-1/3}$, [Fig 8a], in the second case, turbulent flow [Fig 8b], the diffusion layer is confined into the viscous sub-layer of the turbulent boundary layer in the ratio $\delta_d = \delta_v.Sc^{-1/3}$, where δ_v is the viscous sub-layer thickness . Because the viscous sub-layer of a turbulent boundary layer is very thin, for a given value of Cs and Co, a much higher diffusion current can be observed in a turbulent boundary layer.

2.2.5. *Global approach*

The determination of the maximum mass flux is possible calculating the limiting current given using the Nernst approximation : jlim = n.F.N, where the concentration gradient is considered as linear in the diffusion layer. n is the number of electrons involved for the reaction , F the Faraday constant, and N the mass flux. jlim correspond to the limit situation where the EAS concentration at the electrode surface Co is 0 . If we calculate the limiting current jlim for each one of our cases we obtain the following strongly different values :

- electrolysis in cell : jlim ~ 1000 A/m^2,
- electrolysis in flow (Umax = 10 m/s): jlim ~ 4.10^7 A/m^2.

In addition we already mentioned that for practical applications of MHD in sea water, the current range is : 0 - 1000 A/m^2. Consequently it is easy to see that in the first case (electrolysis in cell) the limiting current value can be achieved, which means that the Cl^- species can be totally consummated at the electrode surface. Under these circumstances the oxidation of H_2O species can start because the potential of H_2O/O_2 is the closet from the one of Cl^-/Cl_2. Note that this reaction has no limit because H_2O has an infinite concentration. In the second case (electrolysis in flow) the limiting situation is never achieved because the maximum imposed current (1000 A/m^2) of our study is much lower than the limiting current.

2.2.6. *Local approach*

The transport equation describing the (2D) evolution of the species concentration is the following :

$$u\frac{\partial C}{\partial x} + v\frac{\partial C}{\partial y} = D\frac{\partial^2 C}{\partial y^2}$$

where u and v are the longitudinal and transversal velocity components, x and y the longitudinal and transversal co-ordinates. When the flow is turbulent, we can proceed by analogy with hydrodynamics which introduces an apparent stress τ_{app} as recalled here after:

$$\tau_{app} = \tau + \tau^t$$

where τ is the viscous stress

$$\tau = -\mu\frac{\partial u}{\partial y}$$

and τ^t the turbulent stress which can be expressed using the Reynolds stress as following :

$$\tau^t = -\rho(\overline{u'v'}) = -\rho v^t \frac{\partial u}{\partial y}$$

Regarding now the mass flux of our analysis one an introduce an apparent diffusion mass flux N_{app} :

$$N_{app} = N + N^t$$

composed of the diffusive flux N, defined as previously and a turbulent mass flux N^t defined as following :

$$N^t = -(\overline{v'c'}) = -D^t \frac{\partial \overline{C}}{\partial y}$$

D^t is the turbulent diffusivity which is much higher than molecular diffusivity due to turbulence [14, 15]. The apparent flux of Cl^- species can be more important in turbulent case than in the laminar one. Consequently, in a turbulent flow, we can have a non zero concentration of Cl^- at the electrode surface even for high current density. This means that under these circumstances the second oxidation (H_2O oxidation) is never observed. This analysis demonstrate again how the reaction mechanism can be totally controlled by the flow conditions.

3. Electrolysis bubbles influence on hydrodynamics

The electrochemical reactions described previously generate gas bubbling. (H_2, O_2, Cl_2). In this part, we treat on the electrochemical and hydrodynamic coupling with the aim to determinate if the flow is influenced by the bubbles produced by the electrolysis. In order to show the possible influence of electrolysis micro-bubbles on the hydrodynamics inside the test section, we need to know their distribution, concentration and sizes as well as velocity and turbulent intensity in various axial positions in the flow. Measurements are realised in two axial positions (x = 0.1 m, x = 0.4 m), and for three flow velocities (U = 0.5 - 2 - 10 m/s). Two complementary experimental methods are used :

-1- a Phase Doppler Particle Analyser (PDPA) which measure bubbles sizes and velocity,

-2- a quantitative flow visualisation using a high magnification optics which allow to count the bubbles population and to measure their sizes and concentration.

3.1. VELOCITY AND SIZE MEASUREMENTS

3.1.1. *Velocity profiles*

A first ability of the PDPA is to work as a conventional Laser Doppler Velocimeter (LDV). Figure 9 gives the comparison of velocity profiles with and without bubbles in one case (U = 2 m/s). These are quasi disconcerted with a maximum difference of 6 % corresponding to the measurement system accuracy. The measured velocity is normalised by the bulk flow one given by the flow-meter. The axial u and transversal v components of the velocity are measured. The transversal values are about 10 % of the axial one in each case, which is in agreement with canonical boundary layer [16].

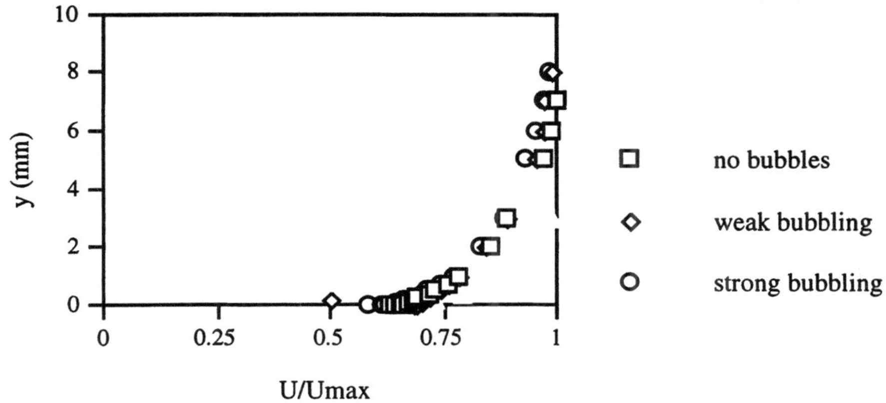

Figure 9 : Velocity profile for a 2 m/s bulk flow velocity in the second axial position.

These measurements have been reproduced in various flow conditions [18]. In any case insignificant differences are observed . It can be summarised saying that the electrolysis bubbles do not modify the boundary layer kinematics in the domain of moderate currents used for sea water MHD.

3.1.2. *Bubbles sizes*

The second ability of the PDPA is to give the bubble size by inter-correlation between signals received by three photo-detectors horizontally placed. The statistical results as regard anodic and cathodic measurements gives mean diameter Φm included in the following ranges :

$$15\ \mu m \leq \Phi m\ (\text{anode}) \leq 25\ \mu m$$
$$20\ \mu m \leq \Phi m\ (\text{cathode}) \leq 40\ \mu m$$

In addition we observed that [18] at a given current, bubbles mean diameter decreases when flow velocity increases. It is in good agreement with previous work [17] and it is probably due to shear stress which pulls bubbles out of the electrode surface. Consequently their shaping mechanism is faster as the velocity is higher. On the other hand we also observed that, at a given flow velocity, the mean bubble diameter increases. This is clearly due to the increasing gas production.

3.2. BUBBLES CONCENTRATION MEASUREMENTS

3.2.1. *Gas fraction*

The PDPA system is not developed for particles counting, and to determine bubbles concentration in the flow. We need a measurement system which permits the count of all the bubbles inside a test volume. For that we conceived a quantitative visualisation system based on shadow-visualisation of bubbles curdle using a high magnification optics and a stroboscopic lighting. The test volume correspond to the flow volume where observed bubbles are focused. The boundary layer thickness is explored (step by step) from the electrode surface to the outer flow. The pictures are digitised by a SYNAPS PC-MAPP board, and stored in a PC computer. Then they are treated with the WIMA software developed in the laboratory TSI from Saint Etienne, France [18]. The images are analysed in term of volume gas fraction α defined as following :

$$\alpha = \frac{Vgas}{Vtest}$$

where V_{gas} represents the volume associated to the observed bubbles and V_{test} the measurement volume (liquid + gas). Using the following hypothesis which idealise the two-phase flow inside the test volume the gas fraction observed can be identified to the void fraction [19, 20] :

- the two-phase flow is locally homogeneous,
- electrolysis bubbles are un-deformable (due to their small size),
- local velocities of bubble and liquid are the same.

As a first result in the working conditions considered, the gas fraction is very low, for instance : a pick of gas fraction ($\alpha = 2.10^{-3}$) is observed for Uflow = 0,5 m/s, j = 800 A/m^2 in the axial position x = 0,4 m [18]. A second result is that α varies with velocity and current. At a given current density, the gas fraction is more important for low velocities. This can be easily understood considering that for a given current density value, the gas produced is evacuated faster from the test volume when increasing velocity. Likewise for a given velocity, α increases with the current density because of the increasing total amount of gas. In fact, the concentration of bubbles inside the boundary layer is much lower than the concentration mentioned in previous work which effectively allowed a measurable modification of the turbulent boundary layer[21, 22]. These measurements give additional confidence to say that, in the working conditions of our study : non-measurable hydrodynamic effect of the electrolysis bubbles can be observed so they act as passive tracers of the flow.

5. Conclusion

The MHD in sea water is related to applications as well for direct propulsion as for electromagnetic flow control. It is the site of an electrolysis due to DC current imposed to the flow. Numerous attempts by previous authors have been realised to find an electrode (anode) material which offers a dominant oxygen production. These materials have been tested in the working conditions of an electrolysis cell and not in real flow conditions. In addition, most of these materials do not offer the reversibility between anode and cathode. We considered that these conditions are not realistic regarding an application to MHD propulsion or flow control. That is why the study of the coupling between electrolysis and hydrodynamics is very important as regard the evolution of the oxidations which appear to the anode. Firstly we studied the coupling between electrochemistry and hydrodynamics using Pt/Ti working electrodes which the same electrochemical behaviour as Pt. This study shows the influence of the flow turbulence on the electrochemical reactions. The electro-active species flux is totally enhanced and controlled by the turbulence. The oxidations evolution is not the same in cell as in flow. In the second case, the Cl^- oxidation (chlorine production) appears in majority. This means that it will be very difficult (if not impossible) to find an anodic electrode material which offers the oxygen production in such working conditions.

Another aspect of the coupling between electrolysis and hydrodynamics is the possible influence of electrolysis micro-bubbles on the boundary layer. The two-phase flow measurements using two methods : Laser Doppler based (PDPA) and high magnification flow visualisation, demonstrate that electrolysis bubbles have no influence on the flow. Bubbles are very small ($\Phi m = 20\ \mu m$), their concentration is very weak ($\alpha_{max} < 2.10^{-3}$) and they are entirely contained in the turbulent boundary layer. The distribution is mainly controlled by the turbulent diffusion. In the imposed current density range, bubbling has no influence on the flow, and micro-bubbles can be considered as passive tracers of the flow.

Acknowledgement

This work is done in the frame of the co-operation between one of the research department of the French Ministry of Defence: "Délégation Générale à l'Armement ", a French company: "Jeumont Industrie" and our institute: "Laboratoire des Ecoulements Géophysiques et Industriels".

References :

1 Mitchell D.L., Gubser D.U. (1988) MagnetoHydroDynamic Ship Propulsion with Superconducting Magnets, *J. of Supercond.*, vol 1, n°4, pp 349-364.
2 Meng J.C.S., Hrubes J.D., Hendricks P.J., Thivierge D.P., Henoch C.W.(1991) Experimental Studies of a Superconducting Electromagnetic Thruster for Sea Water Propulsion, *Proceedings MHDS* , Kobe, Japan, pp 203-210.
3 Meng J.C.S., Hendricks P.J., Hrubes J.D., Henoch C.W (1994) Experimental Studies of a sea water Superconducting Electromagnetic Thruster : A Continuing Quest for Higher Magnetohydrodynamic

Propulsive Efficiency, *Proceedings 2nd Conf. Energy Transfer in MHD Flows*, Aussois, France, pp 491-500.
4 Motora S., Imaichi K., Nakato M., Takezawa S. (1991), An Outline of the R&D Project on Superconducting MHD Ship Propulsion in Japan, *Proceeding MHDS* , Kobe, Japan, pp 53-68.
5 Convert D. (1995) Propulsion MagnétoHydroDynamique en eau de mer, *Ph.D. Thesis*, Université Joseph Fourier, Grenoble.
6 Gailitis A., Lielausis O., Dukure R. (1991) Boundary layer control by means of electromagnetic forces, *Proceedings 1st Conf. Energy Transfer in MHD Flows*, Cadarache, France, pp 5-9.
7 Henoch C.W., Stace J. (1995), Experimental investigation of a salt water turbulent boundary layer modified by an applied streamwise MagnetoHydroDynamic body force, *Phys. Fluids*, vol. 7, n°6, pp 1371-1382.
8 Nosenchuck D., Brown G. (1993) The direct control of wall shear stress in a turbulent boundary layer" Proceeding of the Int. Conf. on Near-Wall Turbulent Flows, Elsevier, pp 689-698.
9 Moreau R. (1990) Magnetohydrodynamics, *Kluwer Academic Publisher*, Dordrecht, The Netherlands.
10 Reed B.C., Lykoudis P.S. (1978) The Effect of a Transverse Magnetic Field on Shear Turbulence, *J. Fluid Mech.*, vol 89, part 1, pp 147-171.
11 Lin T.F., Marks S.P., Gilbert J.B. (1991) Sea Water Conductivity Enhancement by Acid Seeding and the Associated Two-Phase Flow phenomena, *Proceeding MHDS* , Kobe, Japan, pp 367-374.
12 Levitch (1962) Physicochemical Hydrodynamics, *Englewood Cliffs*.
13 Coeuret F., Storck A. (1984) Elements de Génie Electrochimique, *Lavoisier*, Paris, France, 1984.
14 Martemianov S.A., Vorotynsev M.A., Grafov B.M. (1979) Derivation of a non local transport equation of matter in the turbulent diffusion layer, *Soviet Electrochemistry*, vol. 15, n°6.
15 Martemianov S.A., Vorotynsev M.A., Grafov B.M. (1980) Spread of the diffusion boundary layer along the electrode under turbulent flow conditions, *Soviet Electrochemistry*, vol. 16, n°5.
16 Cousteix J. (1989) Turbulence et couche limite, *Cepadus ed.*, Toulouse, France.
17 Glas J.P., Westwater J.W. (1964) Measurements of the Growth of Electrolytic Bubbles, Int. J. Heat Mass Transfer, vol 7.
18 Boissonneau P. (1997) Propulsion MHD en Eau de Mer : Etude des Couplages Hydrodynamique - Electrochimie - Electromagnétisme, *Ph.D. Thesis, Université Joseph Fourier*, Grenoble, France.
19 Gad Hestroni (1982) Handbook of Multiphase Systems, *Mac Graw Hill*, New-York, USA.
20 Delhaye J.M. (1982), Thermohydraulics of two-phase system for industrrial design and nuclear Engineering, *Mac Graw Hill*, New-York, USA.
21 Bogdevitch V.G. (1977) Gas Saturation Effect on Near-Wall Turbulence Characteristics, *Proceedings of 2nd Int. Conf; on Drag Reduction.*
22 Merkle L.M., Deutsch S. (1992) Microbubble Drag Reduction in Liquid Turbulent Boundary Layers, *Appl. Mech. Revue*, vol 45, n°3.

III - MHD IN METALLURGY AND CRYSTAL GROWTH

THERMOELECTRIC MAGNETOHYDRODYNAMIC EFFECTS DURING BRIDGMAN SEMICONDUCTOR CRYSTAL GROWTH WITH A UNIFORM AXIAL MAGNETIC FIELD: LARGE HARTMANN-NUMBER ASYMPTOTIC SOLUTION

Y.Y. KHINE, J.S. WALKER
Department of Mechanical and Industrial Engineering
University of Illinois, Urbana, IL 61801, U.S.A.

This paper treats the thermoelectric magnetohydrodynamic (TEMHD) effects during Bridgman semiconductor crystal growth in a cylindrical ampoule with a strong uniform axial magnetic field. The melt is bounded by a planar crystal-melt interface, the cylindrical ampoule surface and a planar free surface. Inertial effects are negligible everywhere with a sufficiently strong magnetic field, while viscous effects are important only in thin boundary layers. A parabolic temperature variation is assumed at the crystal-melt interface. The thermoelectric current circulates through the crystal and the Hartmann layer adjacent to the crystal-melt interface. There is no current in the core and in the free surface Hartmann layer. The axially uniform azimuthal velocity in the inviscid core region is zero at the centerline and at the ampoule wall, with a maximum at some radial position. The meridional melt motion involves radially inward flow near the crystal-melt interface. The differences between the numerical and asymptotic Bridgman models are discussed.

1. Introduction

Since the values of the absolute thermoelectric power for the solid and liquid, S_s and S respectively, are quite different for most semiconductors, the presence of a radial temperature variation along the crystal-melt interface would drive a meridional circulation of TEMHD currents in the melt and the crystal. This circulation of thermoelectric current is proportional to both the difference between the two values of S and the magnitude of the temperature gradient. The radial thermoelectric current interacts with the externally applied axial magnetic field to drive the important TEMHD melt motion.

During the bulk growth of semiconductor crystals, the buoyant convection in the melt is usually unsteady and often turbulent, resulting in fluctuations in the local growth rate which in turn produce spatial oscillations in the composition of the crystal, called striations [1]. The application of a relatively weak magnetic field, as low as the magnetic flux density B_0 = 0.5 T, can eliminate all unsteadiness in the melt motion since most molten semiconductors are good electrical conductors. However, this also eliminates the desirable turbulent mixing which may result in large compositional variations in the melt. These compositional variations may be incorporated into the crystal, producing undesired macrosegregation [2].

The convection driven by TEMHD effects is smaller than the buoyant convection with weak magnetic fields, but it may be relatively much larger with strong fields, especially in the small residual convective mass transport of dopants and species.

A. Alemany et al. (eds.),
Transfer Phenomena in Magnetohydrodynamic and Electroconducting Flows, 269–278.

For pure materials, such as silicon, there is often faceted growth over a region near the center of the crystal-melt interface and atomically rough growth near the periphery. These different modes of solidification lead to a temperature variation along this interface. The TEMHD effects may be much more pronounced in the magnetically damped growth of alloyed semiconductors, such as 80% mercury telluride and 20% cadmium telluride, due to much larger temperature variations along the crystal-melt interface because of the compositional dependence of the solidification temperature.

The first and most complete discussion of TEMHD was presented by Shercliff [3]. The importance of TEMHD effects during bulk crystal growth with an externally applied magnetic field were first shown by Gorbunov [4] and Gel'fgat and Gorbunov [5]. Recently, Moreau et al. [6] showed that the application of a magnetic field during the dendritic solidification of metallic alloys produces strong TEMHD flows around the dendrites.

A qualitative treatment of the TEMHD effects in the Bridgman process with an applied axial magnetic field for arbitrary Hartmann number (*Ha*) was presented by Khine and Walker [7]. Khine et al. [8] have presented the asymptotic solution for the thermoelectrically driven melt motion for the floating zone process with a strong axial magnetic field. Khine and Walker [9] have also presented numerical results for the TEMHD effects for the floating zone crystal growth process with an axial magnetic field for arbitrary *Ha*.

The TEMHD flows for the floating zone and Bridgman processes are quite different. In the floating zone process, the feed rod has the same electrical conductivity as the crystal and provides an important part of the electric circuit. In the Bridgman process, the melt boundary which is parallel to the crystal-melt interface is an ampoule surface or a free surface, both of which are electrical insulators which limit the thermoelectric currents and the resultant melt circulation, particularly for a strong magnetic field. Since the numerical solution [7] has some limitations for large *Ha* and the actual magnetic field applied during the vertical Bridgman crystal growth is often quite strong (i.e., *Ha* may be quite large), we determine the TEMHD effects for vertical Bridgman process for large *Ha* in this paper. We only consider the case with a free surface parallel to the crystal-melt interface.

A complete model of the TEMHD flows for alloyed crystal growth with magnetic damping would require considering the complex interactions among the thermal field, the compositionally driven buoyant convection and the compositionally dependent solidification temperature. However, in this model, we focus on the basic physical TEMHD phenomena, and therefore we assume (1) that the crystal-melt interface is planar; (2) that S_s and S are uniform absolute thermoelectric powers of the crystal and melt, respectively; (3) that there is a parabolic temperature variation along the crystal-melt interface; and (4) that the flow is steady and axisymmetric. The asymptotic solution of the model problem is presented in the next section.

2. Problem Formulation

There is an induced magnetic field which is produced by the thermoelectric currents in the melt and the crystal, and which is superimposed on the applied uniform axial field produced by the solenoid around the Bridgman furnace. The characteristic ratio of the induced to applied magnetic fields is $\mu_p j_{ac} R / B_o$, where μ_p is the melt's magnetic permeability, j_{ac} is the characteristic magnitude of the thermoelectric current density, and R is the inside radius of the ampoule. For all crystal growth processes, this ratio is very small. Thus, the induced magnetic field is negligible, and the dimensional magnetic field is $\mathbf{B}^* = B_o\hat{z}$, where $\hat{r}$, $\hat{\theta}$, and $\hat{z}$ are unit vectors for cylindrical coordinates with the z axis along the centerline of the ampoule and the origin at the center of the crystal-melt interface.

For an axisymmetric flow, the characteristic ratio of the electromagnetic (EM) body force to the meridional convection of the linear momentum is the interaction parameter, $N = \sigma B_0^2 R/\rho V_m$, where σ and ρ are the melt's electrical conductivity and density, while V_m is the characteristic velocity of the meridional motion (i.e., for radial and axial velocities). For a strong magnetic field, N>>1, and the meridional convection of the linear momentum is negligible. With these assumptions, the governing equations become linear, and the TEMHD flow is decoupled from other melt motions, such as the thermally and compositionally driven buoyant convections, and we only consider the TEMHD flow here. Further, the model problem can be split into two sequential problems.

The first problem governs the azimuthal motion, which is driven by the meridional thermoelectric currents and is decoupled from the secondary motion which is the meridional motion. The dimensionless governing equations for the azimuthal motion in terms of the azimuthal velocity v_θ and the electric stream functions h and h_s for the melt ($z > 0$) and the crystal ($z < 0$), respectively, were derived in our previous paper [7], and are as follows:

$$\frac{1}{r}\frac{\partial h}{\partial z} = Ha^{-2}\left(\frac{\partial^2 v_\theta}{\partial r^2} + \frac{1}{r}\frac{\partial v_\theta}{\partial r} + \frac{\partial^2 v_\theta}{\partial z^2} - \frac{v_\theta}{r^2}\right), \tag{1}$$

$$r\frac{\partial v_\theta}{\partial z} = \frac{\partial^2 h}{\partial r^2} - \frac{1}{r}\frac{\partial h}{\partial r} + \frac{\partial^2 h}{\partial z^2}, \tag{2}$$

$$0 = \frac{\partial^2 h_s}{\partial r^2} - \frac{1}{r}\frac{\partial h_s}{\partial r} + \frac{\partial^2 h_s}{\partial z^2}. \tag{3}$$

Eq. (1) is the azimuthal component of the Navier-Stokes equation without the meridional convection of azimuthal momentum, without a θ derivative of the pressure due to axisymmetry, and with the azimuthal EM body force due to the radial thermoelectric current and the axial magnetic field. Eqs. (2) and (3) are obtained by cross-differentiating the r and z components of the Ohm's law in order to eliminate the static electric field $-\nabla\phi$ and the Seebeck electromotive force $-S\ \nabla T$ or $-S_s\ \nabla T$, where ϕ is the electric potential function (voltage) and T is the temperature.

For Ha >>1, the subregions of the melt are: (c) the inviscid core with $O(1)$ radial and axial derivatives, (t) the Hartmann layer with $O(Ha^{-1})$ thickness adjacent to the free surface at z=b where b is the dimensionless axial length of the melt, (h) the Hartmann layer with $O(Ha^{-1})$ thickness adjacent to the crystal-melt interface at $z = 0$, and (p) the parallel layer with $O(Ha^{-1/2})$ thickness adjacent to the ampoule wall at $r = 1$. Neglecting $O(Ha^{-2})$ terms, $h = 0$ and $v_\theta = v_{\theta c0}(r) + Ha^{-1} v_{\theta c1}(r)$ in both the core and the free-surface Hartmann layer at $z = b$. For the interface Hartmann layer, we introduce the asymptotic expansions $h_h = h_{h0}(r,\zeta) + Ha^{-1} h_{h1}(r,\zeta) + \ldots$ and $v_{\theta h} = v_{\theta h0}(r,\zeta) + Ha^{-1} v_{\theta h1}(r,\zeta) + \ldots$ where $\zeta = Ha\, z$ is the locally stretched axial coordinate. Eqs. (1) and (2) give $h_{h0} = 0$, $\frac{\partial h_{h1}}{\partial \zeta} = r\frac{\partial^2 v_{\theta h0}}{\partial \zeta^2}$ and $\frac{\partial^2 h_{h1}}{\partial \zeta^2} = r\frac{\partial v_{\theta h0}}{\partial \zeta}$. The solution which satisfies the no-slip condition at the crystal-melt interface ($\zeta = 0$), and matches with the core is

$$v_{\theta h0} = v_{\theta c0}(r)\left(1 - \exp(-\zeta)\right), \tag{4}$$

$$h_{h1} = r v_{\theta c0}(r) \exp(-\zeta). \tag{5}$$

At the interface, the electric stream function is continuous ($h = h_s$). The electric potential function or voltage is also continuous across the crystal-melt interface. Therefore, the difference between the radial components of Ohm's law in the crystal and melt at the interface gives a boundary condition relating the jump in radial electric current density to the jump in the radial Seebeck electromotive force. The dimensionless condition is

$$\left(\frac{\sigma}{\sigma_s}\right)\frac{\partial h_s}{\partial z} - \frac{\partial h}{\partial z} = r\frac{\partial T_i}{\partial r}, \tag{6}$$

where σ and σ_s are the electrical conductivities of the melt and crystal, respectively, while $T_i(r)$ is the temperature variation along the crystal-melt interface, normalized by ΔT^*, the dimensional temperature difference between the centerline and ampoule wall at the interface. We have ignored the fact that the actual interface is generally concave into the crystal, and we have assumed that the crystal and melt have two different uniform values S_s and S for the absolute thermoelectric power. The characteristic velocity for our dimensionless v_θ is $(S - S_s)(\Delta T^*)/R B_o$ for $Ha > 1$ [7]. The separation-of-variables solution of Eq. (3) is

$$h_s = Ha^{-1} r \sum_{j=1}^{\infty} C_j J_1(\lambda_j r) \exp(\lambda_j z), \tag{7}$$

where J_k is the Bessel function of the first kind and of order k while λ_j are the roots of $J_1(\lambda_j) = 0$. Eqs. (5) - (7) give

$$\left(\frac{\sigma}{\sigma_s}\right) r \sum_{j=1}^{\infty} C_j \lambda_j J_1(\lambda_j r) + r v_{\theta c}(r) = 2r^2 , \tag{8}$$

assuming $T_i = r^2$.
From $h = h_s$ at the interface with Eqs. (5) and (7),

$$v_{\theta c}(r) = Ha \sum_{j=1}^{\infty} C_j J_1(\lambda_j r) . \tag{9}$$

The crystal and the interface Hartmann layer represent resistors in series, so that an important parameter is the characteristic ratio of their electrical resistances, $\gamma = Ha^{-1} (\sigma/\sigma_s)$ [10]. The crystal has a high electrical resistance because σ_s is considerably smaller than σ for most semiconductors, while the Hartmann layer has a high resistance because it is thin. Our solution assumes that Ha is large enough to exclude current from the core. Each value of γ corresponds to a specific value of Ha for each material, e.g., with $(\sigma/\sigma_s) = 20$ for silicon. Then, Eqs. (8) and (9) give

$$\hat{C}_j = -4 \left[\lambda_j J_0(\lambda_j)(1 + \gamma \lambda_j) \right]^{-1} , \tag{10}$$

where $\hat{C}_j = Ha C_j$.
The meridional motion is governed by the r and z components of the Navier-Stokes equation with the radial centrifugal force and the EM body force, but without the meridional convection of the radial and axial momentum. The characteristic velocity of the meridional motion (radial and axial velocities) is $V_m = \rho V_\theta^2 / R\sigma B_0^2 = \rho (S - S_s)^2 (\Delta T)^2 / R^3 \sigma B_0^4$ [7]. We introduce a dimensionless stream function ψ which satisfies the continuity of velocity, and cross-differentiating the aforementioned Navier-Stokes equation in order to eliminate the pressure gives

$$\frac{\partial^2 \psi}{\partial z^2} - Ha^{-2} \left(\frac{\partial^4 \psi}{\partial r^4} - \frac{2}{r}\frac{\partial^3 \psi}{\partial r^3} + \frac{3}{r^2}\frac{\partial^2 \psi}{\partial r^2} - \frac{3}{r^3}\frac{\partial \psi}{\partial r} + 2\frac{\partial^4 \psi}{\partial r^2 \partial z^2} - \frac{2}{r}\frac{\partial^3 \psi}{\partial r \partial z^2} + \frac{\partial^4 \psi}{\partial z^4} \right)$$
$$= 2 v_\theta \frac{\partial v_\theta}{\partial z} . \tag{11}$$

Eq. (11) shows that only the axial variations of v_θ drive the meridional melt motion since a radial pressure gradient exactly cancels the centrifugal force due to a v_θ which is independent of z. In the core and in the free-surface Hartmann layer, v_θ is independent of z, neglecting $O(Ha^{-2})$ terms, so that there are no axial variations of v_θ to drive any meridional melt motion. The $O(1)$ v_θ changes from $v_{\theta c0}(r)$ to zero across the crystal-melt interface Hartmann layer, but this large axial gradient of v_θ is confined to a thin boundary layer and only produces an $O(Ha^{-1})$ meridional

circulation. The only meridional flow in the core completes the flow circuit for the $O(Ha^{-1})$ flow inside the interface Hartmann layer.
Neglecting $O(Ha^{-2})$ terms, the stream function solution for the interface Hartmann layer which satisfies the no-slip condition at the crystal-melt interface ($\zeta = 0$) and matches with the core is

$$\psi_{h1} = Ha^{-1} v_{\theta c0}^2(r) \left[(\zeta + \tfrac{2}{3}) \exp(-\zeta) + \tfrac{1}{6} \exp(-2\zeta) - \tfrac{5}{6} \right]. \qquad (12)$$

For the free surface Hartmann layer, ψ is zero everywhere as a result of no shear stress at the free surface. In the inviscid core, the solution of Eq. (11) with a free surface at $z = b$ and matching with the interface Hartmann layer gives

$$\psi_{c1} = \psi_{h1}(r,\infty)(1 - z/b). \qquad (13)$$

In the following section, we present the results of these solutions in each region.

3. Results and Discussion

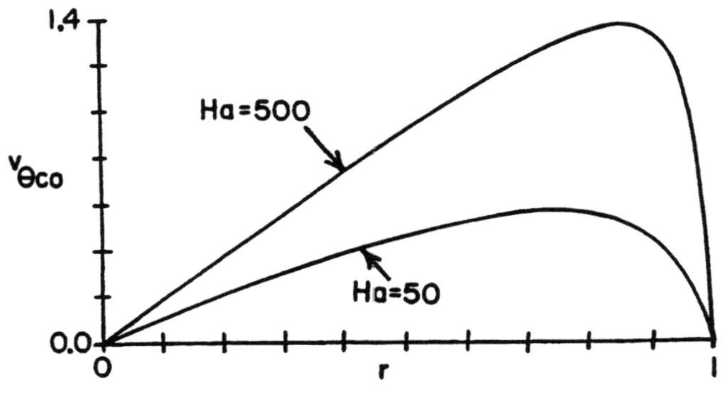

Figure 1. Velocity distribution in the core for different Hartmann numbers.

In *Figure 1*, we present the azimuthal core velocity $v_{\theta c0}(r)$ for $(\sigma/\sigma_s) = 20$, so that $\gamma = 0.4$ and 0.04 for Ha=50 and 500, respectively. For Ha=50 and 500, $v_{\theta c0}$ has maxima of 0.55 at r=0.75 and 1.37 at r=0.85, respectively. The two velocities are zero at the centerline and increase to a maximum and then decrease to zero at the ampoule wall. The magnitude of the buoyant convection varies as B_o^{-2} [11], while the present characteristic azimuthal velocity varies as B_o^{-1}. As B_o is increased so that Ha increases from 50 to 500, the magnitude of the buoyant convection is supposed to be reduced by a factor of one hundred, while the TEMHD azimuthal velocity is only reduced by a factor of four, taking into account the fact that the maximum value of the dimensionless azimuthal velocity increases from 0.55 to 1.37 as γ is reduced from 0.4 to 0.04. In this asymptotic solution, v_θ is independent of z in the core, while the jump in v_θ across the crystal-melt interface Hartmann layer is proportional to the radial electric current inside this layer. Since the radial electric current in this layer is zero at $r=1$, $v_\theta = 0$ at $r=1$ in the core and the parallel layer at $r=1$ has no jump in v_θ to match.

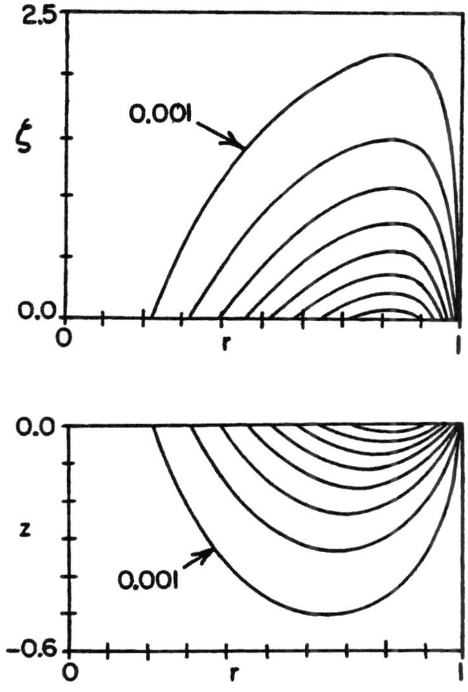

Figure 2. Thermoelectric currents for $Ha = 50$. (a) Contours for $Ha{*}h = 0.001$n,for n = 1 to 8 in the crystal-melt interface Hartmann layer. (b) Contours for $Ha{*}h = 0.001$n for n = 1 to 8 in the crystal.

Figure 2 presents the contours of thermoelectric currents in the crystal and in the interface Hartmann layer for $Ha = 50$. The total current with a maximum $h_s = 0.00874$ in the crystal circulates through the interface Hartmann layer. The amount of total currents produced in the crystal is 0.00874 and the same amount of current circulates in the interface Hartmann layer. There is no current in the core and the free surface Hartmann layer which results in no axial variation of v_θ in the core and the free surface Hartmann layer. This is a unique result of having an electrically insulated free surface at the top. For $Ha = 500$, the contours of h are similar in nature to those for $Ha = 50$, with a maximum of 0.00245 in both the crystal and the interface Hartmann layer.

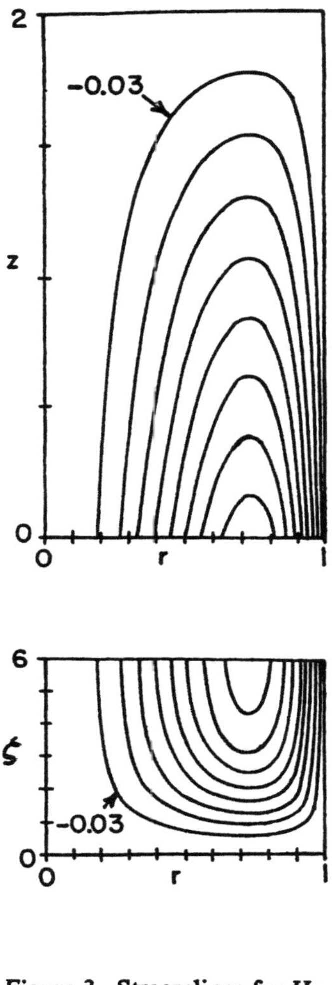

Figure 3. Streamlines for $Ha = 50$.
(a) Contours for $Ha^*\psi = -0.03n$ for n = 1 to 8 in the core.
(b) Contours for $Ha^*\psi = -0.03n$ for n = 1 to 8
in the crystal-melt interface Hartmann layer.

The streamlines in the core and in the interface Hartmann layer for $Ha = 50$ are presented in *Figure 3*. The maximum value of the stream function is -0.261 in both the core and the interface Hartmann layer. The melt flows downward near the ampoule wall, flows radially inward near the crystal-melt interface and moving upward near $r = 0$ in order to complete a loop. The maximum stream function value for $Ha = 500$ is -1.59 and the streamlines have a similar pattern to those for $Ha = 50$. In the interface Hartmann layer, the positive values of $dv_{\theta i}/dz$ (i.e., $v_{\theta i}$ increases as z increases) produce a radially inward flow in this layer. In our previous numerical solution for arbitrary Hartmann number [7], we treated a Bridgman process with an electrically insulated solid surface at $z = b$, i.e., the ampoule. The no-slip condition at $z = b$ led to a jump in v_θ and the associated radial electric currents in the Hartmann layer at $z = b$. The axial electric currents across the core which fed the radial currents in the free-surface Hartmann layer also produced an axial variation of the $O(Ha^{-1})$ v_θ in the core. This axial variation of v_θ produced a meridional circulation with positive stream function, i.e., with radially outward flow near the crystal-melt interface, which totally overwhelmed the radially inward flow associated with the interface Hartmann layer. Thus, the flow with a free surface opposite the interface is very different from that with a solid surface opposite the interface.

4. Conclusions

Our previous numerical solution for arbitrary values of Ha provides accurate results with minimal computational resources for $Ha \leq 30$ [7]. However, as Ha is increased from 30, the computational resources needed to resolve the Hartmann layers for the large dimensionless axial melt-zone lengths b for the actual Bridgman process increase dramatically so that the arbitrary -Ha numerical solution is not practical for $Ha \geq 50$.

In the present asymptotic solution for $Ha >> 1$, the parallel layer adjacent to the ampoule wall at $r = 1$ plays no significant role as long as it is thin. Its dimensionless thickness is $O((b/Ha)^{1/2})$ which is only small for quite large values of Ha. Therefore, both the numerical and asymptotic solutions are only qualitatively correct for $Ha = 50$ and $b = 4$, for example, so that it is surprising that the present results for $v_{\theta c0}$ agree reasonably well with the numerical solutions for $Ha = 50$ [7].

Here, we note that the numerical and asymptotic models are exactly the same except the top boundary condition, one being rigid and the other a free surface. But the stream functions are quite different: a radially outward flow and a radially inward flow near the crystal-melt interface for the solid surface and free surface, respectively. The radially outward flow near the crystal-melt interface is desirable for the mass transport of dopants and species since it has a tendency to balance the radially inward flow due to the compositionally driven buoyant convections. In conclusion, this model problem gives us some physical insights into the TEMHD flows.

Acknowledgments

This research was supported by the U.S. National Aeronautics and Space Administration under Cooperative Research Agreement NCC8-90 and by the U.S. National Science Foundation under Grant CTS 94-19484. Dr. Frank Szofran of the NASA Marshall Space Flight Center provided information about TEMHD effects in crystal growth.

References

1. Garandet, J.P.: Microsegregation in crystal growth from the melt: an analytical approach, *J. Crystal Growth* **131** (1993), 431-438.
2. Ravishankar, P.S., Braggins, T.T., and Thomas, R.N.: Impurities in commercial-scale magnetic Czochralski silicon: axial versus transverse magnetic fields, *J. Crystal Growth* **104** (1990), 617-628.
3. Shercliff, J.A.: Thermoelectric magnetohydrodynamics, *J. Fluid Mechanics* **91** (1979), 231-251.
4. Gorbunov, L.A.: Effect of thermoelectromagnetic convection on the production of bulk single-crystals consisting of semiconductor melts in a constant magnetic field, *Magnetohydrodynamics* **23** (1987), 404-408.
5. Gel'fgat, Y.M. and Gorbunov, L.A.: An additional source of forced convection in semiconductor melts during single-crystal growth in magnetic fields, *Sov. Phys. Dokl.* **34** (1989), 470-473.
6. Moreau, R., Laskar, O., Tanaka, M., and Camel, D.: Thermoelectric magnetohydrodynamic effects on solidification of metallic alloys in the dendritic regime, *Materials Science and Engineering* **A173** (1993), 93-100.
7. Khine, Y.Y. and Walker, J.S.: Thermoelectric magnetohydrodynamic effects during Bridgman semiconductor crystal growth with a uniform axial magnetic field, *J. Crystal Growth* **183** (1998), 150-158.
8. Khine, Y.Y., Ma, N. and Walker, J.S.: Thermoelectrically driven melt motion during floating zone crystal growth with a strong axial magnetic field, *Proceedings of the Seventh AIAA/ASME Joint Thermophysics and Heat Transfer Conference*, June 1998.
9. Khine, Y.Y. and Walker, J.S.: Thermoelectrically driven melt motion during floating zone crystal growth with an axial magnetic field, submitted to *J. Fluids Engineering*.
10. Hjellming, L.N. and Walker, J.S.: Melt motion in a Czochralski crystal puller with an axial magnetic field: isothermal motion, *J. Fluid Mechanics* **164** (1986), 237-273.
11. Hjellming, L.N. and Walker, J.S.: Melt motion in a Czochralski crystal puller with an axial magnetic field: motion due to buoyancy and thermocapillarity, *J. Fluid Mechanics* **182** (1987), 335-368.

EXPERIMENTAL AND NUMERICAL ANALYSIS OF THE INFLUENCE OF A ROTATING MAGNETIC FIELD ON CONVECTION IN RAYLEIGH-BÉNARD CONFIGURATIONS

B. FISCHER, J. FRIEDRICH, C. KUPFER AND G. MÜLLER
Institut für Werkstoffwissenschaften WW6,
Universität Erlangen-Nürnberg, Martensstr. 7,
91058 Erlangen, Germany

AND

D. VIZMAN
Faculty of Physics, West University of Timişoara,
Bd. V. Pârvan 4, 1900-Timişoara, Romania

Abstract. The influence of a rotating magnetic field on convection in a Rayleigh-Bénard configuration was experimentally and numerically investigated. Temperature measurements for different Rayleigh and magnetic Taylor numbers were carried out in a cylindrical test cell filled with liquid gallium. The experimental results were compared to 3D time-dependent numerical simulations. The main result of our study is that it is possible to significantly damp temperature fluctuations caused by time-dependent buoyancy driven convection using a rotating magnetic field.

1. Introduction

The use of magnetic fields is a wide-spread method to control transport processes during semiconductor crystal growth from the melt. In the past mainly steady magnetic fields, which in general have a damping effect on convection, were applied to avoid undesireable crystal properties caused by time-dependent convection, e. g. dopant striations. The possible effects of time-dependent, especially of rotating magnetic fields (RMF), however, have been facing a rapid increase of interest over the recent years, in crystal growth as well as in metallurgy (see e. g. Spitzer *et al.*, 1986).

A. Alemany et al. (eds.),
Transfer Phenomena in Magnetohydrodynamic and Electroconducting Flows, 279–294.

The action of a RMF is not a damping of convection as in the case of steady magnetic fields (see e. g. Baumgartl, 1992), but a deliberate enforcing of a certain flow pattern (Dahlberg, 1972). A lot of mainly theoretical work has been done to describe the flow phenomena occuring in an electrically conducting fluid under the influence of a RMF (for reviews see Dahlberg, 1972, and Gelfgat and Priede, 1995). Fig. 1 illustrates the principal action of a RMF: The Lorentz force generated by a magnetic field (with one pole pair in fig. 1) rotating around the axis of a cylindrical melt volume (infinite in fig. 1) enforces a predominantly azimuthal, rotational flow.

However, in the field of semiconductor crystal growth there are only a few publications of mainly experimental investigations, which demonstrate some effects caused by the use of RMF, e. g. Salk *et al.*, 1994, Dold and Benz, 1995, 1997.

We have studied more systematically the action of a RMF on buoyancy driven convection in a Rayleigh-Bénard (RB) configuration, both numerically and experimentally.

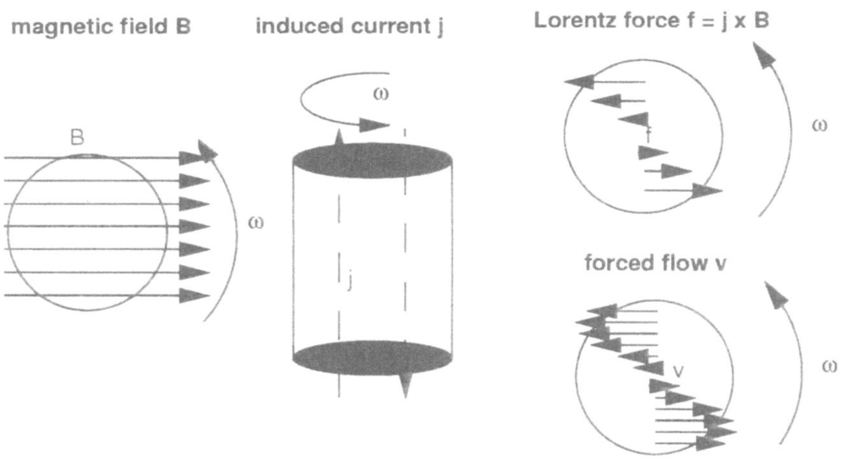

Figure 1. Principal action of the Lorentz force generated by a magnetic field rotating around the axis of a cylindrical melt volume.

2. Experimental Setup

We generate a RMF with two pole pairs by using a 3-phase stator with an inner diameter of 15.5 cm and a height of 17.0 cm, supplied with a 50 Hz AC current. The resulting magnetic field, which is rotating in the horizontal plane of the stator, can – for an arbitrary number of pole pairs p – be written in cylindrical coordinates r, φ, z as:

$$\vec{B}(r,\varphi,z) = B\left(\frac{r}{R}\right)^{p-1}\left[\vec{e}_r \sin(p\varphi - \omega t) + \vec{e}_\varphi \cos(p\varphi - \omega t)\right] \quad (1)$$

$$\text{with} \quad p = 2 \quad \text{in our case.}$$

In eqn. 1 B denotes the amplitude of the magnetic induction at the position $r = R$, where R is the radius of our test cell, ω the frequency of the applied AC current and t the time. Fig. 2 shows the measured magnetic field lines inside the stator, illustrating the quadrupole character of the rotating magnetic field pattern.

A test cell containing liquid gallium ($T_{\text{melt}} = 303$ K) is placed in the center of the stator to examine the effect of the RMF on fluid behaviour in a finite cylindrical cavity with radius $R = 2$ cm, height $H = 4$ cm, and, consequently, aspect ratio $\frac{H}{2R} = 1$. A linear RB temperature profile can be established along the cylinder walls using three independent closed water circuits supplied from thermostats. For more details on this test cell, see Baumgartl *et al.*, 1989.

The temperature difference ΔT between the hot bottom and the cold top of the cylinder was varied from 5 to 40 K, with the magnetic induction B ranging from 0 to about 10 mT in any case. The frequency of the AC current usually was 50 Hz. Temperature measurements were carried out at several positions in the gallium melt.

3. Theoretical Background and Numerical Method

The fluid flow is determined by the well-known set of hydrodynamic equations governing mass, momentum, and heat transport:

$$\vec{\nabla} \cdot \vec{u} = 0 \quad (2)$$

$$\rho_0 \frac{\partial \vec{u}}{\partial t} + \rho_0 \left(\vec{u} \cdot \vec{\nabla}\right) \vec{u} - \rho_0 \nu \vec{\nabla}^2 \vec{u} = -\vec{\nabla} P - \rho_0 \vec{g} \beta (T - T_{\text{ref}}) + \vec{f} \quad (3)$$

$$\frac{\partial T}{\partial t} + \left(\vec{u} \cdot \vec{\nabla}\right) T - \kappa \vec{\nabla}^2 T = 0 \quad (4)$$

We use the Boussinesq approximation for the buoyancy term in the Navier-Stokes equations (3). In eqns. 2–4 $\vec{u}$ is the flow velocity, ρ_0 the reference density, ν the kinematic viscosity, P the pressure, $\vec{g}$ the gravitational acceleration, β the thermal expansion coefficient, T the temperature, T_{ref} a reference temperature, $\vec{f}$ the Lorentz force density, and κ the thermal diffusivity (for gallium, $\rho_0 = 6.1 \cdot 10^3\,\frac{\text{kg}}{\text{m}^3}$, $\nu = 3.126 \cdot 10^{-7}\,\frac{\text{m}^2}{\text{s}}$, $\beta = 1.26 \cdot 10^{-4}\,\text{K}^{-1}$, and $\kappa = 1.512 \cdot 10^{-5}\,\frac{\text{m}^2}{\text{s}}$, see Baumgartl, 1992).

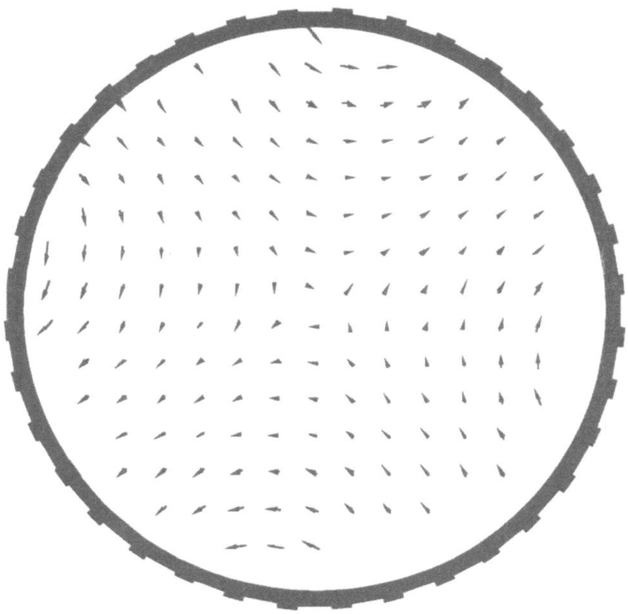

Figure 2. Measured magnetic field lines inside the stator.

The rotating magnetic field $\vec{B}$ induces an electric current in the melt, which leads to a Lorentz force density $\vec{f} = \vec{j} \times \vec{B}$. The current density

$$\vec{j} = -\sigma \vec{\nabla} \Phi - \sigma \frac{\partial \vec{A}}{\partial t} + \sigma \vec{u} \times \vec{B} \tag{5}$$

can be obtained from Ohm's law, where $\vec{A}$ is the magnetic vector potential, Φ the scalar electric potential, and σ the electrical conductivity (for gallium, $\sigma = 3.87 \cdot 10^6 \, \Omega^{-1} \mathrm{m}^{-1}$, see Baumgartl, 1992). The resulting Lorentz force $\vec{f}$ is considered by an additional source term in the Navier-Stokes equations (3).

For our experimental setup the frequency of the RMF is small enough to allow the magnetic field lines to penetrate through the melt so that the skin effect is negligible (Moreau, 1990). Furthermore, we can neglect the influence of the fluid flow on the magnetic field, i. e. omit the $\vec{u} \times \vec{B}$ term in eqn. 5. According to Gelfgat *et al.*, 1993, we use only a time-averaged value of $\vec{f}$ in our numerical simulations, which has only one component in azimuthal direction.

The Lorentz force $\vec{f}$ can be derived analytically for a cylinder with finite length. The magnetic field in eqn. 1 is related to a magnetic vector potential

$$\vec{A} = -\frac{B}{2}\frac{r^2}{R}\cos(2\varphi - \omega t)\,\vec{e}_z\,. \tag{6}$$

As $\vec{\nabla}\cdot\frac{\partial\vec{A}}{\partial t} = 0$, the continuity equation $\vec{\nabla}\cdot\vec{j}$ for the current density yields a Laplace equation

$$\Delta\Phi = 0 \tag{7}$$

for the electric potential, which can be solved analytically, using the boundary condition of electrically insulating walls, i. e. there is no component of $\vec{j}$ normal to the cylinder walls. This means

$$\frac{\partial\Phi}{\partial r} = 0 \quad \text{at the vertical walls } (r = R)\text{, and} \tag{8}$$

$$-\frac{\partial\Phi}{\partial z} - \frac{\partial A_z}{\partial t} = 0 \quad \text{at the horizontal walls } (z = 0,\ z = H). \tag{9}$$

Finally, we get the following solution for the time-averaged Lorentz force density:

$$\begin{aligned}\langle f_\varphi\rangle = {} & \frac{1}{4}\sigma\omega B^2\cdot \\ & \left\{\frac{r^3}{R^2} - r\cdot\sum_n J_2\left(\lambda_n\frac{r}{R}\right)\cdot\frac{2}{\lambda_n}\cdot\frac{J_3(\lambda_n)}{J_2^2(\lambda_n) - J_1(\lambda_n)\,J_3(\lambda_n)}\cdot\right. \\ & \left.\left[\cosh\left(\lambda_n\frac{z}{R}\right) + \frac{1-\cosh\left(\lambda_n\frac{H}{R}\right)}{\sinh\left(\lambda_n\frac{H}{R}\right)}\cdot\sinh\left(\lambda_n\frac{z}{R}\right)\right]\right\}\end{aligned} \tag{10}$$

The J_p are the Bessel functions and the λ_n fulfill the equation

$$J_3(\lambda_n) = \frac{2}{\lambda_n}J_2(\lambda_n)\,. \tag{11}$$

For more details on the modelling of an RMF see Vizman *et al.*, 1998, Gelfgat *et al.*, 1993, Mazuruk *et al.*, 1997, or Barz *et al.*, 1997.

In the following we use the Prandtl Pr, Rayleigh Ra, Grashof Gr and magnetic Taylor number $\mathrm{Ta_m}$ to describe our problem. These numbers emerge in a dimensionless representation of eqns. 2–4 and are defined as follows:

$$\mathrm{Pr} = \frac{\nu}{\kappa},\quad \mathrm{Ra} = \frac{\beta \mathrm{g}\Delta \mathrm{T}\mathrm{R}^3}{\nu\kappa},\quad \mathrm{Gr} = \frac{\mathrm{Ra}}{\mathrm{Pr}} \tag{12}$$

$$\mathrm{Ta_m} = \frac{\mathrm{B^2R^4}\sigma\Omega}{4\nu^2\rho_0}\quad \text{with } \Omega = \frac{\omega}{\mathrm{p}} \tag{13}$$

Here g is the absolute value of the gravitational acceleration, and Ω the rotation frequency of the magnetic field pattern. Ra or Gr are measures for the buoyancy forces, while Ta_m is a measure for the electromagnetic forces.

We have performed 3D time-dependent numerical simulations using the finite volume code STHAMAS (see Vizman *et al.*, 1998) developed at our institute. The computational domain is divided into five blocks using a structured non-orthogonal grid. Within each block, the equations governing mass, momentum and heat transport (eqns. 2–4) are solved using no-slip conditions for the flow velocity and fixed temperatures at the walls as boundary conditions. Further details on the numerical method of STHAMAS are given by Leister, 1994.

4. Results

Fig. 3 shows a horizontal and vertical cross section of numerically calculated velocity fields including isothermal contour plots for $\text{Ra} \approx 2 \cdot 10^4$. When no magnetic field is applied a three-dimensional time-dependent flow results, with a flow pattern being very similar to previous results, e. g. Crespo del Arco *et al.*, 1988. For $\text{Ta}_\text{m} \approx 5 \cdot 10^5$ an axisymmetric solution is obtained with a clearly dominating azimuthal velocity component. In addition, one can recognize a secondary flow consisting of two tori near the caps of the cylinder.

The temperature signal measured on the cylinder axis 12 mm beneath the top of the cell versus time can be seen in fig. 4 for $\text{Ra} \approx 1 \cdot 10^4$. Having started from $B = 0\,\text{mT}$, the magnetic field was increased every 5 minutes. There is a significant decrease of the temperature at the monitor point as well as a strong reduction of the amplitude of the temperature fluctuations after a critical threshold of the applied magnetic induction B. When exceeding an optimum value of the magnetic induction, the amplitude of the temperature fluctuations increases slightly again. One can also notice an increase of the frequency of the fluctuations as well as a slow increase of the temperature at higher values of the magnetic induction B. The same dependence of the temperature fluctuations on the magnetic induction B was observed for other values of Ra. At the same time, the threshold of the applied magnetic induction B, where the strong reduction of temperature fluctuations occurs, increases with increasing Ra.

Fig. 5 shows the average deviation $\langle T - \langle T \rangle \rangle$ of both the experimental and the numerical temperature fluctuations for $\text{Ra} \approx 4 \cdot 10^4$ versus the values of the magnetic induction B. The experimental and calculated curves correspond pretty well.

Fig. 6 shows time-averaged temperature profiles along the cylinder axis for $\text{Ra} \approx 4 \cdot 10^4$ without and with magnetic induction $B = 3.3\,\text{mT}$, or $\text{Ta}_\text{m} \approx$

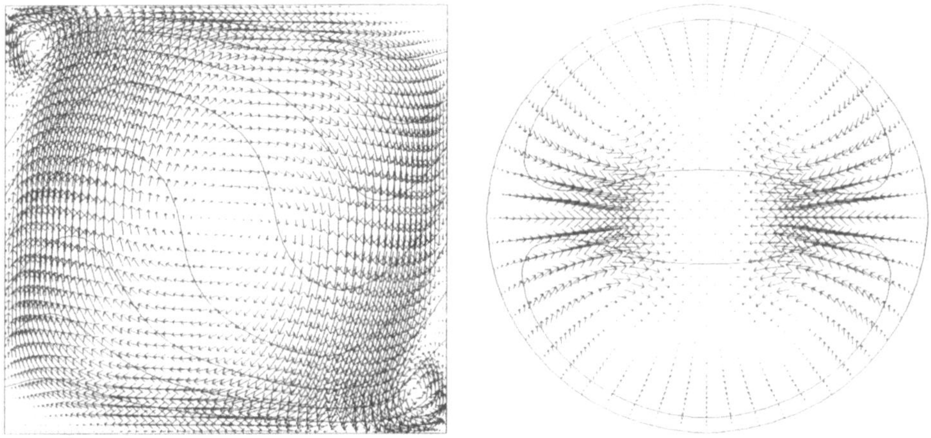

$\mathrm{Ta_m} = 0, \quad v_{\varphi(\max)} = 9\,\frac{\mathrm{mm}}{\mathrm{s}}, \quad v_{z(\max)} = 9\,\frac{\mathrm{mm}}{\mathrm{s}}, \quad v_{r(\max)} = 9\,\frac{\mathrm{mm}}{\mathrm{s}}$

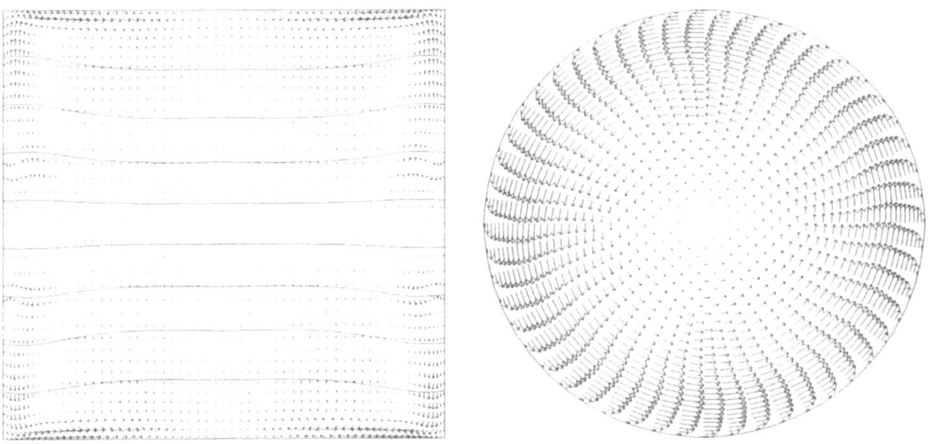

$\mathrm{Ta_m} \approx 5 \cdot 10^5, \quad v_{\varphi(\max)} = 22\,\frac{\mathrm{mm}}{\mathrm{s}}, \quad v_{z(\max)} = 3\,\frac{\mathrm{mm}}{\mathrm{s}} \quad v_{r(\max)} = 7\,\frac{\mathrm{mm}}{\mathrm{s}},$

Figure 3. Numerically calculated velocity fields and isotherms in an axial (left) and horizontal (right) cross section for $\mathrm{Pr} = 0.02$, $\mathrm{Ra} \approx 2 \cdot 10^4$. Top: without magnetic field ($\mathrm{Ta_m} = 0$). Bottom: with rotating magnetic field ($\mathrm{Ta_m} \approx 5 \cdot 10^5$). The lengths of the vector arrows cannot be compared from one plot to another because they are differently magnified in order to achieve a proper presentation. The isotherms are equally spaced ($\Delta T_{\mathrm{iso}} = 1.1\,\mathrm{K}$).

$9 \cdot 10^5$. The symbols are measured values, whereas the lines were obtained from our numerical simulations. The calculated results agree qualitatively with the measurements. Fig. 6 reveals that when switching on a sufficiently strong magnetic field, the profiles undergo a significant change, that can be seen in the linearization of the temperature profile in the core of the melt,

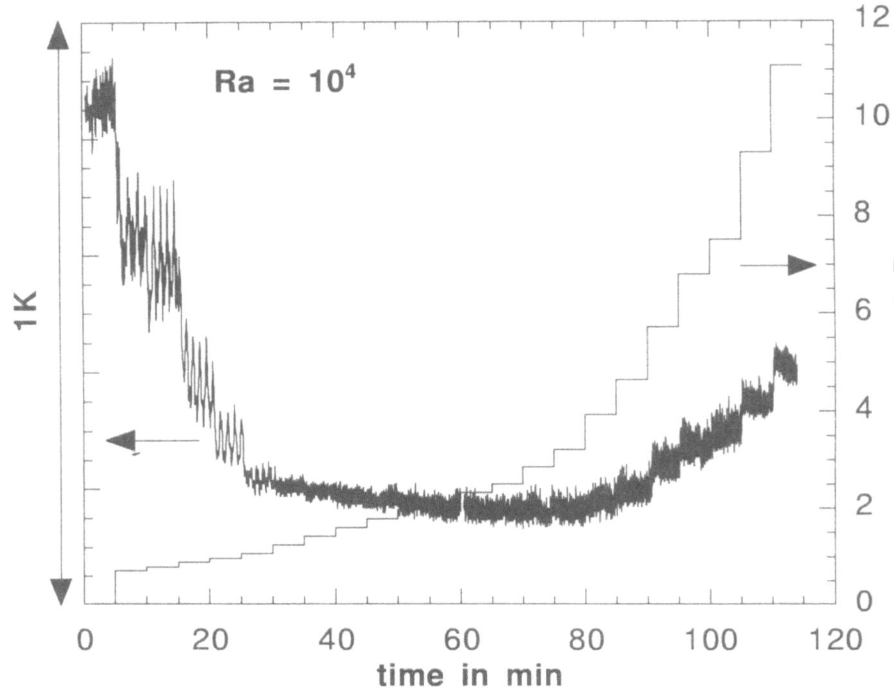

Figure 4. Measured temperature signal on the cylinder axis 12 mm beneath the top of the cell for Ra $\approx 1 \cdot 10^4$. The values of magnetic induction at the vertical wall of the melt cylinder are given as the second ordinate.

which also explains the change of the absolute value of the temperature in fig. 4.

The significant change in the flow pattern occuring when the magnetic induction is increased is also illustrated in fig. 7, where numerically obtained contour lines of the azimuthal velocity are shown in an axial cross section for Ra $\approx 4 \cdot 10^4$ and different values of $\mathrm{Ta_m}$. One can clearly recognize that for sufficiently high values of $\mathrm{Ta_m}$ a predominantly azimuthal flow is established. In addition, for $\mathrm{Ta_m} > 8 \cdot 10^5$ there are small regular patterns at the vertical walls of the cylinder, which come from small vortices at these walls, that can also be found in the velocity vector plot for Ra $\approx 2 \cdot 10^4$ and $\mathrm{Ta_m} \approx 5 \cdot 10^5$ in fig. 3. These patterns can be attributed to so-called Taylor vortices (see Discussion).

For small values of $\mathrm{Ta_m}$ the electromagnetic forces are not yet strong enough to clearly dominate the flow, so that one finds a mixture of flow patterns typical for pure buoyancy and typical for the forced rotation. Fig. 8 shows two horizontal and two axial cross sections of the numerically cal-

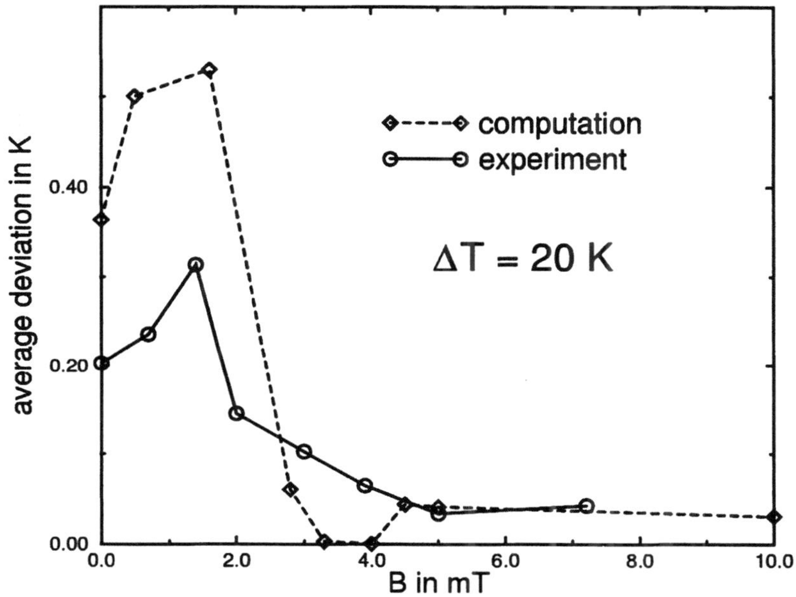

Figure 5. Numerical and experimental average deviation of temperature fluctuations $\langle T - \langle T \rangle \rangle$ versus magnetic induction B for $\mathrm{Ra} \approx 4 \cdot 10^4$.

culated velocity field including thermal contour plots for $\mathrm{Ra} \approx 1 \cdot 10^4$ and $\mathrm{Ta_m} \approx 5 \cdot 10^4$. The fluid rotates around an axis which is tilted with respect to the vertical cylinder axis. Such observations were also made by Volz and Mazuruk, 1996, who experimentally investigated different flow regimes in dependence on Ra and the amplitude of a RMF with one pole pair in a configuration similar to ours, using a gallium melt, too. For weak magnetic fields and exceeding a critical Rayleigh number, they describe the flow as a non-axisymmetric meridional roll originating from buoyancy effects, which rotates around the cylinder axis driven by the RMF.

Our simulations and measurements (Friedrich, Lee *et al.*, 1997) show that the flow pattern precedes around the cylinder axis, resulting in large-scale regular thermal waves. Numerical and experimental results for the corresponding temperature fluctuations are shown in fig. 9 for two points 5 mm from the cylinder wall that are azimuthally seperated by an angle of 90°. Both numerical and experimental results have been obtained for $\mathrm{Ra} \approx 1 \cdot 10^4$ and similar ranges of $\mathrm{Ta_m}$ ($1\text{-}5 \cdot 10^4$). The two signals show a clear phase shift in any case (see also Volz and Mazuruk, 1996). The

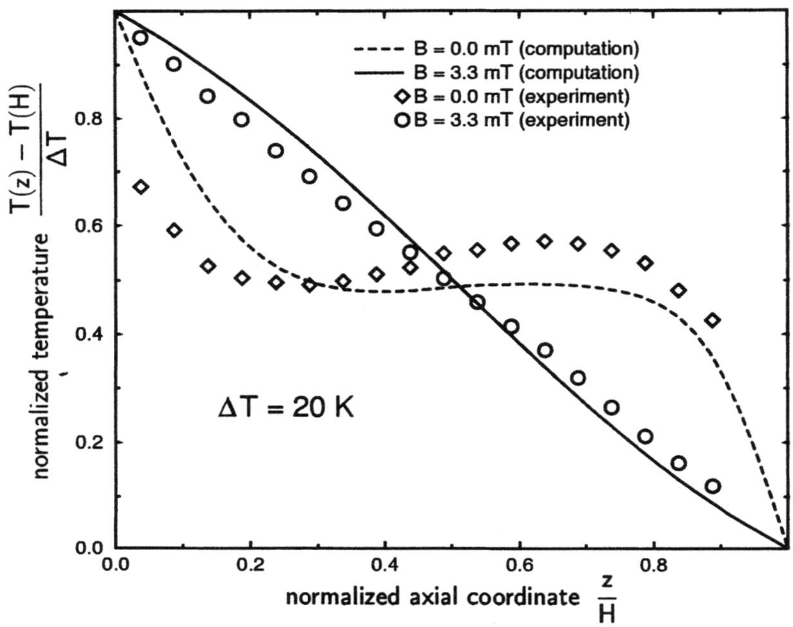

Figure 6. Numerical (lines) and experimental (symbols) temperature profiles normalized by the adjusted temperature difference ΔT versus z normalized by the melt height H at the cylinder axis for Ra $\approx 4 \cdot 10^4$ without and with the influence of a rotating magentic field ($B = 3.3$ mT, $\text{Ta}_\text{m} \approx 9 \cdot 10^5$).

principal effects of increasing Ta_m on frequency and amplitude of the large-scale fluctuations agree quite well for simulation and experiment.

5. Discussion

The action of the RMF results in a damping of temperature fluctuations caused by time-dependent buoyancy driven convection. Our experimental and numerical results confirm the previous work done by Dold and Benz, 1995, 1997. The RMF leads to a stabilization of the fluid motion due to the high azimuthal flow velocity. Exceeding a critical value of the magnetic induction, a transition from a 3D flow with large temperature fluctuations to an axisymmetric flow with only very small ones takes place in dependence on Ra. In fig. 10 we plotted the Grashof number Gr versus the magnetic field value B_{crit}, at which the temperature fluctuations observed in the experiments as well as in the numerical siumlations are reduced to the ar-

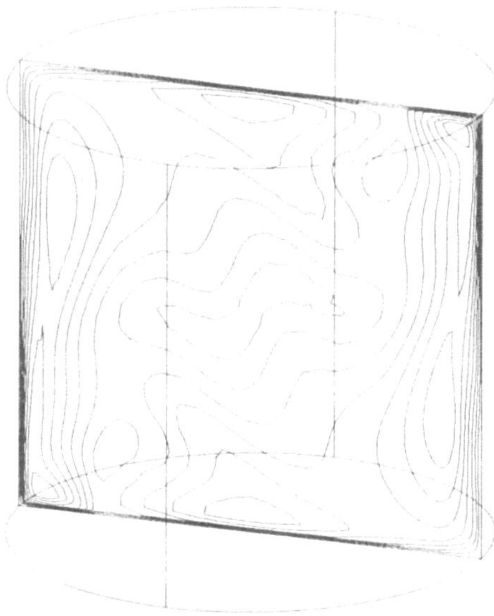

$\mathrm{Ta_m} = 2.1 \cdot 10^5$, $v_{\varphi(\max)} = 18\ \frac{\mathrm{mm}}{\mathrm{s}}$

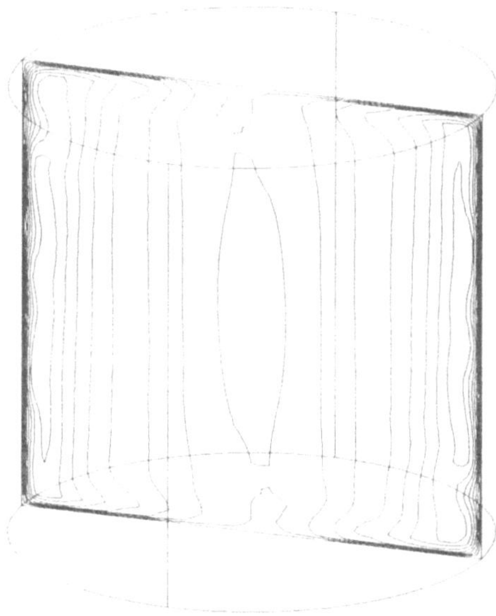

$\mathrm{Ta_m} = 6.4 \cdot 10^5$, $v_{\varphi(\max)} = 27\ \frac{\mathrm{mm}}{\mathrm{s}}$

$\mathrm{Ta_m} = 8.8 \cdot 10^5$, $v_{\varphi(\max)} = 30\ \frac{\mathrm{mm}}{\mathrm{s}}$

$\mathrm{Ta_m} = 2.0 \cdot 10^6$, $v_{\varphi(\max)} = 48\ \frac{\mathrm{mm}}{\mathrm{s}}$

Figure 7. Numerically calculated contour plots of the azimuthal velocity in a horizontal cross section for $\mathrm{Ra} \approx 4 \cdot 10^4$ and different values of $\mathrm{Ta_m}$. The maximum values of the azimuthal velocity v_φ for the whole cylinder are given for each value of $\mathrm{Ta_m}$.

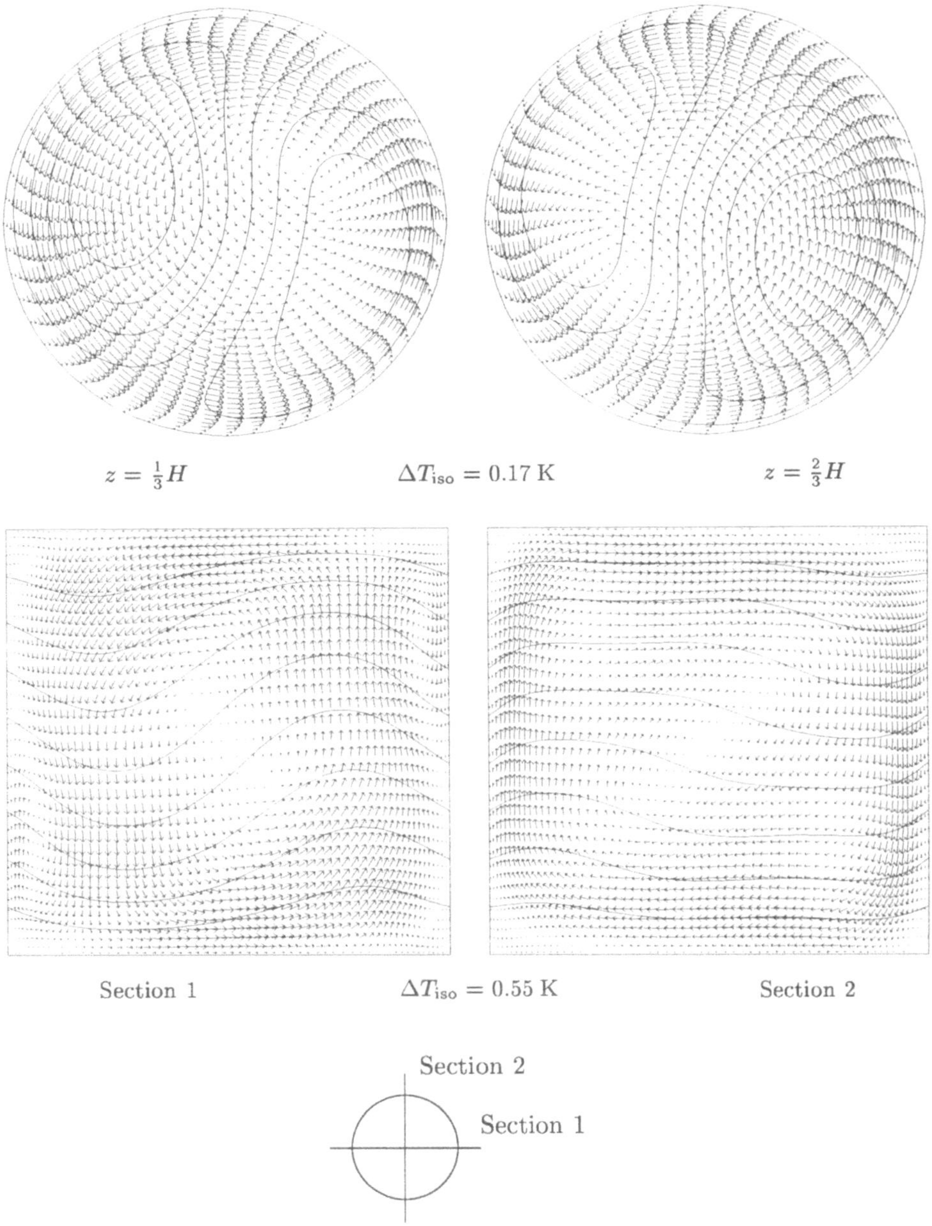

Figure 8. **Numerically calculated velocity fields and isotherms in two different horizontal (top) and two different axial (bottom) cross sections for Pr = 0.02, Ra ≈ $1 \cdot 10^4$ and $\text{Ta}_m \approx 5 \cdot 10^4$. The lengths of the vector arrows cannot be compared from one plot to another because they are differently magnified in order to achieve a proper presentation. The isotherms are equally spaced (ΔT_{iso}).**

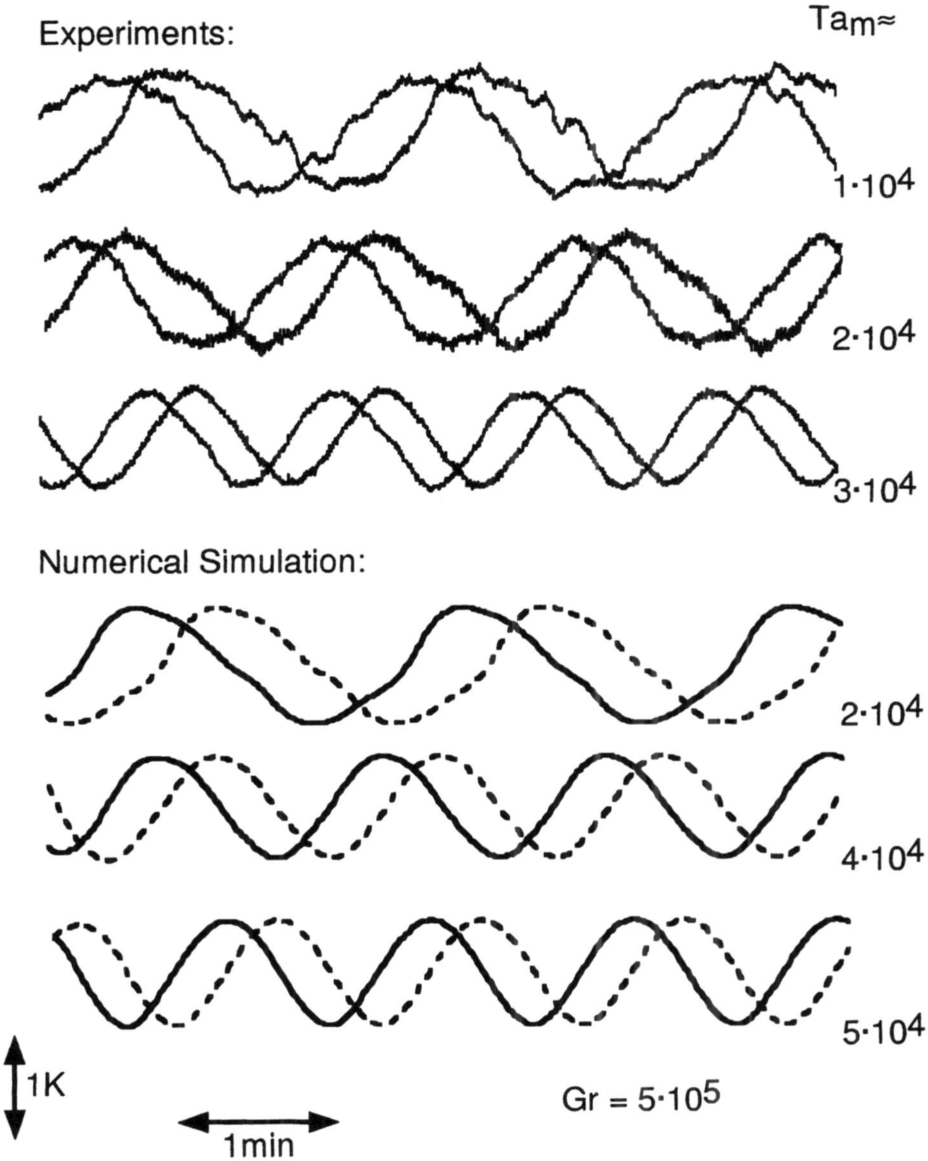

Figure 9. Numerical and experimental temperature signals at two points ($z = \frac{1}{3}H$, $r = R - 5\,\mathrm{mm}$) azimuthally seperated by an angle of 90° for $\mathrm{Ra} \approx 1 \cdot 10^4$ and different moderate values of $\mathrm{Ta_m}$. In any case, there is a clear phase shift between the two signals.

bitrarily chosen fraction of 20% of the initial amplitude without a magnetic field. It is obvious that the value of the necessary magnetic induction increases with increasing Gr. There are two forces driving convection in the melt: buoyancy and the electromagnetic force, the first predominantly inducing meridional flow, the second azimuthal flow. The corresponding flow intensities may be characterized by Gr and Ta_m, respectively. It is a simple assumption that azimuthal convection due to the RMF dominates over buoyancy driven convection for $Gr \leq Ta_m$, and is therefore able to suppress meridional convection due to buoyancy to a sufficient extent. So as a rough basis of comparison the boundary $Gr = Ta_m$ is shown as a solid line in fig. 10. It is obvious that with increasing Ra, i. e. increasing buoyancy force, a larger value of the magnetic field strength is required to achieve a convection regime dominated by the RMF with only small temperature fluctuations.

However, under the influence of the RMF temperature fluctuations are still present in the experiments as well as in the numerical simulations, although here the predicted amplitudes are much smaller than in the experiments. This may be due to the idealized boundary conditions used in the calculations. These oscillations are caused by Taylor instabilities, which result from the steep gradient of the azimuthal flow velocity at the vertical wall of the melt as described by Moffatt, 1965, and Dahlberg, 1972. Such Taylor vortices were also examined in a recent study by Martin Witkowski and Marty, 1997.

The change of the temperature field and therefore of the axial temperature profile is caused by the primary flow acting in the azimuthal direction of the melt. It suppresses the heat flux in the radial and axial direction in the core of the melt volume. Therefore, the isotherms become flat and the axial temperature field looks like a temperature field governed by diffusion only.

For sufficiently high values of Ta_m buoyancy becomes negligible, and we find a slow re-increase of fluctuations and instabilities because of increasing forced rotation. In this limit the flow behaviour is very similar to that in Vertical Bridgman configurations heated from above, where buoyancy has no influence (see Dold and Benz, 1997, and Friedrich, Kupfer *et al.*, 1997).

For very small values of Ta_m ($\frac{Ta_m}{Gr} \leq 0.03$ according to Friedrich, Lee *et al.*, 1997) there exist large scale temperature fluctuations resulting from regular thermal waves travelling azimuthally in the rotation direction of the RMF. For larger values of Ta_m they disappear, and we find only irregular high-frequency oscillations attributed to the instabilities described above. For a summary on these oscillation regimes refer to Lee, 1997 or Friedrich, Lee *et al.*, 1997.

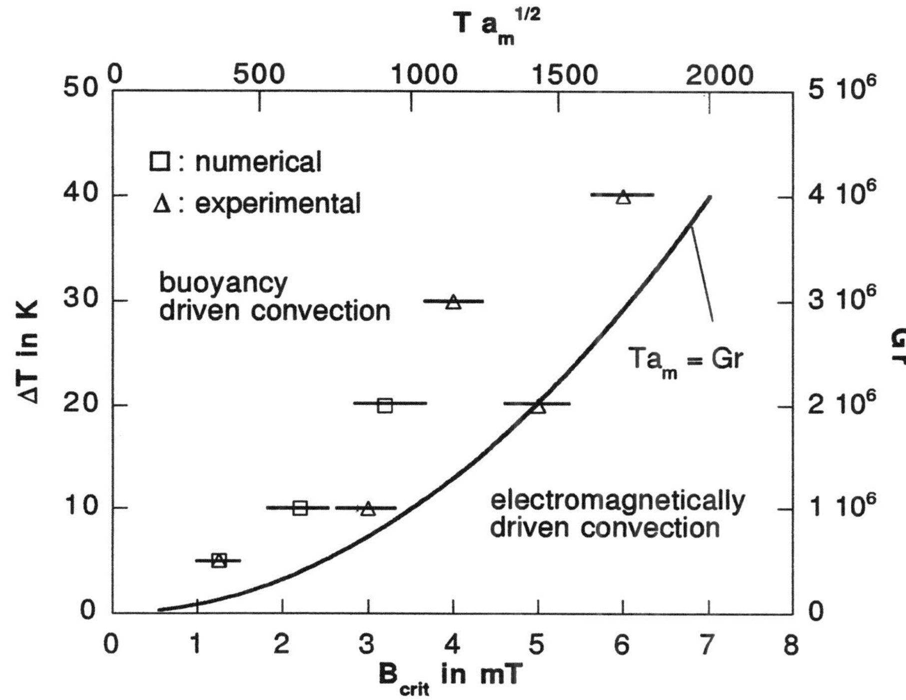

Figure 10. The Grashof number $\mathrm{Gr} = \frac{\mathrm{Ra}}{\mathrm{Pr}}$ versus the magnetic field value B_{crit}, at which the temperature fluctuations observed in the experiments (triangles) as well as in the numerical simulations (squares) are reduced to 20% of the initial amplitude without magnetic field. The solid line divides buoyancy driven convection from the regime where the flow is dominated by the electromagnetic force (see text).

6. Conclusion

We have shown for a low-Pr melt in a vertical cylindrical test cell that it is possible to reduce temperature fluctuations caused by time-dependent buoyancy driven convection using a rotating magnetic field. The strong reduction of the flow oscillations is connected with a transition from a three-dimensional flow to an axisymmetric flow due to the action of the rotating magnetic field. The transition takes place at a critical value of $\mathrm{Ta_m}$ for given Ra, which increases with increasing Ra.

Acknowledgement This work was supported by the german space agency DARA, contract no. 50WM94545.

References

Barz, R.U., Gerbeth, G., Wunderwald, U., Buhrig, E. and Gelfgat, Yu.M. (1997) Modelling of the isothermal melt flow due to rotating magnetic fields in crystal growth, *J. Crystal Growth* **180**, 410–421

Baumgartl, J. (1992), *Numerische und experimentelle Untersuchungen zur Wirkung magnetischer Felder in Kristallzüchtungsanordnungen*, PhD Thesis, Erlangen

Baumgartl, J., Budweiser, W., Müller, G. and Neuman, G. (1989) Studies of buoyancy driven convection in a vertical cylinder with parabolic temperature profile, *J. Crystal Growth* **97**, 9–17

Crespo del Arco, E., Bontoux, P., Sani, R.L., Hardin, G. and Extrémet, G.P. (1988) Steady and oscillatory convection in vertical cylinders heated from below. Numerical simulations of asymmetric flow regimes, *Adv. Space Res.* **8**, (12)281–(12)292

Dahlberg, E. (1972) *On the action of a rotating magnetic field on a conducting liquid*, AB Atomenergi, Sweden, Rep. AE-447

Dold, P. and Benz, K.W. (1995) Convective temperature fluctuations in liquid gallium in dependence on static and rotating magnetic fields, *Cryst. Res. Technol.* **30**, 1135–1145

Dold, P. and Benz, K.W. (1997) Modification of fluid flow and heat transport in Vertical Bridgman configurations by rotating magnetic fields, *Cryst. Res. Technol.* **32**, 51–60

Friedrich, J., Kupfer, C., Fischer, B. and Müller, G. (1997) Influence of rotating magnetic fields on heat and species transport in crystal growth by the Vertical Gradient Freeze method, 3rd Int. Conf. on Transfer Phenomena in Magnetohydrodynamic and Electroconducting Flows, Aussois, France

Friedrich, J., Lee, Y.-S., Fischer, B., Kupfer, C., Vizman, D. and Müller, G. (1997) Experimental and numerical study of Rayleight-Bénard convection affected by a rotating magnetic field, submitted, *Phys. Fluids*

Gelfgat, Yu.M., Gorbunov, L.A. and Kolevzon, V. (1993) Liquid metal flow in a finite-length cylinder with a rotating magnetic field, *Exp. Fluids* **15**, 411–416

Gelfgat, Yu.M. and Priede, J. (1995) MHD flows in a rotating magnetic field (a review), *Magnetohydrodynamics* **31**, 188–200

Lee, Y.-S. (1997), *Experiments on the Oscillatory Convection of Low Prandtl Number Liquid in Czochralski Process with Magnetic Field*, PhD Thesis, Pohang

Leister, H.-J. (1994), *Numerische Simulation dreidimensionaler, zeitabhängiger Strömungen unter dem Einfluß von Auftriebs- und Trägheitskräften*, PhD Thesis, Erlangen

Martin Witkowski, L. and Marty, Ph. (1997) Some peculiar flow patterns in a fluid driven by a rotating magnetic field, 3rd Int. Conf. on Transfer Phenomena in Magnetohydrodynamic and Electroconducting Flows, Aussois, France

Mazuruk, K., Ramachandran, N., Volz, M.P. and Gillies, D. (1997) Study of frequency effects of a rotating magnetic field on fluid flow in vertical cylinders, SPIE, San Diego, vol. 3123, Soc. of Photo-Optical Instrumentation Engineers

Moffatt, H.K. (1965) On fluid flow induced by a rotating magnetic field, *Exp. Fluids* **16**, 411–416

Moreau, R. (1990), *Magnetohydrodynamics*, Kluwer Academic Publishers, Dordrecht

Salk, M., Fiederle, M., Benz, K.W., Senchenkov, A.S., Egorov, A.V. and Matioukhin, D.G. (1994) CdTe and $CdTe_{0.9}Se_{0.1}$ crystals grown by the travelling heater method using a rotating magnetic field, *J. Crystal Growth* **138**, 161–167

Spitzer, K.-H., Dubke, M. and Schwerdtfeger, K. (1986) Rotational electromagnetic stirring in continuous casting of round strands, *Metallurgical Transactions B* **17B**, 119–131

Vizman, D., Fischer, B., Friedrich, J. and Müller, G. (1998), to be published

Volz, M.P. and Mazuruk, K. (1996) Flow transitions in a rotating magnetic field, *Exp. Fluids* **20**, 454–459

NUMERICAL SOLUTIONS OF MOVING BOUNDARY PROBLEM WITH THERMAL CONVECTION IN THE MELT AND MAGNETIC FIELD DURING DIRECTIONAL SOLIDIFICATION

Enthalpy-Porosity method and time dependent melt effect on interface shape

EL GANAOUI M., BONTOUX P. AND MORVAN D.
IRPHE-CNRS
Technopôle de Château Gombert 13451 Marseille, France
ganaoui@mfmjessy.univ-mrs.fr

Abstract. An enthalpy-porosity method is extended to the simulation of crystal growth process via directional solidification. The approach is based on a homogeneous formulation for Navier Stokes and energy equations and involving a two-phases intermediate zone modelled as a porous medium. A finite volume approximation is used with a fixed mesh. The front capturing technique is validated with respect to an interface tracking method. The applications concern the interaction of steady and oscillatory melts with the interface during Bridgman crystal growth. The effect of an axial magnetic field is considered in steady case. The amplification of oscillatory instability in the melt is studied in the case of inverted Bridgman configuration (heated from below). Various solutions (steady symmetric and asymmetric, time-periodic, aperiodic ...) are analyzed. Successive interfaces are considered over a characteristic growth time scale in order to exhibit some typical crystal constitution.

1. Introduction

The improvement of new electronic (or optoelectronic) components requires a better understanding of the physical mechanisms that govern the heat and mass transfer during the solidification. The numerical simulation constitute a complementary approach to experimental studies. They permit to investigate each physical phenomena separately (volume forces induced by gravity or magnetic field, inertial force induced by rotation ...) and their coupling, with varying conditions to determine optimal conditions. They

A. Alemany et al. (eds.),
Transfer Phenomena in Magnetohydrodynamic and Electroconducting Flows, 295–309.

are seen in the purpose of reducing significantly the number of experiments which often need important supports (as for microgravity experiment).
Initially microgravity conditions have been used to avoid the natural convection inside the melting zone. The convective motions are caused by small density variations induced in the liquid material by radial temperature gradient. Unfortunately, under actual microgravity, gravity magnitude is not completely cancelled and generally a small but not negligible thermoconvective flow subsists. The close environment also induces unsteady g-jitter phenomena. An alternative solution is to damp natural convection during a crystal growth by using a magnetic field parallel to the axis of the ampoule. The given Lorentz force induces a significative damping of the flow in the melt. The technique represents a promising way to perform high quality single crystals both in earth and in space in particular when the density gradient drives oscillatory convection as illustrated in the Inverted Vertical Bridgman technique (IVB).
The mathematical model is derived from a continuum formulation, based on the integration of empirical laws and macroscopic descriptions of transport behaviour based on classical mixture theory. The phase change is taken into account by using the enthalpy equation. The jump on the velocity between the liquid and solid phases is accounted in the momentum equation by considering an additional source term given by the solid-liquid interaction force using a permeability Kozeny-Carman relation. This formulation does not require boundary condition at the solidification front. The problem has been solved with a fixed mesh, which reduces the computation time. For a description of the method see references [3,4,8,9,20].
The interactions between the melt and the solidification front during directional solidification are illustrated in the case of the vertical Bridgman configuration. The amplification of oscillatory instability in the melt is studied in the case of the inverted Bridgman configuration (the melt is below the crystal). Different physical and geometrical configurations are investigated and they exhibit several kinds of symmetric, asymmetric and time-dependent behaviours that develop in the melt.

2. Mathematical modelling

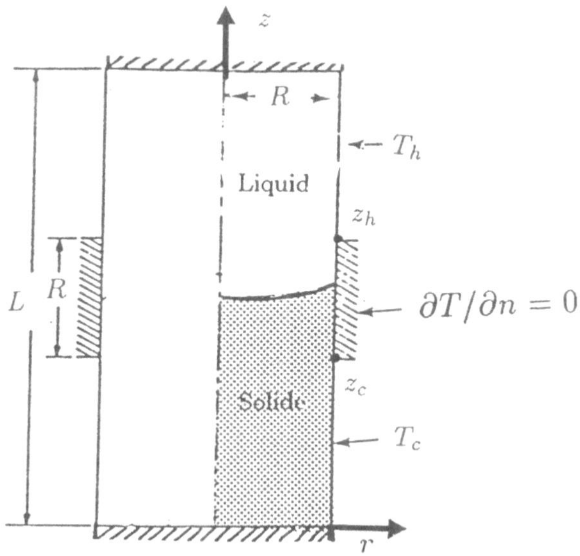

Figure 1. Geometry

A cylindrical ampoule with radius R and length L, contains melt and crystal. The ampoule is translated inside a furnace producing a hot zone, a cold zone and an intermediate insulating zone at a uniform velocity. The Normal Vertical Bridgman technique (NVB), the melt is located above the crystal solidifying. The top and the bottom of the ampoule are respectively at T_h and T_c ($T_h > T_c$). Melt and crystal variables are identified by the subscript l (liquid phase), s (solid phase) and sl ('mushy zone'). We consider the growth of a pure material with equal density, thermal conductivity and heat capacity in the two phases ($\rho_s = \rho_l, k_s = k_l, c_s = c_l$). The variables are dimensionless by using L, α/L, and $c\Delta T = c(T_h - T_c)$ (where α is the melt thermal diffusivity) as length, velocity, and enthalpy scales, respectively that is (where ~ represents the nondimensionless variable) :

$$(r, z) = (\tilde{r}/L, \tilde{z}/L), u = \tilde{u}L/\alpha, h = \tilde{h}/c\Delta\tilde{T} \tag{1}$$

r and z are the radial and the axial coordinates, $u(u_r, u_z)$ are radial and axial velocity components.
We consider the solidification occurs at $T = T_{sol}$, inside a 'mushy region', defined by :

$$I_{sl} = \{x(r, z, t)/T_{sol} - \varepsilon \leq T(x) \leq T_{sol} + \varepsilon\} \tag{2}$$

where ε is a small temperature difference (some % of T_{sol}), and I_{sl} corresponds to the location of the porous region, when the permeability vary between 0 (solid phase) and 1(liquide phase).
f_l is the mass fraction for the liquid phase, the saturation relation gives $f_s = 1 - f_l$. In the solid phase $h_s = T$ and in the liquid phase $h_l = T + L_f/c\Delta T$. The number $L_f/c\Delta T$ is the inverse of the Stefan number (Ste). The enthalpy of the system is defined as :

$$h = f_l h_l + (1 - f_l) h_s = T + f_l Ste^{-1} \tag{3}$$

This use of enthalpy variable avoids the need of the Stefan condition at the solid/liquid interface: $\nabla T_l - \nabla T_s = Ste^{-1} Pe n.e_z$. The Peclet number $Pe = U_g L/\alpha$ is the dimensionless pulling velocity.
The energy equation is written as:

$$\frac{dh}{dt} = \nabla.\nabla(T) - \nabla[u(1 - f_l)Ste^{-1})] \tag{4}$$

The velocity is defined as : $u = f_l u_l$.
For the momentum equation, we consider the mushy zone I_{sl} as a porous medium domain where the interaction between liquid and solid phase is modelled by :

$$F_{sl} = u \times K^{-1}(f_l) \tag{5}$$

where K is a function given by Kozeny-Carman : $K(f_l) = K_0 f_l^3/(1 - f_l)^2$, and the constant K_0 depends on the solidification microstructures. In the solid phase ($f_l = 0$), $K^{-1}(f_l)$ goes to infinity, and gives $u = (F_{sl}/K^{-1}(f_l)) \to 0$, so the velocity in the solid phase goes to zero. In the liquide phase ($f_l = 1$), we have $K^{-1}(1) = 0$ so $F_{sl} = 0$ and the motion is described by the standard Stokes momentum equation.
With these assumptions, the momentum equations are written :

$$\frac{du_r}{dt} = -\frac{\partial p}{\partial r} + Pr\nabla.\nabla u_r - \frac{Pr}{Da} u_r - Pr\frac{u_r}{r^2} \tag{6}$$

$$\frac{du_z}{dt} = -\frac{\partial p}{\partial z} + Pr\nabla.\nabla u_z - \frac{Pr}{Da} u_z + RaPrT \tag{7}$$

where $Pr = \nu/\alpha$, is the Prandtl number, $Ra = g\beta\Delta T L^3/\nu\alpha$, is the Rayleigh number, $Da(f_l) = K(f_l)/L^2$ is a local Darcy number, ν is the melt viscosity and β is the melt thermal expansion coefficient.
When a vertical magnetic field Be_z is added, the effect on the electrically conducting fluid is considered via the Lorentz body force (see assumptions in [16,17,18], $F_B = j \wedge Be_z$ where the induced current density j is derived from Ohm law $j = u \wedge Be_z$. Dimensionless $F_B = -Ha^2 Pr u_r e_r$. Then :

$$\frac{du_r}{dt} = -\frac{\partial p}{\partial r} + Pr\nabla.\nabla u_r - \frac{Pr}{Da}u_r - Pr\frac{u_r}{r^2} - Ha^2 Pru_r \tag{8}$$

$Ha = \sqrt{\sigma BL^2/\mu}$ is the Hartmann number, σ is electrical conductivity.
The thermal boundary condition are:
At $z = 0$ and $z = 1$: $\partial T(r,0)/\partial n = \partial T(r,1)/\partial n = 0$.
At $r = R/L$, the external temperature T_{ex} is imposed under the four of a three zones profile (cold $z < z_c$, hot $z > z_h$ and adiabatic ($z_c < z < z_h$)). $T_{ex} = 0$ for $z < z_c$, $T_{ex} = 1$ for $z > z_h$, and $\partial T_{ex}/\partial r = 0$ for $z_c < z < z_h$.
For the velocity the boundary conditions are :
$u_r(0,z) = u_r(1,z) = u_r(r,0) = (\partial u_z(0,z)/\partial r) = 0$, $u_z(r,0) = u_z(r,1) = -Pe$.

3. Numerical method and validations

The spatial discretisation uses a finite volume method. The convective terms are evaluated using a Quadratic Upwind Interpolation for Convective Kinematics scheme (QUICK). This scheme uses a quadratic interpolation between two upstream and one downstream neighbours in order to estimate the variable considered at any control-volume interface. The formulation of this interpolation depends on the sign of the convective flux on the interface. The QUICK scheme assumes a parabolic shape profile in the interpolation procedure. The tridiagonal linear systems are solved with a classical Tridiagonal Matrix Algorithm (TDMA). It is stable as long as the diagonal dominance of the matrix is ensured. It is also less sensitive to the consistency of boundary treatments than the second order Upwind scheme. The time discretisation uses a second order EULER scheme. The pressure velocity coupling is solved by using a PISO algorithm [13,4,20].

3.1. VALIDATION OF THE ONE PHASE SOLUTION

The comparisons are made in the case of a two dimensional square cavity with L of length submitted to filled with a liquid of $Pr = 0.01$ heated from below (IVB) that was studied elsewhere with a spectral method [2]. The lateral walls are submitted to a temperature profile which is relevant of Bridgman technique application: an isotherm hot region on $3L/4$ and an adiabatic region on $L/4$. The interface is planar and the results at $Ra = 2000$ with a 128×130 resolution correspond to a symmetric solution that compare well qualitatively and quantitatively with the spectral solution obtained with various resolutions, 33×33 and 65×65, and with one domain and two-domains decomposition (which improve the accuracy at the with junction between the isotherm and the insulating parts of the temperature profile). The results are given in table 1 and in figure 2. Above this Ra value, the solution changes into an asymmetric pattern which was first revealed in [2]. The present solution confirms qualitatively the bifurcation and the scenario is detailed by El Ganaoui [20].

TABLE 1. Comparison for Ra = 2000, Pr = 0.01, Pe = 0.01 and $Ste^{-1} = 0$ of ψ_{max} obtained for spectral and finite volumes approaches.

	$N_x \times N_z$	$h_{N_x} = x_N - x_{N-1}$	ψ_{min}	ψ_{max}
Spectral 1 domain	33 × 33	0.00481	−0.309	0.304
Spectral 1 domain	65 × 65	0.00120	−0.308	0.303
Spectral 2 domains	2 × (17 × 17)	0.0048(h_{N_z})	−0.264	0.270
Finite-volumes	128 × 130	0.00781	−0.264	0.264

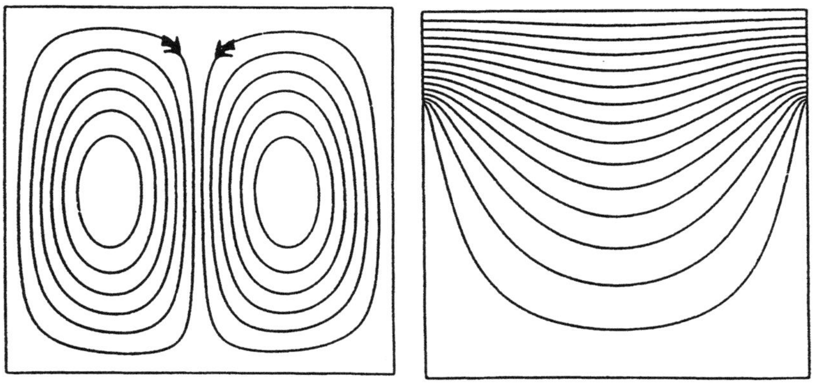

Figure 2. Streamlines and isotherm patterns for a two dimensional fluid phase. $Ra = 2000$, $Pe = 0.01$ and $Ste^{-1} = 0$ of ψ_{max}.

3.2. VALIDATION OF THE TWO PHASES SOLUTION

Solutions were performed in the NVB case, with aspect ratio of the ampoule $L/R = 4$ and with the adiabatic zone $z_h - z_c$ fixed to one radius. Solutions compare well with the ones obtained by using a two-domains adaptive mesh method [15]. The method used in Brown's paper is a Galerkin finite element method, the computations in the liquid and the solid phases are made separatelly and linked via a balance of conductive heat fluxes across the interface. Our enthalpy method is used with 'mushy zone' parameter $\epsilon = 0.05 = 10\% T_{sol}$. The results in the table 2, show that the enthalpy method combined with finite volume approximation is accurate and adapted to compute the front advancement with a fixed mesh approach.

TABLE 2. Present Finite Volume results compared with Brown's Finite Element results for $Pr = 10^{-2}$, $Pe = 10^{-2}$, $Ste^{-1} = 1$ and $10^3 \leq Ra \leq 10^7$.

Rayleigh number	ψ_{max}Brown	ψ_{max} present
10^3	0.00077	0.00078
10^4	0.00601	0.00596
10^5	0.03590	0.03394
10^6	0.12900	0.11912

4. Results

The following results concern steady regimes resulting from the interaction between buoyancy convection and magnetic field effect. The unsteady calculation illustrates the interaction between an oscillatory melt and the solidification interface. Axisymmetric model is used for symmetric solution and 2D model for asymmetric solution. Two models are considered an axisymmetric one and a two-dimensional cartesian one in order to investigate some asymmetric behaviours.

4.1. STEADY NVB REGIMES WITH MAGNETIC FIELD

The results obtained by increasing magnetic field are represented on figure 5, 7 and table 3. The flow pattern is crushed along the ampoule wall and damped by viscous and Lorentz force effect. The figure 3 represents the mutual influence of the parameters (Ra, Ha). We note that for each Ra, there is a level of Hartmann number which ensures the dumping of the intensitiy of convection. The hatched part of the figure 3, shows the domain of Hartmann number which is of practical interest. The crystal growther considers generally values of less than 5000 Gauss in these applications.

TABLE 3. Example of the effect of magnetic field $(0 \leq Ha \leq 10^4)$ on the convection in the melt $Pr = 10^{-2}$, $Pe = 10^{-2}$, $Ste^{-1} = 1$ and $Ra = 10^4$.

Hartmann number	0.0	10	$10^{3/2}$	10^2	10^4
ψ_{max}	0.005960	0.001817	0.000546	0.000517	0.000313

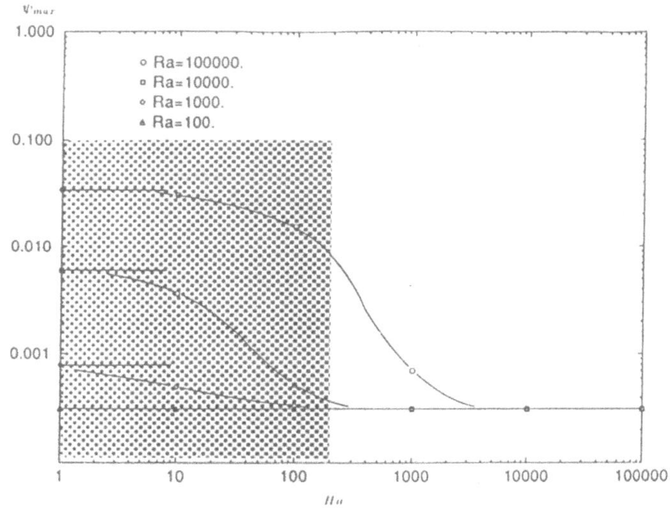

Figure 8. Influence for coupled gravity, magnetic field on the ψ_{max}

4.2. UNSTEADY AXISYMMETRIC MELT FLOW INTERACTING WITH THE SOLIDIFICATION INTERFACE

For IVB configuration, comparison with Brown's results performed in the case of steady state solution have shown a very good agreement qualitatively and quantitatively [20]. The magnitudes of streamfunction are 4.52×10^{-2} and 4.35×10^{-2} respectively for the present solution and for the reference solution in the stiffer case ($Ra = 10^5$). Interface remains quasi-planar in all the steady state configurations. The simulations were carried out above the threshold of oscillatory convection. The first transition at $Ra = 10^6$ corresponds to a periodic flow in the melt which does not bring severe modification to the interface shape. It remains quasi-steady but a slightly increased convex shape. Ra is increased up to $Ra = 10^7$, the melt flow structure becomes very complex and the motion is aperiodic. The driving cell moves progressively with increasing Ra to the bottom of the crucible, with reducing size and increasing strength. The neigbour melt is submitted to the oscillating cell which drives a succession of moving cells which rise up to the top of the melt and affect the mushy region and the interface curvature. Secondary cells are shown at four different stages of the process, to steeply modify the shape of the interface fig. 9, following the time-dependent patterns.

4.3. ASYMMETRIC MELT CONVECTION AND MELT INTERFACE CHANGE

We have represented in figure 4.3 the axisymmetric model studied in the previous section together with twodimensional model (in cartesien coordinate and corresponding to a diameter section of the Bridgman ampoule) that we consider in this section. A twodimensional model with a planar interface and without solid phase is also considered in order to qualify the solution with respect to time dependent behaviour (and to spectral solutions existing in this configuration). This basic cavity is square $L/2$, and the temperature gradient is $\Delta T/2$. In the case of solid liquid model, the lenght is L and the temperature gradient is ΔT. In the axisymmetric configuration the height is still L but the width is $L/4$. The relevant Rayleigh number are deduced from model $M_{\alpha\beta}$ with characteristic length scale $(\beta L/2)$ and temperature difference $(\alpha \Delta T/2)$ as:

$$Ra_{\alpha\beta} = (\alpha\beta^3) Ra_{11}$$

With this notation Ra_{22} is the Rayleigh number used for the axisymmetrical case. Ra_{21} for the 2D-2 phase case and Ra_{11} for the 2D-1 phase case.

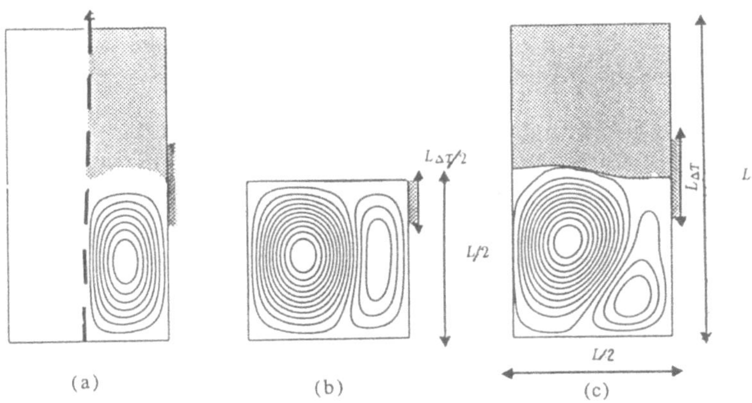

Figure 4. Models configuration used in insteady study. (a) Axisymmetric model (M_{22}), (b) Basic 2D fluid phase model (M_{11}), (c) 2D-2 phases model (M_{21}).

The Ra_{11} is used as a common reference of Rayleigh number for the 3 models. The computation exhibits steady symmetrical solutions up to about $Ra_{11} = 5000$, transition to asymmetrical solutions at $10000 < Ra_{11} < 15000$. The asymmetrical instability is shown to become time-dependent at $25000 < Ra_{11} < 30000$.

The ability of the finite-volume solution to predict such behaviours is assessed in the case of the simplified one-fluid-phase model for which we obtained highly accurate spectral solutions [2]. The finite volume solution exhibits the same flow features : steady symmetrical - asymmetrical - time dependent. The two associate critical Ra ranges are of the same order of magnitude. The first asymmetrical transition occurs at $5000 < Ra_{11} < 7000$ and the time-dependent solution arises at $30000 < Ra_{11} < 40000$. Respectively, the spectral solution gives $2500 < Ra_{11} < 3500$ and $10000 < Ra_{11} < 20000$. The comparisons are shown in the following table 4.

TABLE 4 : Characteristic Rayleigh number for the symmetric, asymmetric and time dependent regime.

Régimes	Finite-volumes			Spectral
	Axis	2 Phases	1 Phase	1 Phase
	6250	5000	3000	1000
Steady / Unsteady		$Sym \rightarrow Asym$ 10000 – 15000	$Sym \rightarrow Asym$ 5000 – 7000	$Sym \rightarrow Asym$ 2500 – 3000
		$Asym$ 25000	$Asym$ 30000	$Asym$ 10000
	62500	30000	40000	20000

4.4. SOME INSIGHTS ON CRYSTAL CONSTITUTION

Crystal constitution process is simulated by considering the successive interfaces during a characteristic growth time scale and for the two axisymmetrical and two-dimensional models. The crystal characterised by its shapes is represented in (figs. 9 and 10) with the complete section of the crystal. The figure exhibits irregularities wich shows similarities with the ones exhibited in experimental result [19]. In this case the occurrence of an asymmetric mode of convection is suggested and the two-dimensional solution exhibits very similar features.

5. Conclusion

The finite volume enthalpy approach has been validated to represent correctely the flow pattern and the heat transfer in crystal growth configuration. The result obtained with an axial magnetic field show that the gravitational convection can be significantly damped using this technique. The interaction between unsteady melt and the solidification interface is studied and we can describe for aperiodic flow at high Rayleigh number ($Ra = 10^7$) strong effects on the shape of the interface. Different models are used to investigate the various symmetric, asymmetric, steady and time-dependent flow solutions. The transitions between regimes have been analysed and we have determined realistic sections of solidified crystals.

6. Acknowldgments

We would like to thank J. Ouazzani from Arcofluid for helpful comments and discussions. We thank also the CNRS for support on CRAY C98 of IDRIS-CNRS Center in Orsay. Supports from the French German CNRS-MENESR-DFG Program 'Numerical Simulation of Flow' and from CNES are also acknowledged.

References

1. G.H. Meyer. Multidimensional Stefan Problem. *J. Num. Anal* 10(3), 522, 1973.
2. P. Larroudé, J. Ouazzani, J.I.D. Alexander and P. Bontoux, Symmetry breaking flow transitions and oscillatory flows in a 2D directional solidification model, European Journal of Mechanics/B, 13(3), pages 353-381, 1994.
3. D. Morvan, M. El Ganaoui, and P. Bontoux. Application of porous medium approach for the resolution of solidification problems. Bulletin ERCOFTAC pages 49-52, 1996
4. D. Morvan, M. El Ganaoui, and P. Bontoux. Numerical simulation of 2D crystal growth problems in a vertical Bridgman-Stockbarger furnace, latent heat effects and crystal-melt in interface morphology. *.Int.J. Heat Mass Transfer* in print, 1998.
5. El Ganaoui M., Bontoux P. et Morvan D. Une méthode de capture du front de changement de phase pour un problème de solidification dirigée C. R. Acad. Sciences Paris, en préparation 1998.
6. El Ganaoui M., Morvan D. and Bontoux P. Numerical simulation of solidification problems in material processing techniques. In 3rd *International Conference Phenomena in Magnetohydrodynamic Electroconducting Flows*, Septembre 1997 Aussois France, pp 343–349, 1997
7. D. Morvan, F. D. Cipriani and Ph. Bournot. Convection thermocapillaire au cours d'une operation de refusion-solidification par laser. *Int. J. Heat Mass Transfer*, 37(14): 1973–1983, 1994.
8. W.D. Bennon and F.P. Incropera. A continuum model for momentum, heat and species transport in binary solid-liquid phase change systems: 1 model formulation. *Int. J. Heat Mass Transfer*, 30(10):2161–2170, 1987. See also 30(10):2171–2187, 1987.
9. V. Voller and M. Cross. Accurate solutions of moving boundary problems using the enthalpy method. *Int. J. Heat Mass Transfer*, 24:545–556, 1980.
10. I. J. Crank. *Free and Moving Boundary Problems*. Clarendon Press, Oxford, U. K., 1984.
11. F.P. Incropera, H.H. Engel, and W.D. Bennon. Numerical analysis of binary solid–liquid phase change with buoyancy and surface tension driven convection. *Numerical Heat Transfer, Part A*, 16:407–427, 1989.
12. C. Beckermann and R. Viskanta. Double–diffusive convection during dendritic solidification of a binary mixture. *Physico–Chemical Hydrodynamics*, 10(2):195–213, 1988.
13. D.S. Jang, R. Jetli, and S. Acharya. Comparison of the PISO, SIMPLER and SIMPLEC algoritms for the treatment of the pressure-velocity coupling in steady flow problems. *Numerical Heat Transfer*, 10:209–228, 1986.
14. D. T. J. Hurle, Crystal Pulling from the Melt, Springer-Verlag Berlin Heidelberg (1993).
15. J. Chang and R.A. Brown. Radial segregation induced by natural convection and melt–solid interface shape in vertical bridgman growth. *J. Crystal Growth*, 63:343–

364, 1983.
16. W.E. Langlois , Ann. Rev. Fluid. Mech., 17,191-215, (1985).
17. I. V. Friazinov, M. P. Marchenko, O. S. Mazhorova. Transient simulation of crystal growth by Bridgman technique under normal and low gravitation. In *Proceeding of International Aerospace Congress (IAC 94)* August 1994, Moscow.
18. L. N. Hjellming, J.S. Walker, J. Fluid Mech. 164, 237,(1986).
19. Jamgotchian Recherches sur l'influence de la convection naturelle sur la morphologie interfaciale lors de la solidification dirigée d'alliages binaires*Phd thesis* U Aix Marseille I, 1987.
20. M. El Ganaoui Modélisation de la convection instationnaire en présence d'un front de solidification déformable. *Phd Thesis* Université de la Méditerranée Aix-Marseille II Octobre 1997.

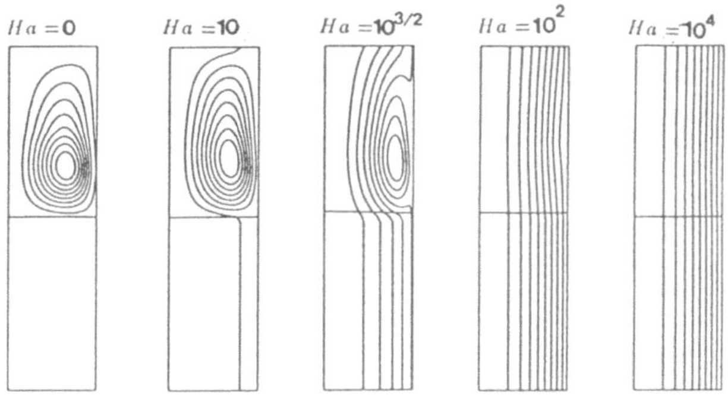

Figure 5. Magnetic field effect on the convection for $Pr = 10^{-2}$, $Pe = 10^{-2}$, $Ste^{-1} = 1$, $Ra = 10^4$ and $0 \leq Ha \leq 10^4$.

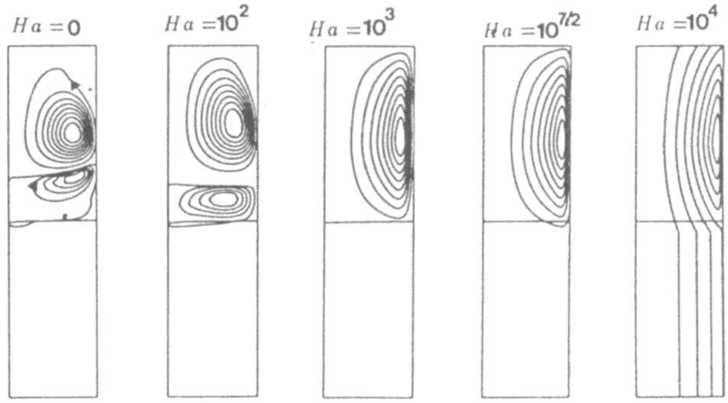

Figure 6. Magnetic field effect on the convection for $Pr = 10^{-2}$, $Pe = 10^{-2}$, $Ste^{-1} = 1$, $Ra = 10^7$ and $0 \leq Ha \leq 10^4$.

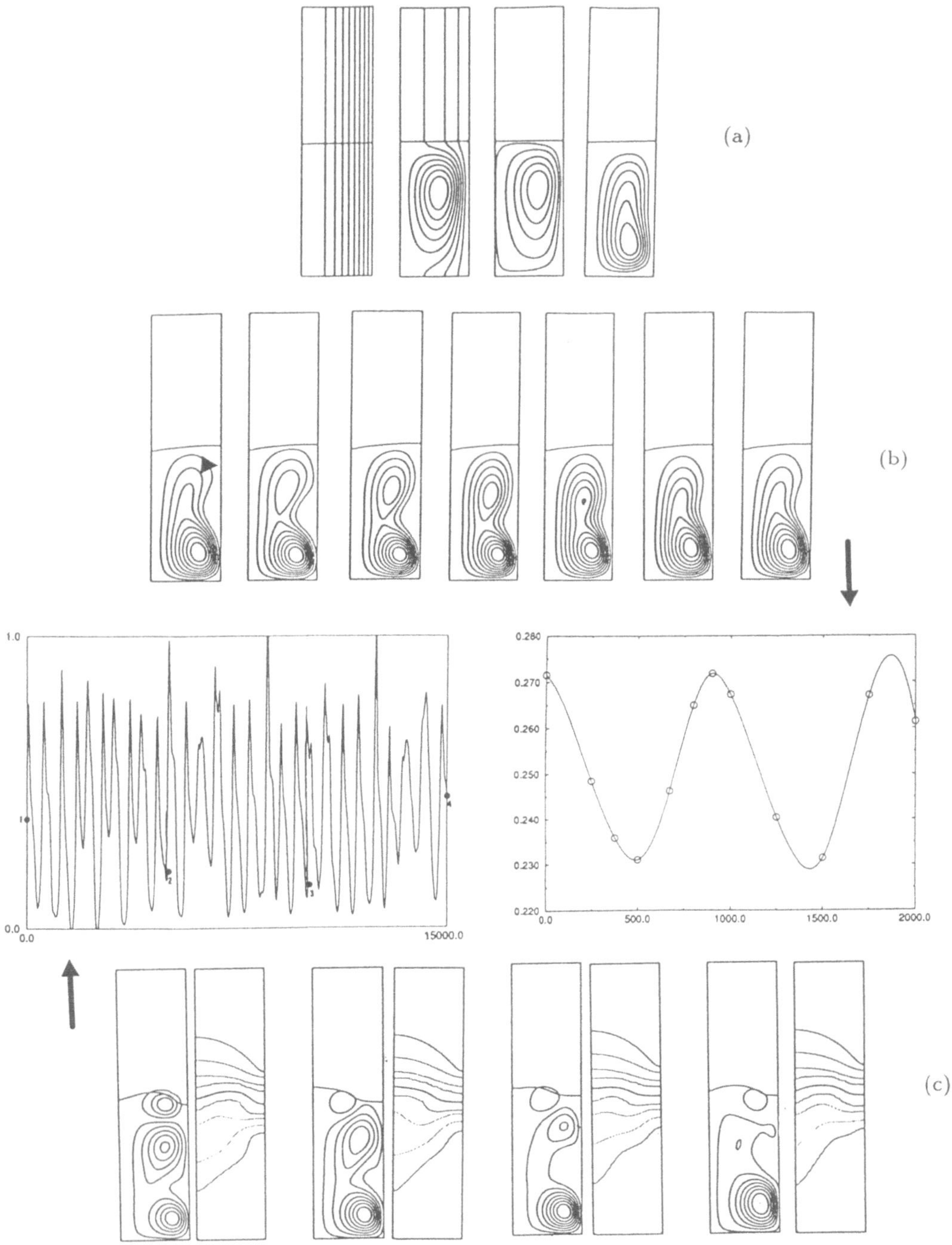

Figure 7. Steady and time-dependent axisymmetric regimes. Streamlines patterns: (a) Steady $Ra_{22} = 0.0, 10^3$, 10^4 and 10^5, (b) Unsteady periodic for $Ra_{22} = 10^6$ with signal of $\psi_{10,10}(t)$, (c) Unsteady aperiodic for $Ra_{22} = 10^7$ with signal of $\psi_{10,10}(t)$. $Pr = 10^{-2}$, $Pe = 10^{-2}$, $Ste^{-1} = 1$, $Ha = 0.0$.

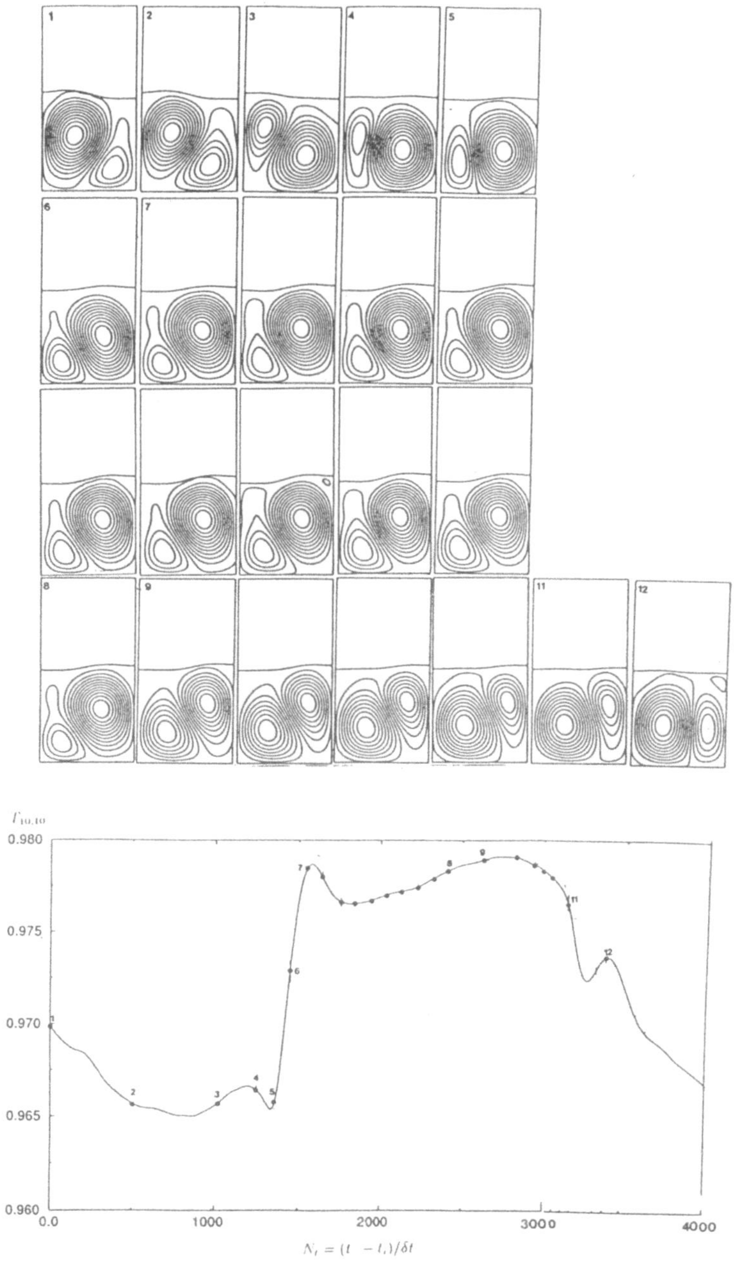

Figure 8. Time-dependent asymmetric régime. Streamlines patterns. $Pe = 5 \times 10^{-3}$, $Ste^{-1} = 0$, $Ha = 0.0$, $Ra_{21} = 80000$.

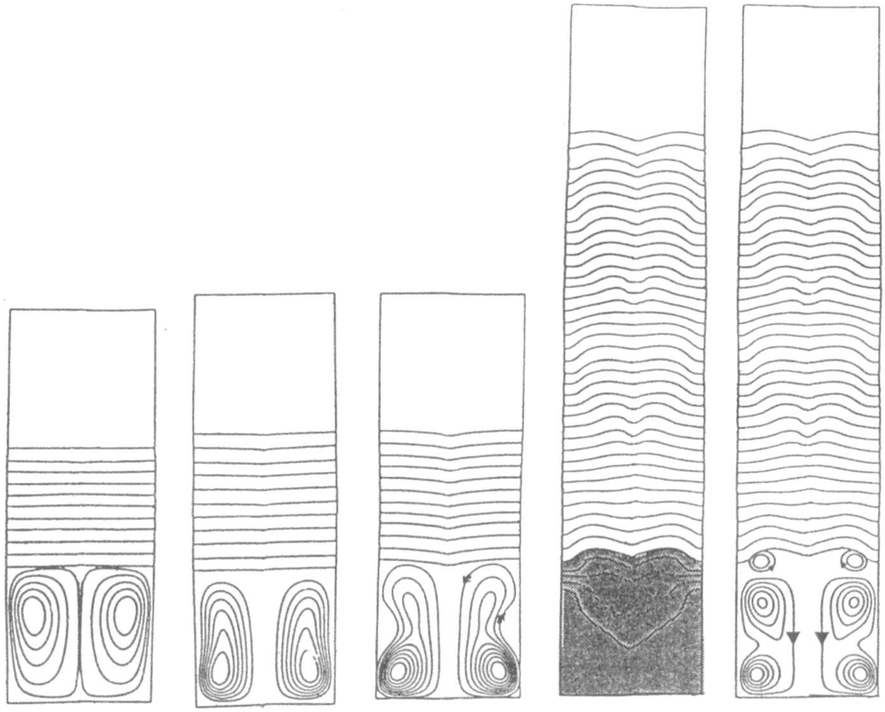

Figure 9. Crystal constitution and streamlines patterns for steady and time-dependent axisymmetric regimes: $Pr = 10^{-2}$, $Pe = 10^{-2}$, $Ste^{-1} = 1$, $Ha = 0$ and $10^4 \leq Ra_{22} \leq 10^7$.

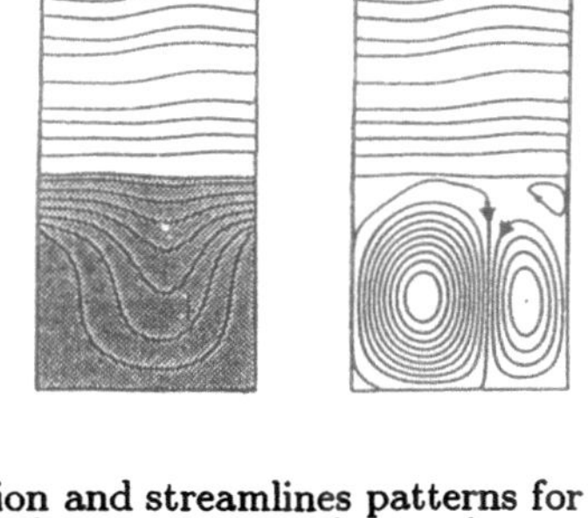

Figure 10. Crystal constitution and streamlines patterns for steady and time-dependent asymmetric regimes: $Pr = 10^{-2}$, $Pe = 5 \times 10^{-3}$, $Ste^{-1} = 0$, $Ha = 0$ and $Ra_{21} = 80000$.

EFFECT OF A STEADY MAGNETIC FIELD AND IMPOSED ROTATION OF VESSEL ON HEAT AND MASS TRANSFER IN SWIRLING RECIRCULATING FLOWS

A Model of Hydrodynamically Guided Solidification

I. GRANTS, Y. M. GELFGAT
Institute of Physics, University of Latvia.
32 Miera, Salaspils LV-2169, LATVIA
ig@sal.lv, yglf@sal.lv

A simplified theoretical model for the solidification interface shape prediction is introduced and tested. We linearised a coupled hydrodynamic-solidification problem about the state with a flat interface. In such a way we split the problem into a hydrodynamic part with a flat solid-liquid front and a solidification part with a calculated heat flux from the liquid phase. The method allows obvious conclusions on optimum heat conditions near the solidification interface providing its flatness and maximum pulling velocity at the same time. Comparison to the results by FLUENT package showed that the method provides a reasonable accuracy even for a noticeably deformed interface shape. Another part of the contribution deals with an influence of artificial swirling-recirculating flow due to a rotating magnetic field with an optional superimposed steady axial field or rotation of crucible. Estimates of necessary forcing are given and illustrated by numerical simulation. The limits of possibilities to control heat regime (consequently, the interface shape and pulling velocity) are discussed and illustrated.

1. Introduction

The current paper concerns in an artificial flow as a means to optimise the solidification interface shape during crystal growth. Traditionally solidification problems are solved numerically on adapting meshes or in enthalpy-porosity formulation on a fixed mesh (Voller & Cross 1980, Bennon & Incropera 1987, Voller & Prakash 1987). Both methods are widely used including commercial codes. However, the numerical simulation of full problem takes much time and gives less understanding of the physical process. The shape of solidification front

A. Alemany et al. (eds.),
Transfer Phenomena in Magnetohydrodynamic and Electroconducting Flows, 311–326.

has a significant influence on the quality of produced crystals. There are optimum growth conditions with a flat growth interface and a strong and stable motion of melt. These conditions are expected to ens⁻re a minimum concentration of dislocations. In our paper we split a hydrodynamic problem, in order to obtain an obvious model of guided solidification. Such model is expected to provide more understanding of the process and, hence, possibilities of its control. Besides, it can be used in estimates as well as for testing of numerical codes. The model is originally intended for the vertical gradient freeze method, but it can be easily generalised.

The shape of solid-liquid interface is closely linked to a heat flux between both phases. If crystals are small, then the heat transfer is diffusive and completely determined by thermal boundary conditions. In recent technologies, however, the size of grown crystals is more and more enlarged. If a properly defined Peclet number is more than unity, then the melt motion can significantly influence the heat transfer near the solid-liquid front and, hence, its shape, too. Thus, there are two tools to control the interface shape: thermal boundary conditions and the artificial flow of melt controlling a convective heat flux. As mentioned, there is an optimum solidification interface that simply has a flat shape. Suppose, both controlling tools allow to obtain an almost flat quasi-steady interface. In order to find corresponding conditions, we can linearise a full problem about this optimum state. It implies neglecting of the influence of actual shape of almost flat interface on the melt motion and heat transfer within it. Having solved a hydrodynamic-thermal problem with an isotermic flat interface, we receive heat flux from the liquid phase. Together with a latent heat flux due to pulling and heat transfer in the solid phase it determines the shape of actual quasi-steady almost flat front. The first main task of our work was to introduce a link between these variables.

Another task concerns a non-isotermic flow of melt. Natural motion occurs due to buoyancy. An artificial swirling-recirculating flow of conducting liquid with a number of favourable properties (Priede 1993) can be induced by a rotating magnetic field (Moffat 1965, Davidson 1992). Superimposed actions such as a steady axial magnetic field (Grants, Priede & Gelfgat 1996, Grants 1998, ch. 3) or imposed rotation of crucible (Priede 1993, 1994, Grants 1998, ch. 4) significantly widen possibilities to control the induced base flow and its stability. A first question arises on necessary strength of imposed forcing to suppress buoyancy, i.e., to make the flow completely artificial. Another question is the influence of considered artificial flow on a heat flux through the solidification interface and its shape.

The obtained method reveals a close relationship among heat flux from the melt, pulling velocity and shape of quasi-steady solidification interface. It follows from this relationship that optimum growth conditions are provided by a minimum heat flux with a uniform distribution, especially in the middle part of new solidified crystal. Our numerical simulation showed that the buoyancy increased this flux in comparison to a purely diffusive regime. A swirling-recirculating flow due to a rotating magnetic field (RMF) of sufficient strength suppresses the buoyancy and changes the direction of meridional flow. As a result, the heat flux can be significantly reduced. An imposed steady magnetic field (SMF) or rotation of crucible promotes a transition to a swirl dominated flow and provides additional options to improve the thermal conditions of growth.

Section 2 introduces the model and governing equations. A link is established for the shape of quasi-steady solidification interface, heat flux conditions on it and pulling velocity in §3. Estimates of necessary forcing are made in §4. Section 5 discusses the influence of artificial flow on the heat transfer. Section 6 proceeds with the main results of numerical simulation. The summary is given in §7.

2. Model and equations

Let us consider a solidification problem in a cylindrical vessel schematically depicted in figure 1. Basic assumptions of our model are:

(i) Melt motion considerably influences the solidification interface $\zeta^*(r)$. In terms of dimensionless criteria it requires the Peclet number more than unity.

(ii) A guided artificial flow allows to transform the interface into almost flat.

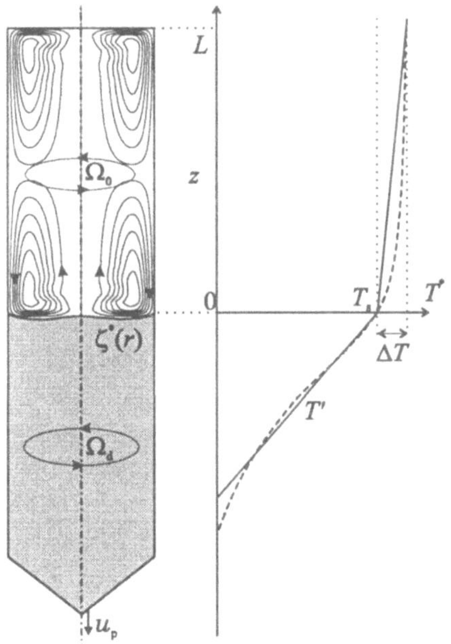

Figure 1. Sketch of the model.

Besides, the solidification interface was considered to be isotermic and quasi-steady. A non-isotermic interface may take place when a binary system solidifies. We neglected a possible morphological instability and consequent dendrite growth with respect to a strong azimuthal flow due to a rotating magnetic field considered for the generation of artificial flow (Priede 1993, 1994, Grants *et al.* 1996, Grants 1998, ch. 1–4). A condition of quasi-steady interface shape requires that its change in time is much slower than pulling velocity $\partial\zeta^*/\partial r << u_p$. This condition is consistent with our task to keep the front as flat as possible. Under these assumptions a full solidification problem can be easily split. First, let us find a heat flux through a flat isotermic interface $\zeta^*(r)=0$ solving a corresponding hydrodynamic-thermal problem in the melt. After that, let us find the shape of actual front by solving the thermal problem in the solid phase.

We assumed fixed temperature profile on crucible surface as schematically shown in figure 1. Besides, we substituted an actual temperature distribution by a stepwise linear function that is characterised by a temperature drop DT in the liquid phase and a temperature gradient T′ on the crystal surface. Of course, such assumption seems inappropriate for a modelling of real process. However, our task was mostly the developing and testing of the method itself rather than the investigation of specified technological process. Thus, to avoid less important details, we restricted ourselves to a minimum set of parameters.

The assumptions for heat transfer in the liquid phase included axial symmetry, laminar motion, lack of volumetric heat sources and Bussinesq approximation for buoyancy. Axial symmetry is ensured by a strong swirling flow. The motion is stabilised and, hence, laminar due to a superimposed steady axial magnetic field or rotation of crucible (Grants 1998, ch. 3–4). A rotating magnetic field is usually generated by AC of industrial frequency that can not rise a considerable heating. Under these assumptions, the Navier–Stokes equations were supplemented by a buoyant body force $\mathbf{f}_b$ that is proportional to a temperature $\mathbf{f}_b=\rho\beta g T^*\mathbf{e}_z$, where $\mathbf{e}_z$ is unity vector in the axial direction, T^* is the temperature, ρ is the density and β is the coefficient of thermal expansion of the melt, but g is the gravity acceleration. According to the assumptions, the temperature field is described by an equation in dimensionless form

$$\frac{\partial T}{\partial t}+(\mathbf{v}\nabla)T=\frac{1}{\mathrm{Pr}}\nabla^2 T, \tag{1}$$

with the boundary conditions

$$T|_{z=1}=1, \qquad T|_{r=R}=z, \qquad T|_{z=0}=0, \tag{2}$$

where $T=(T^*-T_s)/\Delta T$ is the dimensionless temperature, T_s is the solidification temperature, $\mathbf{v}$ is the velocity of melt motion, Pr is the Prandtl number defined as $\mathrm{Pr}=\nu/a$, ν is the kinematic viscosity, a is the thermal diffusivity, $R=R_0/L$ is a dimensionless crucible radius or aspect ratio of liquid zone. The characteristic height of liquid zone L is used as a linear lenghtscale since usually it is a determining scale in a rotating flow (Greenspan 1968).

Having solved that problem, we found an axial heat flux on isotermic flat bottom $q^*(r)=\lambda(\Delta T/L)\partial T/\partial z$ that substituted the actual one on an almost flat front in our model (λ is a thermal conductivity of liquid). Write a heat balance on the solidification interface

$$\left.\frac{\partial T}{\partial z}\right|_{z=\zeta(r)} = \mathrm{St} + q(r), \tag{3}$$

where T is the dimensionless crystal temperature defined as $T = (T^*-T_s)/(T'R_0)$, St is the Stenton number $\mathrm{St} = u_p\rho_s\Lambda/(\lambda_s T')$, ρ_s and λ_s are density and thermal conductivity of crystal, but Λ is the latent heat of solidification. We used another lenghtscale that was the radius of crystal R_0 for a thermal problem in the solid phase since here it was more convenient than the previous one L used in the hydrodynamic part of the problem. Both parts are linked by the heat flux written in dimensionless form $q(r) = q^*(r)/(\lambda_s T')$. We considered a slow pulling with a Peclet number $\mathrm{Pe_p} = u_p R_0/a_s << 1$ and constant thermal conductivity. Actual conditions in technological processes may turn out different. However, they play a secondary role and may be easily incorporated into the present model. Under these assumptions the temperature distribution T in crystal is described simply by the Laplace equation

$$\nabla^2 T = 0, \tag{4}$$

with Dirichlet boundary conditions on the crucible wall

$$T|_S = z. \tag{5}$$

3. Stefan problem with a known heat flux from the liquid phase

There are two boundary conditions on the quasi-steady solidification interface of unknown shape. Obviously, an isothermal solidification front will move, to finally adjust the second boundary condition for the heat flux, too. Suppose, we have found a heat flux q^* from the liquid phase to a flat isotermic bottom that substitutes an almost flat front. Remember that our practical interest was exactly to find the conditions for a flat solidification interface. Of course, even under optimal conditions, the heat flux continuity on the assumed flat front may be unsatisfied. The disbalance in the heat flux continuity will be a measure of shift of actual solidification front from the assumed flat one. This section concerns in a link between these variables.

The shape of quasi-steady solidification interface is described by a solution of (3–5). Let us introduce a temperature correction due to an actual deformed interface as follow

$$Y(r,z)=T(r,z)-T_0(r,z), \qquad (z<\zeta(r)) \tag{6}$$

where $T_0(r,z)$ is the temperature field in crystal with a flat front. From (4–5) it follows that $T_0(r,z)=z$ in our model. Hence, the heat flux continuity (3) on an almost flat interface takes the form of

$$\left.\frac{\partial Y}{\partial z}\right|_{z=\zeta(r)} = \mathrm{St}-1+q(r), \tag{7}$$

The correction $Y(r,z)$ satisfies the Laplace equation and takes zero value on crystal surface with Dirichlet boundary conditions used in our model. Definition (6) establishes a relationship between the correction $Y(r,z)$ and corresponding profile of front $\zeta(r)$

$$Y(r,z)|_{z=\zeta(r)} = T(r,z)|_{z=\zeta(r)} - T_0(r,z)|_{z=\zeta(r)}\,. \tag{8}$$

The solidification interface $z=\zeta(r)$ is isotermic, so the temperature $T(r,z)|_{z=\zeta(r)}=0$ takes zero value on it, but $T_0(r,z)|_{z=\zeta(r)}=\zeta(r)$ in case of the assumed linear temperature profile and slow pulling. Thus, the profile $\zeta(r)$ to be found coincides with a value of temperature correction $Y(r,z)$ on surface defined by this profile itself

$$\zeta(r)=-Y(r,z)|_{z=\zeta(r)}\,. \tag{9}$$

The problem is still defined by the boundary condition on unknown surface $\zeta(r)$. However, it can be easily found iteratively, e.g.,

$$\left.\frac{\partial Y^{n+1}}{\partial z}\right|_{z=\zeta^n(r)} = \mathrm{St}-1+q(r), \quad \zeta^{n+1}(r)=Y^{n+1}\Big|_{z=\zeta^n(r)}, \quad \zeta^0(r)=0. \tag{10}$$

The first approximation $\zeta^1(r)$ is expected to provide a reasonable accuracy for an almost flat solidification interface. It is defined simply by a solution of Laplace equation with homogenous Dirihlet conditions on the crucible wall and Neuman condition (7) on a flat front $z=0$.

Now we can derive some rather evident conclusions, that define certain optimal conditions for the heat flux. A flat quasi-steady front corresponds to zero right hand expression of (7). Hence, fiom the point of view of optimum solidification interface and maximum pulling velocity, a minimal heat flux $q(r)$ is required. The heat flux through the interface may be reduced either by thermal boundary conditions (such as a temperature drop ΔT in melt or gradient T' on the crystal surface) or by a guided motion of the melt. Second, the solidification interface is more sensitive to the heat flux in the central part of new solidified crystal. If this flux has a uniform distribution $q(r)=q_0$ and a magnitude below unity $q_0<1$ there, then an almost flat front can be achieved by an appropriate pulling velocity that follows from (7) $\mathrm{St}=1-q_0$.

4. Estimates of buoyancy suppression

Let us consider an artificial swirling flow of the melt driven by azimuthal body force due to a rotating magnetic field in a rotating vessel. Different rotation of the core region that appears in such flow (Priede 1993, 1994) leads to a secondary recirculation. At a certain threshold of magnetic forcing and imposed rotation of vessel this flow is expected to suppress the natural buoyant one. In order to estimate this limit, let us consider a circulation of centrifugal and buoyant forces over a closed loop *ABCDA* in figure 2. Horizontal parts *AB* and *DC* give no contribution to the circulation of a buoyant force $\mathbf{f}_b=\rho\beta g T^* \mathbf{e}_z$, since it is purely axial

$$\Phi_b = \oint_{ABCDA} \mathbf{f}_b d\mathbf{l} = \rho\beta g\left(\int_B^C T(r,z)dz - \int_A^D T(r,z)dz\right) = \rho\beta g O(\Delta TL). \tag{11}$$

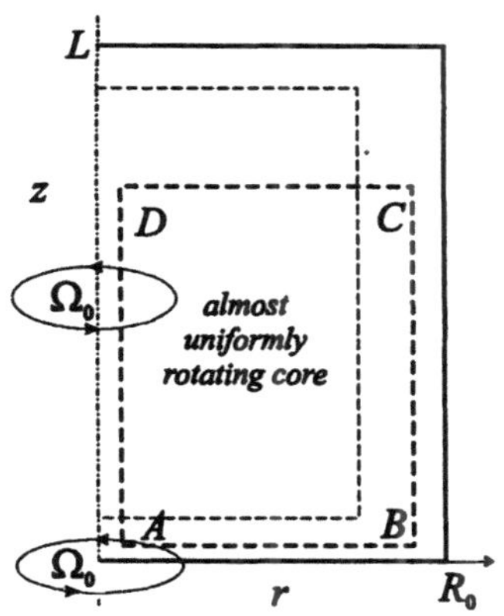

Figure 2. Sketch of swirl dominated flow.

A centrifugal force is $\mathbf{f}_c=\rho\Omega^2 r\mathbf{e}_r$ (Ω is the angular velocity). Let us consider a completely artificial (swirl dominated) flow. Then the circulation of $\mathbf{f}_c$ can be estimated as follow

$$\Phi_c = \oint_{ABCDA} \mathbf{f}_c d\mathbf{l} = \rho\left(\int_A^B \Omega^2(r,z) r dr - \int_D^C \Omega^2(r,z) r dr\right) \approx \frac{\rho R_0^2}{2}\left(\Omega_d^2 - \Omega_0^2\right), \tag{12}$$

where Ω_d and Ω_0 are the angular velocities of the crucible and the core, respectively. Suppose, the buoyancy is suppressed, if $\Phi_c >> \Phi_b$. Then, the threshold forcing can be estimated comparing (11) and (12).

First, consider a sole rotating magnetic field driven flow $\Omega_d=0$. Grants (1997) numerically found that the core angular velocity (for moderate Reynolds numbers $\mathrm{Re} \leq O(10^4)$) is approximately $\Omega_0 \approx \nu / L^2 \mathrm{T}^{2/3} R^{1/2}$, where $R=R_0/L$ is the aspect ratio of liquid zone and T is the Taylor number of forcing proportional to a driving magnetic torque $\mathrm{T}=\dfrac{ML^4}{0.5mR_0^2\nu^2}=\dfrac{M}{M_0}\dfrac{\sigma\omega_0 B_0^2 L^4}{2\rho\nu^2}$. A factor M/M_0 expresses the ratio of actual magnetic driving torque M in a truncated cylinder to the idealised one M_0 acting on a corresponding piece of infinite cylinder; m is the mass and σ is the conductivity of enclosed liquid; ω_0 is the angular frequency and B_0 is the induction of rotating field. Substituting in (12) and comparing to (11) yields the following estimate of strong forcing:

$$\mathrm{T} \sim (\mathrm{Gr} R^{-3})^{3/4}. \tag{13}$$

The Grashoff number Gr is defined here as $\mathrm{Gr}=\dfrac{\beta g \Delta T L^3}{\nu^2}$. Estimate (13) differs from the one given by Fischer *et al.* (1997). They simply compared $O(\mathrm{T})$ and $O(\mathrm{Gr})$. One should notice that both these parameters characterise the body forces that act in perpendicular directions, so such straightforward comparison hardly justifies itself. On the other hand, results of such comparison agreed with the experimental and numerical results on the stabilising action of rotating field. In our case, however, we are interested mostly in the effect on heat flux and, thus, characteristics of base flow. To verify (13), we performed a numerical simulation with a forcing rate ranging about the estimated threshold value. The results for different aspect ratios are depicted in figure 5.

We restricted our study of action of superimposed SMF to a "small" induction. From practical point of view the imposed steady magnetic field is expected to be "hardly strong" that is with about a minimum strength providing a stable flow. The angular velocity of such flow is of order of RMF alone driven flow. Hence, the above estimate (13) is sufficient for such cases. In case of a strong steady field, a considerable angular velocity drop occurs along the height of $O(\mathrm{Ha}^{-1/2})$ side layer. Besides, this layer may have its own velocity scale of a higher order of magnitude. As a result, an axial pressure gradient appears driving the meridional flow that vanishes slower in a strong steady field (Grants *et al.* 1996, Grants 1993, ch. 3). Hence, estimate (12) is not suitable any more.

Now, consider an almost rigidly rotating flow ($\Omega_0 \approx \Omega_d$). Then the circulation of centrifugal forces (12) can be estimated as

$$\Phi_c \approx \frac{\rho R_0^2}{2}\left(\Omega_d^2 - \Omega_0^2\right) \approx \rho R_0^2 \Omega_d \Delta\Omega . \tag{14}$$

The differential rotation $\Delta\Omega$ of such flow is (Priede 1993, Grants 1998, ch. 4)

$$\Delta\Omega \approx \nu / L^2 \mathrm{T} |\mathrm{E}|^{1/2}, \tag{15}$$

where $\mathrm{E}=\nu/(\Omega_d L^2)$ is the Ekman number of crucible rotation. Then the comparison of (11) and (14) yields:

$$\mathrm{T}|\mathrm{E}|^{-1/2} \sim \mathrm{Gr} R^{-2}. \tag{16}$$

The condition of almost rigidly rotating flow requires $|\mathrm{E}|^{-1} \gg \mathrm{T}^{2/3}$ (Priede 1993).

Notice that expression (11) overestimates the contribution of buoyancy and it is valid only for large Peclet numbers, when the recirculation significantly deforms a diffusive temperature field. The circulation of buoyant forces (11) vanishes as the Peclet number tends to zero in our model, since no radial temperature gradient occurs according to boundary conditions (2).

5. Heat transfer

In the previous section we estimated a limit of necessary forcing to suppress a buoyancy driven flow. Now let us consider heat transfer in swirl dominated flow as a limiting case of non-isotermic motion. An azimuthal body force due to a rotating magnetic field induces a swirling flow that provides axial symmetry of time-averaged characteristics. However, the main role belongs to an induced secondary recirculation that influences the heat transport in the normal direction to the solidification interface.

Flows of all three types considered in the current paper have a single vortex structure as schematically depicted in figure 3. From the point of view of heat transfer through the solid-liquid interface this vortex can be characterised by two parameters the core axial velocity v_z and the location of vortex eye r^*. We see in §3 that the solidification interface is more sensitive to the heat flux in its central part. The direction of core axial velocity determines whether the heat flux is reduced or increased here in comparison to a purely diffusive regime. The magnitude of core axial velocity H_0 determines the role of convection. Peclet number $\mathrm{Pe}=H_0 L/a$ should be larger than unity according to our basic assumption on a hydrodynamically guided solidification. The radial location of vortex eye determines the place where the influence of convection on heat transfer is inverted. Thus, it marks out the place, where a significant non-uniformity of convection dominated heat flux is expected. From the practical point of view the vortex eye should be pushed towards the side wall as far as possible, since exactly such configuration could increase the size of flat part of the heat flux profile and, hence, the solidification interface.

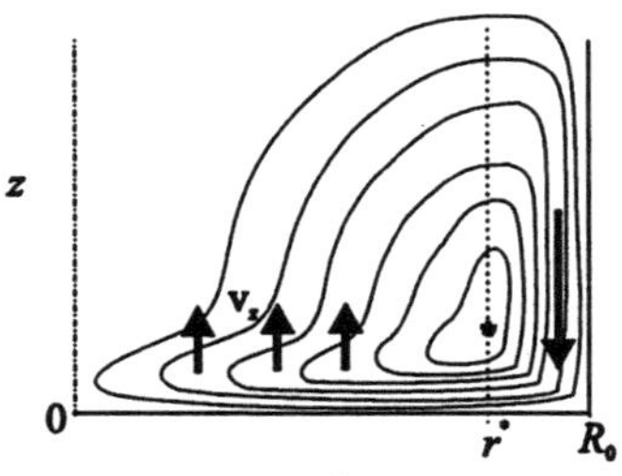

Figure 3. Characteristic pattern of swirl dominated motion near the solidification interface.

Let us consider a sole rotating field driven flow. According to Grants' (1997) numerical simulation of moderate Reynolds number flow in an elongated cylinder (the aspect ratio $1/8<R<1$), the core angular velocity is $\Omega_0 \approx 0.52\nu/L^2 T^{2/3} R^{1/2}$. This expression for $R=1$ coincides with the corresponding similarity solution (Davidson 1992, Ungarish 1997). Employing a Bödewadt's solution (e.g., Greenspan 1968), we obtained the core axial velocity and Peclet number in form

$$H_0 \approx 1.35(\nu\Omega_0)^{1/2} = \nu/L 0.97 T^{1/3} R^{1/4},\ Pe \approx 0.97 T^{1/3} R^{1/4} Pr\,. \tag{17}$$

Evidently, the location of vortex eye is closely connected to the thickness of side layer. Grants (1998, ch. 2) found that the relative thickness of this layer is $(\delta_\tau/R_0) \sim 0.075 R^{-2/3}$ for moderate Reynolds numbers near the onset of Taylor–Götler instability and the aspect ratios ranging $1/8<R<1$. Here the side layer was defined via the location of maximum in the azimuthal velocity profile. Correspondingly, the "thickness of side layer" in an infinite cylinder is $(\delta_\tau/R_0)=1-3^{-1/2} \approx 0.422$ (Richardson 1974). The current investigation revealed that such widening of side layer persisted in elongated cylinders for the Reynolds numbers far beyond the onset value as well. One can see that in case of elongated liquid zone that is characteristic for the initial stage of the vertical gradient freeze method, the vortex eye moves away from the side wall. It reflects in the heat flux distribution so that the central uniform part is less expressed (cf. figure 5 *a* and *b* for forcing rate $T=3T_0$). The thickness of critical side layer is depicted here by a vertical dashed line.

Under the magnetic field with a large Hartmann number the core angular velocity of flow greatly depends on a relative conductance of new solidified crystal $\alpha_n=\sigma_s L_s/(\sigma L)$, where σ_s is the conductivity of crystal and L_s is its height. In case of semiconductor growth it is of order of few percents (Hjelming & Walker 1986). Such seemingly negligible conductance, however, is much larger than that of adjacent Hartmann layer if the imposed field is strong $\alpha_n >> Ha^{-1}$; $Ha=(\sigma/(\nu\rho))^{1/2} LB$

is the Hartmann number; B is the induction of steady magnetic field. As a result, the swirling flow is braked more efficiently than in case of insulating walls (Grants *et al.* 1996, Grants 1998, ch. 3). Besides an indirect braking via the reduction of swirl, the meridional flow is suppressed directly as well. The core axial velocity scales as

$$H_0 = \frac{L^3}{\nu}\Omega_0^2 \mathrm{Ha}^{-3}, \quad \text{where} \quad \Omega_0 = \frac{\nu}{L^2}\frac{\mathrm{T}}{\mathrm{Ha}^2}\left(\frac{0.5+\alpha_n}{\alpha_n + \mathrm{Ha}^{-1}}\right). \tag{18}$$

One can see that a strong SMF brakes recirculation drastically and, thus, it is appropriate to control the heat transfer and solidification in a narrow range beyond the threshold of strong field that is

$$\mathrm{Ha}=O(\mathrm{T}^{1/4}\alpha_n^{-1/4}) \quad \text{and} \quad \mathrm{Ha}=O(\mathrm{T}^{1/3}) \tag{19}$$

for poorly conducting and insulating walls, respectively.

A rapid imposed rotation rises an almost rigidly rotating flow with the axial velocity sign as that of crucible angular velocity (Priede 1993)

$$H_0=\nu/L\mathrm{TE}. \tag{20}$$

Priede predicted that there was a second branch of steady similarity solution at counter-rotation with a larger axial velocity and lower core angular velocity. However, Grants (1998, ch. 4) showed that this solution was not feasible due to an unstable force balance in the core. Thus, both imposed actions reduce the core axial velocity. Consequently, they can not reduce the heat flux more than a sole rotating field does. A rapid counter-rotation of crucible changes the direction of normal to bottom flow towards it. As a result, the heat flux increases and, consequently, an optimum pulling velocity decreases. There is the so-called Stewartson layer near the curved side wall in an almost rigidly rotating flow. It has an exponentially rising profile, a thickness of $O(L|\mathrm{E}|^{1/4})$ and an inner more complicated layer of $O(L|\mathrm{E}|^{1/3})$ order that smoothes out a discontinuity in the primary axial velocity profile (Stewartson 1957). Besides, the radial profile of core axial velocity is no more uniform but fits the driving magnetic body force distribution averaged over the height (Grants 1998, ch. 4). Minding that (20) expresses a softer damping law than (18), the above features of almost rigid flow make it more suitable for the controlling of heat flux distribution.

6. Numerical results

First, let us consider a buoyancy dominated flow (figure 4). The melt moves upward at a warmer side wall and is directed towards the solidification interface in its middle part. As a result, the buoyancy considerably increases the heat flux from the liquid phase in comparison to a diffusive regime (figure 5).

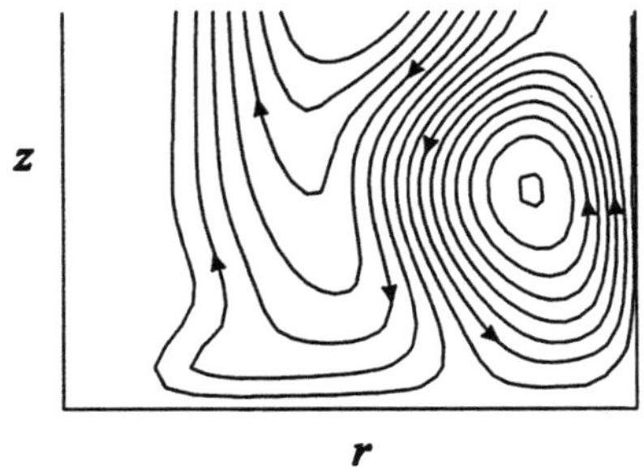

Figure 4. Characteristic pattern of buoyancy dominated motion near the solidification interface.

A magnetic forcing of sufficient strength changes the pattern of recirculation similar to the one depicted in figure 3. In such a way the heat flux is reduced. Figure 5 illustrates the above said. The heat flux from the liquid phase is depicted here in several regimes of forcing about estimated threshold value (13) for aspect ratio R=1/2 and R=1/8. The Grashoff number is Gr=$4\times10^6\times R^{-3}$ with corresponding threshold value $T_0=3.2\times10^7$ and $T_0=10^9$ for aspect ratios 1/2 and 1/8, respectively.

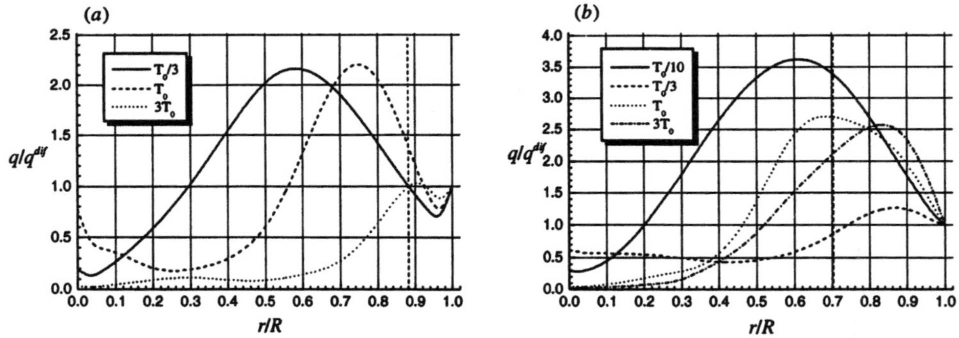

Figure 5. The heat flux distributions in the range at transition from buoyancy to swirl driven motion. (*a*) the aspect ratio R=1/2, the Grashoff number Gr=$4\times10^6R^{-3}$=32×10^6 and estimated rate of threshold forcing $T_0=2\times10^6$. (*b*) R=1/8, Gr=$4\times10^6R^{-3}$=2×10^9, $T_0=10^9$. A constant q^{dif} characterises purely diffusive heat flux.

One can see that a transition between buoyancy and swirl dominated regimes really occurs as predicted by scaling. The numerical simulation showed that the heat flux could be more reduced in a shorter cylinder (figure 5 *a*). Transition regime with an almost uniform heat flux was detected for an elongated cylinder (figure 5 *b*). The effect of sidelayer widening together with a relative height of liquid zone (§5) restricts possibilities to reduce the heat flux from an elongated liquid zone.

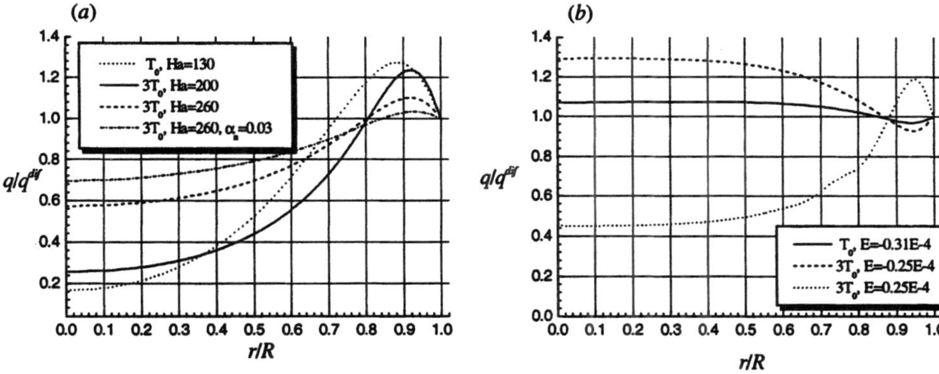

Figure 6. The heat flux profiles under imposed: (*a*) steady magnetic field or (*b*) rotation of crucible. The aspect ratio of liquid zone R=1/2, Gr=32×10^6, estimate of forcing threshold rate T$_0$=2×10^6. A constant q^{dif} characterises a purely diffusive heat flux.

Figure 6 depicts the characteristic heat flux distributions under a superimposed steady magnetic field or rotation of crucible. The aspect ratio is R=1/2, the Grashoff number is Gr=32×10^6 and a corresponding threshold forcing is T$_0$=2×10^6. An imposed steady magnetic field promotes a transition to the heat regime characteristic for a magnetically driven flow. Figure 6 (*a*) illustrates the role of seemingly negligible conductance of grown crystal α_n=3% as well.

The imposed rotation of vessel (figure 6 *b*) provided a more uniform heat flux distribution. A counter-rotation of magnitude about $\Omega_d \approx -1.5\nu/L^2T^{2/3}$ (Priede 1993, 1994, Grants 1998, ch. 4) changed the direction of core axial velocity towards the solidification front. As a result, the heat flux grows. Contrary to the motion due to buoyancy, the axial velocity distribution is more uniform. Besides, its magnitude is much lower, so the heat flux hardly increases in comparison to the diffusive regime.

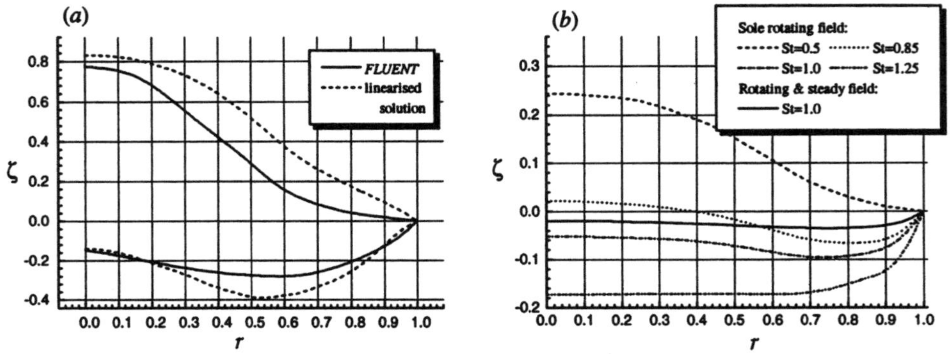

Figure 7. (*a*) comparison of front shape calculated by simplified model and *FLUENT*; zero pulling velocity; Gr=0.31×10^9; R=1/3. Lower and upper curves correspond to buoyancy and swirl (T=0.2×10^9) dominated flow, respectively. (*b*) an example of solid-liquid interface optimisation. Dashed curves depict sole rotating field guided profiles at different pulling velocities. Solid curve corresponds to rotating (T=0.2×10^8) and steady (Ha=510) magnetic field driven regime.

In order to verify the method, we calculated a case modelling a 100 mm diameter GaAs crystal growth process by the vertical gradient freeze method (for parameters of GaAs see, e.g., Liang & Lan 1997: λ=7 W/(Km), λ_s=14 W/(Km), $a=6.45\times10^{-6}$ m^2/s, $\nu=3.45\times10^{-7}$ m^2/s, $\Lambda=0.726\times10^6$ J/kg, $\rho=\rho_s=5.17\times10^3$ kg/m^3, $\beta=1.16\times10^{-4}$ K^{-1}). We assumed a temperature drop ΔT=10 K in the melt, a temperature gradient on the crystal surface T'=10 K/cm, a height of liquid zone L=150 mm, the Grashoff number Gr=$\beta g\Delta TL^3\nu^{-2}=0.32\times10^9$. Figure 7 (*a*) depicts a steady solidification interfaces calculated by a simplified method as well as by commercial hydrodynamic code *FLUENT*. We considered two cases of buoyancy or swirl dominated flows and zero pulling velocity. The melt motion influenced the interface shape drastically, since the crystal diameter was large and the ratio of solid-liquid thermal conductivity was poor. One can see that our simplified method provided a reasonable accuracy even for a large front deformation of crystal radius order. In case of a buoyancy driven motion, the front is noticeably concave to the melt even for zero pulling velocity and favourable thermal boundary conditions used. According to (7) and its consequences, interface convex to the melt, that occurs for a swirl driven motion, can be always smoothed out by an appropriate pulling velocity (see, figure 7 *b*). However, a non-uniformity of heat flux distribution always manifested itself in a corresponding profile of the front (compare figure 7 *b* with the results by Friedrich *et al.* 1997). An imposed steady magnetic field reduced this non-uniformity and, thus, improved the growth interface.

7. Summary

This paper concerns in the solid-liquid interface between a grown crystal and the melt in the vertical gradient freeze method. We considered a lenghtscale large enough to significantly influence the heat transport by forced motion of melt. We linearised the coupled hydrodynamic-solidification problem about the optimum state simply with a flat growth interface. Thus, the hydrodynamic problem is to be solved in a fixed simplified geometry. After that a corresponding quasi-steady interface shape can be calculated solving the Laplace equation with the previously found flux source in the solid phase. Main benefits of the method are:

(i) The solution of a more complicated hydrodynamic part once calculated can be used repeatedly to find the corresponding optimum pulling velocity and thermal conditions on the crystal surface;

(ii) An evident relation between the heat flux from the melt and the shape of solidification interface allows obvious conclusions on an optimum pattern of melt motion;

(iii) Originally intended for the vertical gradient freeze technique with a lot of additional assumptions, the model can be easily generalised and supplemented;

(iv) The method provides a reasonable accuracy even for a significantly deformed interface.

We considered an artificial swirling-recirculating flow of conducting melt forced by a rotating magnetic field under an action of either steady magnetic field or rotation of crucible. The estimates of necessary forcing to suppress a natural buoyant flow are given and illustrated by the results of numerical simulation for different aspect ratios of the liquid zone. A sole rotating field driven flow could significantly reduce the heat flux from the liquid phase in comparison to a purely diffusive one. Hence, the optimum pulling velocity can be increased. An imposed steady field or co-rotation of crucible control the heat flux in the range between sole rotating field driven and purely diffusive regimes. At the same time, the above actions allow to manipulate its distribution. Therefore the front shape can be noticeably improved with an additional benefit to the pulling velocity. The goal of steady field or crucible rotation is to provide a certain compromise between the magnitude of heat flux and its uniformity (or, in other words, the optimum velocity of pulling and flatness of the front). The imposed counter-rotation increases the heat flux from the liquid phase in comparison to diffusive regime. Thus, from the point of view of solidification interface shape and pulling velocity, it is not an optimal case. However, the heat flux is reduced in comparison to the buoyancy driven flow. Besides, exactly such a configuration with a uniform flow towards a new solidified crystal may have benefits in quality characteristics that depend on mass transfer.

References

Bennon, W. D. and Incropera, F. P. (1987) *Int. J. Heat Mass Transfer* 30, 2171

Davidson, P. A. (1992) *JFM* 245, 669-699

Fischer, B., Friedrich, J., Kupfer, C., Müller, G. and Vizman, D. (1997) In: Proc. on 3rd Int. Conf. on Energy Transfer in MHD Flows. Vol. 2, p. 337. Aussois, France.

Friedrich, J., Kupfer, C., Fischer, B., Müller, G. (1997) In: *Proc. on 3rd Int. Conf. on Energy Transfer in MHD Flows.* Vol. 2, p. 439. Aussois, France.

Grants, I. (1998) *Magnetically driven swirling flow guided by steady magnetic field and imposed rotation.* Thesis, Salaspils, Latvia

Grants, I. (1997) In: *Proc. on 3rd Int. Conf. on Energy Transfer in MHD Flows.* Vol. 2, p. 409. Aussois, France.

Grants, I., Priede J., Gelfgat Y. M. (1996) *Magnetohydrodynamics* 33, #3, 281

Greenspan, H. P. (1968) *The Theory of Rotating Fluids.* Cambridge University Press.

Hjelming, L. N. and Walker, J. S. (1986) *JFM* 164, 237

Liang, M.C. and Lan, C. W. (1997) *J. Cryst. Growth* 180, 587

Moffat, H. K. (1965) *JFM* 22, 521

Priede, J. (1993) *Theoretical study of a flow in an axisymmetric cavity of finite length, driven by a rotating magnetic field.* Thesis, Salaspils

Priede, J. (1994) In: *Proc. on 2nd Int. Conf. on Energy Transfer in MHD Flows.* Vol. 1, p. 87, Ausois, France

Richardson, A. T. (1974) *JFM* 63, 593

Stewartson, K. (1957) *JFM* 3, 17

Ungarish, M. (1997) *JFM* 347, 105

Voller, V. R. and Cross M. (1980) *Int. J. Heat Mass Transfer* 24, 545

Voller, V. R. and Prakash, C., (1987) *Int. J. Heat Mass Transfer* 30, 1709

ON THE STABILITY OF ROTATING MHD FLOWS

MARTY PH, MARTIN WITKOWSKI L,
TROMBETTA P AND TOMASINO T
Laboratoire LEGI
BP 53 X, 38041 Grenoble cedex, France
e-mail:Philippe.Marty@hmg.inpg.fr

AND

GARANDET J.P.
CEA-Grenoble
38054 Grenoble cedex 9, France

We present a numerical study of the flow induced by a rotating magnetic field on a liquid metal which fills a cylindrical container. Using a low frequency approximation and assuming axisymmetry, a finite difference technique is employed for the calculation of the flow field. Two different cases are considered in order to show that using a rotating magnetic field requires a detailed knowledge of its interaction with the flow. In the first situation, which is isothermal, it is shown that increasing the field intensity leads to the occurrence of Taylor-Couette type centrifugal instabilities depending upon the aspect ratio of the cavity. In the second case, which includes a heat transfer problem, it is shown that applying a very moderate rotating field to an initially unstable thermally driven convection is able to restore the flow stability.

1. Introduction

It is now admitted that applying a rotating magnetic field (RMF) may have, under particular circumstances, the effect of stabilizing an initially unstable flow. This property is of great interest in many respects, and particularly in crystal growth technology. Recent experimental works support this idea like the one made in Riga by Professor Gelgat's group (Gelfgat *et al*, 1996) or in Germany (see Dold and Benz 1995). As an example, in figure 1 published by these authors, one clearly show the effect of a limited magnetic field on the dopant striations in the solidified crystal.

A. Alemany et al. (eds.),
Transfer Phenomena in Magnetohydrodynamic and Electroconducting Flows, 327–343.

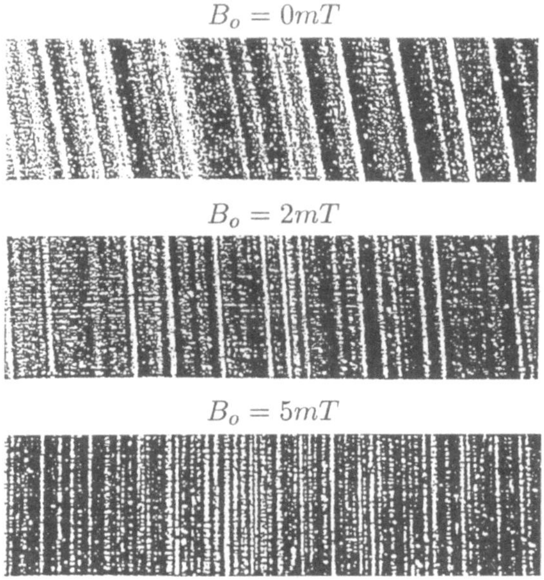

Figure 1. Decrease of the dopant striations when a rotating magnetic field is applied. Experimental results from P.Dold *et al* 1995 (the size of the samples is approx. $160 \times 630\ \mu m$)

These results have given a renew of interest to the topic of RMF for at least two reasons: on the one hand, creating a RMF device requires only a small amount of room and of electrical energy (what can be of great interest for space applications); on the other hand, the recent works which have been done to control a flow by means of a *dc* magnetic field have made clear the point that a rather heavy device had to be used in order to reach the required magnetic field. Up to now, no clear theoretical explanation or numerical demonstration have been given to explain or to illustrate the detailed action of a RMF on a fluid or thermal instability. It is precisely the aim of this paper to study the stability of a rotating flow driven by a RMF of limited frequency. Section 2 describes the model which has been used. The method for solving the electromagnetic force field is described in section 3. Section 4 gives the base flow solution in an isothermal situation. The results allow to determine the hydrodynamic stability diagram as a function of the magnetic Taylor number, T_m, and the aspect ratio of the cylindrical container. This is described in section 5. Section 6 studies a situation which is directly connected to the real crystal growth technology where unavoidable radial temperature defects are responsible for a thermally driven convection loop which instabilities can be damped by the use of a RMF.

2. Mathematical model

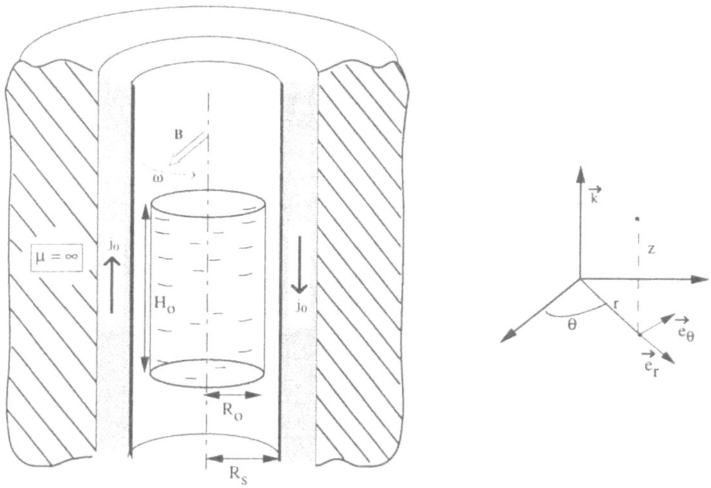

Figure 2. Geometry

We consider an isothermal, incompressible, viscous, electrically conducting fluid, contained in an electrically insulating truncated circular tube (radius R_o, height H_o) as shown in Fig. 2. The ratio H_0/R_0 is denoted by h. The properties of the fluid represented by the density ρ, the kinematic viscosity ν, the permeability μ and the conductivity σ are assumed to be constant. The fluid is submitted to the magnetic field created by an infinitely long surface current sheet located on the cylinder surface in $r = R_s > R_o$: $\vec{j} = -j_o \cos(p\theta - \omega t)\vec{k}$ where j_o is the lineic current density, $2p$ is the number of poles of the field and ω stands for the electric pulsation. The screening (or shielding) parameter $R_\omega = \mu_0 \sigma \frac{\omega}{p} R_0^2$ is small compared to unity. The region $r > R_s$ is filled with a material of infinite permeability. We use the cylindrical coordinates (r, θ, z) associated to units vector $(\vec{e_r}, \vec{e_\theta}, \vec{k})$. The following characteristic scales have been used: $l_o = R_o$ for length, $\varphi_o = \frac{\omega}{p} B_o R_o^2$ for electric potential, $f_o = \sigma \frac{\omega}{2p} R_o B_o^2$ for electromagnetic forces. B_o is constructed as $B_o = \mu j_o$. The velocity scale is denoted by U^* and its value will be precised later on according to the nature of the problem to be solved. The time scale t_o is R_o/U^*.

3. Electromagnetic force field

In the low frequency approximation ($\mu\sigma\frac{\omega}{p}R_o^2 \ll 1$) the magnetic field is unperturbed by the load and its expression:

$$\vec{B} = B_0 \left(\frac{r}{R_0}\right)^{p-1} (\sin(p\theta - \omega t)\vec{e_r} + \cos(p\theta - \omega t)\vec{e_\theta}) \quad (1)$$

satisfies:

$$\Delta\vec{B} = 0 \quad (2)$$

In (1) B_0 is the maximum intensity of the magnetic field in $r = 1$. $\vec{B}$ is two-dimensional and the vector potential $\vec{A}$, such as $\vec{B} = \vec{curl}\vec{A}$, has only a z component which is:

$$A_z = -\frac{r^p}{p}\cos\left(p(\theta - t)\right) \quad (3)$$

In this dimensionless expression B_0R_0 and p/ω have been used as characteristic values for the vector potential and for the time. Moreover, the electric field $\vec{E}$ is classically written:

$$\vec{E} = -\vec{\nabla}\varphi - \frac{\partial\vec{A}}{\partial t} \quad (4)$$

Assuming the velocity of the fluid being much smaller than the velocity of the field, Ohm's law reduces to:

$$\vec{j} = \sigma\vec{E} \quad (5)$$

Taking the divergence of (5) together with equation (3) yields the dimensionless equation of the electric potential:

$$\Delta\varphi = 0 \quad (6)$$

where $\frac{\omega R_0^2 B_0}{p}$ is used to make φ dimensionless. Looking for solutions such as:

$$\varphi = \Phi(r, z)\sin\left(p(\theta - t)\right) \quad (7)$$

gives:

$$\frac{\partial^2\Phi}{\partial r^2} + \frac{1}{r}\frac{\partial\Phi}{\partial r} - \frac{p^2}{r^2}\Phi + \frac{\partial^2\Phi}{\partial z^2} = 0 \quad (8)$$

The boundary conditions at the cylinder boundaries are:

$$\Phi)_{r=o} = 0 \qquad \left.\frac{\partial\Phi}{\partial r}\right)_{r=1} = 0 \qquad \left.\frac{\partial\Phi}{\partial z}\right)_{z=0} = \left.\frac{\partial\Phi}{\partial z}\right)_{z=h} = r^p \quad (9)$$

These conditions express the cancelling of the normal electric current at the cylinder boundaries. A separation of variables solution for Φ gives :

$$\Phi = \sum_{n=1}^{+\infty} \{C_{1n} cosh(\lambda_n z) + C_{2n} sinh(\lambda_n z)\} J_p(\lambda_n r) \tag{10}$$

where J_p are the Bessel functions of the first kind and where the λ_n are such as $J_p'(\lambda_n) = 0$ whereas *sinh* and *cosh* stand for the hyperbolic sinus and cosinus, respectively.

The boundary conditions in $z = 0$ and $z = h$ give:

$$r^p = \sum_{n=1}^{+\infty} \lambda_n C_{2n} J_p(\lambda_n r) = \sum_{n=1}^{+\infty} \lambda_n \{C_{1n} sinh(\lambda_n h) + C_{2n} cosh(\lambda_n h)\} J_p(\lambda_n r) \tag{11}$$

From (11) one obtains :

$$C_{1n} = \frac{C_{2n}(1 - cosh(\lambda_n h))}{sinh(\lambda_n h)} \tag{12}$$

and, after some manipulations one finds :

$$C_{2n} = \frac{2p}{\lambda_n (\lambda_n^2 - p^2) J_p(\lambda_n)} \tag{13}$$

The final solution for Φ becomes :

$$\Phi = \sum_{n=1}^{+\infty} \frac{2pJ_p(\lambda_n r)}{\lambda_n (\lambda_n^2 - p^2) J_p(\lambda_n)} \quad \frac{cosh(\lambda_n z) - cosh(\lambda_n (h - z))}{sinh(\lambda_n h)} \tag{14}$$

From Ohm's law one deduce the time and space averaged azimuthal force :

$$f_\theta = r^{p-1} \left(-\frac{\partial \Phi}{\partial z} + r^p \right) \tag{15}$$

which can be written :

$$f_\theta = r^{p-1} \left(r^p - \sum_{n=1}^{+\infty} \frac{2pJ_p(\lambda_n r)}{(\lambda_n^2 - p^2) J_p(\lambda_n)} \frac{sinh(\lambda_n z) + sinh(\lambda_n (h - z))}{sinh(\lambda_n h)} \right) \tag{16}$$

This expression extends the work from (Mazuruk *et al.*, 1997) which is valid for $p = 1$ only.

4. Base flow solution in the isothermal case

The flow equations are described in terms of vorticity $\Omega \cdot \vec{e_\theta} = \vec{\nabla} \wedge \vec{V}$, stream function ψ defined by $\vec{\nabla} \wedge (-\frac{\psi}{r} \cdot \vec{e_\theta}) = (u, 0, w)$ and angular momentum $\Gamma = rv$:

$$\frac{\partial \Omega}{\partial t} + \frac{\partial(u\Omega)}{\partial r} + \frac{\partial(w\Omega)}{\partial z} - \frac{\partial}{\partial z}\left(\frac{\Gamma^2}{r^3}\right) = \frac{1}{T_m^{1/2}}\left(\frac{\partial^2 \Omega}{\partial r^2} + \frac{1}{r}\frac{\partial \Omega}{\partial r} - \frac{\Omega}{r^2} + \frac{\partial^2 \Omega}{\partial z^2}\right) \tag{17}$$

$$\frac{\partial^2 \psi}{\partial r^2} - \frac{1}{r}\frac{\partial \psi}{\partial r} + \frac{\partial^2 \psi}{\partial z^2} = r\Omega \tag{18}$$

$$\frac{\partial \Gamma}{\partial t} + \frac{\partial(u\Gamma)}{\partial r} + \frac{u\Gamma}{r} + \frac{\partial(w\Gamma)}{\partial z} = \frac{1}{T_m^{1/2}}\left(\frac{\partial^2 \Gamma}{\partial r^2} - \frac{1}{r}\frac{\partial \Gamma}{\partial r} + \frac{\partial^2 \Gamma}{\partial z^2}\right) + rf_\theta \tag{19}$$

where $T_m = (\frac{U^* R_o}{\nu})^2$. The velocity U^* is chosen equal to $\sqrt{\sigma\omega/2(p\rho)}B_o R_o$ and results from an equilibrium between the inertial forces and the electromagnetic forces. Then, the Taylor number becomes $T_m = \frac{\sigma\omega B_o^2 R_o^4}{2p\rho\nu^2}$. The equations (17) to (19) are solved numerically by using a second order finite difference scheme derived from (Sørensen *et al.*, 1989). The azimuthal force also contains a periodic component but we assume that it doesn't affect the flow as far as the fluid velocity is negligible compared to the velocity of the magnetic field (Davidson *et al.*, 1987) when the flow is dominated by inertia or as far as $\frac{\omega R_o^2}{p\nu} \gg 1$ when the flow is viscous dominated (see (Priede, 1993)).

Consequently, only the time-averaged value of the force (eq 16) is introduced in the momentum equation. This assumption can also be understood as follows: due to the equilibrium between the mean value $\sigma\frac{\omega}{2p}R_0 B_0^2$ of the electromagnetic forces and the inertial forces, the flow has a mean value U^* such as: $\sigma\frac{\omega}{2p}R_0 B_0^2 \approx \frac{\rho U^{*2}}{R_0}$. During a time interval of one electric period, $\frac{\omega}{p}^{-1}$, the flow experiences a fluctuation of the driving force of order $O(\sigma\frac{\omega}{2p}R_0 B_0^2)$. Again, a simple balance between this force and inertia writes: $\sigma\frac{\omega}{2p}R_0 B_0^2 \approx \rho\frac{\delta U^*}{(\omega/p)^{-1}}$ where δU^* stands for the fluctuating part of the velocity. Consequently, assuming the flow to be steady despite the fluctuating nature of the forces can also be written $\frac{\delta U^*}{U^*} << 1$ or, introducing an interaction parameter, N:

$$N = \frac{\sigma B_0^2}{\rho\omega} << 1 \tag{20}$$

This situation is quite at the opposite of that recently considered by (Walker, 1998) who studies the behavior of a flow driven by a RMF in a truncated liquid metal cylinder with $Ha >> 1$ and $N >> Ha$. Figure 3 shows the electric current lines for $H_0/R_0 = 0.66$. Their path is roughly vertical in the mid-height region , thus creating in this part a maximum value of the azimuthal force.

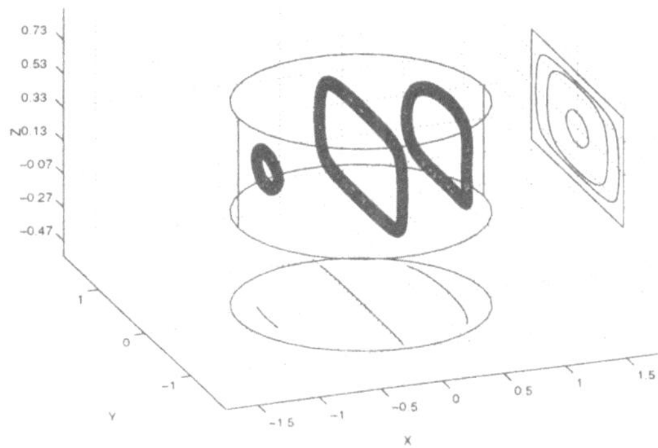

Figure 3. Electric current lines for $H_0/R_0 = 0.66$. At the instant when this plot has been made, the magnetic field is normal to y = constant planes.

The distributions of the azimuthal force f_θ is shown in Fig 4 for different aspect ratio of the cylinder. One notes that for $H_0/R_0 > 3$, the force distribution in the mid-plane becomes almost equal to that obtained in an infinitely long cylinder. Nevertheless, H_0/R_0 has to be greater than approximately 10 to allow the existence of a region near the mid-plane having an almost z-independent plateau.

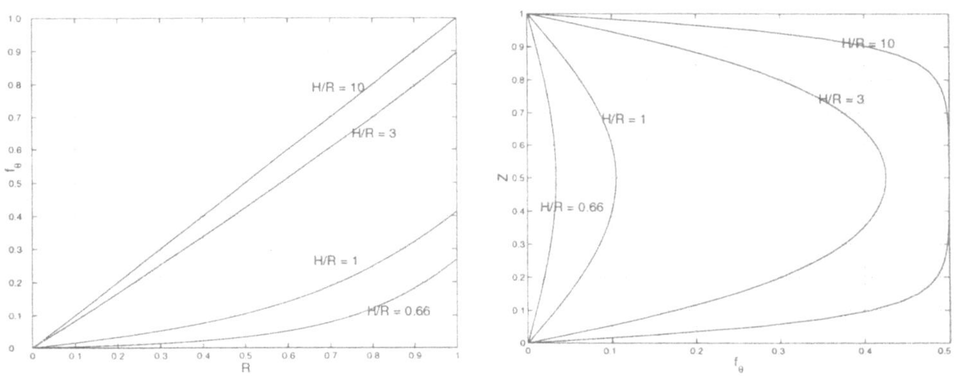

(a) (b)

Figure 4. Radial and vertical distributions of the azimuthal force for different aspect ratio. a) $z = 0.5$ and b) $r = 0.5$.

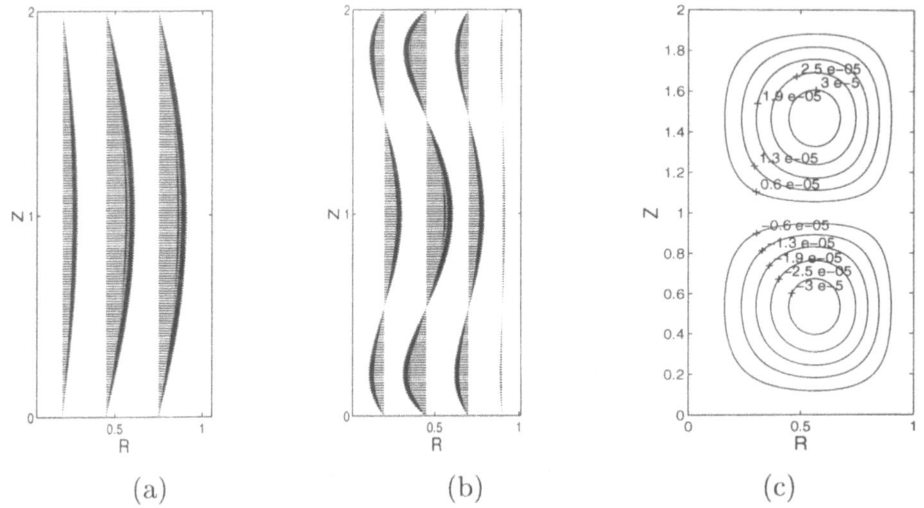

Figure 5. Velocity profiles at different radius and streamlines for a magnetic Taylor number $Tm = 10$ and an aspect ratio $H_o/R_o = 2$ (a) :Azimuthal velocity, v - (b) : Radial velocity, u (the length used for drawing the arrows is roughly 400 times larger than in (a)) - (c) : Meridional streamlines.

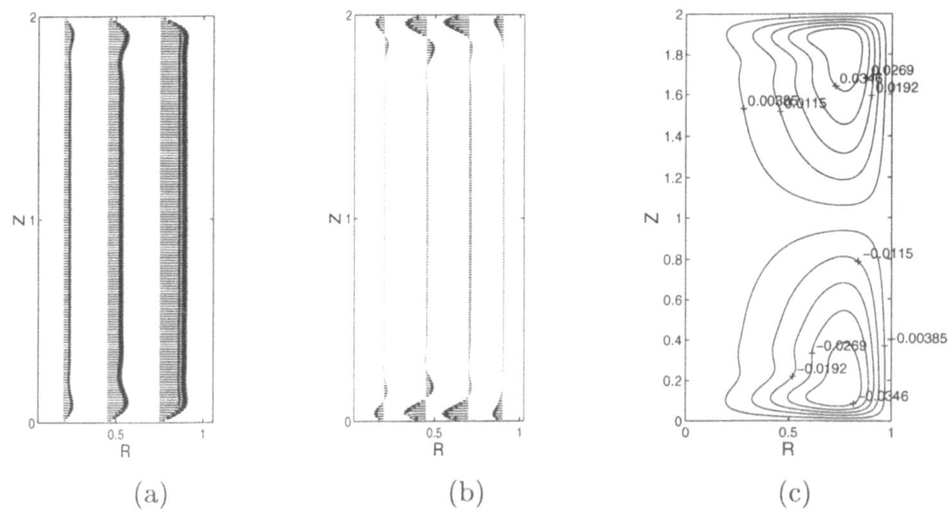

Figure 6. Velocity profiles at different radius and streamlines for a magnetic Taylor number $Tm = 10^5$ and an aspect ratio $H_o/R_o = 2$ (a) :Azimuthal velocity, v - (b) : Radial velocity, u (the length used for drawing the arrows is roughly the same as in (a)) - (c) : Meridional streamlines.

Different flow regimes can be observed. For an aspect ratio $H_o/R_o = 2$, the influence of T_m on the flow has been studied in details. As T_m increases three flow regimes occur successively: a viscous regime, an inertial regime, an unstable regime. Gelfgat (1994) predicted a fourth regime where the flow again becomes quasi stationary and stable but this prediction has not been observed in our numerical calculations.

In the viscous regime, the recirculating flow in the meridional plane has no effect on the angular momentum distribution which is controlled by an equilibrium between viscosity and the magnetic force. This can be written: $\mu v_\theta/R_0^2 \sim \sigma \frac{\omega}{2p} R_0 B_0^2$ or, in dimensionless form: $\frac{v}{U^*} \sim {T_m}^{1/2}$. The flow is shown on Fig 5. Both v or u have smooth profiles and no boundary layers appear. The meridional streamlines are almost circular.

In the inertial regime, the recirculating flow is strongly coupled with the azimuthal flow and by virtue of an analysis detailed in (Davidson, 1992) the order of magnitude of azimuthal velocity can be found: $\frac{v}{U^*} \sim {T_m}^{1/6}$. This situation is characterized by the existence of thin Ekman layers along the end walls in $z = 0$ and $z = h$ (see Fig 6). Most of the meridional recirculating flow passes through these layers where it looses the angular momentum picked up in the core due to the effect of its interaction with the electromagnetic forces. The thickness of the Ekman layers has the well-known dimensionless value $\left(\frac{vR}{\nu}\right)^{-1/2}$ what can be written in dimensionless units:

$$\delta_{Ekman} \sim O(T_m^{-1/3}) \tag{21}$$

The radial velocity in these layers scales as the azimuthal velocity in the core, ie, $O(T_m^{1/6})$. The mass conservation of the recirculating flow gives the value of the meridional velocity in the core: $v_m(core) \sim O(T_m^{-1/6})$. Finally, on Fig 6, we can note an increase in v as one move from the core to the end walls. This is clearly a convective effect of the meridional flow which carries towards the center, fluid particles having a higher azimuthal velocity.

Figure 7 shows the evolution of the maximum value of the angular momentum in the whole fluid as a function of T_m. The slope found by the above calculations of order of magnitude matches reasonably well with our numerical calculations. It is clearly 1/2 in the viscous regime for $T_m < 10^3$ and 1/6 in the full inertial regime for $T_m > 10^5$. It is worth noting the existence of an intermediate region between $T_m = 10^3$ and $T_m = 10^5$ where v scales as $T_m^{0.3}$ but where the flow is still stable. This region probably allows a transition of the flow between the viscous and the inertial regimes. This latter becomes rapidly unstable if T_m increases. In figure 7, the last three points are already in the unstable regime. It is then surprising that the occurrence of flow instabilities does not affect the validity of an order

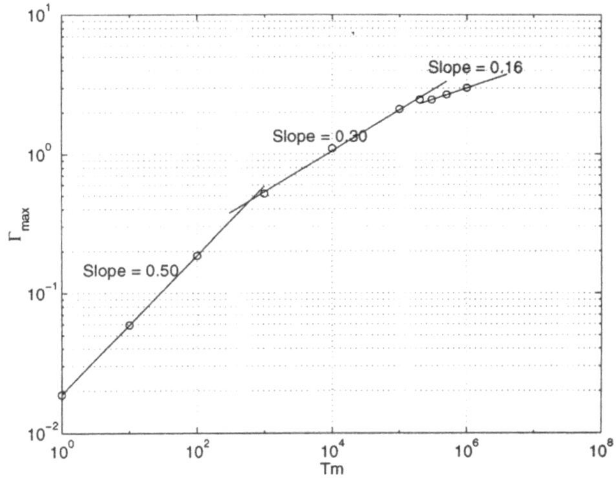

Figure 7. $H_o/R_o = 2$. Evolution of the maximum of angular momentum Γ_{max} in $[0, R_o] \times [0, H_o]$ vs magnetic Taylor number T_m

of magnitude analysis which has nevertheless been made for a stable flow. The general tendency of these results is in agreement with those recently obtained by other authors(see (Priede, 1993) and (Davidson, 1992)).

5. Departure from stability due to the RMF action

The first stability study was performed by Richardson (Richardson, 1974). He showed the Taylor Couette like nature of the instability in the case of an infinite height. Gelfgat (Gelfgat *et al.*, 1994) found that the finite length of the cylinder was a stabilizing factor. For an aspect ratio $H_o/R_o = 2$, Gelfgat predicts the first transition for $T_m > 3.2 \cdot 10^5$ while for an infinite model the calculation gives $T_m = 2917$. In order to understand the mechanism of appearance of the instability and the effect of top and bottom end walls, we choose a sufficiently enhanced cylinder (aspect ratio $H_o/R_o = 10$). The exact determination of the transition threshold by means of direct simulation is a difficult task as the first perturbations are very weak and take a long time to appear. For the aspect ratio considered, we found that the critical T_m ranges in $[4050, 7200]$. The lower bound is obtained by decreasing T_m from the unstable regime till a stable regime is obtained while the upper bound is obtained doing the route backward. To illustrate the unsteady case, we choose therefore a regime where the oscillations are not too weak. On figure 8, we see the time evolution of the streamlines over one period and of a probe signal measuring the azimuthal velocity. As predicted by Richardson, we see the appearance of Taylor Couette vortices in

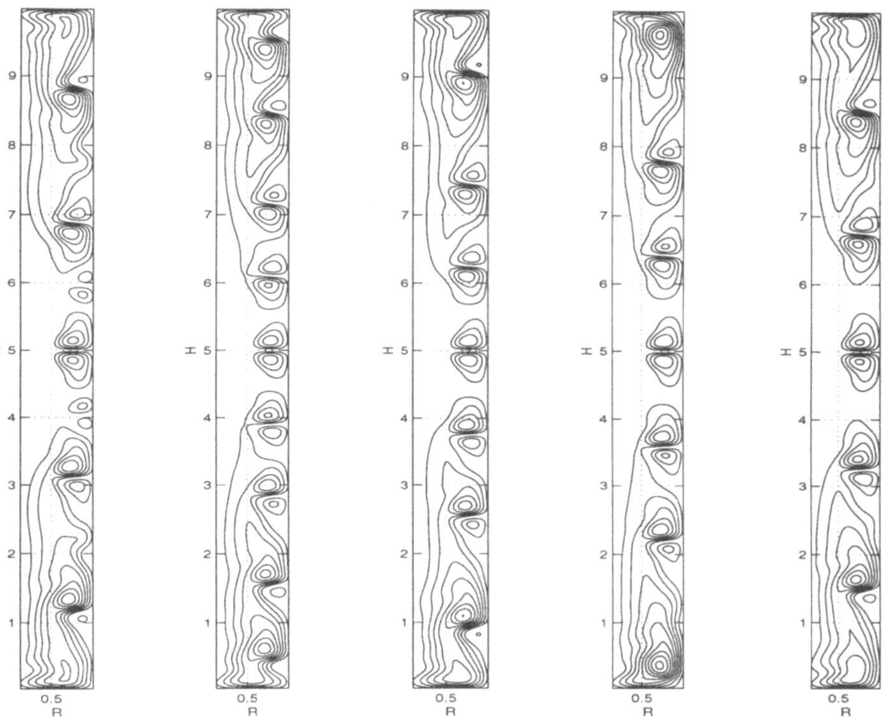

(a) T_o $\quad T_o + 0.2 \cdot \Delta T \quad T_o + 0.4 \cdot \Delta T \quad T_o + 0.6 \cdot \Delta T \quad T_o + 0.9 \cdot \Delta T$

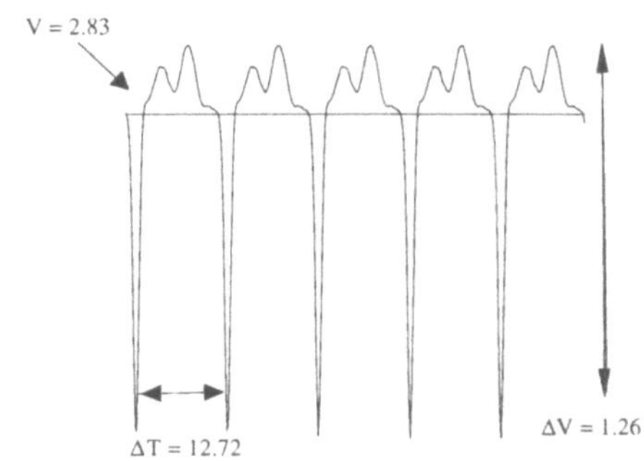

(b)

Figure 8. (a) Evolution of distribution of the stream function ψ over a period $\Delta T = 12.72$ for $T_m = 16200$, and $H_o/R_o = 10.(\psi \in [-0.157, +0.157]$ with 20 isovalues equally spaced). One can follow the evolution of a pair of vortices created at $z = 6$ ascending over $3 \cdot \Delta T$ toward the top wall and note the appearance of a pair of vortices at $z = 8.3$ for $T_o + 0.2 \cdot \Delta T$. (b) Signal of a probe measuring the azimuthal component of the velocity at $r = 0.85, z = 7.5$.

the meridional plane created by the unstable part of the azimuthal velocity profile near the cylinder wall. The major difference between infinite case and the configuration of finite height is the fact that when unstable the flow is unsteady. The plane $z = H_o/2 = 5$ remains a plane of symmetry. The vortices are generated in the quasi z-independent zone of the cylinder and move from the plane of symmetry toward the extremities where they are destroyed. This is due to the presence of the Ekman pumping created by the end walls that attract the vortices. The signal shows this time evolution, each peak corresponding to a vortice passing in front of the probe.

It is interesting to draw the stability diagram of the evolution of the critical Taylor number vs the aspect ratio H_0/R_0. This is done in Fig 9. It is most probable that $(T_m)_{critical}$ will have an asymptotic dependence in $(H_0/R_0)^{-4}$ when $H_0/R_0 \to 0$. As a matter of fact, the -4 exponent eliminates the influence of the radius R_0, what is expectable when the cylinder height becomes vanishingly small compared to the radius.

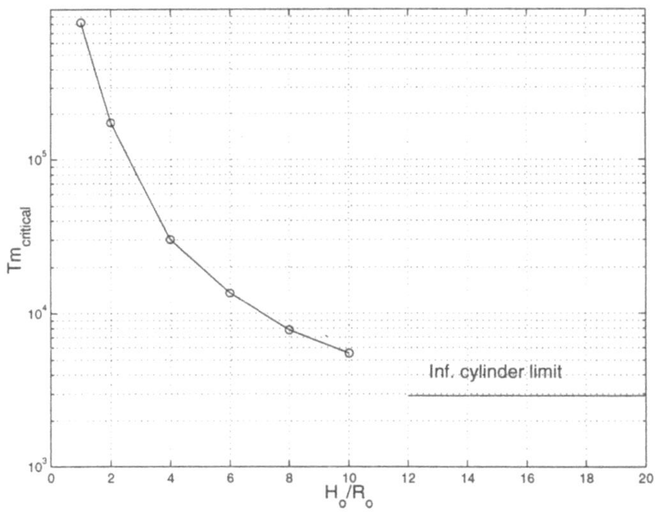

Figure 9. Evolution of the critical value of the magnetic Taylor number vs the aspect ratio H_0/R_0.

6. Recovering stable conditions with a RMF

The aim of this part is to explain the detailed mechanism by which a RMF may, under particular circumstances, contribute to stabilize an initially unstable flow. As many of the potential applications of this type of magnetic fields are connected with the crystal growth technology, it has appeared interesting to consider the situation described in Fig 10 where a cylindrical cavity is filled with a liquid metal which is possibly submitted to a rotating magnetic field. This situation is then close to the vertical Bridgman configuration.

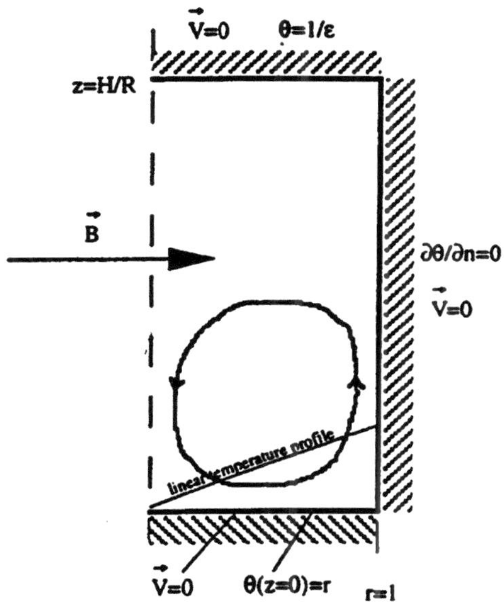

Figure 10. Test case for creating an initially unstable flow: a radial temperature distribution is imposed along the bottom wall of a cylindrical cavity having an adiabatic vertical wall and an isothermal top wall.

The top boundary is at temperature $T_0 + \Delta T$ whereas a radial temperature defect due in practice to a difference between the thermal conductivities of the liquid and of the solid phase, is maintained along the bottom boundary in $z = 0$ where T is linearly increasing from T_0 in $r = 0$ to $T_0 + \epsilon\Delta T$ in $r = R$. The vertical wall is thermally insulated. Changing the parameter ϵ then allows to control the intensity of the thermally driven meridional flow in the (r, z) plane.The scale velocity U^* results from an equilibrium between buoyancy and inertial forces: $\rho g\beta\epsilon\Delta T \sim \rho U^{*2}/R_0$ where $\epsilon\Delta T$ is considered as being the significant quantity to estimate the buoyant effect and β is the thermal expansion coefficient. As a result, the scale velocity U^* becomes equal to $\sqrt{g\beta\epsilon\Delta T R_0}$. Introducing a dimensionless temperature $\Theta = \frac{(T-T_0)}{\epsilon\Delta T}$, equations (17) to (19) and the energy equation become:

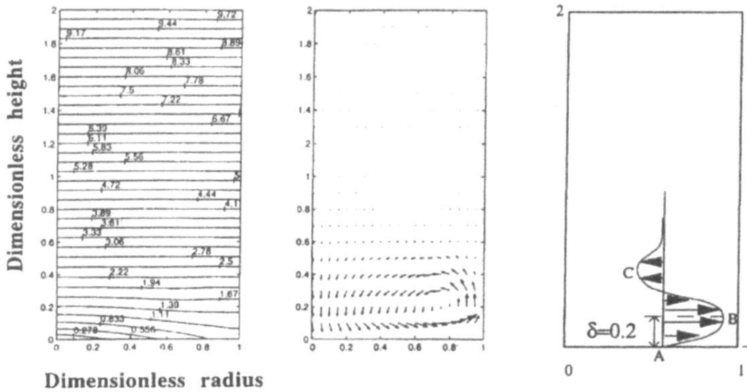

Figure 11. Temperature (a) and velocity (b) distributions for $\epsilon = 0.1$, $Pr = 0.01$, $Gr = 2.9\ 10^6$ and $T_m = 0$.

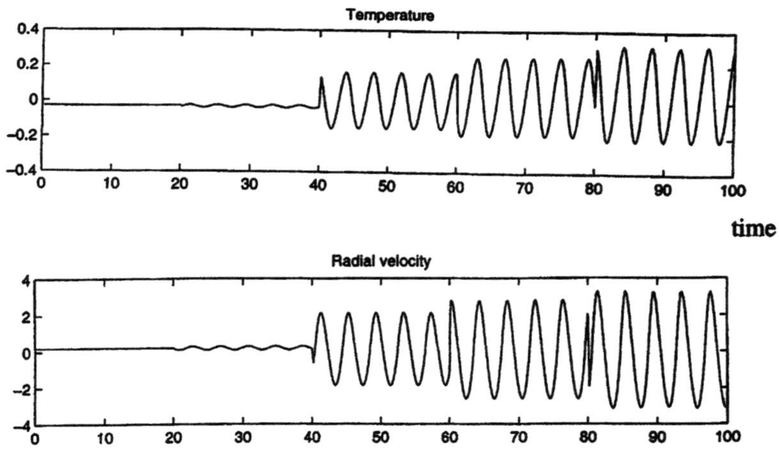

Figure 12. Time evolution of the temperature for $Gr = 3.1$, 3.2, 3.5, 3.8 and 4.1 10^6 (Gr is increased in time every $\Delta t = 20$; the transient period between two successive values of Gr is not shown). $T_m = 0$, $H/R = 2$, $Pr = 0.01$ and $\epsilon = 0.1$.

$$\frac{\partial \Omega}{\partial t} + \frac{\partial (u\Omega)}{\partial r} + \frac{\partial (w\Omega)}{\partial z} - \frac{\partial}{\partial z}\left(\frac{\Gamma^2}{r^3}\right) = \frac{1}{Gr^{1/2}}\left(\frac{\partial^2 \Omega}{\partial r^2} + \frac{1}{r}\frac{\partial \Omega}{\partial r} - \frac{\Omega}{r^2} + \frac{\partial^2 \Omega}{\partial z^2}\right) - \frac{\partial \Theta}{\partial r} \tag{22}$$

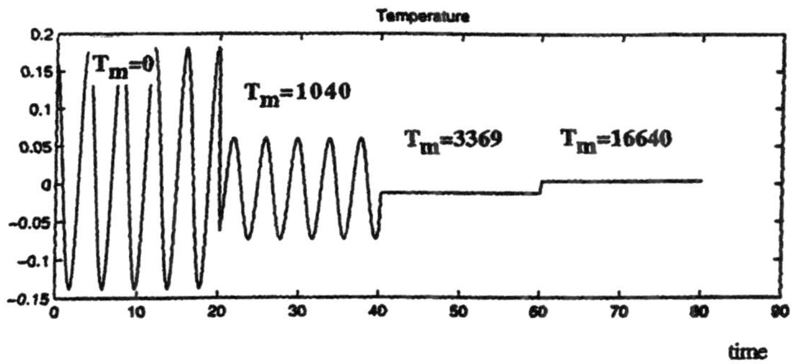

Figure 13. Temporal evolution of the temperature fluctuations in $r = 0.25$, $z = 0.2$ vs time for different T_m and for $Pr = 0.01$, $\epsilon = 0.1$, $Gr = 3.5\ 10^6$, $H/R = 2$. The transient period between two successive values of T_m is not shown.

$$\frac{\partial^2 \psi}{\partial r^2} - \frac{1}{r}\frac{\partial \psi}{\partial r} + \frac{\partial^2 \psi}{\partial z^2} = r\Omega \tag{23}$$

$$\frac{\partial \Gamma}{\partial t} + \frac{\partial (u\Gamma)}{\partial r} + \frac{u\Gamma}{r} + \frac{\partial (w\Gamma)}{\partial z} = \frac{1}{Gr^{1/2}}\left(\frac{\partial^2 \Gamma}{\partial r^2} - \frac{1}{r}\frac{\partial \Gamma}{\partial r} + \frac{\partial^2 \Gamma}{\partial z^2}\right) + \frac{T_m}{Gr} r f_\theta \tag{24}$$

$$\frac{\partial \Theta}{\partial t} + u\frac{\partial \Theta}{\partial r} + w\frac{\partial \Theta}{\partial z} = \frac{1}{Pr\sqrt{Gr}}\left(\frac{\partial^2 \Theta}{\partial r^2} + \frac{1}{r}\frac{\partial \Theta}{\partial r} + \frac{\partial^2 \Theta}{\partial z^2}\right) \tag{25}$$

The Grashof and Prandtl numbers have their standard definitions:

$$Gr = \frac{g\beta\epsilon\Delta T R_0^3}{\nu^2} \qquad Pr = \frac{\nu}{\alpha} \tag{26}$$

The equations are solved using the same finite differences technique as that described in section 4.

When $T_m = 0$, Fig 11 shows the temperature and velocity distribution for $Gr = 2.9\ 10^6$, $Pr = 0.01$ and $\epsilon = 0.1$. The value 0.01 is a characteristic value of the Prandtl number of liquid metals whereas $\epsilon = 0.1$ corresponds to a radial temperature difference of $2K$ when a vertical temperature difference of $20K$ is applied between the top and bottom boundaries. The flow is confined at the bottom of the cell where the radial inhomogeneity of the temperature is important. It consists of a stationary radial jet along the bottom wall having an almost constant thickness $\delta \approx 0.2$. The flow stability is lost when Gr is increased (Fig 12). The amplitude A of the flow oscillations vs $Gr - Gr_c$ displays a Hopf bifurcation behavior. For $Gr = 3.5\ 10^6$, Fig 13 shows the damping of the temperature fluctuations when the magnetic Taylor number, T_m, is progressively increased from 0 to 16640.

A detailed study of the flow around the critical Grashof number Gr_c has shown that the bottom jet along the wall in $z = 0$ was responsible for the instability (Marty *et al.*, 1998). More precisely, the inflexional nature of its velocity profile is the dominant ingredient of this instability. For an unbounded geometry, such a profile is known to be unstable even for rather low Reynolds numbers (Villermaux, 1998). In our case, the dimensional confinement in the radial direction imposed by the vertical wall has the effect of preventing the perturbations having a wave-length greater than R to develop. This low-pass filtering is consequently at the origin of a critical Reynolds number for the jet to which corresponds an associated critical Grashof number. Adding a rotation to the flow through electromagnetic azimuthal forces will generate a radial Ekman recirculating inward flow along both end walls. Along the bottom wall, this Ekman flow will then be opposite to the thermally driven flow and will thus lower the jet Reynolds number to a value situated below the critical threshold. This, in turn, will restore the flow stability.

7. Conclusion

A method of calculation of the flow induced by a low-frequency rotating magnetic field in a truncated liquid metal cylinder has been presented. In an isothermal cavity, the flow is found in agreement with calculations of other contributors: it consists of a viscous flow when the magnetic Taylor number, T_m, is lower than approximately 10^3 and of a flow dominated by inertia when T_m is greater than 10^5. In the viscous regime, the azimuthal velocity scales as $T_m^{1/2}$ whereas it scales as $T_m^{1/6}$ in the inertial regime. For various aspect ratios, the critical value of T_m which produces Taylor Couette instabilities in the flow has been found. The most important results of this study concerns the possibility to use a RMF to stabilize an initially unstable and time-dependent flow. This is of high practical interest for crystal growth technology. Although all these results show the promising future of RMF, they also illustrate how careful one has to be when this type of fields is considered for industrial applications. It is also likely that specific experimental studies on these flows should be undertaken in addition to the numerical approach.

References

DAVIDSON, P.A. and HUNT, J.C.R. Swirling recirculating flow in a liquid-metal column generated by a rotating magnetic field. *Journal of Fluid Mechanics*, 185:67–106, 1987.

DAVIDSON, P.A. Swirling flow in an axisymmetric cavity of arbitrary profile, driven by a rotating magnetic field. *Journal of Fluid Mechanics*, 245:669–699, 1992.

DOLD, P., and BENZ, K.W. Convective temperature fluctuations in liquid Gallium in Dependence on static and rotating magnetic fields, *Crystal Research Technology*, Vol.30, n.8, pp.1135–1145, 1995.

DOLD, P. , KAISER, Th., BOSCHERT, S. and BENZ, K.W., Modification of the heat transport by MHD-Flows in the vertical Bridgman crystal growth process, 8th Beer-Sheva Seminar on MHD Flows and Turbulence, 1995, Beer Sheva Conf., Israel.

GELFGAT, Y.M., GORBUNOV, L.A. and KOLEVZON, V. Liquid metal flow in a finite-length cylinder with a rotating magnetic field. *Experiments in Fluids*, 15:411–416, 1993.

GELFGAT, Y.M. and GORBUNOV L.A. Effect of alternating magnetic field on melt hydrodynamics in a cylindrical vessel with free surface. *Magnetohydrodynamics*, 30(3):237–247, 1994.

GELFGAT, Y.M., KRUMINS J., and ABRICKA A., On hydrodynamics, heat and mass transfer control in a cylindrical vessel by rotating magnetic field, *Proceedings of the 8th Bat-Sheva International Seminar*, 1996,Beersheva, Israel.

MARTY, Ph., TROMBETTA, P., AND GARANDET, J.P. Contrôle de la stabilité d'un écoulement thermoconvectif par un champ magnétique tournant *Comptes Rendus de l'Académie des Sciences, Paris, Série IIb*, t.326, p. 185-190, 1998.

MAZURUK, K., RAMACHANDRAN, N., VOLZ, M. and GILLIES, D. Study of frequency effects of a rotating magnetic field on fluid flow in vertical cylinders. *Soc. of Photo-optical Instrum. Eng.*, Vol.3123, San Diego, 1997.

PRIEDE, J. Theoretical study of a flow in an axisymmetric cavity of finite length, driven by a rotating magnetic field. *PhD thesis, Institute of Physics, Latvian Academy of Science*, Salaspils, 1993.

RICHARDSON, A.T. On the stability of a magnetically driven rotating fluid flow. *Journal of Fluid Mechanics*, 63:593–605, 1974.

SØRENSEN, J.N. and PHUOC LOC, TA.. High-order axisymmetric Navier-stokes code: Description and evaluation of boundary conditions. *International Journal For Numerical Methods In Fluids*, 9:1517–1537, 1989.

VILLERMAUX, E. On the role of viscosity in shear instabilities, *Physics of Fluids*, Vol. 10, n.2, 1998.

WALKER, J.S. Bridgman crystal growth with a strong, low-frequency, rotating magnetic field, *to appear in J. of Crystal Growth*, 1998.

DYNAMICS OF AN AXISYMMETRIC ELECTROMAGNETIC 'CRUCIBLE' MELTING

V.BOJAREVICS, K.PERICLEOUS, M.CROSS
University of Greenwich,
School of Computing and Mathematics,
Wellington Str., London SE18 6PF, UK

Abstract. Different industrial induction melting processes involve free surface and melt-solid interface of the liquid metal subject to dynamic change during the technological operation. Simulation of the liquid metal dynamics requires to solve the non-linear, coupled hydrodynamic-electromagnetic-heat transfer problem accounting for the time development of the liquid metal free boundary with a suitable turbulent viscosity model. The present paper describes a numerical solution method applicable for various axisymmetric induction melting processes, such as, crucible with free top surface, levitation, semi-levitation, cold crucible and similar melting techniques. The presented results in the cases of semi-levitation and crucible with free top surface meltings demonstrate oscillating transient behaviour of the free metal surface indicating the presence of gravity-inertial-electromagnetic waves which are coupled to the internal fluid flow generated by both the rotational and potential parts of the electromagnetic force.

1. Introduction

The magnetic induction melting procedures were theoretically and numerically studied for many years using different quasi-magnetostatic and boundary layer approximations, e.g. [1-4]. A limited skin layer penetration effect for the external electromagnetic field of a high frequency is the key argument in these theories. In the case of pure levitation melting, relatively high frequencies are typically used (10^4-10^5 Hz), which justifies the use of different electromagnetic skin-layer based theories [2,3]. However, a typical frequency used in the semi-levitation melting process [1] is an order of magnitude lower and the magnetic field penetration is significant (approximately 10% of the charge radius in the computed example case, see Fig. 1. The magnetic field lines of the current induced in the metal are closing within the internal part of the liquid, and the

A. Alemany et al. (eds.),
Transfer Phenomena in Magnetohydrodynamic and Electroconducting Flows, 345–357.

resulting electromagnetic force is rotational, in general, generating a fluid motion.

In addition to the rotational electromagnetic force generated fluid flow, a significant fluid motion is created because of the dynamic external boundary shaping. When the metal is being melted, the magnetic confining force pushes the liquid radially inwards to the symmetry axis and upwards until a balance with gravity force is reached. With inertial effects this leads to an oscillating interface motion and waves, the damping of which depends on the effective turbulent viscosity. In addition to the fluid flow and electromagnetic field coupling with the liquid metal shape change, there is a coupling to the temperature field. Joule heat generation depends on the coil position and the interface deformation at a given time moment. The heat transfer is significantly affected by the turbulent nature of the fluid flow, the melting front motion and the associated latent heat, and the temperature dependence of material properties including the solid metal region.

The numerical modelling presented in this paper is an attempt to put together all these dynamic dependencies, and to simulate at the best possible accuracy the time development of the melting and confinement by the magnetic induction effect.

2. Problem statement

The most significant approximation for the magnetic induction melting and the resulting fluid flow is the assumption that the problem can be considered as axisymmetric. In reality there are always three dimensional perturbations present, especially with the waves at the free surface. However, to certain accuracy, in a time average sense the problem can be considered as axisymmetric, what makes feasible the numerical solution for this highly non-linear and different physical effects coupled problem. Let us consider an axisymmetric volume of metal which initially is solid and is gradually being melted.

In general, the governing equations for the incompressible turbulent flow can be represented in the following rotational form:

$$\begin{aligned} &\partial_t \mathbf{v} - \mathbf{v}\times\mathbf{w} = -\nabla(\rho^{-1}p + \tfrac{1}{2}\mathrm{v}^2) + \nabla\cdot(\nu_e(\nabla\mathbf{v}+\nabla\mathbf{v}^T)) - K_D\mathbf{v} + \rho^{-1}\mathbf{J}\times\mathbf{B} + \mathbf{g}, \\ &\nabla\cdot\mathbf{v} = 0, \end{aligned} \qquad (1)$$

where $\mathbf{v}$ is the velocity vector, p the pressure, ρ the density, $\nu_e(\nabla\mathbf{v} + \nabla\mathbf{v}^T)$ is the effective turbulent viscosity according to the Boussinesq hypothesis and the velocity deformation rate tensor, K_D is Darcy coefficient representing the mushy zone and melting front modelling [8], $\mathbf{f} = \mathbf{J} \times \mathbf{B}$ is the electromagnetic force, **g** the gravity. There can be two types of boundary conditions for velocities at different portions of the external boundary:

1) at solid walls (no slip for velocities)

$$u_n = 0, \quad u_\tau = 0; \tag{2}$$

2) at the free surface boundary normal stress is compensated by the surface tension, tangential stress is zero, and the kinematic condition states that a material fluid particle moves with the interface location:

$$\Pi_{nn} = \gamma K \; ; \; \Pi_{n\tau} = 0 \; ; \quad \mathbf{e_n} \cdot \partial_t \mathbf{R} = \mathbf{e_n} \cdot \mathbf{v} \, , \; (\mathbf{R} = \mathbf{e_R} R(\theta, t)) \, , \tag{3}$$

where Π is stress tensor, $\mathbf{e}_n$ and $\mathbf{e}_\tau$ - normal and tangent unit vectors at the free surface, γ the surface tension coefficient, K the local surface mean curvature, **R** the position vector at the interface.

The effective viscosity is represented as a sum of molecular and turbulent viscosity ($\nu_e = \nu + \nu_T$), which is computed according to a generalised mixing length hypothesis:

$$\nu_T = l_m^2 \tfrac{1}{2}\sqrt{[(\nabla\mathbf{v}+\nabla\mathbf{v}^T):(\nabla\mathbf{v}+\nabla\mathbf{v}^T)]}, \tag{4}$$

$$l_m = \max\left\{ l_{m-} \, ; \; \min\left\langle f_l \frac{|\mathbf{v}|}{\frac{1}{2}\sqrt{[(\nabla\mathbf{v}+\nabla\mathbf{v}^T):(\nabla\mathbf{v}+\nabla\mathbf{v}^T)]}} ; \; l_{m+} \right\rangle \right\},$$

where the mixing length l_m is expressed by dimensional considerations, limited from below by the minimum grid size $l_{m-} = \Delta R_{min}$ and from above by a largest turbulent vortex size assumed as $l_{m+} = 0.1R_0$ (R_0 is a typical radial dimension of the flow domain). There are no special adjustments for the empirical constant f_l =1 in the present calculations, except at an initial stage when f_l = 8 in order to damp the first very high free surface oscillation amplitude. This model for the turbulent viscosity was found to be the most satisfactory between other tried variants of mixing length expressions and $k - \varepsilon$ type models commonly used in engineering simulations, and which failed to give stable, converging results for the free surface evolution.

The heat transfer equation with the variable effective thermal diffusivity $\alpha_e(\mathbf{R},t)$ and Joule heating $|\mathbf{J}|^2/\sigma$ is

$$C_p^*(\partial_t T + \mathbf{v}\cdot\nabla T) = \nabla\cdot(C_p\alpha_e\nabla T) + \rho^{-1}|\mathbf{J}|^2/\sigma, \tag{5}$$

where T is the temperature, $\sigma(T)$ the electrical conductivity, $C_p(T)$ the specific heat, and C_p^* the modified specific heat function accounting for the latent heat as the melting front progresses according to the simple linear solid fraction function [8]. The turbulent thermal diffusivity is assumed to be equal to the turbulent viscosity ν_T.

A realistic heat transfer at the chill block requires an empirical heat transfer coefficient $h(T)$ to be used with the heat flux boundary condition:

$$-\rho C_p\alpha\,\mathbf{e_n}\cdot\nabla T = h(T - T_w), \tag{6}$$

where T_w is wall or ambient air temperature. Unfortunately the data for $h(T)$ are available in the case of solid aluminium cylinder within the temperature range up to 540°C only [4], therefore we used an approximate extrapolation for this empirical coefficient which permitted us to obtain reasonable results for the melting front propagation:

$$h = \left\{\begin{matrix} 30 & ;\text{if } T \le 120\ ^\circ C \\ 30 + 0.93(T-120) & ;\text{if } 120 < T < 540\ ^\circ C \\ 420 & ;\text{if } T \ge 540\ ^\circ C \end{matrix}\right\} \left(\frac{W}{m^2 K^o}\right). \tag{7}$$

At the free surface of liquid metal the adiabatic condition $h=0$ is used (neglecting radiation losses).

3. Numerical solution method

The numerical solution method is a significant modification of the spectral collocation procedure, as described, e.g. in [6], in combination with a co-ordinate transformation dynamically adapting to the interface change. The initial shape is cylindrical; however, we prefer to use spherical polar co-ordinates because during the melting process the boundary of liquid metal can assume a shape which is more easily described as a deformed sphere. The spherical co-ordinates (R,θ,ϕ), with the origin in the middle of the melted body, are transformed to computationally convenient new co-ordinates:

$$\overline{R} = R_0^{-1}(1+s(\mu,t))^{-1}R\,, \qquad \overline{\mu} = \mu = \cos\theta\,, \quad \overline{t} = t\,, \tag{8}$$

where the continuous function $s(\mu,t) = s(\overline{\mu},\overline{t})$ depends on the polar angle and time, R_o is a reference radius which in our case is the initial cylinder radius. With the transformation (8) the deformed external boundary $R{=}R_0(1{+}s(\mu{,}t))$ is always at $\overline{R}=1$. The equations and boundary conditions (1)-(6) are first expressed in spherical co-ordinates and then transformed to the new co-ordinates (8) with the appropriate expressions for the derivatives:

$$\partial_R = \frac{1}{\mathrm{R}_0(1+s)}\partial_{\overline{R}}\,, \quad \partial_\mu = \partial_{\overline{\mu}}\mathrm{v} - \frac{\overline{R}\partial_{\overline{\mu}}s}{1+s}\partial_{\overline{R}}\,, \quad \partial_t = \partial_{\overline{t}} - \frac{\overline{R}\partial_{\overline{t}}s}{1+s}\partial_{\overline{R}}\,. \tag{9}$$

The expressions for the second derivatives can be obtained by applying the operators (9) repeatedly.

The velocity vector components, the pressure and the temperature are represented as series of Chebyshev polynomials $T_n(\overline{R})$ and Legendre functions $P_n(\mu)$, $P_n^1(\mu)$:

$$\begin{aligned} V_R &= u = \sum_{n=0}^{N}\sum_{m=0}^{M} u_{1nm} P_n(\mu) T_{2m}(\overline{R}), \\ V_\theta &= \mathrm{v} = \sum_{n=0}^{N}\sum_{m=0}^{M} u_{2nm} P_{n+1}^1(\mu) T_{2m}(\overline{R}), \\ p &= \sum_{n=0}^{N}\sum_{m=0}^{M} u_{3nm} P_n(\mu) T_{2m}(\overline{R})\,; \end{aligned} \tag{10}$$

$$T = \sum_{n=0}^{N} \sum_{m=0}^{M} T_{nm} P_n(\mu) T_{2m}(\overline{R}) . \tag{11}$$

The transformed equations and boundary conditions are discretised on the grid:

- extrema and endpoints for $T_{2M}(\overline{R})$ on the interval $\overline{R} \in (0,1]$:

$$\overline{R}_i = \cos\left[\tfrac{\pi}{2}\left(1 - \frac{i-1}{M}\right)\right] \; ; \quad i = 2,3,\ldots.\ M + 1 , \tag{12}$$

and one additional collocation point close to the origin: the extremum of $T_{4M}(\overline{R})$;

- $P_{N-2}(\mu_j), (j = 2,3,\ldots N)$ extrema points on the interval $\mu \in (-1,1)$, which are computed numerically by the algorithm given in [6] modified for this application to achieve higher accuracy. Two additional collocation points are chosen as the extrema of $P_{2(N-2)}(\mu_j), (j = 1 \text{ and } j = N+1)$ close to the negative and positive z-axis, respectively.

For the Navier-Stokes equations (and in a similar way for the heat transfer problem) the second order implicit time discretization and iterative linearisation for non-linear terms is implemented according to the following scheme:

$$\begin{aligned} &\frac{3\mathbf{v}^k - 4\mathbf{v}^{k-1} + \mathbf{v}^{k-2}}{2\Delta t} - \mathbf{v}^{k-1/2} \times (\nabla \times \mathbf{v}^{k-1/2}) = \\ &-\nabla(\rho^{-1} p^k + \tfrac{1}{2} \mathrm{v}^{k-1/2} \mathrm{v}^{k-1/2}) + \nabla \cdot (\nu_e^{\,k-1} (\nabla \mathbf{v}^k + \nabla \mathbf{v}^{k^T}) - K_D \mathbf{v}^k + \rho^{-1} \mathbf{J}^{k-1} \times \mathbf{B}^{k-1} + \mathbf{g}, \end{aligned} \tag{13}$$

where the differential operators depending on the local co-ordinate transformation are computed at the 'k-1' level when the interface is updated. Typically 2-4 iterations are necessary for the non-linear terms at each time step to reach convergence. Occasionally the first order Euler scheme is applied when the iterations are not converging (to reduce the time step) or when the solution is sufficiently smooth to increase the time step. The procedure is stable for typical time steps Δt=2.5 ms. After the iterations for the velocity field have converged, a new interface location is found explicitly solving the kinematic condition (3). The derivatives of the function $s(\mu,t)$, defining the co-ordinate transformation, are computed using the third order local monotone Hermite interpolation procedure.

4. Electromagnetic field calculation

The electromagnetic force is recalculated at each time step according to the following procedure implemented with the same grid as the fluid dynamic equations, which ensures a high resolution within the external boundary layer because of the dense Chebyshev grid in this region. The metal charge is surrounded by a coaxial coil carrying a given AC electric current. We are interested in the electric current and magnetic field distributions within the metal, which in the axisymmetric situation are described by means of vector

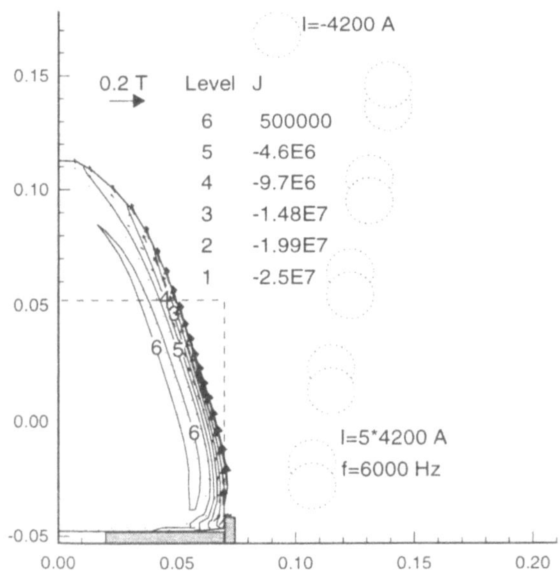

Figure 1. The axisymetric semi-levitation melting device configuration (half of the axial cross-section), and the computed magnetic field and electric current distribution within the liquid metal (Al) at the quasi-stationary final stage of melting.

potential azimuthal component $A_\varphi(R,\theta,t)$ related to the magnetic field as $\mathbf{B}=\nabla\times\mathbf{A}$.

In the AC case all functions can be expressed by means of complex exponents for the time dependence: $Q(x,t)=\mathrm{Re}[Q(x)e^{i\omega t}]$. The electric current density in this case is given by Ohm's law:

$$\mathbf{J}=-\mathrm{i}\omega\sigma\mathbf{A}\ , \tag{14}$$

and Ampere's law:

$$\nabla\times\mathbf{B}/\mu_0=\mathbf{J}, \tag{15}$$

gives the equation to determine the vector potential:

$$\nabla^2\mathbf{A}=-i\omega\mu_0\sigma\,\mathbf{A}, \tag{16}$$

which is solved by dividing the conducting region in finite cross-section rings according to the prescribed grid. Then the solution can be represented analytically as a superposition of the remote infinitely thin rings' mutual inductances, the finite section rings' self inductances M and the source coil contributions M_s , which all can be expressed by means of elliptic integrals F and E [7]. The expressions for the inductances were re-derived taking into account the local grid element areas ΔS_{mn} and the temperature dependent conductivity $\sigma_{mn}(T)$. Finally we have to solve the set of algebraic equations (in cylindrical co-ordinates r, z):

$$\begin{aligned} A_\varphi(r_i,z_j) &= \sum_{m,n} M(r_i,z_j;r_m,z_n)\, A_\varphi(r_m,z_n) \\ &+ \sum_s M_s(r_i,z_j;r_s,z_s)\, I_s(r_s,z_s)\ , \end{aligned} \tag{17}$$

where $\mathrm{M} = i\frac{\mu_0}{4\pi}\omega\sigma_{mn}\Delta S_{mn} L_{ijmn}$, $\qquad \mathrm{M}_s = \frac{\mu_0}{4\pi} I_s L_{ijss}$;

if $m \neq i,\ n \neq j,\quad L_{ijmn} = \frac{4}{\kappa}\sqrt{\frac{r_{mn}}{r_{ij}}}[(1-\frac{\kappa^2}{2})F(\kappa) - E(\kappa)]$,

$$\kappa = \sqrt{\frac{4 r_{ij} r_{mn}}{(r_{ij} + r_{mn})^2 + (z_{ij} - z_{mn})^2}};$$

$$L_{ijij} = 6\ln 2 - 3 - \ln\frac{\Delta S_{ij}}{\pi} + 2\ln r_{ij}\,.$$

r_{ij} is the distance between the discretised ring elements, I_s are the amplitude values for the total current in the source coil, which are presently assumed to be fixed.

The magnetic field is expressed analytically in terms of the elliptic integrals $F(\kappa)$and $E(\kappa)$ [7] without the need for numerical differentiation. A typical electromagnetic field distribution calculated with this model is shown in Fig. 1 for the conical design semi-levitation device [9]. Note, the current induced in the core region (level 6) is of the opposite sign to the current in the skin-layer.

The volume integral method was tested against the analytical solution for a spherical conducting body surrounded by a coaxial current loop [7]. The self-induction effects were found to be of importance in explaining such details of the field as a core region of induced current opposite to the current in the boundary skin-layer (Fig. 1) and to ensure the smoothness of the tangential magnetic field within the skin-layer.

3. Results of computation

The calculated results are presented for the case of pure Al material properties, including the temperature dependence for electrical conductivity and the heat transfer coefficient [4]. The numerical solution was tested by varying the grid resolution and by checking the total volume conservation at each time step. A typical calculation used the grid with M=15 (or 20), N=30 (or 40).

The same grid of a high resolution at the electromagnetic boundary layer, owing to the dense Chebyshev nodes distribution, was used for the electromagnetic field calculations. It was found necessary to update the

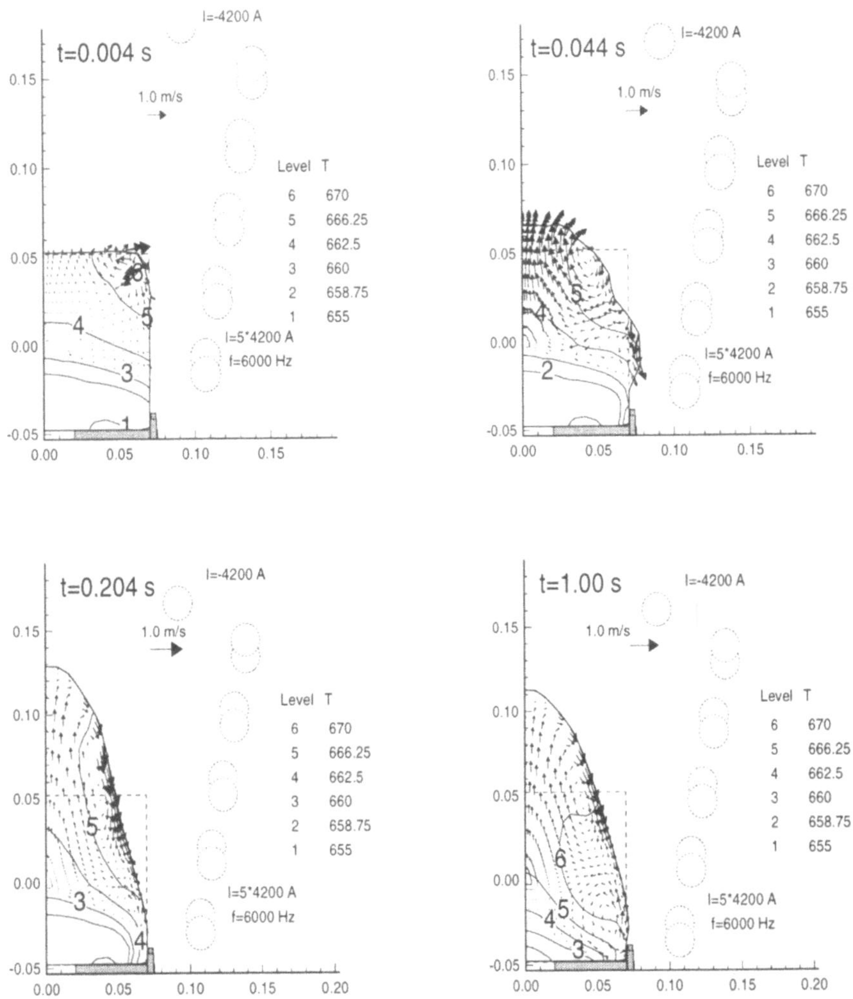

Figure 2. The computed development of the free surface, melting front, velocity and temperature fields for the semi-levitation melting (half of the axial cross-section). The source coil moves vertically with the melt front (isotherm 3) propagation.

electromagnetic field at each time step to prevent unstable situations when the metal tends to "leak out".

The calculation of the temperature field was started at room temperature letting the solid cylinder to be heated by Joule heat. At this stage the coil was located stationary with the bottom positioned at the middle of the billet. The heating stage was run with large time steps (1 s) until approximately a half of the metal at the top part has reached the temperature exceeding the Al melt temperature (660 deg C).

The hydrodynamic calculation was initiated, and the starting flow and temperature fields are shown in the first picture of Fig.2. The coil was moved with the melt front propagation. The initial time steps for hydrodynamics calculations were 2 ms.

The second frame shows a dangerous initial situation when the liquid metal bulged out at the bottom. However, the electromagnetic field was strong enough to prevent a spill as the liquid interface comes closer to the source coil and since the coil was moved, too. It is remarkable that the bottom perimeter corner was melted quite fast because of the vicinity of the source coil ensuring the holding force.

The bottom bulging was fully recovered at later stages and a dramatic increase in the dome height followed. Owing to inertial effects, the top liquid was pushed too high, then it fell back in a series of oscillations detectable at the top and side surface. The largest velocities were observed at the time moment when the gravity induced downflow was superposed onto the rotational electromagnetic force induced circulation. Since the tangential stress is zero by the boundary condition at the free surface, an intense shear layer was developed along the side surface.

At the final quasi-stationary stage the double vortex structure induced by the rotational electromagnetic force was observed. The last solid unmelted metal (see isotherm 3) remained in the middle of the bottom up to the last moments.

The second considered case is for a cylindrical crucible with free top surface for the aluminium experiment case where the measured surface contour data are available [5]. The flow field and the final stationary interface shape computed with the present spectral code are shown in Figure 5 where the measured interface locations are shown for comparison. The correspondence is quite satisfactory.

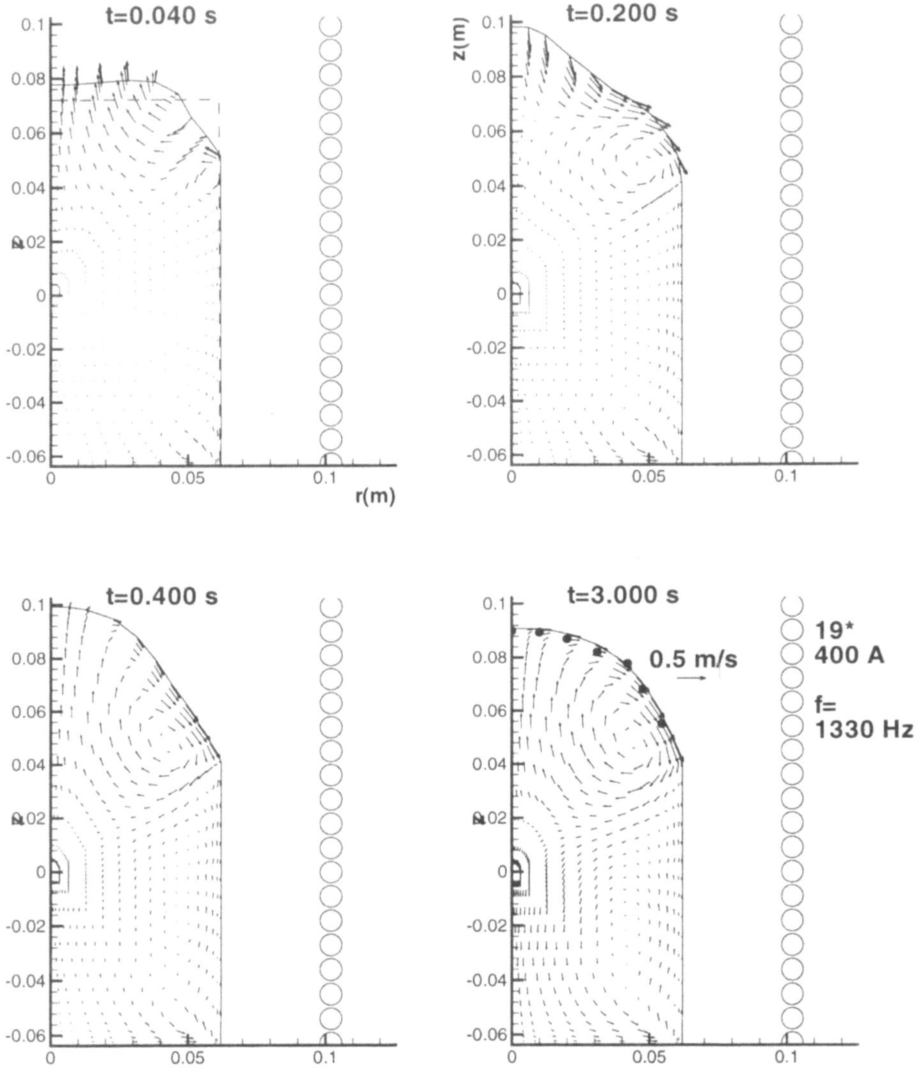

Figure 3. The computed velocity field and the free surface for the induction melting in cylindrical crucible (half of the axial cross-section) as compared to the measured liquid metal dome shape [5]. o – measured position.

The most difficult numerical problem here is the soft unknown contact position of the free surface at the solid side wall. The situation is quite similar to the electromagnetic caster problem where the liquid metal is supported by the magnetic field above the solidification line, without a contact to the mould wall. The initial dynamics of the interface demonstrate gravity-inertial wave oscillations: the inertia dominated high dome shape followed by a reflected wave from the "electromagnetic side wall".

References

1. Bhamidipati, J.R. and El-Kaddah, N. (1991) Calculation of Electromagnetic Field and Melt Shape in the Magnetic Suspension Melting Process, in J.Szekely et al. (eds.), *MHD in Process Metallurgy*, Minerals, Metals&Materials Society, pp. 69-74.
2. Sneyd, A.D. and Moffatt, H.K. (1982) Fluid Dynamical Aspects of the Levitation-Melting Process, *J.Fluid Mech.* **117**, 45-70.
3. Mestel, A.J. (1982) Magnetic Levitation of Liquid Metals, *J.Fluid Mech.* **117**, 27-43.
4. Bhamidipati, J.R. (1995) *Mathematical Modelling of Electromagnetic Field, Free Surface, Heat Transfer, and Fluid Flow in Electromagnetic Confinement Systems,* Ph.D.Th., University of Alabama.
5. Lavers, J.D. and Ahmed, M.R. (1988) Mathematical Modelling of Electromagnetic Confinement, in Y.Sahai et al. (eds.), *Casting of Near Net Shape Products*, The Metallurgical Society, pp.395-410.
6. Canuto, C., Hussaini, M., Quarteroni, A., and Zang, T. (1988) *Spectral Methods in Fluid Dynamics*, Springer, Berlin.
7. Smythe, W.R. (1989) *Static and Dynamic Electricity,* Hemisphere Publishing Corp., New-York.
8. Voller, V.R., Cross M., and Markatos N.C. (1987) An enthalpy method for convection/diffusion phase change, *Internat. J. Numer. Meth. Eng.* **24**, 271-284.
9. Harding, R. A. and Zhu, X. R. (1997) The future development of the semi-levitation melting technique for the production of castings in reactive alloys, in *Proceedings of the International Congress of Electromagnetic Processing of Materials* **1**, Centre Francais de l'Electricite, Service Diffusion de la Documentation, Paris, pp. 165-170.

MEASUREMENT OF SOLUTE DIFFUSIVITY IN ELECTRICALLY CONDUCTING LIQUIDS

T. ALBOUSSIÈRE
Department of Engineering
University of Cambridge
Trumpington Street, Cambridge CB2 1PZ

J.P. GARANDET
Commissariat à l'Énergie Atomique, DTA/CEREM/DEM
LRBS au CENG, 38054 Grenoble Cedex 9

AND

P. LEHMANN AND R. MOREAU
Laboratoire EPM-Madylam, CNRS UPR A 9033
ENSHMG BP 95, 38402 Saint Martin d'Hères

Abstract. The use of a magnetic field in the measurement of diffusion coefficients in liquid metals or semi-conductors is analyzed. Previous studies have demonstrated the disturbing influence of buoyancy-driven convection on the species transport in the shear-cell technique. A theory of effective diffusivity was proposed to account for the convective part of the transfer (GBD95). The idea developed here is to combine this effective diffusivity with the Lorentz electromagnetic damping of convection. The expected influence of the magnetic field is that the convective contribution should scale as Ha^{-4}. Experiments were carried out and, for a small convection level, this magnetic influence was confirmed. It is suggested that a strong steady magnetic field could enable the precise measurement of liquid metal alloys diffusivities. The present analysis gives the required magnetic field magnitude.

1. Introduction

For two independent reasons, it is important to measure accurately the chemical diffusivities in liquid metals. Firstly, some fundamental studies

A. Alemany et al. (eds.),
Transfer Phenomena in Magnetohydrodynamic and Electroconducting Flows, 359–372.

of the structure of liquid metal could use precise measurements to select one diffusion theory among all of them. Secondly, precise values are now needed to feed the powerful modelling methods in crystal growth or metallurgy. However, due to the extremely small values of these solute diffusivities of order $10^{-9}m^2.s^{-1}$, any attempt to measure them is a challenge: during an experiment of diffusion where the liquid is supposed to be at rest, a small reminiscent motion is likely to contribute more to the mass transfer than pure diffusion itself (GBD95; ARR96; HR86). The method of the long capillary seems to be the most widely used and this paper addresses the question of the influence of parasitic convection during its application. The compositions being assumed small[1], the undesirable motion takes its origin in thermal buoyancy forces. A careful thermal design of the experimental device helps reducing these forces. Nevertheless, it has been necessary to carry out measurements in the microgravity environment to obtain reasonably precise diffusivity values for some alloys. Another approach is investigated here, relying on the damping of the movements in an electrically conducting liquid that a steady magnetic field should induce. After a presentation of the experimental technique in section (2), the model of effective diffusivity (first presented in (GBD95)) aiming at modelling the mass transport induced by convection will be recalled (section 3). How a magnetic field damps the convective contribution to mass transport is analyzed in (4) in the framework of the effective diffusivity. Section (5) is devoted to our recently started experimental program and section (6) to a discussion of the use of magnetic fields for the purpose of measuring precise values of solute diffusivities.

2. The shear cell technique

In a purely diffusive transfer, the mass transfer density field $\mathbf{J}$ is modelled by Fick's law in the limiting case of dilute alloys: $\mathbf{J} = -\rho D \nabla c$, where ρ, D, c denote the constant and uniform density, solute diffusivity and solute concentration expressed in mass fraction respectively. A configuration is settled in such that a one dimensional diffusion is expected. The initial state is a step function of concentration along the axis where diffusion will proceed, which we denote the x-axis. The conservation of solute mass ρc along with the expression of the diffusion flux $\mathbf{J}$ leads to a local equation of conservation in the x direction: $\rho\, \partial c/\partial t + \partial J_x/\partial x = 0$, where J_x is the single non zero component of $\mathbf{J}$. The solution of this local equation is well

[1]This is the so-called dilute approximation

known since the 19th century results in heat transfer and is an erf function

$$c(x,t) = c_1 + \frac{c_2 - c_1}{2}\left[1 + erf\left(\frac{x}{2\sqrt{Dt}}\right)\right], \qquad (1)$$

when the initial condition is $c = c_1$ for $x < 0$ and $c = c_2$ for $x > 0$. The knowledge of the composition at a time T enables to derive the diffusivity.

The originality of the shear cell technique (see figure 1) initially proposed at the CEREM (Pra89), is aimed at creating the conditions of one-dimensional diffusion from a step function and at providing the possibility of "freezing" the composition function at a time T, determined by the operator. A thin capillary of diameter of order 1 mm is formed by the alignment of a number of discs where an eccentric circular hole has been drilled . The material of the discs forms then the boundary of the long capillary. All the discs are liable to rotate around a common central axis. Three configurations are needed during a measurement experiment. Firstly, the long capillary is split in two half capillaries, allowing their filling with the alloy at different compositions. Secondly, the full capillary should be formed, creating the initial step function. Thirdly, after a duration T of diffusion prescribed by the operator, all fractions of the capillary should be isolated, *e.g.* by rotating every other disk. How these steps can be performed is a matter of subtle technology. After the last step, the temperature is decreased to room temperature and the solid elementary alloy samples contained in each disk are globally analyzed. The mean value of the axial solute distribution on each disc length after a diffusion duration T is obtained. By comparison with the solution (1), an experimental value of diffusivity D is derived.

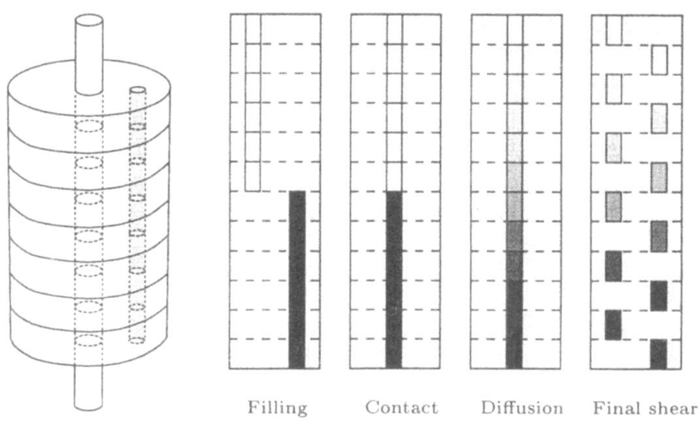

Figure 1. Shear cell sketch and principle

The figure 2 shows the evolution of the purely diffusive concentration profile given by the expression 1. Typically, the distance over which the composition gradient is spread scales as $\sqrt{Dt}$ with time.

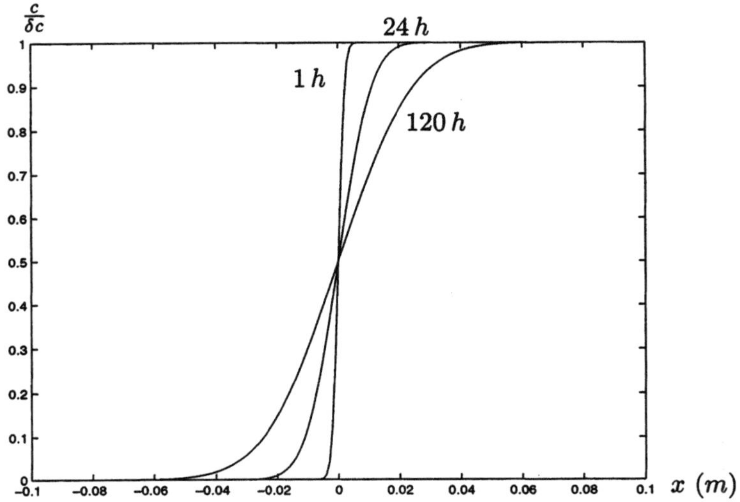

Figure 2. Pure diffusion profiles for SnBi after 1 hour, 24 hours and 120 hours ($D = 1.5 \times 10^{-9}\, m^2.s^{-1}$)

3. Convection and effective diffusivity

The previous section describes the ideal purely diffusive case. In this section, the effect of a weak convection on mass transport is addressed. A steady divergence free velocity field $\mathbf{v}$ is supposed to take place in the capillary with a zero normal component at the boundaries. In comparison with earlier works in the field, it is not necessary to suppose that the flow is one-dimensional. Nevertheless its variations in the direction of the x axis are assumed to take place on a larger length scale than the capillary diameter.

Fick's law for solute transfer holds in the reference system where matter is at rest. When expressed in the laboratory system of reference, it becomes $\mathbf{J} = -\rho D \nabla c + \rho c \mathbf{v}$, in the presence of convection. It remains possible and enlightening to express a local solute conservation equation along x. The basic experimental observable is the mean value of the concentration in the plane orthogonal to the x-axis at a given coordinate x, denoted $c_0(x,t)$, while the variations of concentration within this plane are represented by $c_1(x,y,z,t) = c(x,y,z,t) - c_0(x,t)$. The conservation of species along x takes the form $\rho\, \partial c_0/\partial t + \partial < J_x >/\partial x = 0$, where $< >$ denotes the average in the plane perpendicular to the x axis. A zero net global mass flux being assumed in the x direction, because the fluid is confined in a closed cavity, it is not difficult to derive the following form $< J_x > = -\rho D\,(\partial c_0/\partial x) + \rho < v_x c_1 >$, where v_x is the x component of the velocity field. The solute equation along x can now be written

$$\frac{\partial c_0}{\partial t} = D\frac{\partial^2 c_0}{\partial x^2} - \frac{\partial < v_x c_1 >}{\partial x}. \tag{2}$$

It is clear from equation (2) that the field c_1 has to be considered in order to obtain a proper equation for c_0. The local point-wise solute conservation equation is $\rho\,\partial c/\partial t + div\,\mathbf{J} = 0$, which can be expanded as $\partial c/\partial t + \mathbf{v}.\nabla c = D\,\Delta c$. Hopefully, the difference between this isotropic equation and the x equation (2) should essentially involve c_1, the variation of the composition in the plane perpendicular to the x axis

$$\frac{\partial c_1}{\partial t} + \mathbf{v}.\nabla c_0 + \mathbf{v}.\nabla c_1 - \frac{\partial}{\partial x} < v_x c_1 > - D\,\Delta c_1 = 0. \tag{3}$$

Our primary goal being the analysis of the departure of species transfer from the purely diffusive case, we will treat convection as a small parameter. We can anticipate that the perturbation $\mathbf{v}$ is at the origin of a field c_1 of the same order, and both $\mathbf{v}.\nabla c_1$ and $(\partial/\partial x) < v_x c_1 >$ are second order terms in (3) compared to $\mathbf{v}.\nabla c_0$. For a weak convection it is also safe to neglect the first term $\partial c_1/\partial t$ in comparison with $D\,\Delta c_1$: these terms should be comparable if only axial variation was looked at, but by definition of c_1, and due to the aspect ratio of the cell, its variations in the cross-section are very large compared to its axial variation[2]. The first order disturbance form of (3) is

$$\Delta_S c_1 = D^{-1} v_x \frac{\partial c_0}{\partial x}, \tag{4}$$

where Δ_S denotes the Laplacian operator in the cross section and where ∇c_0 was replaced by its single non zero component $\partial c_0/\partial x$. The field c_1 must satisfy a zero normal gradient at the circular boundary. The solution of equation (4) is $f\,D^{-1}\partial c_0/\partial x$, where f is solution of $\Delta_S f = v_x$ and admits a zero normal gradient on the boundary. Using that result, the term $< v_x c_1 >$ takes the form $< v_x f > D^{-1}\,\partial c_0/\partial x$ and the basic equation for c_0 (2) can be written

$$\frac{\partial c_0}{\partial t} = \frac{\partial}{\partial x}\left[\left(D - \frac{< v_x f >}{D}\right)\frac{\partial c_0}{\partial x}\right]. \tag{5}$$

[2]Here is also made use of the assumption that the axial scale of variation of convection is much larger than the diameter.

The term in brackets is called an effective diffusivity (GBD95), $D_{eff} = D - < v_x f > /D$, by analogy of (5) with a diffusion equation. In the case of a fully-established convection along the axis, the effective diffusivity is independent of x and the equation governing c_0 becomes strictly identical to that of pure diffusion: given an experimental measurement of a diffusivity, it is impossible to guess what part of it is due to convection (see (GPVVA97)).

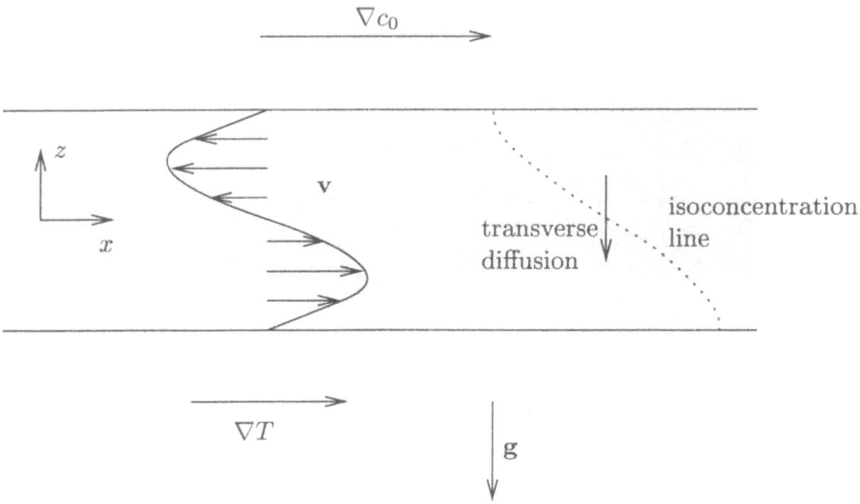

Figure 3. Convection in the capillary and the associated curvature of the iso-concentration lines

Some characteristics of the effect of convection on the apparent diffusivity (see equation 5) are now discussed (see reference (GBD95) for the original presentation of the effective diffusivity). It should be noted first that the term $< v_x f >$ is quadratic in the velocity. Indeed, because of the simple equation $\Delta_S f = v_x$, the function f depends linearly on the velocity v_x. A simple physical explanation can be given for the reason why the convection dimensionless parameter intervenes to the square in the expression of the effective diffusivity. The velocity field acts linearly on the mean axial concentration gradient $\partial c_0 / \partial x$ to produce a lateral concentration difference. The convective transport contribution $< v_x c_1 >$ is then the product of the convective velocity by the lateral concentration. It is also striking that the convective contribution is inversely proportional to the "true" diffusivity D. This is related to how the lateral concentration difference is created. The interaction between the velocity field and the axial concentration gradient is a production term of lateral concentration, whereas lateral diffusion tends to limit the lateral concentration differences. So the equilibrium lateral concentration is inversely proportional to D and so is the convective transport contribution to the overall transport.

4. Effect of a magnetic field

A simple configuration is now specified: the capillary of diameter H is assumed to be horizontal and an axial uniform temperature gradient G is supposed to exist (see figure 3). It has been shown in (GAM92), that this configuration was analogous to a vertical capillary in a transverse temperature gradient, since only the curl of the buoyancy driving force is to be considered. Let us denote z the direction opposite to the gravity vector and y the orthogonal direction to the x and z axis. The solution of the fully-established[3] buoyancy driven convective flow in the cylindrical cell is (Bej84)

$$v_x = \frac{Gr\,Sc}{32}\left[4\,Z^3 + 4\,Z\,Y^2 - Z\right]\frac{D}{H}, \tag{6}$$

where $Z = z/H$, $Y = y/H$ and $Gr\,Sc = (\beta\, g\, G\, H^4)/(\nu\, D)$ can be considered as the product of a Grashof number and a Schmidt ν/D number (β, g and ν denote the volume expansion coefficient, gravity and the kinematic viscosity, respectively). Following the previous section analysis, the disturbance c_1 is derived analytically as the solution of a 2D harmonic equation and the resulting effective diffusivity is determined analytically , and the mean value of its product with v_x is exactly calculated

$$c_1 = \frac{\partial c_0}{\partial x} H \frac{Gr\,Sc}{768}\left[4\,Z^5 + 8\,Y^2 Z^3 - 3\,Z^3 + 4\,Y^4 Z - 3\,Y^2 Z + Z\right]. \tag{7}$$

Equation (5) admits the following expression

$$\frac{\partial c_0}{\partial t} = D\left[1 + \frac{7\,(Gr\,Sc)^2}{11796480}\right]\frac{\partial^2 c_0}{\partial x^2}. \tag{8}$$

$$D_{eff} = D\left[1 + \frac{7\,(Gr\,Sc)^2}{11796480}\right]. \tag{9}$$

[3]The question of the stability of this flow is not relevant since inertia is likely to be negligible. We are indeed interested in regimes for which the effective diffusivity is not massively larger than the molecular diffusivity $D_{eff} \sim D$ and it was shown (GPVVA97) that the convective contribution to D_{eff} is of order DPe^2, where $Pe = VH/D$ is the solutal Péclet number. The kinematic viscosity being much larger than the species diffusivity for liquid metals and semiconductors, the Reynolds number VH/ν will be much smaller than unity if we restrict the analysis to Péclet numbers of order unity or less.

The increase of the effective diffusivity is quadratic in the governing non dimensional parameter $Gr\, Sc$.

In the same configuration, a transverse uniform vertical magnetic field is applied. When its magnitude B is such that the Hartmann number $Ha = \sqrt{\sigma/(\rho\nu)}BH$ (where σ denotes the electrical conductivity) is very large compared to unity, the convective velocity profile for electrically insulating boundaries tends towards (AGM93)

$$v_x = -\frac{Gr\, Sc}{Ha^2}\, 2\, Z\, \frac{D}{H}, \tag{10}$$

except in the small Hartmann layer of typical thickness $Ha^{-1}H$ at the boundary. and the analytical derivation of D_{eff} yields now

$$c_1 = \frac{\partial c_0}{\partial x}\, H\, \frac{Gr\, Sc}{16\, Ha^2}\left[4\, Z^3 + 4\, Y^2 Z - 3\, Z\right]. \tag{11}$$

The equation of effective diffusivity (5) becomes now

$$\frac{\partial c_0}{\partial t} = D\left[1 + \frac{7\, (Gr\, Sc)^2}{384\, Ha^4}\right]\frac{\partial^2 c_0}{\partial x^2}. \tag{12}$$

$$D_{eff} = D\left[1 + \frac{7\, (Gr\, Sc)^2}{384\, Ha^4}\right]. \tag{13}$$

The magnetic field reduces the convective part of the effective diffusivity as Ha^{-4}, which makes it a quite promising tool to achieve good measurements on earth with the shear cell technique.

5. Our experimental program

In EPM-Madylam, an experimental device, EURIDICE, has been designed for the measurement of diffusivities using the shear cell technique under magnetic fields up to $0.75\, Tesla$. Contrary to many other experiments where thermal buoyancy forces are made as small as possible, a small constant temperature gradient (of order $0.5\, K.cm^{-1}$) is created so that the driving force is known. In order to achieve both temperature homogeneity or constant axial temperature gradients, the cells are placed into a graphite cylinder heated by three heating elements. An insulating material is placed around the graphite (Microtherm, $K = 0.03\, W.m^{-1}.K^{-1}$ at $500\, ^oC$) and the external envelope is maintained at a constant temperature by a thermally regulated water circulation. The heating element in the middle compensate radial losses and the elements at each end prevent from axial losses.

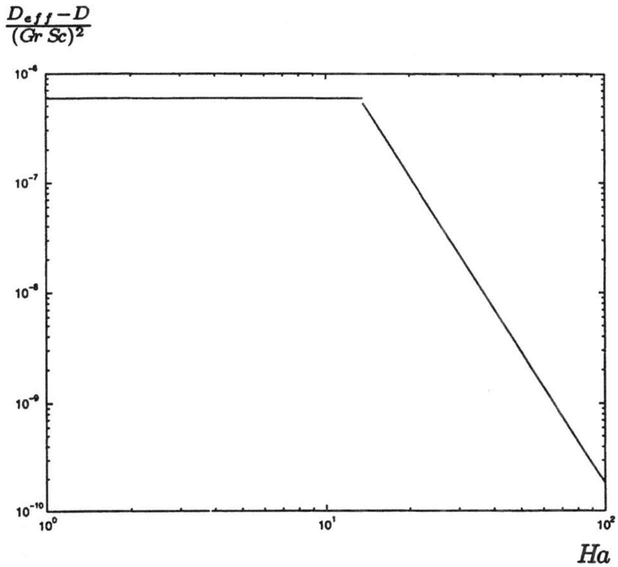

Figure 4. Dependence of the effective diffusivity on the Hartmann number. The transition between expression 9 without magnetic field and expression 13 for strong magnetic fields, occurs for a Hartmann number around 13.

Temperature measurements are made with thermocouples passed through both the graphite cylinder and a hole in the shear cells. When equilibrating carefully the three heating elements, a $0.03\,^oC$ homogeneity is possible at $250\,^oC$ along the $20\,cm$ long capillaries. On one side, a piece of metal links the graphite cylinder to the water cooled external envelope. It allows heat losses : a perfect axial and linear temperature gradient can then be created by increasing the regulation temperature on one side and decreasing it on the other. The whole device is held by a vibration insulated support and placed in an electromagnet. Vertical magnetic fields up to $0.75\,T$ can then be applied.

The tin bismuth alloy, with low concentration in bismuth, has been one of the studied alloys (see figure 5). A concentration of $0.5\,at.\%$ of bismuth was put on one half and pure tin on the other half capillary. Despite this low concentration, it was found that solute-buoyancy effects were more important than thermal-buoyancy forces by a factor ten. In that case, the previous analyses does not apply directly, because the driving force of the flow depends on the axial solute gradient, which evolve in time when diffusion proceeds. Here follows a modification of the analysis developed in section 3 to this case. Concerning the concentration analysis, the equations are unchanged up to equation 4. The only difference is that the velocity appearing is due to solute buoyancy forces rather than thermal effects. Let

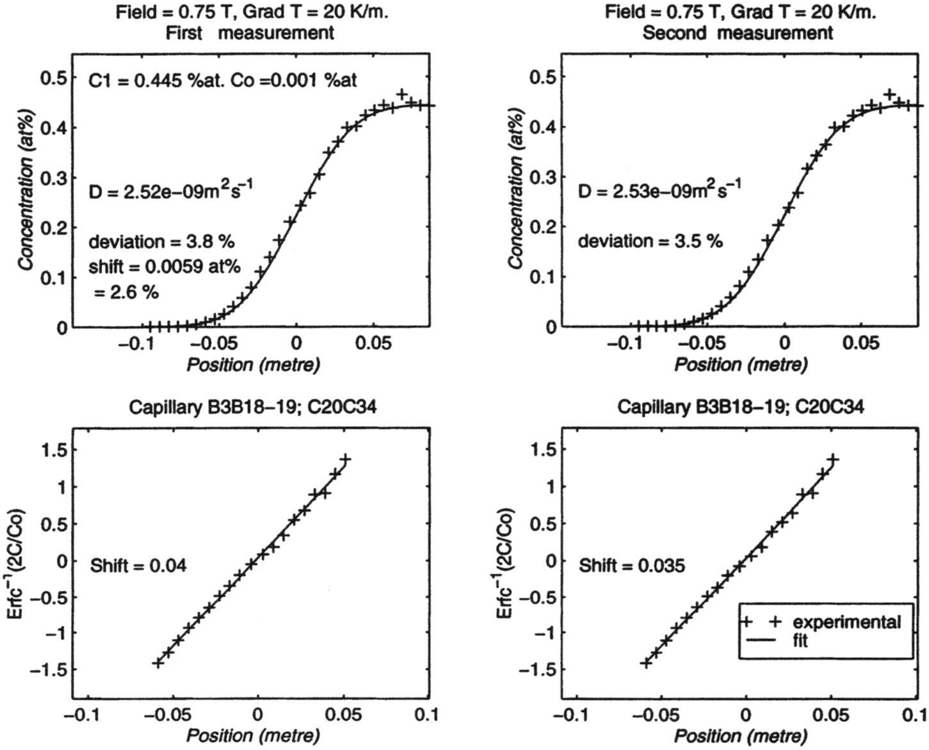

Figure 5. Concentration profile and associated erf function for SnBi at $0.5\,at\%$ in the horizontal configuration under a vertical magnetic field of magnitude $0.75\,T$.

us introduce a solute Grashof number,

$$Gr_c = \frac{\beta_c g \frac{\partial c_0}{\partial x} H^4}{\nu D}$$

where β_c is the compositional expansion coefficient of the alloy. This Grashof number depends on the axial concentration gradient, that depends on the axial position and on time. Nevertheless, the velocity driven by solute buoyancy forces takes the same form as previously where Gr_c is substituted to Gr. Equation 5 is still valid, but the convective contribution will now depend implicitly on the axial coordinate x through $\partial c_0/\partial x$ A difference appears when the corresponding expression to 8 is derived. A multiplicative factor 3 results from the variation of Gr_c with x. Without magnetic field, it is found that the axial concentration satisfies the following equation:

$$\frac{\partial c_0}{\partial t} = D\left[1 + \frac{21\,(Gr_c\,Sc)^2}{11796480}\right]\frac{\partial^2 c_0}{\partial x^2}. \tag{14}$$

and under the condition of a large Hartmann number, we find the expression:

$$\frac{\partial c_0}{\partial t} = D\left[1+\frac{21\,(Gr_c\,Sc)^2}{384\,Ha^4}\right]\frac{\partial^2 c_0}{\partial x^2}. \qquad (15)$$

These two equations are more difficult to handle because they are non-linear ($\partial c_0/\partial x$ appears in Gr_c) and can not be characterized by a uniform and constant effective diffusivity. Instead, the effective diffusivity varies along the capillary (being maximum at the center where composition gradients are maximum) are evolves in time (decreasing in time because the initial composition step is smoothed). Nevertheless, the results showed clearly an effect of the magnetic field. The effective diffusivity measured seems almost uniform and decreased with increasing the magnitude of the magnetic field. The Gr number can be roughly estimated by a sort of averaged value. We have access to a composition gradient at the end of the experiment (which is the minimum over the experiment), but we can consider it at the middle of the capillary (where it is maximum). It is hoped that this composition gradient will be characteristic of the typical solute buoyancy driven forces during the whole experiment. It is estimated that the thermal Grashof number is $Gr \sim 5$, whereas the solute Grashof number is $Gr \sim 44$. Clearly, solute buoyancy forces dominate. In the absence of magnetic field, the associated effective diffusivity is enormous, and indeed the experiments showed a complete homogeneity of the composition. With $B \sim 0.75\,T$ (corresponding to a Hartmann number of 52 for a $2\,mm$ capillary), the effective diffusivity estimated is (with $Sc \sim 150$)

$$D_{eff} = 1.33 \times D. \qquad (16)$$

Considering that the reference value is around $D = 1.5 \times 10^{-9} m^2.s^{-1}$ for SnBi, we predict an effective diffusivity around $2.26 \times 10^{-9} m^2.s^{-1}$, not far from the measured $2.52\times 10^{-9} m^2.s^{-1}$, on figure 5. On figure 5 is also plotted the erf^{-1} function that shows that the compositional profile is not far from an erf function, despite the fact that the effective diffusivity is not strictly a constant.

On the figure 6 is plotted the value of the measured diffusivity coefficient versus the different values of the solutal Grashof-Schmidt product squared divided by the Hartmann number to the power four. The measured diffusivity increases with the this convective parameter. When this parameter is relatively small, the effective diffusivity theory developed above can be applied and a linear fit of the first points allows us to prolongate the curve at the origin, which should correspond to the purely diffusive case. The value found $D = 1.66 \times 10^{-9} m^2.s^{-1}$ is very close to the reference micro-gravity

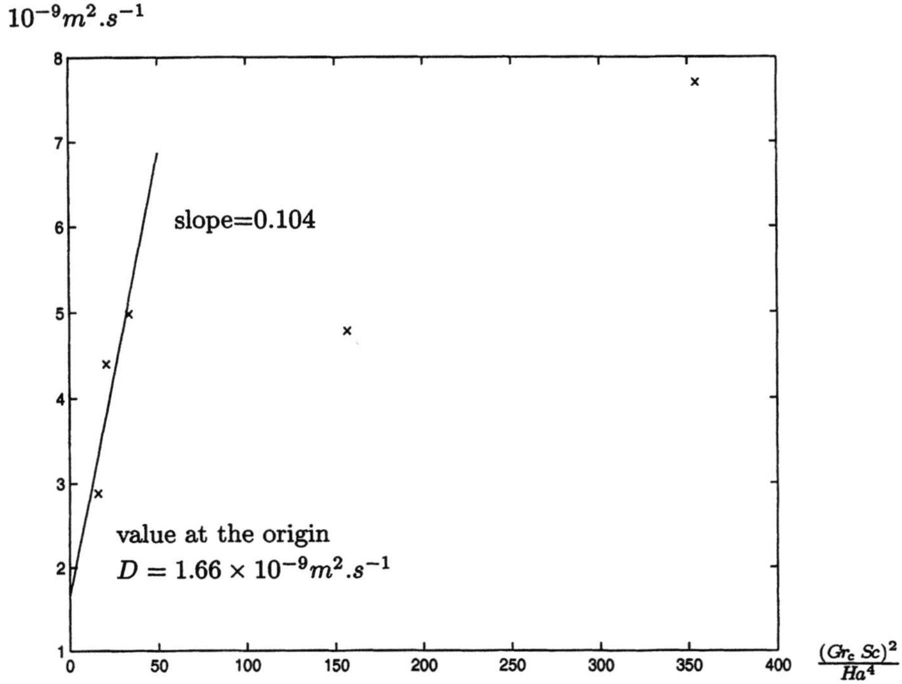

Figure 6. Measured diffusivities for different convective coefficients $\frac{(Gr_c\ Sc)^2}{Ha^4}$

value $1.5 \times 10^{-9}m^2.s^{-1}$. It should be noted that for this solute buoyancy-driven case, the measured diffusivity becomes smaller than the predicted effective diffusivity at high value of the parameter $Gr\ Sc/Ha^4$. An explanation to this behaviour is still to be found and might be connected to some solutal stratification effect that would damp the flow, somehow similar to the stratification mechanism in the vertical configuration (Har71).

Other experiments have been carried out with another alloy, SnIn. An interesting property of this alloy is that its density does not depend very much with composition. Only the thermal buoyancy forces should act on the flow. Nevertheless, some preliminary results show that even if the solutal contribution is *a priori* small, it affects significantly the overall transfer. Further investigations are necessary to clarify this aspect.

6. Discussion

We have compared an analytical model predicting the influence of buoyancy convection on the measurement of diffusivities in the long capillary technique with experimental results. The model seems to provide a good understanding of these effect and moreover, it has been seen that a steady magnetic field could help significantly to reduce convective effects. In the thermal buoyancy driven case, an exact analytical solution (erf function) can be found to the axial effective diffusion problem. In the solute buoyancy driven case, a one-dimensional unsteady non linear equation should be solved numerically in general.

The experimental results shown for the SnBi alloy confirm the analysis. A significant reduction of the apparent solute transfer has been demonstrated when a magnetic field is applied. For small convective levels, the convective contribution to the effective diffusivity varies as $(Gr\,Sc)^2/Ha^4$. This model can be useful in practice, to determine what magnetic field magnitude is necessary to provide experimental results with negligible convective contribution. With a $0.75\,T$ magnetic field, the Hartmann number achieved is 55 for a $2\,mm$ diameter capillary. At this Hartmann number, it is believed that the measured diffusivity is close to the actual diffusivity to a factor 1.3 for the SnBi alloy. According to the Ha^{-4} law, a slight increase in the magnitude of the magnetic field should permit us to measure correctly the diffusion coefficient.

Other configurations should be studied, like the vertical capillary (BG96). In that case, solute buoyancy forces can provide a stabilizing effect, instead of the driving effect here. This vertical configuration is extensively used in practice. The question to address is what should be the combined effect of the solute stabilization and a possible magnetic damping.

A general aspect concerning the use of a magnetic field to damp convective effects is related to the thermoelectric effect (MGKF96). If the crucible is not electrically insulating, electric currents of thermoelectric origin exist and generate a driving force when a magnetic field is applied.

Finally, it was suggested (YCC64) that a magnetic field could change the diffusivity as a physical property; however, it is not clear whether the experiments mentioned showed a true variation of D or a magnetic damping of an initially unexpected convection.

Acknowledgment: this work has received support from the CNES under the contract 97/CNES/071/6881.

References

T. Alboussière, J.P. Garandet, and R. Moreau. Buoyancy-driven convection with a uniform magnetic field. Part I. Asymptotic analysis. *J. Fluid Mech.*, **53** :545–563, 1993.

J.I.D. Alexander, J.F. Ramus, and F. Rosenberger. Numerical simulations of the convective contamination of diffusivity measurements in liquids. *Microgravity science and technology*, **9** :158–162, 1996.

A. Bejan. *Convection Heat Transfer*. Wiley and sons, 1984.

C. Barat and J.P. Garandet. The effect of natural convection in liquid phase mass transport coefficient measurements: the case of thermosolutal convection. *Int. J. Heat Mass Transfer*, **39**:2177–2182, 1996.

J.P. Garandet, T. Alboussière, and R. Moreau. Buoyancy driven convection in a rectangular enclosure with a transverse magnetic field. *Int. J. Heat and Mass Transfer*, **35**(4):741–748, 1992.

J.P. Garandet, C. Barat, and T. Duffar. On the effect of natural convection in mass transport measurements in dilute liquid alloys. *Int. J. Heat Mass Transfer*, **38**:2169–2174, 1995.

J.P. Garandet, J.P. Praizey, S. Van Vaerenbergh, and T. Alboussière. On the problem of natural convection in liquid phase thermotransport coefficients measurements. *Phys. Fluids*, **9**(3):510–518, 1997.

J.E. Hart. On sideways diffusive instability. *J. Fluid Mech.*, **49**:279–288, 1971.

D. Henry and B. Roux. Three-dimensional numerical study of convection in a cylindrical thermal diffusion cell: its influence on the separation of constituents. *Phys. Fluids*, **29**:3562–3572, 1986.

G. Mathiak, A. Griesche, K.H. Kraatz, and G. Frohberg. Diffusion in liquid metals. *Journal of non-crystalline solids*, **207**:412–416, 1996.

J.P. Praizey. Benefits of microgravity for measuring thermotransport coefficients in liquid metallic alloys. *Int. J. Heat Mass Transfer*, **32**(12):2385–2401, 1989.

W.V. Youdelis, D.R. Colton, and J. Cahoon. On the theory of diffusion in a magnetic field. *Canadian Journal of Physics*, **42**:2217–2237, 1964.

MAGNETIC CONTROL OF CONVECTION IN LIQUID METAL HEATED FROM ABOVE

O. ANDREEV, Yu. KOLESNIKOV
Institute of Physics, University of Latvia
Salaspils-1, LV-2169 Latvia

A. THESS
Department of Mechanical Engineering
Dresden University of Technology
01062 Dresden, Germany

ABSTRACT. We investigated experimentally the buoyancy driven convection induced by localized central heating from above in cylindrical tanks filled with liquid metal eutectic alloy (low-Prandtl-number liquid). The experiments were carried out for two aspect ratios A = (radius/height) = 4.125 and 0.333. Time-averaged local temperature distributions and radial and azimuthal velocity components in the liquid at different distances from heater axis were measured. It is shown that maximum temperature gradient is centered in vicinity heater. Temperature difference in radial direction creates pressure gradient between heater and cold wall regions. This gradient leads to upraising of liquid nearly heater and its down in vicinity side cold wall. The Nusselt number tends to a nearly constant value at small aspect ratio and increases considerably at large aspect ratio in the range of heat flux density up to 400 W/cm^2. A static homogeneous magnetic field, in the case of large aspect ratio cell, decreases convective heat transfer intensity by 30% for small enough magnetic field B = 0.1 T because the convective motion in the plane parallel to the field undergoes the most dumping effect.

1. Introduction

In some applications, such as electron beam evaporation of refractory metals [1], where heat convection in low-Prandtl-number fluids dominates over heat conduction, there is a requirement to control and damp convective heat transfer. The goal of the present communication is to report a series of model experiments performed in a simple, yet realistic, prototype geometry in order to understand the structure of temperature fields and heat transfer mechanisms and to develop efficient techniques for their magnetic control [2]. More precisely, we study buoyancy driven convection in a cylindrical pool with a localized heater at the free surface and cold lateral walls.

The objective of the present study, namely the *reduction* of heat transfer, is opposite to that of earlier experimental work [3, 4], focussed on heat transfer *intensification* in a cylindrical closed flow under axial magnetic field using electrically conducting inserts and obstacles on the walls perpendicular and parallel to the magnetic field.

A. Alemany et al. (eds.),
Transfer Phenomena in Magnetohydrodynamic and Electroconducting Flows, 373–380.

2. Experimental Device, Measurement Method, and Parameters

The experimental device, sketched in Figure 1, consists of either a large-aspect-ratio (diameter $2R$ = 165 mm, height H = 20 mm) or a small-aspect-ratio ($2R$ = 60 mm, H = 90 mm) cylindrical tank. The walls of the tanks are made from copper. The working liquid is an eutectic InGaSn alloy with the following properties at 20°C. The density ρ = 6.36×10^3 kg/m^3, electrical conductivity σ = 3.3×10^6 (Ohm·m)$^{-1}$, thermal expansion coefficient β = 1.32×10^{-4} K^{-1}, heat conductivity λ = 39 W/m·K, heat capacity c_p = 366 J/kg·K, kinematic viscosity ν = 3.4×10^{-7} m^2/s, the Prandtl number Pr = 2.03×10^{-2}. The side and bottom walls contain water-cooling channels whose temperature is kept at 20±0.5°C using a thermostat. A special cylindrical dc electrical heater with diameter $2r_d$ = 12 mm was submerged into the working liquid on the depth of 0.5 mm. The heating power Q can be increased up to 0.5 kW/cm^2. The heater was thermoinsulated by compact mineral wadding and was fixed on the cover on the center of device with a position precision of 0.1 mm. The ratios between the radii of heater and side wall r_d/R are 0.017 and 0.006 for the large-aspect-ratio A_1 = R/H = 4.125 and the small-aspect ratio A_2 = R/H = 0.333 cells, respectively.

Time-averaged local temperature distributions in the liquid are obtained by a thermocouple. Thermocouple with a sensor 0.3-mm in diameter and sensitivity of 40 mV/K was made of copper and constantan wires. A specially manufactured co-ordinate device with spatial accuracy of 0.1 mm is used to displace the thermocouple in radial and vertical directions. The temperatures of the heater and of the cooling-wall are measured by the same thermocouples with sensors placed into the orifices of the heater and cold walls. The thickness of wall between thermocouples and working liquid was 0.2 mm. For measuring time-averaged temperatures, the signal from the sensor is fed into a dc voltmeter with a measurement range from 3 μV to 10V. The experimental device is subjected to a vertical magnetic field whose strength can be varied in the range B = 0-0.1 T.

We used electrical potential probe to measure the radial and the expected azimuthal velocity components. A sensor consisted of a cylindrical permanent magnet and four copper electrodes 0.25 mm in diameter electrically insulated by vanish along its length excepting electrically sensitive ends. The magnet is 4 mm in diameter and 5 mm in height. The magnetic field is perpendicular to flat surface of the magnet. Because of the velocity in the fluid is proportional to an electric potential difference created due to a

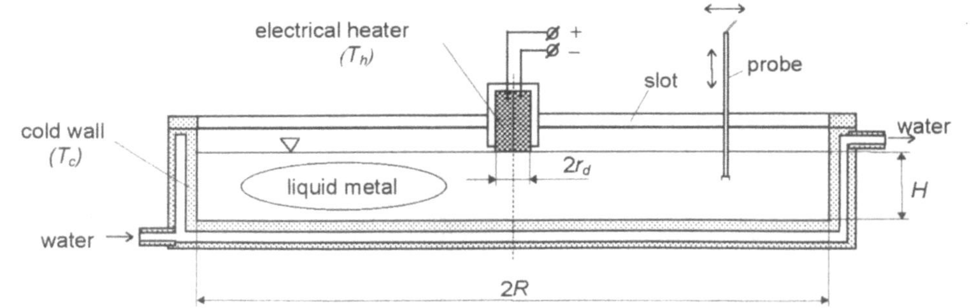

Figure 1. Sketch of the experimental device.

local interaction of moving fluid and the magnetic field, the velocity components were measured by two pairs of electrodes in intercross directions. A sensitivity of this probe was 1.2 μV·s/cm. The probe was preliminary calibrated in an isothermal flow with controlled velocity in a special stand. To exclude the influence of temperature difference (thermoelectrical current) in signal of the potential velocity probe, the small copper-constantan thermocouples with sensors 0.1-mm in diameter were glued near the ends of the potential electrodes. Coefficient of the thermal electric moving force of the pair of copper-eutectic alloy InGaSn is 2.12 μV/K.

A heat transfer from the heated zone to the flow is characterized by the Nusselt number

$$\mathrm{Nu} = \frac{\alpha l}{\lambda},$$

where α is a heat exchange coefficient

$$\alpha = \frac{Q}{\pi r_h^2 \Delta T},$$

Q the heater power, ΔT the temperature difference of the submerged heater surface and the cooled walls. The characteristic spatial scale l of the heated zone is estimated as

$$l = r_h \left(1 - \frac{\pi}{4} \frac{r_h}{} \sqrt{\left(\frac{1}{R^2} + \frac{1}{H^2} \right)} \right)$$

at $r_h < (R$ and $H)$. This scale depends on all spatial parameters of container and allows presentation of experimental results obtained in devices with moderate aspect ratios. In the device with large aspect ratio A_2 the scale l is 5.83 mm. For small aspect ratio A_1, this scale is 4.55 mm.

For each experiment we prescribe Q and B and determine the resulting temperature difference $T_0 = T_h - T_c$, and the Nusselt number $\mathrm{Nu}(Q, B) = \alpha l/\lambda$, where $\alpha = Q/(\pi r_h^2 T_0)$, $q = Q/(\pi r_h^2)$ is beat flux density. The so defined Nusselt number is proportional to $T_0/\Delta T$. The Grashof number (which is unknown *apriori!*) is evaluated as

$$\mathrm{Gr} = \frac{g\beta \Delta T l^3}{\nu^2}.$$

The influence of the magnetic field on the liquid motion can be characterized by the Hartmann number

$$\mathrm{Ha} = BH\sqrt{\frac{\sigma}{\nu\rho}}\,.$$

3. Experimental Results and Discussion

3.1. TEMPERATURE AND VELOCITY DISTRIBUTIONS

At small heat flux (q = 5 W/cm^2, ΔT = 6.3°C), measurement of the time-averaged local temperature field showed that the shape of isotherms is nearly spherical in both cells (i. e. the effective heat conductivity is closed to the molecular heat conductivity of fluids). An increase of heat flux intensity leads to a considerable deformation and extension of temperature isolines in radial direction. If there were no convection, then the temperature of the heater T_h, would increase up to its conductive value leading to a temperature difference $T_0 = T_h - T_c = cQr_h/\lambda$, where c is a geometry parameter of order unity which can be determined numerically. Due to convective heat transfer, however, the temperature of the heater decreases, and the ratio $T_0/\Delta T$ becomes a measure for the convective heat losses (where ΔT is temperature difference at convective motion development).

Time-averaged local temperature distributions in the liquid measured by thermocouple in the middle vertical plane (q = 110 W/cm^2, ΔT = 100° C) are shown in Figure 2 for large A_1 aspect ratio (temperature distributions at small A_2 aspect ratio were similar). The temperature step ΔT_s between ithotherms is 0.05 that defined as $\Delta T_s/T_h$. Indeed, such distribution of isotherms essentially differs from the spherical one at small heat flux produced by the heater. Extension of isotherms in radial direction is due to excitation of convective motion for large temperature difference ΔT at large aspect ratio. Maximum temperature gradient is centered in vicinity heater because the liquid heated from above undergoes tendency to density stratification. This difference leads to upraising of liquid nearly heater and its down in vicinity side cold wall. As a result, pressure gradient emerges between these regions in radial direction. It is that impels liquid, typical of that occurring in buoyancy-dominated situations of natural convection.

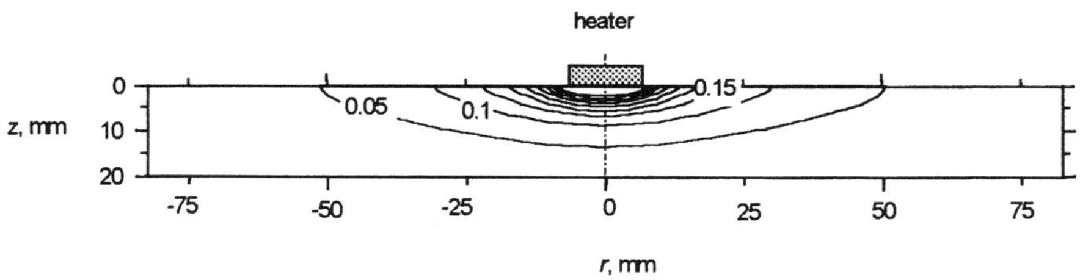

Figure 2. Time-averaged temperature distributions in cross-section of tank with a large aspect ratio A_1 = 4.125.

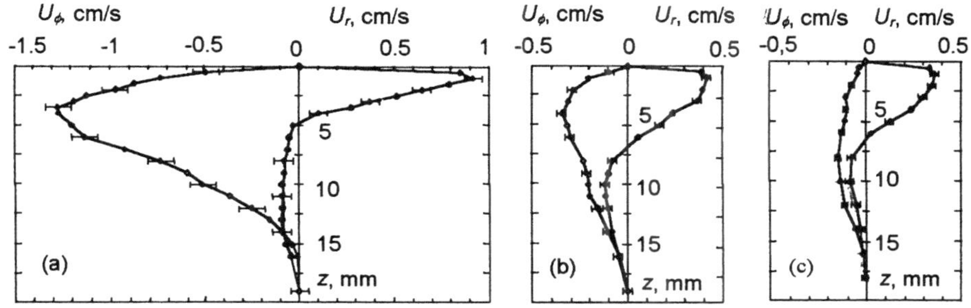

Figure 3. Profiles of radial and azimuthal velocity components in dependence on vertical coordinate at different radii $r_i = r/R$ from the heater. (a) $r_1 = 0.25$; (b) $r_2 = 0.5$; (c) $r_3 = 0.75$.

Let us analyze the measurement of velocities and excited convective motion at the same temperature regime ($\Delta T = 100°C$) at large-aspect-ratio cell A_1. Figure 3 shows a sequence of distributions of radial and azimuthal velocity components measured by the conductance anemometer in dependence on vertical coordinate z, $0<z<H$, at different distances from heater axis. We can note the existence of two horizontal layers in which positive and negative radial velocity components take place. The flow with the positive velocity is centered in upper thin layer and towards the cold wall; its maximum values lie in immediate vicinity liquid surface (zero value of velocity at $z = 0$ is caused by acidification of liquid metal on its free surface). The flow with negative radial velocity is closed in low-lying layer and towards the heater. Azimuthal velocities in measuring sections at the same coordinates are positive everywhere along z. Both velocity components diminish with increasing distance from heater. As we can see from these graphs the motion at large heat flux possesses complicated three-dimensional character.

3.2. HEAT TRANSFER WITHOUT MAGNETIC FIELD

Figure 4 shows change of the Nusselt number at large and small aspect-ratios cells as a function of the heat flux density q. Because of mentioned above spherical form of

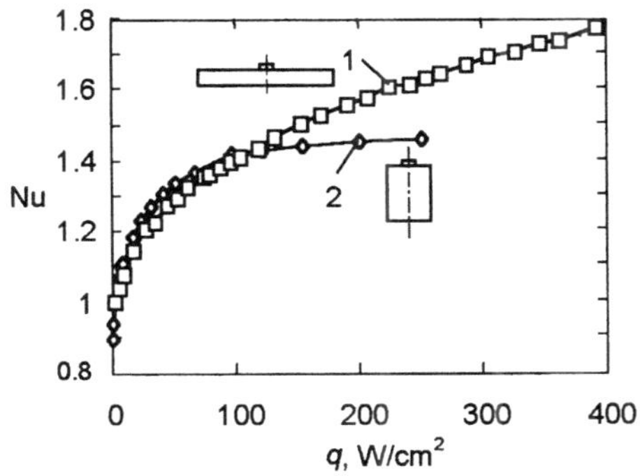

Figure 4. Dependence of the Nusselt number on heat flux density at aspect ratios $A_1 = 4.125$ (curve 1) and $A_2 = 0.33$ (curve 2).

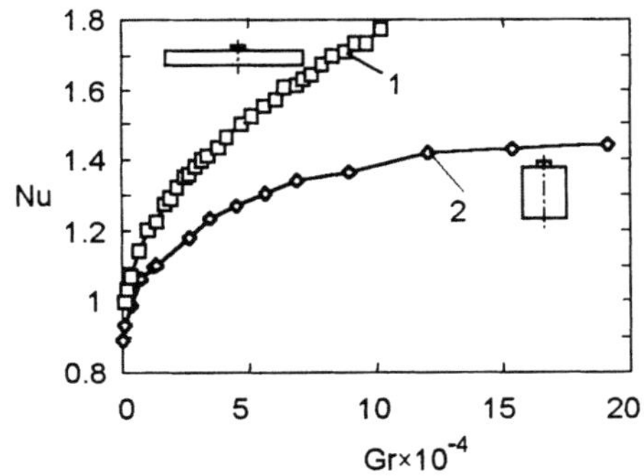

Figure 5. Dependence of the Nusselt number on the Grashof number $A_1 = 4.125$ (curve 1) and $A_2 = 0.33$ (curve 2).

isotherms occurs at both aspect ratios at small heat flux, the Nusselt number does not practically differ from unit in these cases (i. e. the effective heat conductivity is closed to the molecular heat conductivity of the fluid). With heat flux growing, the Nusselt number value asymptotically tends to the value of Nu = 1.45 at small-aspect-ratio (curve 2). If the aspect ratio is large enough (when the cold wall is far away from heater), then the Nusselt number becomes considerable, Nu = 1.8, (curve 1). Considerable value of the Nusselt number is reached due to the fact that large flow structures occur in the liquid (Figure 3). Growth of Nusselt number (and expansion of temperature isolines in radial direction shown in Figure 2) is caused by convective motion in whole liquid due to approximity of cold bottom to the heater.

Figure 5 presents the dependencies of heat transfer intensity on buoyancy effect in liquid for both cases cells in coordinates Nusselt number and Grashof number in range of heat flux density up to 400 W/cm^2. It is seen from these graphs that heat transfer with the Nusselt numbers of 1.4 for large and small aspect ratios is provided by discrepancy of buoyancy effect 3-fold. We assume that the difference of buoyancy effect is specified by nearness of bottom and far side wall in the first cell and nearness of side wall, and far bottom in the second cell, with respect to the heater. Appearing forcible horizontal temperature gradient at small aspect ratio provides most buoyancy effect then that is in large aspect ratio cell. Furthermore, nearness of thermostatic cold bottom to the heater at latter case allows tendency of liquid to achieve a stable horizontal stratification.

3.3. INFLUENCE OF MAGNETIC FIELD ON HEAT TRANSFER

Next we turn to the influence of a magnetic field on heat transfer. In this part of we consider a possibility of an influence of external magnetic field on the convective flow in the ranges of the heat parameters. The dependence of the Nusselt number on the heat flux, shown in Figure 6 for aspect ratio $A_1 = 4.125$, indicates that increasing magnetic field leads to a damping of heat transfer intensity. Convective motion produced by temperature gradient in parallel plane to the magnetic field induces electric current. Interaction of that current and magnetic field creates electromagnetic force

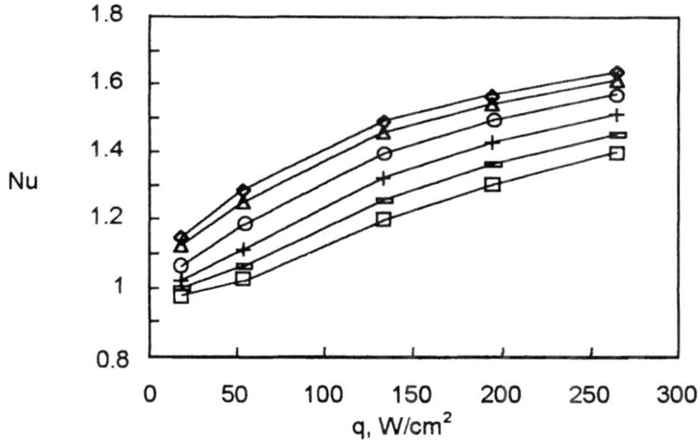

B, T: ◊, 0; Δ, 0.02; O, 0.04; +, 0.06; ▭, 0.08; □, 0.1

Figure 6. Dependence of the Nusselt number on heat flux for different magnetic field values

counteracting the convective motion. Apparently, this tendency admits suppression of convective motion at a large magnetic field.

In studies [3] at aspect ratio A = 0.71, natural convection was suppressed under 0.35 T. That heat transfer decrease is also confirmed by the data presented in Figure 6 at various values of heat flux. It is well known that using of the magnetic field allows transition of a liquid metal flow to two-dimensional (2D) state in perpendicular plane to the field [3, 4]. Motion in the plane parallel to the field undergoes the most dumping effect. It follows to point out that the magnetic field does not disable development of 2D perturbations in plane perpendicular to the field. Figure 6 shows that increasing magnetic field (the Hartmann number growth) up to 0.1 T leads to a dumping of the heat transfer intensity. Such convective motion (see Figure 3) produced by horizontal

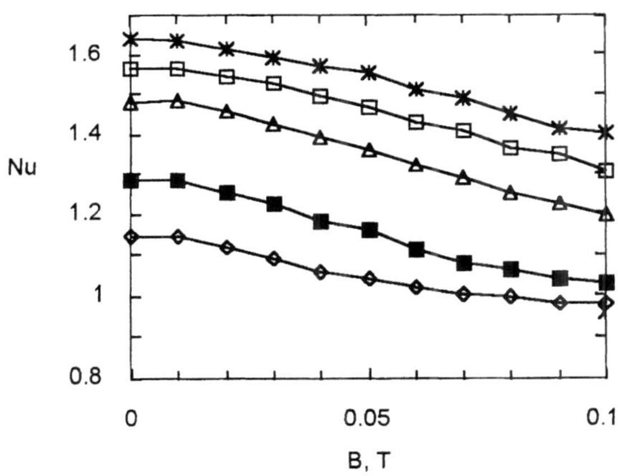

q, W/cm^2: ∗, 300; □, 220; Δ, 150; ■, 60; ◊, 20

Figure 7. Dependency of the Nusselt number on magnetic field at different heat flux.

temperature gradient in the plane perpendicular to the magnetic field induces electric current component perpendicular to the field. Interaction of that component and magnetic field creates electromagnetic force ($\mathbf{F}_{E.M.} = \mathbf{j}\times\mathbf{B}$) counteracting convective motion. As a result, heat transfer intensity decreases by 30% at even small enough magnetic field used in our experiments (B = 0.1 T). Apparently, this tendency admits suppression of convective motion at large magnitude of the magnetic field. It should be noted that vertical orientation of the magnetic field used in the experiment does not counteract development of azimuthal motion shown in Figure 3.

4. Conclusions

Heat transfer produced by buoyancy driven convection between heater, localised above the liquid, and cold walls increases with growing heat flux. The Nusselt number tends to a nearly constant value at small aspect ratio and increases considerably at large aspect ratio in the range of heat flux densities available in our experiment. Increasing axial magnetic field suppresses convective motion and thereby decreases the Nusselt number.

It should be noted that melt flow in real applications is a combination of buoyancy driven convection and surface-tension driven convection, the latter being at least as important as the former. The present study has been devoted only to buoyancy driven flow; future work will be necessary to elucidate the mechanisms of turbulent thermocapillary flows and possibilities of their magnetic control.

5. Acknowledgement

This work is supported by the Deutsche Forschungsgemeinschaft in the framework of the "Innovationskolleg Magnetofluiddynamik". The authors are grateful to Ch. Karcher, R. Schaller, Th. Boeck and Ch. Metzner for interesting discussions.

6. References

1. Schiller, S., Heisig, U. and Panzer, S. (1982) Electron Beam Technology, Wiley & Sons, New York.
2. Kolesnikov, Yu.B., Andreev, O.V. and Thess, A. (1997) Laboratory modelling of hermoconvection in a liquid metal heated from above, Workshop: Hydrodynamic Aspects of Electron Beam Technologies, Dresden, March 16-18.
3. Andreev, O.V. and Kolesnikov, Yu.B. (1994) Magnelohydrodynamic heat transfer intensification to the problems of fusion blankets, in H. Branover and Y. Unger (eds.), *Progress in Turbulent Research, Progress in Astronautics and Aeronautics,* 162, AIAA, Washington, DC, pp. 179-188.
4. Kolesnikov, Yu.B. and Andreev, O.V. (1998, in press) Heat transfer intensification promoted by vortical structures in closed channel under magnetic field, *J. Exp. Fluid Sci.*.

IV - ENERGETIC APPLICATIONS

CHANNEL DESIGN INFLUENCE ON STABILITY AND WORKING CHARACTERISTICS OF INDUCTION MHD PUMP

J.VALDMANIS, I.BUCENIEKS, Y.W.CHO*
Institute of Physics of Latvian University, 32 Miera St., Salaspils, LV-2169, LATVIA
**Korea Institute of Science and Technology, PO Box 131, Cheongryang, Seoul 130-650, KOREA*

Abstract. An induction magnetohydrodynamic (MHD) pump based on the principle of creating an alternating traveling magnetic field by system of rotating permanent magnets has been designed for pumping of liquid Wood's metal alloy through a vertical loop. The results of modeling experiments using eutectic alloy In-Ga-Sn showed that there is a good agreement between the estimated characteristics and the working parameters of real pump.

1. Introduction

For practical purposes a MHD pump must have stabilized working characteristics in a definite region of pressure and flow rate. To design an optimized MHD induction pump, it is imperative to generate maximum electromagnetic force in the channel and to minimize hydrodynamics losses (the hydraulic resistance of the pump channel). But a small hydrodynamic resistance of the channel can stimulate inner motions in it, which results in decreasing of developed flow rate and electromagnetic pressure. In the present work we have investigated the influence of channel design on the stability of working characteristics of the induction pump using a simple physical model in which alternating traveling magnetic fields were generated by permanent magnets fixed on rotating ferromagnetic octagonal columns. The idea of using rotating permanent magnets for creating a traveling magnetic field is not a new one. It was described already many years ago [1] and the example of practical application was reported recently [2]. The advantage of using permanent magnets is the possibility of creating more stronger alternating traveling magnetic field in the air gap at much lower energy consumption in comparison with linear

A. Alemany et al. (eds.),
Transfer Phenomena in Magnetohydrodynamic and Electroconducting Flows, 383–394.

inductors requiring relatively more higher power supply for their feeding to produce the same magnetic field strength.
An induction pump with rectangular channel was used in the model experiments. Several channel assemblies were tested: with and without electrically conducting side bars made from copper, as well as with wetted and not wetted inner walls. In the model experiments eutectic alloy In-Ga-Sn having low melting temperature (T_M = 9 ^{0}C) was used. Another pump, based on the same principle used in model experiments, was designed and made for pumping Wood's metal (T_M = 69 ^{0}C) through closed vertical loop. Results obtained from model experiments can be used to evaluate the pressure developed by real pump.

2. Model experiments

It is well known, that instabilities arise in the channel of induction MHD pump at large values of non-dimensional parameter ε_s (when $\varepsilon_s >> 1$), which is defined as $\varepsilon_s = 2\sigma\mu_o f \tau^2 s/\pi$, where σ is electrical conductivity of liquid metal, μ_o - magnetic permeability of vacuum, f – frequency of alternating magnetic field, τ - pole length (half length of wave of traveling electromagnetic field), and s is a slip. The slip $s = 1 - V_o / V_f$, where V_o is the velocity of liquid metal flow in channel and V_f is velocity of a traveling electromagnetic field. Using the relative velocity $V = V_f - V_o$, we can get another expression for parameter $\varepsilon_s = \sigma\mu_o \tau V/\pi$.
To maximize electromagnetic forces, hydraulic resistance of channel and inner motions must be minimized. Inner motions in the channel reduce the developed electromagnetic pressure and the efficiency of the pump. The detrimental inner motions are intensified both by electromagnetic interaction and low hydraulic resistance of pump's channel. At large value of ε_s instabilities arise and result is non-uniform pressure and hysteresis. It is assumed that, at large ε_s , the characteristic of pump ($P(Q)$) is determined by pressure at slip parameter $s = 1$. From this view the increment of flow rate does not accompany increment of pressure in any region. To investigate the processes, partly related to the mentioned phenomenon, a simple model experiments were carried out. Experimental facility is shown in Fig.1. Traveling magnetic field was created by two ferromagnetic rotators. Eight NbFeB permanent magnets were fixed on each rotator. By placing magnets with alternating poles, alternating magnetic field was made. This corresponds to four poles machine. The cross sectional dimension and the length of the magnets were 12.5 mm x 25 mm and 50 mm, respectively. The overall outer diameter of each assembled rotator was 95 mm.
Rotators were rotated synchronously by a speed-controlled motor. It was possible to fix their axes in different positions. There were two types of position.

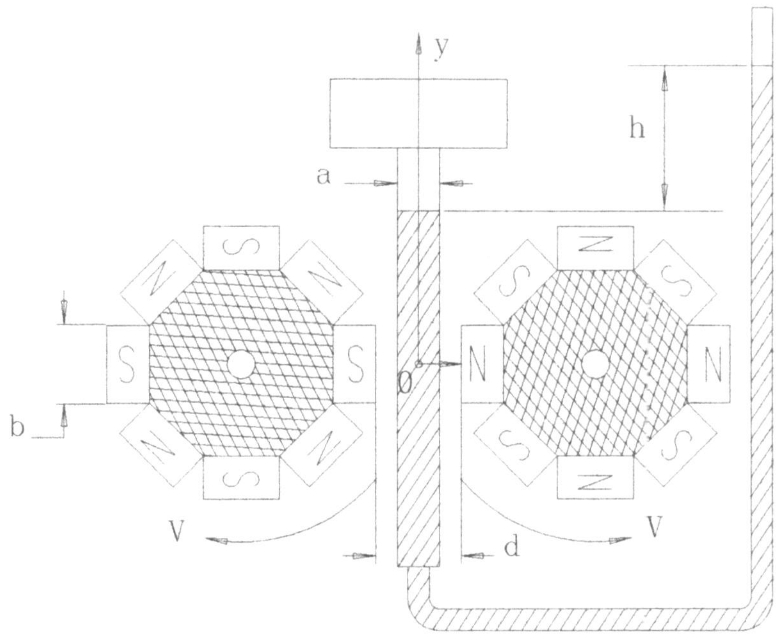

Figure 1. Experimental device for model experiments.

First, when in the gap between rotators opposite each other, there were poles of opposite polarity N – S, which corresponds to equal connection of linear inductors of induction pump, and second, when the poles of each rotator have the same polarity N - N or S – S, which corresponds to opposite connection of inductors. The length of rotators was 100 mm. Two permanent magnets were fixed along the side surfaces of octagonal column. The air gap width between two rotators and the rectangular cross-section channel dimensions were d and $a \times l$, respectively. Where a is the thickness of the channel and l is its width, which was equal to the length of rotators. Because the dimensions of magnetic bricks are relatively small, the strength of magnetic field in the air gap is not uniform and, naturally, depends on d. For the case of opposite polarity (N-S), measured magnetic field distribution between rotators is shown in Fig.2 and Fig.3. The maximum strength of constant magnetic field between not rotating rotators was approximately 1 Tesla(T) and was measured at minimum gap. In this model experiments the average value of constant magnetic field strength in the center of gap was about 0.5 T at air gap width d = 16mm. The distribution of alternating traveling magnetic field created by rotating permanent magnets along the symmetry axis y is represented on figure 3.

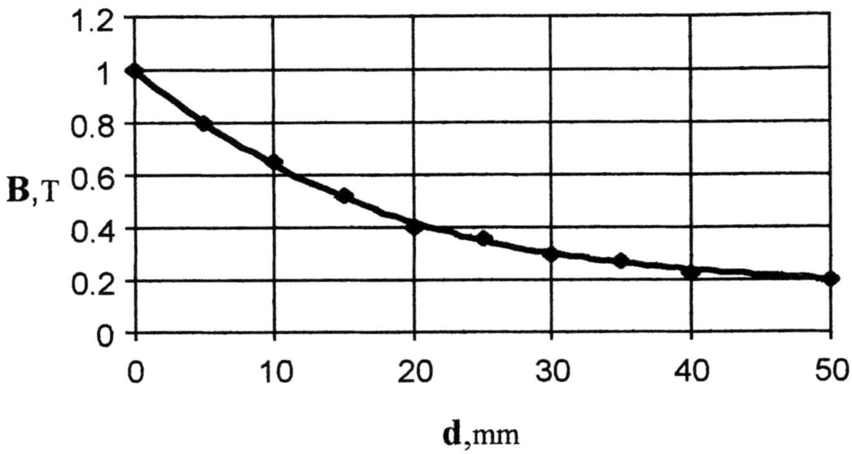

Figure 2. Strength of a constant magnetic field in the center between rotators.

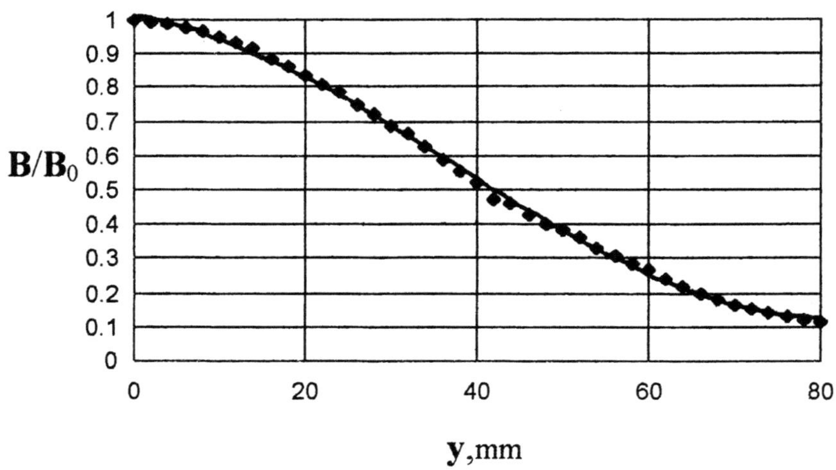

Figure 3. Distribution of alternating traveling magnetic field in the air gap between rotators.

The parameters of the channel were changed during experiments. The inner thickness of channel a (filled with liquid metal) was 10 mm and the width of liquid metal in channel l was a little smaller than the length of magnetic field (the length of rotators). The walls of channel of pump were made from transparent acrylic. The inner motions of liquid metal were observed directly through them. For measuring the developed pressure a differential manometer was connected to the vertical channel. The upper part of the channel had a wider cross-section, which served as a reservoir for overflow. Influence of static pressure on the intensity of inner motions was investigated by varying the level of liquid metal in the reservoir. As the vertical liquid metal loop was not closed, only electromagnetic force and pressure difference developed by pump could be measured. In this case velocity of metal (V_0) in the channel is zero and velocity of traveling magnetic field (V_f) corresponds to above mentioned relative velocity V, which equals just to the linear velocity of rotating magnets.

To achieve maximum electromagnetic force action in the liquid metal with relatively low electrical conductivity, side bars with high electrical conductivity are frequently used [3]. At the presence of side bars the distribution of induced currents and, consequently, electromagnetic force become more uniform in all cross-section of channel of pump. These side bars exclude the formation of the short-circuiting of induced current in the liquid metal near end walls. At the absence of conducting bars the electromagnetic force does not make contribution to develop pressure gradient along channel near end walls. It is clear from Fig.2 that, at the presence of side bars, the pump pressure developed by electromagnetic action is sufficiently higher. In both situations the electromagnetic action corresponds to the linear velocity of a traveling magnetic field that is proportional to the rotation speed of rotators and their radius. All gained experimental results shown in Fig. 4 can be divided into two categories: 1) rotation of axes with opposite direction and opposite magnetic poles (equal linear inductors), and 2) rotation of axes with opposite direction and the same poles (opposite linear inductors). 1) and 2) are related to series 3,4,5 and 8, and series 1,2,6 and 7, respectively. No conducting bars were supplied in the experiments of series 1, 2, 3 and 4 whereas series 5, 6, 7 and 8 were supplied. By changing rotation direction the influence of external pressure on the value of developed electromagnetic pressure was obtained. The downward electromagnetic force rises the level of liquid metal in the other part of loop with a small cross-section and changes the external pressure, which in turn prevents development of intensive inner motions. The external pressure i.e. electromagnetic pressure was read from the manometer. By rotating axes in another direction, upward electromagnetic force was acquired. The upward force pumped the liquid metal up to the reservoir. In this case, as the cross-section of reservoir was much greater in comparison with the cross-section of the pump's channel and other part of the loop, which was not closed, there was not enough volume of liquid metal for increasing the level of metal in the reservoir and for increasing the static pressure in the channel. That is why in this case the inner motions were more intensive. Series 2, 4, 7 and 8 and series 1,3,5 and 6 are downward pressure and upward pressure, respectively. In all cases the inner motions are responsible for the decreasing of the developed electromagnetic force in the channel at fixed rotation speed.

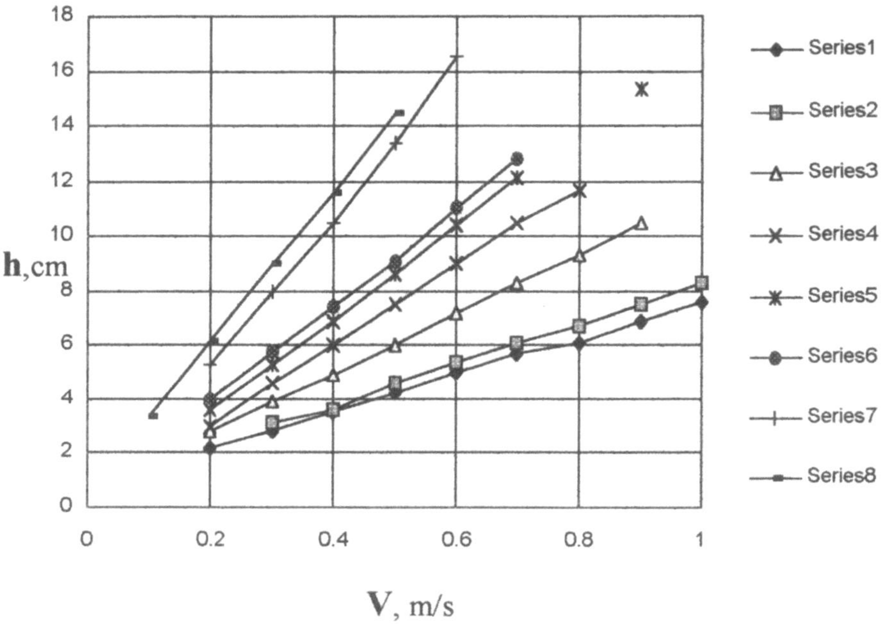

Figure 4. Measured developed static pressure versus velocity of traveling magnetic field for 100mm wide channel

In all cases, the backward flows were arising near end walls. Backward flows at presence of copper side bars were induced due to decreasing of magnetic field strength by 10% - 15% in the gap at the ends of rotators. At the absence of side bars more intensive backward flows were observed due to absence of transversal current component in the region near side walls. It should be noticed that working metal in the channel must wet the short-circuiting side bars completely. No wetting or partial wetting of the side bars leads to sufficient decreasing of the developed electromagnetic pressure in the regions near side walls of channel.

In order to exclude the influence of decreasing of magnetic field strength at the ends of rotators the width of channel of pump was reduced up to 80 mm. On Fig. 5 curves for developed electromagnetic pressure depending from velocity of traveling magnetic field are represented for this channel. Curve 1 corresponds to the case, when the channel of pump is fully filled with liquid metal and electromagnetic force is directed downwards. In this case in comparison with the previous channel, which had a greater width equal to 100 mm, the inner motions were sufficiently less intensive. Curve 2 on Fig. 5 corresponds to the same case, but side bars are short-circuited outside the channel using external copper bar. As we can see, in this case the developed pressure is higher in comparison with curve 1. It is associated with the fact that

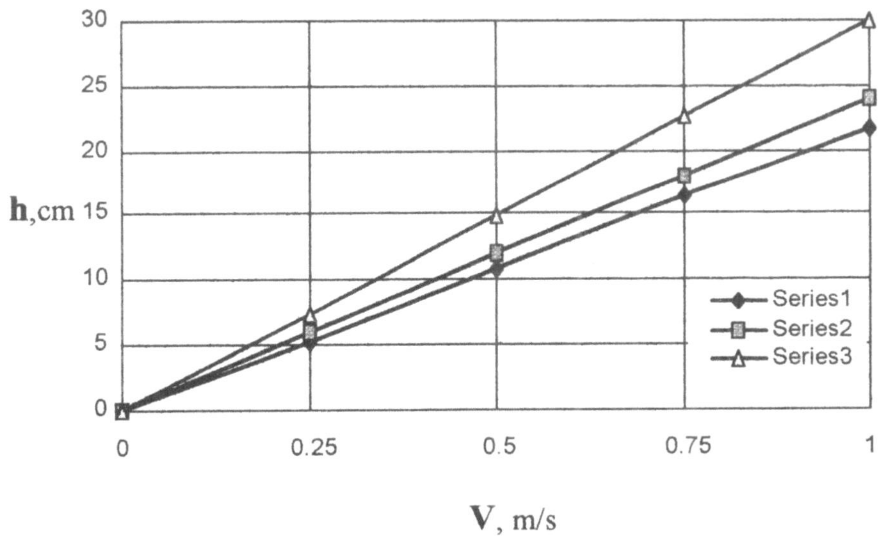

Figure 5. Experimentally measured developed static pressure depending on velocity of traveling magnetic field for the channel of pump having width 80 mm.

rotating permanent magnets practically do not produce active action on channel with even number of poles pairs. The potential difference between end bars is arising and at external short-circuiting the additional electric current arises, which leads to the increasing of developed electromagnetic pressure. Curve 3 on Fig. 5 corresponds to the case, when the channel of pump is filled with liquid metal only partly. In this case, as one can see, the developed electromagnetic pressure grows sufficiently. Magnetic field in this case as if has a possibility to be accelerated before acting on the free surface of liquid metal. May be this effect in some way can be used in induction pumps with inductors.

The influence of wetting of all inside walls of channel on the character of inner motions was also investigated. All experimental results previously demonstrated on Fig. 4 and 5 correspond to the cases, when side walls of channel of pump were not wetted by working liquid metal. The experiments also were carried out, when side walls of channel were perfectly wetted using amalgamated copper foil glued inside to the channel walls. The wetting of side walls sufficiently damps inner motions, but practically has no influence on developed electromagnetic pressure. This obviously is related with the fact, that working electrical currents are shunted through electrically conducting layers, which in the sum results in small change of developed pressure.

At opposite rotation (N - N; S - S) the decreasing of developed electromagnetic pressure was observed, as in this case the main electromagnetic action is concentrated near the side walls of the channel. From visual observations it follows, that except above mentioned effects it is important also to provide a hydraulic seal of side bars in such a way, that there would be contact between side bars and liquid metal only from one side just directed towards liquid metal in the channel. The leakage between side bars and side walls leads to intensive inner jet -like motions decreasing developed pressure.

In conclusion it is worth to mention, that all observed effects depending on the constructive elements of channel of pump are linear with electromagnetic action in rather wide range of it. It is justified also by evaluating of non-dimensional parameter ε_s, which for the range of parameters in our modeling experiments was less than unit ($\varepsilon_s < 1$). As particular measurements showed, at high rotation speeds (high linear velocity of traveling magnetic field) the linear relations showed on Fig. 4 and 5 are valid for the developed electromagnetic pressure at least up to 0.5 atmosphere. Furthermore, this linearity in the case of rotating permanent magnets can be expected in more wide range in comparison with the case of linear inductors. It is because the permeability of permanent magnets is lower than that of ferromagnetic yoke in linear inductors and in the case of rotators demagnetization effect due to electrical currents induced in liquid metal is reduced.

3. Evaluation of the order of pressure being developed by system of rotating magnets

Gained experimental data mentioned above in previous paragraph allow to make an approximate evaluations of the developed by pump electromagnetic pressure, which can be expected using principle of rotating permanent magnets for creating alternating traveling magnetic field. For evaluation of pressure developed by a system of rotating permanent magnetic bricks let us refer to Fig.1. Permanent magnetic bricks having height b = 25 mm in the air gap between rotators, where the channel of pump having thickness a was located, create a magnetic field the distribution of which in the gap is not uniform due to relatively small dimensions of magnetic bricks and relatively big air gap. So for evaluation of pressure we should take some averaged value of magnetic field **B** in the air gap. System of these rotating permanent magnets produce alternating traveling magnetic field close to sinusoidal. This alternating magnetic field induces an electromotive force $\mathbf{E} = \mathbf{V} \times \mathbf{B}$, where **V** is a linear velocity of created traveling magnetic field (linear velocity of magnetic bricks at rotation). In accordance with Ohms' law in the liquid metal in the channel of pump the electric current is induced having density $\mathbf{j} = \sigma\mathbf{V} \times \mathbf{B}$, where $\sigma = 3.46 \cdot 10^6$, $(\text{om}\cdot\text{m})^{-1}$ is the electrical conductivity of liquid metal used in modeling experiments (eutectic alloy In-Ga-Sn). As a result of interaction between this electric current and magnetic field the electromagnetic force is generated, which specific value acting on the volume unit is $f_{em} = \mathbf{j} \times \mathbf{B} = \sigma(\mathbf{V}\times\mathbf{B}) \times \mathbf{B}$. So integral electromagnetic force acting on the part of liquid metal in the channel between magnetic bricks will be $\mathrm{F}_{em} = f_{em} \cdot alb$, where l is a length of the cross-section of the channel of pump, and pressure developed in the channel will be $\mathrm{P} = \mathrm{F}_{em}/(al) = \sigma \mathrm{VB}^2 b$, which is not dependent on geometry of channel. In other terms this pressure corresponds to the head of static pressure (which just was measured in modeling experiments) $h = (\sigma \mathrm{VB}^2 b)/(\rho g)$, where the density of liquid metal $\rho = 6.4 \cdot 10^3$,kg/m^3. So for above mentioned parameters and effective magnetic field strength B in our experiments about 0.5 T and velocity of traveling magnetic field V = 1 m/s the height of static pressure should be about 0.34 m, which corresponds to pressure 0.21 atmosphere. The experimentally measured static pressure for the same parameters was smaller: 0.27 m, so discrepancy is only about 20% and these data can be used for the rough evaluations of the order of pressure being developed by the induction pump using the concept of system of rotating permanent magnets for creating alternating traveling magnetic field. Such a pump (Fig. 6) was designed and constructed for supplying a flow rate in a Wood's metal loop. This pump having the same cross-section of channel (10mm × 100mm) and copper side bars as in modeling experiments had much longer effective zone of electromagnetic action due to half-circular geometry, but smaller averaged magnetic field about 0.35 T (using the same permanent magnetic bricks) in the air gap, as from other side of channel there was only passive Fe-yoke. Maximal magnetic field strength in the gap between rotator and channel of pump was 0.45 T and 0.3 T in the gap between outer channel wall and ferromagnetic yoke. The diameter of ferromagnetic rotor of pump was 200 mm, twelve rows of permanent magnetic bricks were fixed around rotator, so it was a six pole machine, the length of rotator was 100 mm and equal to the width of channel of pump. The active length of channel of pump was approximately about 300 mm. Corresponding to the evaluations this pump can develop static pressures over one atmosphere and supply flow rate up to one liter per second, which was proved during its operation in real conditions in the vertical loop for pumping eutectic Wood's metal alloy having melting temperature 69^0 C. The flow rate in the closed loop at fixed pressure naturally is defined by hydraulic resistance of channel of pump and resistance of loop. Power consumption at operating this pump was relatively small (from 1 kW up to 3 kW) and was supplied by conventional 3-phase motor. So such a design concept of magnetohydrodynamic induction pump can be used successfully and rather effectively for different practical applications.

Similar induction pump using also the same design concept and apparently the same very close to our case the order of parameters, but different construction, in which alternating traveling magnetic field was created by permanent magnets fixed on rotating discs, have been reported in [4]. The channel of pump having U-shaped geometry was located between these discs. Efficiency of this pump also was proved both theoretically and experimentally.

4. Potential ways of improving the characteristics of pump

The operating characteristics of a pump with rotating permanent magnets can be easily advanced. First of all, developed static pressure and flow rate could be much higher at increased rotation speed by supplying higher power to motor, as from estimations of parameter ε_s we have a reserve in respect to the velocity of traveling magnetic field V, which can be increased without violating condition

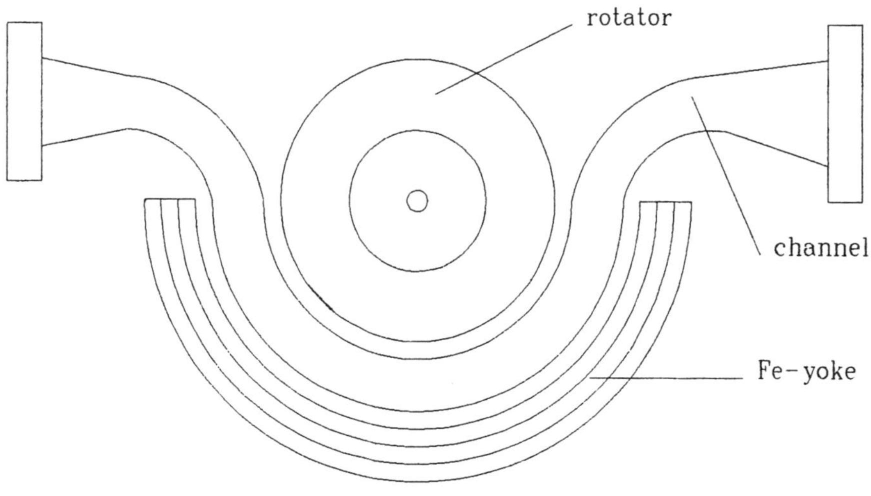

Figure 6. Induction pump design concept.

for parameter $\varepsilon_s \leq 1$, still remaining in the region of linear relation. Second possibility: a grater developed by pump pressure approximately two times bigger can be achieved at the same rotation speed simply by prolonging the channel of pump making it not a half circle, as it was in our case (see fig. 6), but around all rotator. Third possibility: flow rate can be increased by increasing of cross-section of channel of pump, preferably increasing its width and consequently the length of rotator. Increasing of the thickness of channel of pump is not desirable, as it will lead to decreasing of averaged magnetic field strength and more non-uniform distribution of electromagnetic forces across the channel, which results in worsening of working characteristics of pump, and can promote more intensive inner motions and arising of instabilities in the channel. The developed by pump pressure, as it was shown earlier, does not depend on the geometry of channel. So increasing of the cross-section of channel of pump is reasonable only in the case when higher flow rates should be provided.

Another possibility of improving operation characteristics of induction pump with rotating permanent magnets is based on the possibility of increasing the frequency of alternating traveling magnetic field. Investigations carried out on linear induction pump demonstrated that increasing of frequency of feeding current improves characteristics of induction pump [3]. So developed by pump static pressure at fixed feeding current grows linearly with frequency in the range up to $f = 200$ Hz and efficiency of pump also grows with frequency and has a maximum value at frequency about $f = 150$ Hz.

In practice linear induction pumps are fed only by standard frequency current as a resulting efficiency falls drastically due to increasing of impedance of winding and losses in the ferromagnetic yoke with frequency growth. But in the case of induction pump with rotating permanent magnets these limitations do not exist: there is no winding at all and no losses in the main ferromagnetic yoke - rotating cylinder, in which magnetic field is constant, losses exist only in the outer non-moving ferromagnetic yoke , but it is not a problem to provide its cooling. Generating of traveling field with higher frequency in the case of induction pump with rotating permanent magnets can be performed easily by increasing the number of poles around rotator. From the energetic point of view these facts are the main advantage of such an induction pump in comparison with linear inductors.

This pump concept has also some other advantages, which may be considered at designing a pump for liquid metal for specific applications. For example, the longitudinal end effect in the linear inductor is more distinctly expressed and has a higher negative influence on working characteristics of pump at relatively short inductor as near the ends of inductor instead of alternating traveling magnetic field there is only pulsating field. In the case of induction pump with rotating magnets, as it was demonstrated in our modeling experiments, the length of active part where electromagnetic action takes place may be very short and nevertheless the pump operates efficiently. The same can be said concerning the influence of higher harmonics of alternating magnetic field on the characteristics of induction pump. Again, in the linear inductor the presence of higher harmonics has a negative influence on characteristics of pump, as they interfere with a main frequency. In the case of induction pump with rotating permanent magnets the presence of higher harmonics has no negative influence on characteristics of pump as all harmonics are moving with the same velocity as main field (rotation speed of rotator) and make a positive contribution in generating electromagnetic forces and pressure gradient.

Conclusion

The results of theoretical evaluations and experimental modeling showed the effectiveness of induction pump construction basing on the concept of using a system of rotating permanent magnets for generating alternating traveling magnetic field. The main advantages of such a concept are following: it is simple in design and production, parameters of pump can be evaluated rather simply and easily changed, simple control during operation by adjusting rotation speed of rotator using conventional motor. The pump needs significantly much lower power supply in comparison with induction pump using linear inductors for providing the same order of pressure and flow rate. Such pumps may be used in different applications for pumping of liquid metals.

References

1. Kalnins , T.K. (1969) *MHD pumps with distinct poles*, Zinatne, Riga (in Russian).

2. Vives, C.N. (1994) *New electromagnetic manufacturing process for semisolid alloys and metal matrix composites synthesis*, Proceedings of International Symposium on Electromagnetic processing of materials, October 25-28, Nagoja, Japan, pp. 223-228.

3. Lielpeteris, J.J. (1969) *Liquid metal induction MHD machines*, Zinatne, Riga (in Russian).

4. Davies, O.W., Ellis, P.J., McIntosh, R.L., Owen, P.J. (1997) *The development of non-contact high-head electromagnetic induction pump*, Proceedings of International Congress "Electromagnetic processing of materials", May 27 - 29, Paris, la Defence, v. 1, pp. 143-148.

CONTRAST STRUCTURES AND ROTATING STALL IN MHD FLOWS.

Y.A.POLOVKO, E.P.ROMANOVA AND E.A.TROPP
Ioffe Physico-Technical Institute
26 Polytekhnicheskaya, St.-Petersburg, Russia

Abstract. On the basis of the suggested two-dimensional model the possibility of arising of non-stationary induction MHD flow in wide channels or channels of a ring cross-section is discovered. This secondary flow has the features of "rotating stall", which is an example of non-stationary contrast structure. The existing experimental data validate the possibility to observe this phenomenon, which may be treated as a new type of stability loss. The carried out linear analysis proved that at certain relation between the parameters the system had less stability factor as related to non-stationary perturbations. The calculated region of instability for homogeneous stationary flow turned to be wider than that obtained with the help of one-dimensional model. In particular, the stability of the stationary homogeneous flow may be lost at the descending part of the external characteristic. The first numerical results of the study on arising secondary stationary and non-stationary flows far from branching points are obtained.

1. Introduction.

In connection with increasing interest in self-organization processes in general, physical phenomena, whereby contrast structures arise, attract investigators' attention. (By contrast structures we mean narrow internal regions, where parameters of a process change abruptly). So, in recent years the theory of construction of asymptotic solutions for corresponding mathematical models, as well as methods for stability investigations are being worked out [1,2].

The authors of the present paper have been studying contrast structures in inductive MHD-flows for a long time. Stationary secondary flows which

A. Alemany et al. (eds.),
Transfer Phenomena in Magnetohydrodynamic and Electroconducting Flows, 395–408.

arose after homogeneous flow had lost its stability were investigated first. The analysis was based on so-called one-dimensional turbulent model, applied in [3] for the first time. The main results were presented in details in [4-7]. It was discovered, that the behavior of the system had analogy with the phase transition, which took place for example in a liquid-gas system of contrast structure. Van der Vaals isotherm related to coordinate system "average density of both phases - pressure" is similar to S-form external characteristic of MHD induction device, which corresponds to homogeneous flow. Analogous current contrast characteristics also arise in semiconductors with negative differential conduction and characteristic S-form current density dependence of field [8,9].

However, one-dimensional jet model turned to be insufficient for study of all possible contrast structures. In experiments on flows in MHD induction channels, transverse velocities, which resulted in formation of helical motion were discovered [10,11]. As it was emphasized in [12], it was hardly possible to determine definitely the cause responsible for transverse velocities due to the fact, that in the experiments the initial field has been already asymmetrical along azimuth (transverse coordinate). Thus, were the transverse motions caused by that asymmetry or did they arise as a result of instability of symmetric solution - these questions were still open ones, and to answer them, a study on stability of at least two-dimensional motion is required.

An investigation carried out in [4-7] showed that the stationary secondary flows had much in common with secondary flows in high pressure axial compressors, which arose in so-called "rotating stall" [13]. "Rotating stall" is a well-known dangerous phenomenon, described in the theory of turbo-devices, which lies in the fact that the flow, homogeneous along the azimuth looses its stability and the secondary nonhomogeneous flow with contrast structure arises. The nonhomogeneous flow features a rotating cell, where the flow velocity differs from that of the main flow. This similarity concerns external integral characteristics, such as an ability to operate as an "ideal pressure source", as well as the flow structure (the existence of internal shear layers, separating a "cell", where the velocity differs much from the velocity of the flow core or even may be of opposite direction). The similarity of that kind points to the fact, that the mathematical models of these two different physical phenomena have similar properties. In addition, the secondary flows in compressors are non-stationary, of helical character and the appeared "velocity cell" moves across the channel. The observed similarity of these phenomena suggests that the flows with similar structure may occur in MHD induction flows too.

In the present paper a two-dimensional mathematical model in a long-wave approximation is considered and it is shown that the "rotating stall"

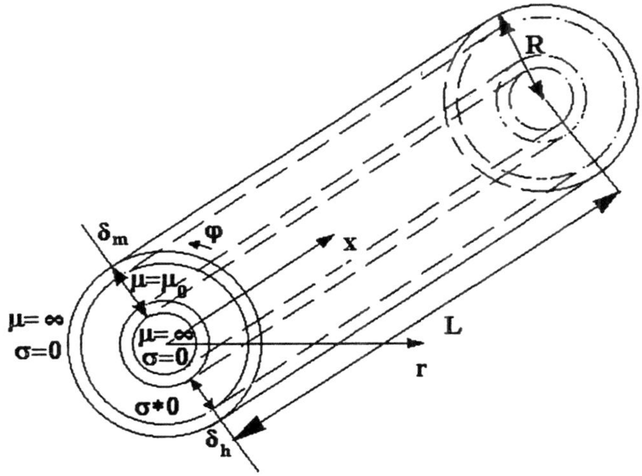

Figure 1. Plan-view of an idealized inductive MHD-channel

type of instability does occur in MHD induction channels. The carried out linear analysis allows to make conclusions on the ways of "rotating stall" formation depending on parameters of the process. On the basis of the newly developed numerical method and the two-dimensional model, the first results of investigation of arising secondary flows far from the branching point are obtained.

2. Mathematical Model

The flow of conductive fluid in an idealized inductive MHD-channel is investigated, its plan-view presented in Fig.1. In such device liquid metal flows through the channel of ring cross-section, formed by two co-axial cylinders. The surfaces of the inductor, facing an operating gap are smooth and an infinite thin current layer, which is an ideal "winding" is spaced on them. The linear density of the current j_0 has the only one component and is a sinusoidal travelling wave $j_0 = A\sqrt{2}\cos(\alpha x - \Omega t)$. Magnetic permeability of the magnetic circuit is taken infinite, and its specific conductance is equal to zero. The thickness of the conducting layer and its specific conductance are taken constant, and its magnetic permeability is equal to that of the vacuum. It is assumed that the dimensions of the device satisfy the following relationships:

$$\pi/\alpha = \tau << L\,,\ \delta_h \approx \delta_m = \delta << \min(R, \tau)\ .$$

Here τ is the pole pitch, L is the device length, δ_h is the hydraulic gap width, δ_m is the width of the magnetic gap between the internal and the external magnetic circuit, R is the radius of the device. These geometrical relations are characteristic for actual cylindrical inductive MHD-devices of high capacity, that makes it possible further to use Cartesian coordinates instead of cylindrical ones with no regard for the curvature of the channel. On this assumption end effects, higher harmonics of the external magnetic field, related to the gear operating surfaces of the stator and discrete winding distribution, as well as the saturation of the magnetic circuit due to the finite value of the magnetic permeability of electrotechnical steel are not considered. The assumptions accepted allow to use the so-called model of plane-parallel field [14, p.155]. It is considered that, there is only one component of the magnetic field, normal to the channel wall

$$\mathbf{B} = \left(0, 0, \tilde{\mathbf{b}}\,(\mathrm{x}, \mathrm{y}, \mathrm{t}) \exp[\mathrm{i}(\alpha \mathrm{x} - \Omega \mathrm{t})]\right), \quad \tilde{\mathbf{b}}\,(\mathrm{x}, \mathrm{y}, \mathrm{t}) = \mathrm{b_a} + \mathrm{i b_r}, \tag{1}$$

which corresponds to the principal mode of the external electromagnetic field travelling along the channel's axis $B_0 = B_m \sin(\alpha x - \Omega t)$. This approximation seems natural for channels of small height and with high magnetic permeability of walls. Unlike so-called "jet model", it is accepted that the azimuth component of the velocity exists despite the fact, that the machine is axially symmetric, i.e. it is considered that $\mathbf{v} = (u, v, 0)$.

After (1) has been substituted in the equation of induction

$$\Delta \mathbf{B} - \mu\sigma \frac{\partial \mathbf{B}}{\partial t} + \mu\sigma \cdot rot(\mathbf{V} \times \mathbf{B}) = -rot j_0, \tag{2}$$

the following equations for dimensionless complex amplitude of magnetic field induction $\mathbf{b} = \tilde{\mathbf{b}}\,/B_m$ are obtained (the curvature of the channel is disregarded):

$$[\mathbf{A}_1 + \mathbf{A}_2]\,\mathbf{b} = i, \quad \mathbf{A}_1 = \frac{\partial^2}{\partial y^2} - \varepsilon \frac{\partial}{\partial t} - 1 + i\varepsilon(1 - u),$$

$$\mathbf{A}_2 = \frac{\partial^2}{\partial x^2} + 2i \frac{\partial}{\partial x} - \varepsilon u \frac{\partial}{\partial x} - \varepsilon v \frac{\partial}{\partial y}, \tag{3}$$

where $\varepsilon = \mu\sigma\Omega$ is magnetic Reynolds number. Here $\mathbf{A}_1$ is an operator, corresponding to a jet flow model, $\mathbf{A}_2$ is an additional operator caused by azimuth component of velocity v and by the ability of field amplitude to alter along axis x. Note that in (2) j_0 is a density of the external current:

$$rot j_0 = -\frac{\partial^2 B_0}{\partial^2 x} = -\alpha^2 B_m i \exp[i(\alpha x - \Omega t)]$$

The equations of motion are averaged over the period and the wave length of the external travelling electromagnetic field. After the averaging operator $\frac{1}{T}\int_0^T \bullet dt \; \frac{1}{2\tau}\int_0^{2\tau} \bullet dx$ has been applied (here $T = 2\pi/\Omega$, $\tau = \pi/\alpha$), the equations of motion for the main velocity component, which varies slowly as compared to the alteration of the external field, become:

$$\begin{aligned} z(\frac{\partial u}{\partial \tau} + u\frac{\partial u}{\partial x} + v\frac{\partial u}{\partial y}) &= -\frac{\partial p}{\partial x} + \frac{j^2}{\varepsilon}(b_a - b_a\frac{\partial b_a}{\partial x} - b_r\frac{\partial b_r}{\partial x}) - u\sqrt{v^2+u^2}, \\ z(\frac{\partial v}{\partial \tau} + u\frac{\partial v}{\partial x} + v\frac{\partial v}{\partial y}) &= -\frac{\partial p}{\partial y} - \frac{j^2}{\varepsilon}(b_a\frac{\partial b_a}{\partial y} + b_r\frac{\partial b_r}{\partial y}) - v\sqrt{v^2+u^2} \\ \frac{\partial u}{\partial x} + \frac{\partial v}{\partial y} &= 0, \end{aligned} \qquad (4)$$

where j is dimensionless density of feeding current and z is hydraulic inductance of the pump. As in the case of the jet model, only turbulent friction against the walls is taken into account (the last terms of the first and the second equations in (4)), because of the small height of the channel.

Equations (3)-(4) are supplemented by periodic boundary conditions over intervals $[0, L], [0, S]$ (L and S are the dimensionless length and width of the channel respectively).

$$\begin{aligned} &\mathbf{b}\,|_{x=0} = \mathbf{b}\,|_{x=L}, \quad \mathbf{b}\,|_{y=0} = \mathbf{b}\,|_{y=S}, \\ &\frac{\partial \mathbf{b}}{\partial x}\,|_{x=0} = \frac{\partial \mathbf{b}}{\partial x}\,|_{x=L}, \quad \frac{\partial \mathbf{b}}{\partial y}\,|_{y=0} = \frac{\partial \mathbf{b}}{\partial y}\,|_{y=S}, \\ &\mathbf{v}\,|_{x=0} = \mathbf{v}\,|_{x=L}, \quad \mathbf{v}\,|_{y=0} = \mathbf{v}\,|_{y=S}\ . \end{aligned} \qquad (5)$$

They may be considered as boundary conditions for cylindrical inductive toroidal MHD-channel or ordinary boundary conditions for wide and long channels. In (3-5) the velocity of the external travelling field $[v] = \Omega/\alpha$ is taken as a velocity scale, $[t] = 1/\Omega$ and $[b] = B_m$ are the scales of time and induction respectively. It should be noted, that the averaging procedure restricts the field of application for mathematical model (3)-(5), as it demands variation of $\mathbf{b}$ and $\mathbf{v}$ within small limits over a length of the order of 2τ and in a time of the order of T, i.e. long-wave approximation.

3. Linear Analysis. Equation System for Neutral Surfaces

Solution, corresponding to the stationary flow, which is homogeneous along both coordinates, satisfies equation system (3)-(5):

$$u = u_0, \quad v_0 = 0, \quad b_{r0} = -\frac{1}{\varepsilon^2(1-u_0)^2+1}, \quad b_{a0} = -b_{r0}\varepsilon(1-u_0)\,,$$

$$\left(\frac{\partial p}{\partial x}\right)_0 = \frac{j^2 b_{a0}}{\varepsilon} - u_0 \left|u_0\right| \tag{6}$$

The solutions, branching from homogeneous solution (5) are sought in the following form

$$\vec{\mathrm{U}}=\vec{\mathrm{U}}_0 + \delta\,\vec{\mathrm{U}}=\vec{\mathrm{U}}_0 + \vec{\mathrm{C}} \exp\left[\mathrm{i}\left(2\pi \mathrm{my/S} + 2\pi \mathrm{nx/L} - \omega \mathrm{t}\right)\right], \tag{7}$$

$$\vec{U}= (u, v, b_a, b_r, p)^T, \vec{U_0}= (u_0, v_0, b_{a0}, b_{r0}, p_0)^T, \vec{C}= (c_1, c_2, c_3, c_4, c_5)^T.$$

Equations (3)-(5) are linearized in the vicinity (6), expressions (7) are substituted in (3)-(5) and a system of linear equations in components of vector $\vec{C}$ and with determinant Δ is obtained. Determinant Δ is as follows:

$$\begin{vmatrix} \begin{matrix} i\varepsilon\omega - k_1^2 - k_2^2 \\ -1 - i\varepsilon k_2 u_0 \end{matrix} & \begin{matrix} -\varepsilon(1-u_0) \\ -2ik_2 \end{matrix} & \varepsilon b_{r0} & 0 & 0 \\ \begin{matrix} \varepsilon(1-u_0) \\ +2ik_2 \end{matrix} & \begin{matrix} i\varepsilon\omega - k_1^2 - k_2^2 \\ -1 - i\varepsilon k_2 u_0 \end{matrix} & -\varepsilon b_{\alpha 0} & 0 & 0 \\ -\frac{j^2}{\varepsilon}(1 - ik_2 b_{\alpha 0}) & \frac{ij^2}{\varepsilon} b_{r0} k_2 & \begin{matrix} izu_0k_2 + \\ 2\left|u_0\right| - i\omega z \end{matrix} & 0 & ik_2 \\ \frac{ij^2}{\varepsilon} k_1 b_{\alpha 0} & \frac{ij^2}{\varepsilon} b_{r0} k_1 & 0 & \begin{matrix} izu_0k_2 + \\ \left|u_0\right| - i\omega z \end{matrix} & ik_1 \\ 0 & 0 & ik_2 & ik_1 & 0 \end{vmatrix}$$

Here new designations are introduced: $k_1 = 2\pi m/S$ $k_2 = 2\pi n/L$ - wave numbers, which express dimensionless density of arising waves across and along the channel respectively. Condition of existence for nontrivial solutions results in a system of two non-linear equations $Re\,(\Delta) = 0$, $Jm\,(\Delta) = 0$, from which branching point u_0 and frequency of the rotating stall ω are determined. The expanded form of this system is as follows:

$$R_3 \left|u_0\right| \left(R_1 + k_1^2\right) - zR_1R_2R_4 - k_1^2 j\left[b_{a0}\varepsilon\left(1 - u_0\right) + b_{r0}\left(R_1 + 1\right)\right] = 0,$$

$$R_4 \left|u_0\right| \left(R_1 + k_1^2\right) + zR_1R_2R_3 - k_1^2 j\left[2b_{a0}k_2 + \varepsilon b_{r0} R_2\right] = 0, \tag{8}$$

where $R_1 = k_1^2 + k_2^2$, $R_2 = k_2 u_0 - \omega$, $R_3 = \left(R_1 + 1\right)^2 + \varepsilon^2 \left(1 - u_0\right)^2 - \varepsilon^2 R_2^2 - 4k_2^2$, $R_4 = 2\varepsilon\left[R_2\left(R_1 + 1\right) + 2\left(1 - u_0\right)k_2\right]$.

4. Analysis of the System, which Determines Branching Points

As it has been already mentioned, the accepted mathematical model is valid only when $k_2 \ll 1$, $\omega \ll \Omega$, because of the procedure, used for averaging of motion equations. In this connection , the form of asymptotics for ω and u_0 at small values of k_2 seems to be interesting. After the series

$$\omega = \omega_0 + \omega_1 k_2 + o(k_2)$$
$$u_0 = u_0^1 + u_0^{(2)} k_2^2 + o(k_2^2)$$

are substituted in (8) and the terms with equal degree are equated, in the first approximation the following expressions are obtained:

$$\omega_0 = 0,$$
$$2\left|u_0^1\right| G(G + K^2 - 1) - j^2(G - K - 1) = 0, \tag{9}$$

where $K = 1 + k_1^2$, $G = 1 + \varepsilon^2 s^2$, $s = 1 - u_0^1$.

Expression (9) is a well-known equation, from which boundary points of stationary solutions are determined [3]. Thus, the stationary flows studied before make a subset of solutions for the two dimensional model, which branching points are determined from (8) at $k_2 = 0$. When the terms of the order of k_2 in the second equation and the terms of the order of k_2^2 in the first equation are set equal respectively, the following expressions are obtained.

$$\omega_1 = \frac{G u_0^1[4\varepsilon(K u_0^1 + 2s) + z(K^2 + \varepsilon^2 s^2)] - \varepsilon j^2(2s - u_0^1)}{G[4\varepsilon u_0^1 K + z(K^2 + \varepsilon^2 s^2)] + \varepsilon j^2}$$

$$u_0^{(2)} = -\frac{G[(K^2+\varepsilon^2 s^2)u_0^1 + 2k_1^2 u_0^1(2K - \varepsilon^2 W^2 - 4) - 2z\varepsilon k_1^2 W(KW + 2s)] + k_1^2 j^2}{2k_1^2[(K^2+\varepsilon^2 s^2)(G - 2u_0^1\varepsilon^2 s) - 2G u_0^1 \varepsilon^2 s + j^2 \varepsilon^2 s]},$$

where $W = u_0^1 - \omega_1$. It is seen, that the phase velocity of long waves propagation along the channel $c = w/k_2 = w_1$ does not depend on the wave length.

In the case of high feeding currents $j \to \infty$, which is also of practical interest, the asymptotic formulas are as follows:

$$u_0 = 1 - \frac{\sqrt{k_1^2 + k_2^2 + 1}}{\varepsilon} + \delta(j);\;\; w = k_2(3u_0 - 2) + \delta(j),$$

where $\delta(j) \to 0$ when $j \to \infty$.

In general, system (8) was solved numerically with the help of Newton method. The major part of calculations was fulfilled for MHD device under pump operation mode $0 < u_0 < 1$. The study on flow in a toroidal closed

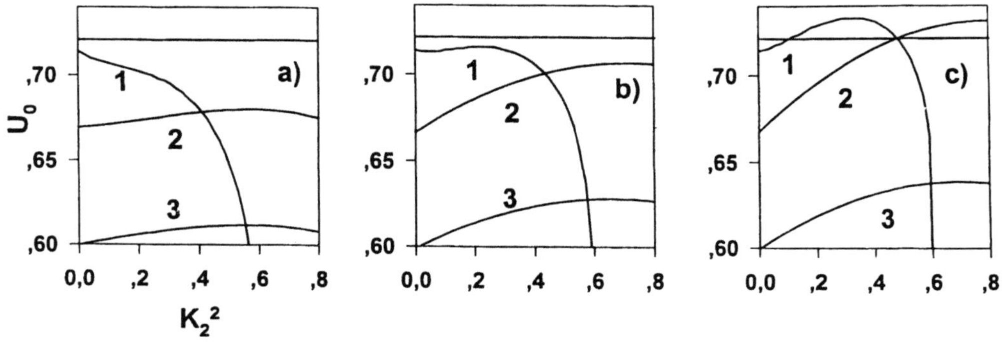

Figure 2. Neutral curves u_0 as a function of k_2^2 at $æ = 0.1$, $\varepsilon = 4$, $j^2 = 40$. and *a*) $z = 5$,) $z = 10$, *c*) $z = 20$ under pump mode for different modal numbers $n = 1 - 1$; $n = 2 - 2$; $n = 3 - 3$.

type channel is probably of no practical interest, so it is better to analyze flow in an infinite tube with circular cross-section. This model corresponds to continuous variation of k_2 and a discrete set of $k_1 = 2\pi n/L = \sqrt{æ}n$, $n = 1, 2, 3...$

In Fig.2 neutral curves, which depict u_0 as a function of k_2^2 in the branching point at different values of z and modal number n , are presented. The fact, that function $u_0(k_2^2)$ is nonmonotonous at sufficiently great values of z, is fundamentally important for the analysis. It means, that for the flows of rotating stall type, the region of existence may be wider, than for secondary stationary flows, for which $k_2 = 0$. It is just non-stationary flows, that are to be observed experimentally at these parameters, because the system's stability margin relatively such perturbations is minimal.

In Fig.3 the critical values of $u^* = \max(u_0)$, ω, k_2^2 are shown as functions of z. The vertical line in Fig.3a and the horizontal one in Fig.3b correspond to the maximal possible u_0 for the stationary homogeneous flow (6) under the pump mode. The intersection of these lines with curves $u^*(z)$, $\omega(z)$, $k_2^2(z)$ points to the possibility of stability loss at the descending branches of the external characteristic. As for the secondary stationary flows, the behavior like that is impossible [4-7]. It is known, that the branching external characteristics observed in the experiments are more smooth than those, predicted by the one-dimensional theory, they lack a characteristic peak in the vicinity of the branching point. The possibility of branching for the external stationary characteristic at its descending section may result in smoothing of this peak. This problem can be solved definitely only after the non-linear analysis and the complete numerical calculation of external characteristics of non-stationary secondary flows are carried out.

The limiting case, when the fluid flows between parallel plates and two

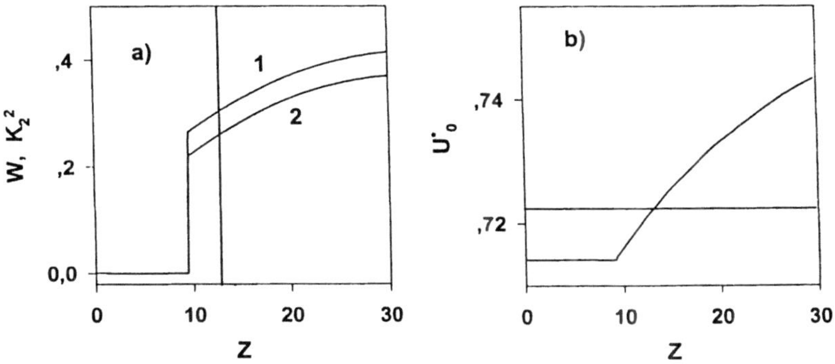

Figure 3. Critical value of angular frequency ω (a, curve 1), squared wave number k_2^2 (a, curve 2) and u_0^* (b) as functions of hydraulic inductance z at $k_1 = 0.1$, $\varepsilon = 4$, $j^2 = 40$. Functions are plotted for maximal pressure values under pump mode. Vertical (a) and horizontal (b) lines correspond to the maximum of external characteristics of homogeneous flow.

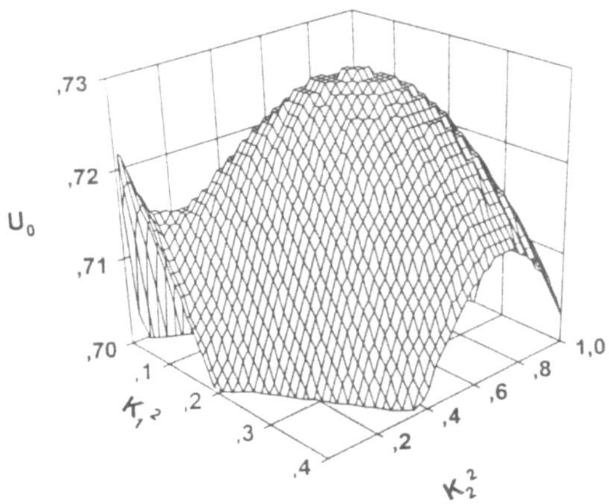

Figure 4. Neutral surface $u_0\left(k_1^2, k_2^2\right)$ at $z = 11.3$, $\varepsilon = 4$, $j^2 = 40$.

wave numbers k_1 and k_2 continuously vary, is of special interest. In Fig.4 neutral surface $u_0(k_1^2,\, k_2^2)$, corresponding to zero decrement of perturbation damping, is given as an example.

It is seen, that at these parameters the surface has maximum, corresponding to the critical point of stability loss. The instability arises at the descending section of the external characteristic, however the sufficiently great values of k_2^2 - $(0.4 - 0.7)$ in the critical point make us doubt, whether the model used for the study of stability of the flow between the plates is valid at least for quantitative evaluations. To accomplish the rigid analysis, the investigation of stability for the system not averaged over time and wave length is required. It makes necessary to solve a problem on eigenvalues for the equation with variable coefficients, which is much more complicated.

It is necessary to note, that axial symmetry makes all directions of stall rotation equally probable, i.e. small perturbations with both wave numbers k_1 and $-k_1$ are possible. So, as it directly results from equation system (8), k_2 is not sign-variable, i.e. the direction of movement of the wave, travelling along x-axis, always coincides with the direction of the external field movement.

5. On the Evolution of Finite Perturbations

The results obtained show, that arising of secondary non-stationary inductive MHD-flows of rotating stall type is in principle possible. However, at least two important problems can be solved only after finite, while small perturbations are examined. First, it is not clear, if the instability investigated can result in twisting of axisymmetric flow, and secondly, if arose waves (7) can suppress each other and result not in "travelling cell", but in "standing wave" in the azimuth direction. Indeed, as it follows from the linear analysis, when stability is lost, the main part of the perturbation of the homogeneous solution in critical point is as follows:

$$\delta \vec{U} = \eta j_1 \vec{U}_1 \exp[i(k_2x+k_1y-wt)]+\eta j_2 \vec{U}_1 \exp[i(k_2x-k_1y-wt)]+c.c.+o(\varepsilon),$$

where c.c are conjugate complex numbers. Amplitudes, j_1, j_2 are determinated from the conditions of solvability of the equations for the next terms of the expansion series in supercriticality $\eta^2 = u_0^* - u_0$. The so-called system of branching equations is composed of these conditions and is of the same form as in the case of stability loss for a submerged jet [15].

As it follows from the main principles of symmetry and as the carried out preliminary analysis shows, the system of branching equations has the solutions of the following form:

$$j_1 = 0, \ |j|_2^2 = j^2; \tag{10}$$

$$j_2 = 0, \ |j|_1^2 = j^2, \tag{11}$$

which correspond to formation of the rotating stall with the sell travelling in opposite directions, and the solution

$$j_1^2 = j_2^2 = \beta^2 \tag{12}$$

corresponding to the wave standing on azimuth and travelling in longitudinal direction

Spiral waves, corresponding to solutions (10),(11) already in the next approximation in ε result in deformation and twisting of the average flow due to Reynolds stresses $(\vec{u}_1\ \nabla)\ \vec{u}_1$ and nonlinear friction in (4), while standing wave (13) does not cause rotating. Which of these two regimes is realized , with twisting (11),(12) or without it (13), this problem is solved by means of analysis of stability for these solutions in the vicinity of the neutral curve. The stability of solutions (10)-(12) depends on the correlation between j and β, which are derived from the parameters of the initial system (3),(4). This analyses is the next step of the investigation.

It should be noted, that the stationary flows, produced in the result of evolution of perturbations (7) at $k_2 = 0$ and their external characteristics are fully constructed in [4], and the stability of these flows is obtained within the frames of one-dimensional jet flow model in [5]. However, it is not clear, if these flows are also stable when analyzed with the help of two-dimensional model (3)-(5). The shift of branching points of the non-stationary solutions to the right towards the region of descending static external characteristic of the homogeneous flow results to the fact, that some of stationary solutions (4) may not exist within sufficiently wide range of variation of the external load, when the stability of the homogeneous flow is lost.

6. Numerical Solution of the Problem. Computational Method

The behavior of the solutions beyond the branching point has been analyzed numerically. To compute non-stationary two-dimensional problem the numerical method suggested in [16] has been used. The method has been generalized for the case, when the boundary conditions are periodic. The set of equations has been supplemented with the induction equation. A staggered arrangement of variables has been used, i.e. pressure and induction have been placed in the cell centres $x(i+\frac{1}{2}), y(j+\frac{1}{2})$, while the velocity components $u_{i,j}$ and $v_{i,j}$ are in the centres of the corresponding faces $x(i+1)$, $y(j+\frac{1}{2})$; $x(i+\frac{1}{2})$, $y(i+\frac{1}{2})$. An advantage of this arrangement is that no boundary condition for pressure is needed. Space derivatives have been approximated by central differences while time approximation has been written as

$$\frac{\mathbf{u}_{ij}^{n+1}-\mathbf{u}_{ij}^{n}}{\Delta t} + \alpha\mathbf{G}^{n+1} + (1-\alpha)\mathbf{G}^{n} = 0, \tag{13}$$

where $\mathbf{G}$ is a space operator, governed by equations (3-4). From the stability considerations α coefficient in (13) is greater or equal to 0.5. The scheme is the second order accurate in time, when $\alpha = 0.5$ (Crank-Nicolson scheme), otherwise it is of the first order.

Equations have been re-written in terms of decrements and in accordance with [16]. The solution of the non-stationary system

$$\frac{\partial \mathbf{u}}{\partial \tau} + \mathbf{u} + \gamma \mathbf{G}(\mathbf{u}^n + \mathbf{u}, p^n + p) = -\gamma \mathbf{G}(\mathbf{u}^n, p^n); \quad \phi \frac{\partial p}{\partial \tau} + \nabla \cdot \mathbf{u} = 0 \quad (14)$$

with corresponding boundary and initial conditions has been sought at each time step. Here $\gamma = \Delta t \alpha$, τ- is so-called pseudo-time and ϕ is an additional parameter chosen experimentally to accelerate the convergence. To solve this problem, the maximum possible pseudo-time step $\Delta \tau$ has been chosen.

Different explicit and difference schemes may be used to solve equations (13)-(14). In the present paper the implicit scheme of special structure has been exploited. The elimination of the pressure reduces the method to a number of scalar tridiagonal systems. When ϕ is chosen correctly, the convergence of the method does not depend on the pseudo-time step. An optimal choice of ϕ is considered in [16].

It should be noted, that periodical boundary conditions on both variables together with axial symmetry of the problem deteriorate the convergence of the method. This is related to the fact, that the finite-difference solution is not single in this case. In reality, a series of n solutions, which differ in the shift of coordinate origin by the value multiple of the step, evolves in the phase space. To improve the convergence, special methods have been applied. One of the existing solutions is selected according to some additional criteria. One more factor, that badly complicates the numerical solution is the presence of contrast structures, i.e. the strongly pronounced moving shear layers, in the developed rotating stall. It makes necessary to apply an adaptive grid, which automatically becomes denser at high solution gradients.

On the basis of the algorithm described above, the first numerical results of the analysis of the developed contrast structure of rotating stall type far from branching points are obtained. In Fig.5 velocity profiles (upper pictures) and magnetic field isolines (lower pictures) for the developed rotating stall at $\varepsilon = 4, j^2 = 40, k_1^2 = 0.1, k_2^2 = 0.032, z = 1$ for the case of closed channel (zero flowrate) are shown. It is discovered, that there are domains of parameters, where such flow is unstable and evolves to the stationary contrast structure. The intermediate flow structure, which arises during this process is shown in Fig.5 (right). The next step of numerical investigation is the study on the stability of arising secondary stationary and non-stationary flows at various parameters and construction of their external characteristics $p(q)$.

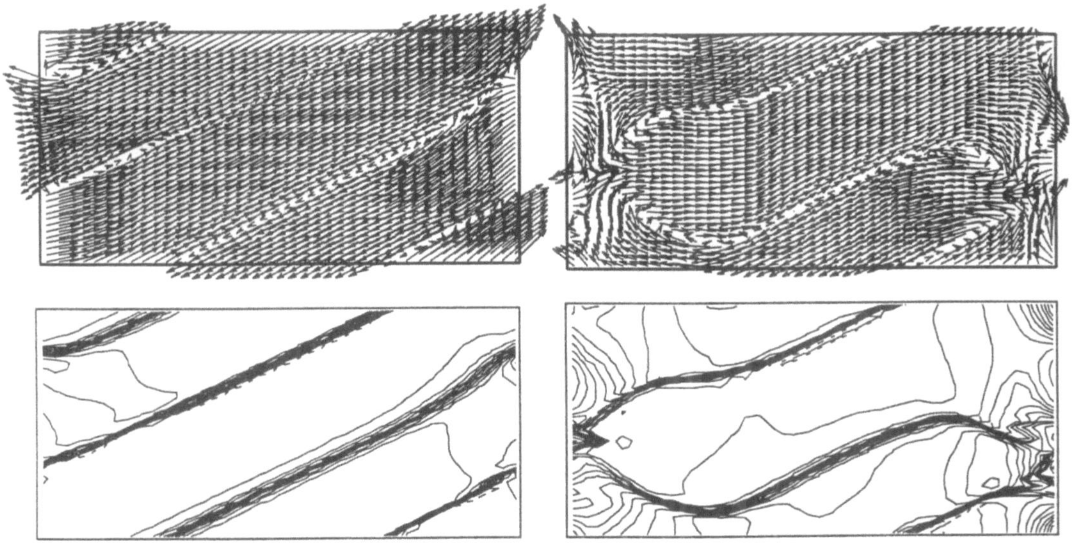

Figure 5. Examples of velocity field and magnetic field isolines with shear layers for developed rotating stall at different moments.

7. Conclusions

The main conclusions, which follow from the carried out analysis:

1.The rotating helical flows of travelling wave type may occur in MHD induction channels alongside with stationary secondary flows, studied before.

2.Arising of such type of secondary flows is similar to so-called rotating stall, discovered in high pressure compressors.

3.At sufficiently great values of hydrodynamic inductance the region of existence for such rotating flows may be wider than for the secondary stationary flows.

4. At sufficiently great values of hydrodynamic inductance rotating flows may arise at the descending branches of static external characteristics under pump and generator regimes, which is impossible for the secondary stationary flows.

5.The numerical experiments showed, that the developed rotating stall could disintegrate and the stationary contrast flow structures could be formed.

6. The carried out analysis probably explains twisting of flows in channels of cylindrical induction MHD devices, formerly observed and smoothing of external characteristics of secondary flows, as compared with those, predicted by the one-dimensional jet flow model.

References

1. Vasilieva, A.B.(1997) On Contrast Structures for the Systems of Singularly Perturbed Equations, *Jurnal Vichislitelnoy Matematiki i Matematicheskoy Fisiki* v **37, N1**, 74-84.
2. Bakaleinikov, L.A. and Polovko, Yu.A. (1993) Asymptotics of Similar to Discontinuous Periodic Solutions of Singularly Perturbed Autonomous Second Order Systems of Differential Equations under Bifurcation Conditions, *Mathematical Physics, Applied Mathematics and Informatics* Ed.Yu.I.Koptev, Nova Sci.Publ., Washington, 207-230.
3. Gailitis, A.K. and Lielausis, O.A (1975) Instability of Homogeneous Velocity Distribution in Inductive MHD-Flows, *Magnitnaya Gidrodinamika* **1**, 106-111.
4. Polovko, Yu.A. and Tropp, E.A. (1986) Asymptotic and Numerical Investigation of One-dimensional Turbulent Model of the Flow in Inductive MHD-pump, *Magnitnaya Gidrodinamika* **4**, 106-113.
5. Polovko, Yu.A. (1989) Study of Operation Stability for Cylindrical Inductive MHD-pump under External Pneumohydravlical Load, *Magnitnaya Gidrodinamika* **3**, 81-88.
6. Polovko, Yu.A., Romanova, E.P. and Tropp, E.A. (1996) A Study on Stability of Induction Flows in Wide MHD Channels,*Technical Physics* **4**, 36-44.
7. Polovko, Yu.A., Romanova, E.P. and Tropp, E.A. (1997) Cylindrical MHD Induction Device in the "Ideal Pressure Source" Regime, *Technical Physics* **6**, 591-594.
8. Volkov,A.F. and Kogan, Sh.M. (1968) Physical Phenomena in Semiconductors with Negative Differential Conductivity, *Uspekhi Fisicheskikh Nauk* **v.96, N4**, 633-672.
9. Gorbatyuk, A.V. and Rodin, P.B. (1997) Current Filamentation in Bistable Semiconductor Systems with Two Global Constraints. Z. Phys. B. **v.104**, pp. 45-54.
10. Klavinya, A.P., Lielausis, O.A and Riekstin'sh, V.V. (1975) Heterogeneous Flow in the Channel of Cylindrical Inductive Pump, *VIII Rigas Conference on Magnetic Hydrodynamics. Abstracts. Riga* **2**, 79-81, (in Russian).
11. Kirillov, I.R., Ogorodnikov, A.P. and Ostapenko, V.P. (1980) Experimental Investigations of Heterogeneous Flow in Cylindrical Linear Inductive Pump, *Magnitnaya Gidrodinamika* **2**, 107-113.
12. R.A.Valdmane, R.A., Valdmanis, Ya.Ya. and Ulmanis, L.Ya. (1986) On Two-dimensional Motions in Cylindrical MHD-pumps, *Magnitnaya Gidrodinamika* **1**, 103-106.
13. Greitzer, E.M. (1980) Review - Axial Compressor Stall Phenomena, *Journal of Fluids Engineering* **102**, 134-151.
14. Vol'dek, A.I. (1970) *Inductive Magneto-hydrodynamic Machines with Liquid-metal Operation Media*, Energia Publishing House, Moscow,(in Russian).
15. Jdanova, A.I. (1985) Auto-oscillations and Non-stationary Transient Phenomena in Axisymmetric Jets, *Stability and Turbulence* Ed. M.A.Goldshtik and V.N.Shtern Publishers, Novosybirsk, 71-81 (in Russian).
16. Soh, W.Y. and Goodrich, J.W. (1988) Unsteady Solution of Incompressible Navier-Stokes Equations, *J. Comput. Physics* **79**, 113-134.

NONEQUILIBRIUM PLASMA MHD POWER GENERATION WITH FUJI-1 BLOW-DOWN FACILITY

Y.OKUNO, T.OKAMURA, K.YOSHIKAWA, T.SUEKANE, K.TSUJI, T.MAEDA, T.MURAKAMI, S.KABASHIMA, H.YAMASAKI AND S.SHIODA
Department of Energy Sciences, Tokyo Institute of Technology 4259 Nagatsuta, Midori-ku, Yokohama 226, Japan

AND

Y.HASEGAWA
Mechanical Engineering Laboratory 1-2 Namiki, Tsukuba, Ibaraki 305, Japan

Abstract

Experimental studies on closed cycle MHD power generation with "Fuji-1" blow-down facility at Tokyo Institute of Technology are presented. Recently, a new disk generator (Disk-F4) has been installed and a new seed injection system has been introduced from IVTAN (Institute of High Temperature in Russia). The design concept of the new generator channel is focused mainly on the reliability of high power generation. In the new seeding system, a melted seed material is pushed by a piston dozator, instead of gas-pressure-driven in the previous system. In the power generation experiments with the new components, the maximum power output of 502kW and the enthalpy extraction ratio of 16.7% have been obtained. These values at the present stage are lower than the maximum values previously achieved in the facility. However, the reproducible high power generation has been demonstrated with the new generator. The behavior of plasma and fluid under strong MHD interaction taking place in the generator channel is discussed.

1. Introduction

One possibility of highly efficient electrical power generation is an MHD power generation with nonequilibrium plasma realized in a alkali metal

A. Alemany et al. (eds.),
Transfer Phenomena in Magnetohydrodynamic and Electroconducting Flows, 409–420.

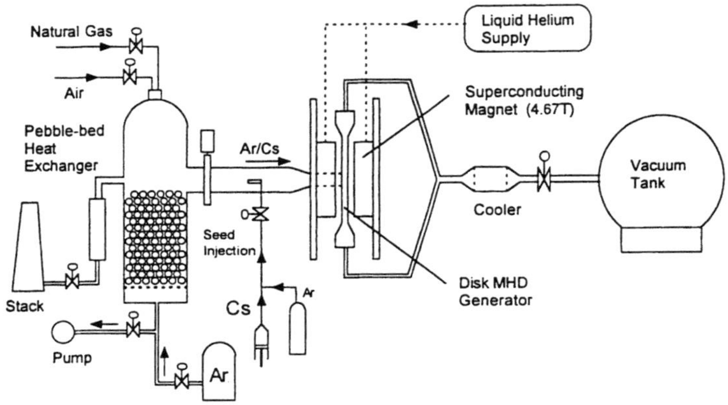

Figure 1. Schematic diagram of FUJI-1 blow-down facility

seeded noble gas(Rosa, 1968). Studies on the nonequilibrium plasma MHD power generation have been conducted with the Fuji-1 blow-down facility at Tokyo Institute of Technology, since 1983(Okamura *et al.*, 1994). One of the main objectives in the Fuji-1 experiments is to demonstrate high enthalpy extraction ratio (electrical output/thermal input) with a disk shaped generator for a period of several tens of seconds under low seed fractions. For this purpose, attentions have been paid to the realization of stable and uniform nonequilibrium plasma with full seed ionization under the condition of strong MHD interaction.

Schematic diagram of the Fuji-1 facility is shown in Fig.1. Argon gas is heated up to ~2000K in a pebble bed type regenerative heat exchanger and cesium as a seed material is mixed with the heated gas in the hot duct through a seed injection nozzle. The working gas of cesium seeded argon is introduced into the disk generator, where the magnetic flux density of ~4.6T is applied by a Helmholtz type superconducting magnet. The working gas after passing through the generator is cooled in a pebble bed cooler, and introduced into a vacuum tank equipped to get a large pressure ratio of the generator inlet to the exit.

The cross sectional views of disk generators installed in the Fuji-1 facility are shown in Fig.2. In a disk MHD generator, the electromotive force is induced by the radial plasma flow and the magnetic field applied perpendicularly to the flow, and yields the azimuthal current (Faraday current). The interaction between the Faraday current and the magnetic field causes the radial Hall electric field. Thus, the electric power is extracted with a load resistance connected between an inner (anode) and an outer (cathode) electrodes. The maximum enthalpy extraction ratio of 15.7% and the corresponding output power of 404kW were obtained with Disk-F3a channel

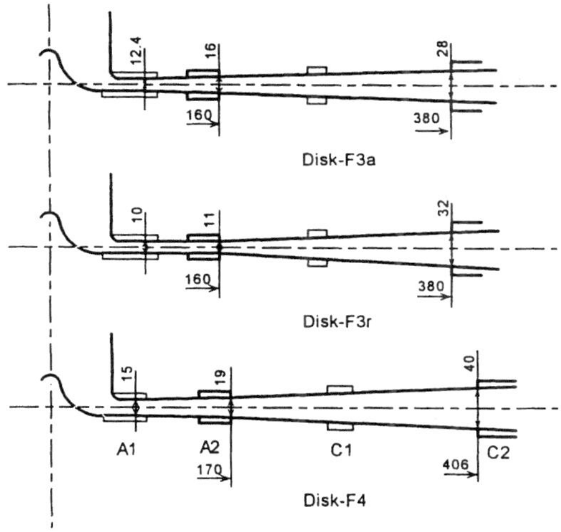

Figure 2. Cross sectional views of disk MHD generators installed in Fuji-1 facility

in 1989(Harada *et al.*, 1990). It was suggested from the Disk-F3a experiments that the increase in the ratio of generator exit area A_{exit} to the inlet A_{inlet} should provide high enthalpy extraction ratio because the low static pressure can be maintained in the channel. On the basis of this suggestion, Disk-F3r channel as shown in Fig.2 was introduced, where the area ratio A_{exit}/A_{inlet} was increased up to 6.9 from 4.1 (Disk-F3a). With the Disk-F3r channel, the maximum enthalpy extraction ratio of 18.0% and the corresponding output power of 297kW were achieved in 1992(Harada *et al.*, 1993). As has been expected, the larger area ratio surely yielded higher enthalpy extraction ratio.

The reproducibility of high power generation with Disk-F3r, however, was not possible. In a sense of the reliability of high power generation, the Disk-F3a channel would be superior to the Disk-F3r channel. One of the reasons could be ascribed to the low Mach number (2.36) and small height at the generator inlet of Disk-F3r generator. Thus, non-equilibrium plasma with a high electrical conductivity could not be realized frequently.

Furthermore, in the experiments with Disk-F3a and -F3r channels, the influence of the fluctuation in the seed fraction on the generator performance was investigated(Harada *et al.*, 1993). The results showed that the large fluctuation with a low frequency of several Hz, particularly for low seed fraction of $\sim 10^{-4}$, caused the fluctuation of output power with the frequency of the same order of that in the seed fraction. It was pointed out at the same time that the generator performance could be markedly improved, if the fluctuation of seed fraction is suppressed.

On the basis of the suggestions above, a new disk shaped generator

(Disk-F4) has been installed and a new seed injection system has been introduced from IVTAN (Institute of High Temperature in Russia) in the Fuji-1 facility. These improvements resulted in large output power and high enthalpy extraction ratio with well-reproducibility. In the present paper, first of all, the design concept of the new generator channel Disk-F4 and the new seed injection system are described. Next, the typical results of power generation experiments are presented and finally the problems for the further improvement of generator performance will be discussed briefly.

2. Improved Disk Generator and Seeding System

The design concept of the new generator channel Disk-F4 was based mainly from the viewpoint of the reliability for high power generation. For this purpose, the Mach number at the generator inlet and the thermal input are slightly increased up to ~2.8 and ~3.0MW (at P_0=0.42MPa, T_{g0}=1950K), respectively. As shown in Fig.2, the geometry of the Disk-F4 channel is roughly similar to that of the Disk-F3a channel which provided the well-reproducibility of power generation. The high Mach number, the large thermal input and the large inlet height (see Fig.2) should contribute to the production of a nonequilibrium plasma with a high electrical conductivity at the inlet. The cross sectional view of the Disk-F4 generator channel is shown in Fig.3. The channel configuration is almost same as that of Disk-F3a and -F3r channels, and consists of a supersonic nozzle (r<140mm), MHD channel (170mm<r<406mm) and a diffuser (r>441mm). Two ring-typed anodes, first (A1) and second anode (A2) and two cathodes, first (C1) and second cathode (C2) are put on each disk wall. The external load resistance is connected between the second anode (A2) and the second cathode (C2). First anode (A1) and second anode (A2) are electrically shorted. The insulator walls between electrodes are made of Si_3N_4 having high mechanical strength. Electric probes, static pressure taps and optical windows are equipped for measurements.

In Fig.4 is shown the schematic view of the new seeding system in which melted cesium as a seed material is pushed by a piston dozator, instead of gas-pressure-driven in the previous system. The main components such as the piston and the cesium store tank are kept in a room where the temperature is maintained constant at ~60°C. The cesium pushed off is mixed with a heated argon gas at a pressure higher than that in the main flow in the hot duct, and introduced into the duct through a multi-hole seed injection nozzle. The seed fraction was found to be more controllable in the new system and particularly the time response to the setting value was markedly improved, in comparison with the corresponding results in the previous system. Furthermore, the large fluctuation of seed fraction

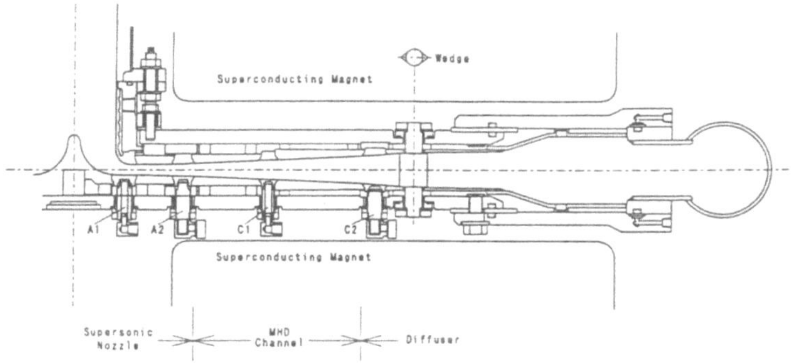

Figure 3. Cross sectional view of Disk-F4 MHD generator

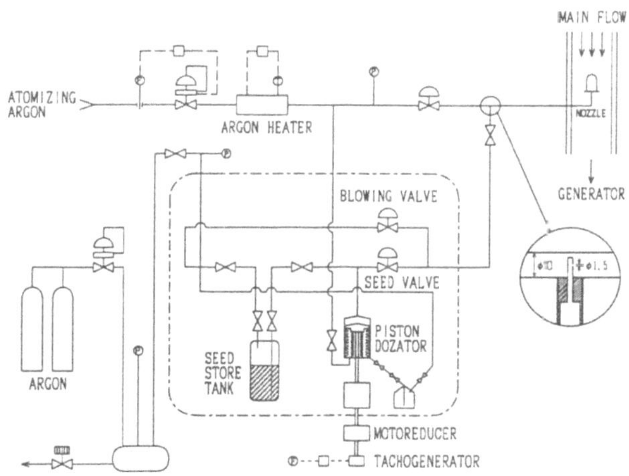

Figure 4. Improved Seeding System

with a low frequency of several Hz, which was observed in the previous system and caused the fluctuation of output power as mentioned before, was diminished. This fact should be attributed to the smooth supply of seed material by the mechanical piston motion.

3. Experimental Results and Discussion

Thirteen runs of power generation experiments have been carried out with the Disk-F4 generator channel and the new seed injection system so far. In this paper, the results of the run number A4105, where the maximum enthalpy extraction ratio and output power were obtained with the Disk-F4

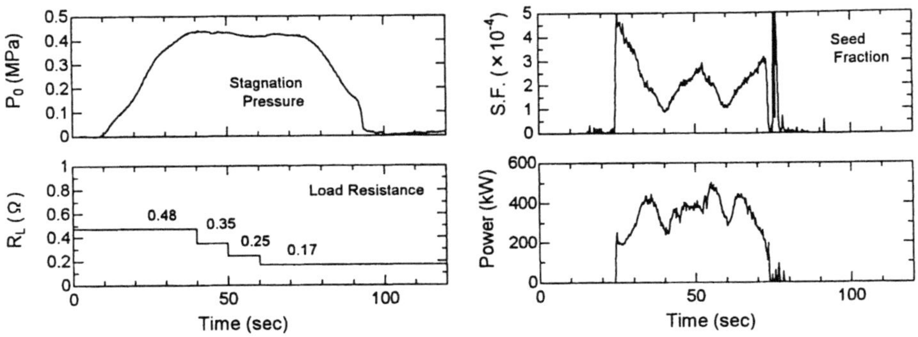

Figure 5. Time sequence of power generation experiment (RUN A4105)

channel, will be shown, as a typical result. The working conditions in this run are listed in Table 1 and the time sequence of the experiment is shown in Fig.5, where the time variations of the stagnation pressure P_0, the load resistance R_L, the seed fraction $S.F.$, and the output power $Power$ are drawn. The time of 0 sec refers to that when the main blow starts. As shown in this figure, the seed fraction was continuously changed in the range of $1.0{\sim}3.0\times10^{-4}$ for four different load resistances, in order to clarify the effect of seed fraction on generator performance.

TABLE 1. Working conditions (RUN A4105)

Working gas	Ar+Cs	Thermal input	3.01 MW
Stagnation pressure	0.42 MPa	Seed fraction	$1.0 \sim 3.0 \times 10^{-4}$
Stagnation temperature	1950 K	Load resistance	0.48, 0.35, 0.25, 0.17 Ω

As is seen from the time variation of output power in Fig.5, the maximum power of 502 kW is obtained at the load resistance of 0.25Ω and seed fraction of $\sim2.0\times10^{-4}$. The corresponding enthalpy extraction ratio is 16.7%. This value at the present stage does not reach the maximum value of 18.0% previously achieved with Disk-F3r generator. It can be emphasized, however, that large output power and high enthalpy extraction ratio are simultaneously obtained. The variation of output power at the load resistance of 0.35Ω (period of 40∼50sec) seems to be somewhat irregular. The phenomena concerned with this fact will be shown later. It is noted that the output power takes a maximum clearly at each load resistance except for 0.35Ω. This fact suggests that there exists an optimum seed fraction giving the largest output power for each load resistance.

The output powers for load resistances of 0.25 and 0.48Ω are plotted against the seed fraction in Fig.6(a). It is found from this figure that the optimum seed fraction is about 2×10^{-4} for both load resistances. Thus, the output power is decreased at the seed fractions lower and higher than this optimum value. In this figure, the fluctuations of output voltage V_{out} for load resistances of 0.25 and 0.48Ω are also plotted as a function of seed fraction. The fluctuation is defined here as the standard deviation divided by the mean value. It can be seen that the fluctuation becomes minimal for the seed fraction of about 2×10^{-4} giving the maximum power output. Thus, the high output power is realized with the small fluctuation.

The electron temperature measured in the upstream region of the generator channel is shown against the seed fraction in Fig.6(b). The measuring port is located at almost middle point between A2 and C1 (r=195mm). The temperature was estimated from the intensities of three lines of cesium (6p-8d:621.3nm, 6p-7d:672.3nm, 5d-6f:722.8nm) detected by a spectrum multichannel analyzer. It is found that the electron temperature decreases nonlinearly with the increase in the seed fraction. This result, which has been also observed in shock-tube experiments(Tsuji *et al.*, 1994), can be due to the fact that high seed fraction enhances the Lorentz force, and results in the deceleration of working fluid and the decrease in the Joule heating. It is noted that the electron temperature is in the range from ~3500 to ~5000K at the seed fraction of 2×10^{-4} giving the maximum power output. This fact suggests that the fully ionized seed plasma with the electron temperature of 3500~6000K is dominantly realized in the generator. Then, the fluctuation of the temperature was relatively small. Extremely high electron temperature observed at low seed fractions could originate from the ionization instability attributed to the noble (argon) gas ionization. At high seed fractions, on the contrary, partially ionized seed plasma is considered to be dominant, because of low gas velocity and low Joule heating. Consequently, at the optimum seed fraction, the fully ionized seed plasma can be dominant in the generator, resulting in large output power with its small fluctuation as well as that in the electron temperature. These results indicate that the precise control of seed fraction should be required to get desirable plasma properties and high generator performance.

Figures 7(a) and 7(b) show Hall potential and static pressure profiles along the generator channel at different seed fractions. The gradient of the potential, that is, electric field seems to be roughly constant along the generator channel (~2.2kV/m), and no negative potential appears in the inlet region of the generator. One of the reason of no negative potential can be attributed to the high Mach number at the inlet of the Disk-F4 genera-

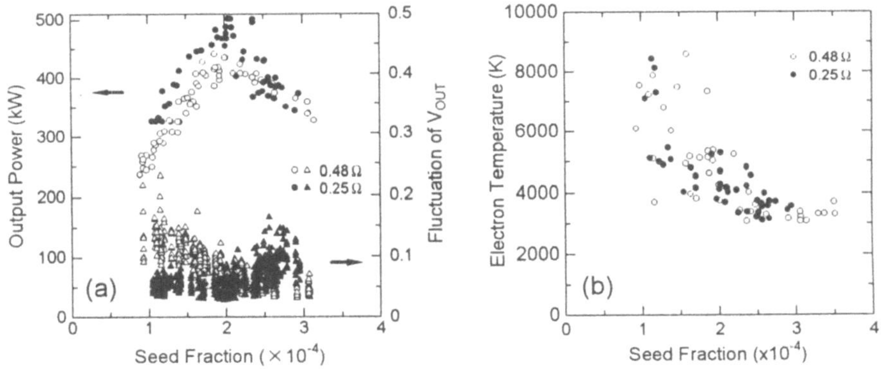

Figure 6. **Dependence of (a) output power and fluctuation of output voltage, (b) electron temperature, on seed fraction**

tor channel. The Hall potentials for low (1×10^{-4}) and high (3×10^{-4}) seed fractions are relatively low probably owing to low electrical conductivity in the plasma and to low gas velocity, respectively. It is seen from Fig.7(b) that static pressures increase with the seed fraction in almost the entire region of the generator, which is ascribed to the increase in Lorentz force. It should be noted that the static pressures at the second anode (A2) are higher than the values expected according to the isentropic relation (dotted line). Furthermore, the deviation is enhanced with the increase in the seed fraction. This phenomenon has not been observed in the previous experiments with Disk-F3a and F3r generators, although confirmed in shock-tube experiments(Tsuji *et al.*, 1994). The pressure increase at A2 is considered to be ascribed to the Lorentz force in both regions of nozzle and generator. Since the Mach number at the inlet of Disk-F4 generator is set to be high enough to produce the plasma with high electrical conductivity mentioned before, the Lorentz force in the nozzle region could be dominant.

Total pressures have been measured at different heights in the diffuser region in order to clarify the uniformity of fluid flow in the height direction (magnetic flux direction). Figure 8(a) shows the time variations of the total pressures measured at five different heights. During the seed injection (25~75sec), the measured pressures almost coincide, whereas the pressures do not coincide for the period of no seed injection (<25sec, and >75sec). This fact can imply that the uniform flow along the height is realized by MHD interaction. Thus, the stronger Lorentz force can be caused in the higher velocity part of fluid flow, which can make the flow uniform.

Figure 8(b) shows the time variations of swirl measured in the diffuser region and Hall current. Here, the swirl is defined as the ratio of the gas

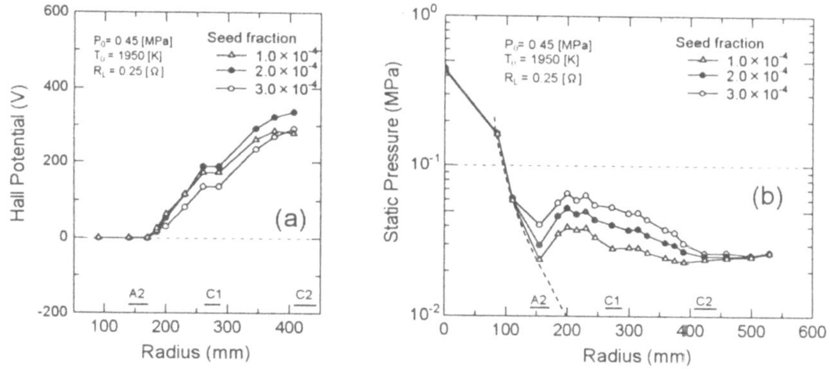

Figure 7. **Radial profiles of (a) potential and (b) static pressure ($R_L = 0.25\Omega$)**

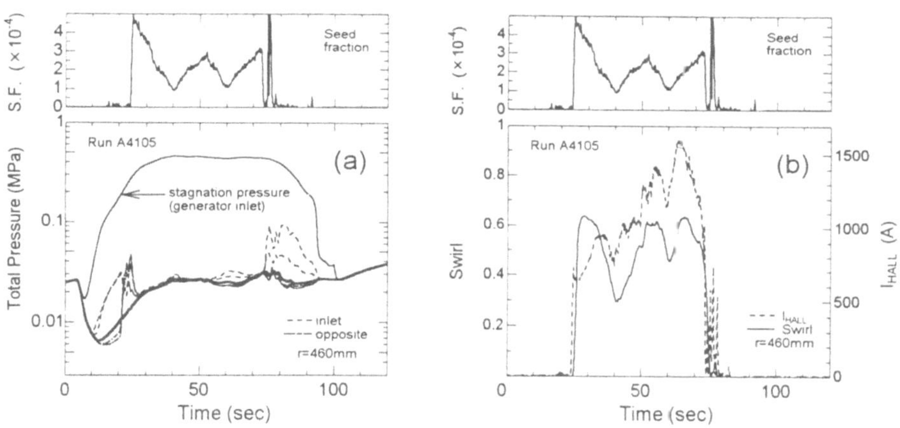

Figure 8. **Time variations of (a) total pressure at different heights, and (b) swirl in the diffuser region and Hall (output) current**

velocity in the azimuthal direction to that in the radial direction. In a disk generator, the swirl is directly coupled with Hall current, because the azimuthal gas velocity originates from Lorentz force of $j_r \times B$. It is found from this figure that the swirl well responds to the Hall current except for the periods of $<\sim$35sec and 40 $\sim$ 50 sec. Since in the period of $<\sim$35sec the gas flow is transient as seen in the time variation of stagnation pressure shown in Fig.5, the swirl measured in this period can be meaningless. The swirl in the periods of 40 $\sim$ 50 sec, when the load resistance is 0.35Ω and the output power does not take a maximum value clearly against the seed fraction, seems to depend not on the Hall current but on the seed fraction. This fact might suggest that the Hall current as the output current is no longer equal to the radial current in the generator. Thus, a part of the

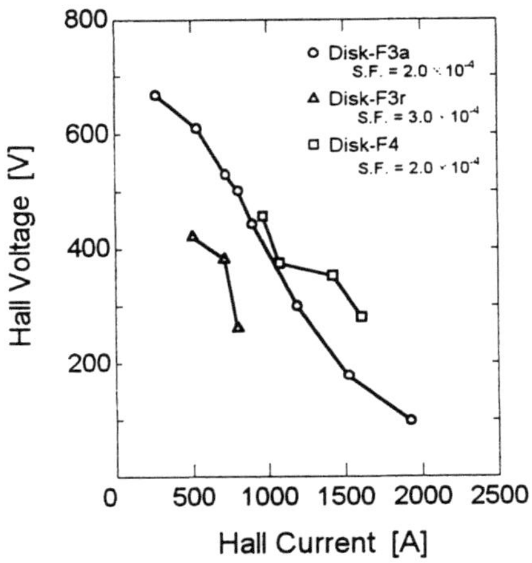

Figure 9. Typical Voltage-Current Characteristics of Disk-F3a, F3r and F4 Generators

radial current might flow, for instance, on the surface of insulator wall, and result in the reduction of output power, although there is no other evidence to verify this phenomenon.

TABLE 2. Typical power generation experiments

Generator channel Run No.	Disk-F3a 6208	Disk-F3r A8108	Disk-F4 A4105
Working gas		Ar+Cs	
Area ratio of outlet to inlet	4.1	6.9	5.0
Stagnation pressure (MPa)	0.46	0.42	0.42
Stagnation temperature (K)	1850	1930	1950
Thermal input (MW)	2.57	1.65	3.01
Seed fraction	2.0×10^{-4}	3.0×10^{-4}	2.0×10^{-4}
Load resistance (Ω)	0.62	0.51	0.25
Output power (kW)	404	297	502
Enthalpy extraction ratio (%)	15.7	18.0	16.7

The results of power generation experiments with Disk-F4 generator channel are roughly compared with those with Disk-F3a and F3r channels. The working conditions and results when the maximum enthalpy extraction ratio was achieved in each channel are summarized in Table 2, and the typical voltage-current characteristics of these generators are shown in Fig.9. Only in the experiments with Disk-F3a channel, the load resistance was widely changed under the fixed seed fraction. It seems that the voltage-current characteristics of Disk-F4 channel are similar to those of Disk-F3a channel. Thus, neglecting the point at R_L=0.35Ω where the voltage-current characteristics shrink and the behavior of plasma and fluid was irregular as has been described, the voltage-current curve can be roughly linear, in comparison with that of Disk-F3r channel. This can be due to the shape of Disk-F4 channel modified on the basis of Disk-F3a channel.

The maximum output power of 517kW was obtained at the large thermal input of 3.4MW (stagnation pressure of 0.6 MPa) with Disk-F3a generator channel, although the enthalpy extraction ratio was 15.2% slightly lower than 15.7% in Run No. 6208. In the power generation experiments with Disk-F4 generator channel, the maximum power is slightly lower than 517kW with Disk-F3a channel, and the maximum enthalpy extraction ratio does not reach 18.0% with Disk-F3r channel. It can be emphasized, however, that large output power and high enthalpy extraction ratio are simultaneously obtained in the experiment with Disk-F4 channel. Furthermore, the reproducibility in the high power generation was surely improved and large fluctuation of output voltage as has been pointed out in the experiments with Disk-F3r channel(Harada *et al.*, 1993) was not observed. In almost other runs with Disk-F4 channel, the enthalpy extraction ratios of 15∼16% were successfully achieved. These improvements are ascribed to the high Mach number at the inlet of the Disk-F4 generator channel, which is suitable for the production of non-equilibrium plasma with high electrical conductivity under the present working gas conditions, and to the seed injection without large fluctuation in the new seeding system, as has been expected.

Some problems, however, have remained for the further improvement of the generator performance in the Fuji-1 facility. One of them is to reduce the impurity contamination in the working gas, particularly H_2O, since the contamination will deteriorate the generator performance(Dellinger and Gray, 1978). The fraction of H_2O roughly measured was of the order of hundreds ppm, which was comparable to the seed fraction of the order of 10^{-4}. Moreover, a fine modification in the shape of the generator channel including the diffuser to meet the present gas pressure condition in the downstream parts of the blow-down facility, should be required. It is important to realize the properly supersonic flow under the strong MHD interaction. Finally, numerical models and codes should be developed, where complicated structure and behavior of plasma and fluid flows in the generator channel can be simulated and the performance can be predicted successfully. In the simulations, the influences of water vapor contamination, back pressure, separation of boundary layer, etc. should be taken into account simultaneously. All the problems mentioned above are inherently concerned with the long time power generation in the blow-down facility.

4. Conclusions

Experimental results on nonequilibrium plasma MHD electrical power generation with the Fuji-1 blow-down facility were presented. In the experiments with a new disk generator (Disk-F4) and a new seed injection system, the maximum power output of 502 kW and the enthalpy extraction ratio of 16.7% have been obtained. These values at the present stage do not reach the maximum values previously achieved in this facility. However, large output power and high enthalpy extraction ratio were simultaneously obtained with Disk-F4 channel. Furthermore, the reproducibility in the high power generation was surely improved and large fluctuation of output voltage was not observed. These improvements are resulted from the high Mach number at the inlet of the Disk-F4 generator channel and from the seed injection without large fluctuation in the new seeding system.

Furthermore, the experimental results indicate that the fully ionized seed plasma can be dominant in the generator, resulting in large output power with its small fluctuation at the optimum seed fraction. Moreover, the static pressure increase at the second anode, which is considered to be ascribed to the Lorentz force in the nozzle region, was clearly observed. Thus, these facts, which have been already pointed out in the experiments of power generation during several milliseconds with shock-tube facilities, were confirmed in the Fuji-1 blow-down experiments.

Acknowledgment

The authors wish to express great thanks to Mr. S.Ikeda and Mr. H.Mukai (Toshiba Corporation), Mr. H.Takahashi and all of the graduated students of the MHD group for their assistance in the experiments.

References

Dellinger, T.C. and Gray, E.L. (1978) Kinetic Model for Molecular Contamination Effects in Closed Cycle, Nonequilibrium MHD Generators, *Proc. 17th Symp. Engineering Aspects of MHD*, pp. H.1.1–H.1.9.

Harada, N., *et al.* (1990) High Enthalpy Extraction Demonstration with Closed Cycle Disk MHD Generators, *Proc. 28th Symp. Engineering Aspects of MHD*, pp. IX3.1–IX3.8.

Harada, Nob., *et al.* (1993) Improvement of Generator performance in FUJI-1 CCMHD Blow-down Experiment, *Proc. 31st Symp. Engineering Aspects of MHD*, pp. IVa1.1–IVa1.9.

Okamura,T., *et al* (1994) Review and New Results of High Enthalpy Extraction Experiments at Tokyo Institute of Technology, *Proc. 32nd Symp. Engineering Aspects of MHD*, Session 11.

Rosa, R.J (1968) *Magnetohydrodynamic Energy Conversion*, McGraw-Hill, New York.

Tsuji, K., *et al.* (1994) Highest Enthalpy Extraction from CCMHD Disk Generator with Cesium Seeded Argon, *Proc. 32nd Symp. Engineering Aspects of MHD*, Session 11.

Authors Index

Mechanics

FLUID MECHANICS AND ITS APPLICATIONS

Series Editor: R. Moreau

21. J.P. Bonnet and M.N. Glauser (eds.): *Eddy Structure Identification in Free Turbulent Shear Flows.* 1993 ISBN 0-7923-2449-8
22. R.S. Srivastava: *Interaction of Shock Waves.* 1994 ISBN 0-7923-2920-1
23. J.R. Blake, J.M. Boulton-Stone and N.H. Thomas (eds.): *Bubble Dynamics and Interface Phenomena.* 1994 ISBN 0-7923-3008-0
24. R. Benzi (ed.): *Advances in Turbulence V.* 1995 ISBN 0-7923-3032-3
25. B.I. Rabinovich, V.G. Lebedev and A.I. Mytarev: *Vortex Processes and Solid Body Dynamics.* The Dynamic Problems of Spacecrafts and Magnetic Levitation Systems. 1994 ISBN 0-7923-3092-7
26. P.R. Voke, L. Kleiser and J.-P. Chollet (eds.): *Direct and Large-Eddy Simulation I.* Selected papers from the First ERCOFTAC Workshop on Direct and Large-Eddy Simulation. 1994 ISBN 0-7923-3106-0
27. J.A. Sparenberg: *Hydrodynamic Propulsion and its Optimization.* Analytic Theory. 1995 ISBN 0-7923-3201-6
28. J.F. Dijksman and G.D.C. Kuiken (eds.): *IUTAM Symposium on Numerical Simulation of Non-Isothermal Flow of Viscoelastic Liquids.* Proceedings of an IUTAM Symposium held in Kerkrade, The Netherlands. 1995 ISBN 0-7923-3262-8
29. B.M. Boubnov and G.S. Golitsyn: *Convection in Rotating Fluids.* 1995 ISBN 0-7923-3371-3
30. S.I. Green (ed.): *Fluid Vortices.* 1995 ISBN 0-7923-3376-4
31. S. Morioka and L. van Wijngaarden (eds.): *IUTAM Symposium on Waves in Liquid/Gas and Liquid/Vapour Two-Phase Systems.* 1995 ISBN 0-7923-3424-8
32. A. Gyr and H.-W. Bewersdorff: *Drag Reduction of Turbulent Flows by Additives.* 1995 ISBN 0-7923-3485-X
33. Y.P. Golovachov: *Numerical Simulation of Viscous Shock Layer Flows.* 1995 ISBN 0-7923-3626-7
34. J. Grue, B. Gjevik and J.E. Weber (eds.): *Waves and Nonlinear Processes in Hydrodynamics.* 1996 ISBN 0-7923-4031-0
35. P.W. Duck and P. Hall (eds.): *IUTAM Symposium on Nonlinear Instability and Transition in Three-Dimensional Boundary Layers.* 1996 ISBN 0-7923-4079-5
36. S. Gavrilakis, L. Machiels and P.A. Monkewitz (eds.): *Advances in Turbulence VI.* Proceedings of the 6th European Turbulence Conference. 1996 ISBN 0-7923-4132-5
37. K. Gersten (ed.): *IUTAM Symposium on Asymptotic Methods for Turbulent Shear Flows at High Reynolds Numbers.* Proceedings of the IUTAM Symposium held in Bochum, Germany. 1996 ISBN 0-7923-4138-4
38. J. Verhás: *Thermodynamics and Rheology.* 1997 ISBN 0-7923-4251-8
39. M. Champion and B. Deshaies (eds.): *IUTAM Symposium on Combustion in Supersonic Flows.* Proceedings of the IUTAM Symposium held in Poitiers, France. 1997 ISBN 0-7923-4313-1
40. M. Lesieur: *Turbulence in Fluids.* Third Revised and Enlarged Edition. 1997 ISBN 0-7923-4415-4; Pb: 0-7923-4416-2

Kluwer Academic Publishers – Dordrecht / Boston / London

Mechanics

FLUID MECHANICS AND ITS APPLICATIONS

Series Editor: R. Moreau

41. L. Fulachier, J.L. Lumley and F. Anselmet (eds.): *IUTAM Symposium on Variable Density Low-Speed Turbulent Flows.* Proceedings of the IUTAM Symposium held in Marseille, France. 1997 ISBN 0-7923-4602-5
42. B.K. Shivamoggi: *Nonlinear Dynamics and Chaotic Phenomena.* An Introduction. 1997 ISBN 0-7923-4772-2
43. H. Ramkissoon, *IUTAM Symposium on Lubricated Transport of Viscous Materials.* Proceedings of the IUTAM Symposium held in Tobago, West Indies. 1998 ISBN 0-7923-4897-4
44. E. Krause and K. Gersten, *IUTAM Symposium on Dynamics of Slender Vortices.* Proceedings of the IUTAM Symposium held in Aachen, Germany. 1998 ISBN 0-7923-5041-3
45. A. Biesheuvel and G.J.F. van Heyst (eds.): *In Fascination of Fluid Dynamics.* A Symposium in honour of Leen van Wijngaarden. 1998 ISBN 0-7923-5078-2
46. U. Frisch (ed.): *Advances in Turbulence VII.* Proceedings of the Seventh European Turbulence Conference, held in Saint-Jean Cap Ferrat, 30 June–3 July 1998. 1998 ISBN 0-7923-5115-0
47. E.F. Toro and J.F. Clarke, *Numerical Methods for Wave Propagation.* Selected Contributions from the Workshop held in Manchester, UK. 1998 ISBN 0-7923-5125-8
48. A. Yoshizawa: *Hydrodynamic and Magnetohydrodynamic Turbulent Flows.* Modelling and Statistical Theory. 1998 ISBN 0-7923-5225-4
49. T.L. Geers (ed.): *IUTAM Symposium on Computational Methods for Unbounded Domains.* 1998 ISBN 0-7923-5266-1
50. Z. Zapryanov and S. Tabakova: *Dynamics of Bubbles, Drops and Rigid Particles.* 1999 ISBN 0-7923-5347-1
51. A. Alemany, Ph. Marty and J.P. Thibault (eds.): *Transfer Phenomena in Magnetohydrodynamic and Electroconducting Flows.* 1999 ISBN 0-7923-5532-6

Kluwer Academic Publishers – Dordrecht / Boston / London

MIX
Papier aus verantwortungsvollen Quellen
Paper from responsible sources
FSC® C105338

If you have any concerns about our products,
you can contact us on
ProductSafety@springernature.com

In case Publisher is established outside the EU,
the EU authorized representative is:
Springer Nature Customer Service Center GmbH
Europaplatz 3, 69115 Heidelberg, Germany

Printed by Libri Plureos GmbH
in Hamburg, Germany